SECOND EDITION

Excursions in Modern Mathematics

Peter Tannenbaum

Robert Arnold

California State University–Fresno

Prentice Hall
Englewood Cliffs, NJ 07632

Editorial Director: Timothy Bozik
Editor-in-Chief: Jerome Grant
Acquisitions Editor: Sally Denlow
Managing Editor: Jeanne Hoeting
Production Project Manager: Carol A. Bollati
Marketing Manager: Evan Hanby
Assistant Editor: Audra Walsh
Manufacturing Buyer: Trudy Pisciotti
Creative Director: Paula Maylahn
Cover Designer: Anthony A. Inciong
Interior Design and Page Layout: Judith A. Matz-Coniglio
Cover Photo: Dennis O'Clair/Tony Stone Worldwide
Photo Editor: Lorinda Morris-Nantz
Photo Researcher: Rona Tuccillo
Editorial Assistant: JoAnne Wendelken
Copy Editor: Lynn Buckingham
Text Compostion: Bi-Comp Incorporated
Color Inserts: New England Book Components

© 1995 by Prentice-Hall, Inc.
A Simon & Schuster Company
Englewood Cliffs, New Jersey 07632

Printed in the United States of America
10 9 8 7 6 5 4 3 2 1

ISBN 0-13-309600-9

Prentice-Hall International (UK) Limited, *London*
Prentice-Hall of Australia Pty. Limited, *Sydney*
Prentice-Hall Canada Inc., *Toronto*
Prentice-Hall Hispanoamericana, S.A., *Mexico*
Prentice-Hall of India Private Limited, *New Delhi*
Prentice-Hall of Japan, Inc., *Tokyo*
Simon & Schuster Asia Pte. Ltd., *Singapore*
Editora Prentice-Hall do Brasil, Ltda., *Rio de Janeiro*

To my parents, Nicholas and Anna,
and my wife Sally

PT

To my wife Rachael,
and my son Craig

RA

Contents

PART II   **Management Science**

6 HAMILTON JOINS THE CIRCUIT
The Traveling Salesman Problem   187

7 SPANNING TREES AND STEINER TREES
Minimum Network Problems   217

8 DIRECTED GRAPHS AND CRITICAL PATHS
Scheduling Problems   259

PART III   Growth and Symmetry

9 THE GOLDEN RULER
Spiral Growth and Fibonacci Numbers   297

10 THERE IS STRENGTH IN NUMBERS
The Growth of Populations   327

11 MIRROR, MIRROR, OFF THE WALL
Symmetry of Motion   357

12 FRACTALLY SPEAKING
Symmetry of Scale and Fractals   393

PART IV   Statistics

13 CENSUSES, SURVEYS, AND CLINICAL STUDIES
Collecting Data   423

Preface

> *To most outsiders, modern mathematics is unkown territory. Its borders are protected by dense thickets of technical terms; its landscapes are a mass of indecipherable equations and incomprehensible concepts. Few realize that the world of modern mathematics is rich with vivid images and provocative ideas.*
>
> IVARS PETERSON, *The Mathematical Tourist*

Excursions in Modern Mathematics is, as we hope the name might suggest, a collection of "trips" into that vast and alien frontier that many people perceive mathematics to be. While the purpose of this book is quite conventional—it is intended to serve as a textbook for college-level liberal arts mathematics course—its contents are not. We have made a concerted effort to introduce the reader to a view of mathematics entirely different from the traditional algebra-geometry-trigonometry-finite math curriculum which so many people have learned to dread, fear, and occasionally abhor. The notion that general education mathematics must be dull, unrelated to the real world, highly technical, and deal mostly with concepts that are historically ancient is totally unfounded.

The "excursions" in this book represent a collection of topics chosen to meet a few simple criteria.

■ **Applicability.** The connection between the mathematics presented here are down-to-earth, concrete real-life problems is direct and immediate. The often heard question, "What is this stuff good for?" is a legitimate one and deserves to be met head on. The often heard answer, "Well, you need to learn the material in Math 101 so that you can understand Math 102 which you will need to know if you plan to take Math 201 which will teach you the real applications," is less than persuasive and in many cases reinforces students' convictions that mathematics is remote, labyrinthine, and ultimately useless to them.

■ **Accessibility.** Sophisticated mathematical topics need not always be highly technical and built on layers of concepts. (Most of us are painfully familiar with the experience of missing or misunderstanding a concept in the middle of a math course and consequently having trouble with the rest of the course material.) In general the choice of topics is such that a heavy mathematical infrastructure is not needed. We have found Intermediate Algebra to be sufficient prerequisite. In the few instances in which more advanced concepts are unavoidable we have endeavored to provide enough background to make the material self-contained. A word of caution—this does not mean that the material is easy! In mathematics, as in many other walks of life, simple and straightforward is not synonymous with easy and superficial.

■ **Age.** Much of the mathematics in this book has been discovered in this century, some as recently as 20 years ago. Modern mathematical discoveries do not have to be only within the grasp of experts.

■ **Aesthetics.** The notion that there is such a thing as beauty in mathematics is surprising to most casual observers. There is an important aesthetic component in mathematics and, just as in art and music (which mathematics very much resembles), it often surfaces in the simplest ideas. A fundamental objective of this book is to develop an appreciation for the aesthetic elements of mathematics. It is not necessary that the reader love everything in the book—it is sufficient that he or she find one topic about which they can say, "I really enjoyed learning this stuff!" We believe that anyone coming in with an open mind almost certainly will.

OUTLINE

The material in the book is divided into four independent parts. Each of these parts in turn contains four chapters dealing with intrrelated topics.

■ **Part I** (Chapters 1 through 4). *The Mathematics of Social Choice.* This part deals with mathematical applications in social science. How do groups make decisions? How are elections decided? How can power be measured? When there are competing interests among members of a group, how are conflicts resolved in a fair and equitable way?

■ **Part II** (Chapters 5 through 8). *Management Science.* This part deals with methods for solving problems involving the organization and management of complex activities—that is, activities involving either a large number of steps and/or a large number of variables (building a skyscraper, putting a person on the moon, organizing a banquet, scheduling classrooms at a big university, etc.). Efficiency is the name of the game in all these problems. Some limited or precious resource (time, money, raw materials) must be managed in such a way that waste is minimized. We deal with problems of this type (consciously or unconsciously) every day of our lives.

■ **Part III** (Chapters 9 through 12) *Growth and Symmetry.* This part deal with nontraditional geometric ideas. What do sunflowers and seashells have in common? How do populations grow? What are the symmetries of a pattern? What is the geometry of natural (as opposed to artificial) shapes? What kind of symmetry lies hidden in a cloud?

■ **Part IV** (Chapters 13 through 16). *Statistics.* In one way or another, statistics affects all of our lives. Government policy, insurance rates, our health, and our diet are all governed by statistical laws. This part deals with some of the basic elements of statistics. How are statistical data collected? How are they sumarized so that they say something intelligible? How are they interpreted? What are the pattern of statistical data?

EXERCISES

We have endeavored to write a book that is flexible enough to appeal to a wide range of readers in a variety of settings. The exercises, in particular, have been assigned to convey the depth of the subject matter by addressing a broad spectrum of levels of difficulty—from the routine to the challenging. For convenience we have classified them into three levels of difficulty: **Walking** (these are straightforward applications of the concepts discussed in the chapter, and it is intended that all readers be able to do them); **Jogging** (these are exercises that are not difficult per se, but require a little thinking on the part of the reader); and **Running** (these are the really challenging exercises and are intended for those readers who like a little challenge in life).

TEACHING EXTRAS

Because this course can be taught in a myriad of styles, we have included material which will increase the text's flexibility. Bibliography listings at the end of every chapter can be used by students and instructors as a source of additional readings which provide more indepth or tangentially related information.

In addition, Prentice Hall and the *New York Times* are sponsoring "A Contemporary View," a program which will provide both students and instructors with access to information which further illustrates the relevance and currency of mathematics. Through this program, at the instructor's request, each student will receive a supplement containing recent articles from the *New York Times* which pertain in one way or another to the themes of our excursions in this book. These articles may be used to prompt classroom discussion, to suggest topics for term papers and research projects, or merely to show that the contents of this course have relevance to a well-informed person. Instructors are encouraged to call their Prentice Hall representative for more information.

To ease the instructor's transition to a new book, we have prepared an instructor's manual based on our own experience using the manuscript in class. The instructor's manual contains a test bank consisting of four hundred multiple choice questions as well as answers to all exercises (including worked out solutions to many), helpful notes about teaching emphasis, suggested topic ordering, etc.

A FINAL WORD

This book grew out of the conviction that a textbook of this type should provide much more than just a battery of technical skills. Our ultimate purpose is to instill in the reader something that is more subjective but at the same time more long-lasting—an appreciation of mathematics as a discipline and an exposure to a few of its global paradigms: algorithms as solutions to mathematical problems, the concept of recursion, the idea of complexity and the trade-offs that it entails, the difference between conceptually difficult and procedurally complicated, the idea that mathematics is one of those rare disciplines that can define its own boundaries (impossibility theorems), etc. Last, but not least, we have tried to show that mathematics can be fun.

THE SECOND EDITION

This second edition of *Excursions in Modern Mathematics* retains the topics and organization of the first edition. Our major focus in producing this edition has been to improve (and update where appropriate) the presentation of the individual chapters. In doing so, we have benefited from a full range of excellent suggestions and miscellaneous input coming from many different sources: instructors, students, reviewers, and editors. A few chapters (such as Chapters 4, 8, and 9), have undergone major reconstructive surgery; many of the others have had just a nose job; a few have been left essentially unchanged. In most cases, the changes have been in the direction of adding a few more things, so that, somewhat regretably, just like the authors themselves, the book has gotten thicker with age. For the most part, however, the additions have been aimed at improved understanding and readability, with many more and new types of examples added. In addition, the exercise sets at the end of each chapter have been refined and expanded (this edition has over 750 exercises which, like those in the first edition, are organized into three different levels of difficulty).

ACKNOWLEDGMENTS Several people have made significant contributions to the second edition. Our editor Jerome Grant, guided us through this revision. Carlos Valencia helped with the exercise sets, and our production editor, Carol Bollati kept everything running smoothly. The following mathematicians reviewed the manuscript and made many invaluable suggestions:

Carmen Artino, *College of Saint Rose*
Ronald Czochor, *Rowan College of New Jersey*
Kathryn E. Fink, *Moorpark College*
Dr. Josephine Guglielmi, *Meredith College*
William S. Hamilton, *Community College of Rhode Island*
Tom Kiley, *George Mason University*
Jean Kirchbaum, *Broome Community College*
Thomas O'Bryan, *University of Wisconsin–Milwaukee*
Daniel E. Otero, *Xavier University*
Matthew Pickard, *University of Puget Sound*
Kathleen C. Salter, *Eastern New Mexico University*

Theresa M. Sandifer, *Southern Conneticut State University*
Paul Schembari, *East Stroudsberg University of Pennsylvania*
William W. Smith, *University of North Carolina at Chapel Hill*
John Watson, *Arkansas Tech University*
Donald Beaton, *Norwich University*
Terry L. Cleveland, *New Mexico Military Institute*
Leslie Cobar, *University of New Orleans*
Harold Jacobs, *East Stroudsburg University*
Lana Rhoads, *William B aptist College*
David E. Rush, *University of California at Riverside*
David Stacy, *Bellevue Community College.*

We would like to extend special thanks to Professor Benoit Mandelbrot of Yale University who read the manuscript for Chapter 12 and made several valuable suggestions.

We gratefully acknowledge Vahack Haroutunian, who class-tested some of the earliest versions of the material and Ronald Wagoner who helped with the development of many of the exercises

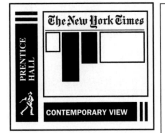

I The Mathematics of Social Choice

CHAPTER 1

The Paradoxes of Democracy

The Mathematics of Voting

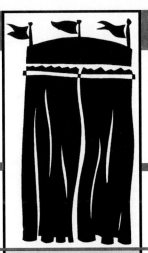

We are often reminded that the right to vote, one of the fundamental tenets of a democracy, is something that we should not take for granted. Voting is the vehicle by which a democratic society makes decisions, and studying the process by which the many and conflicting opinions of the individuals are consolidated into the single choice of the group is what **voting theory** is about. Voting theory? Why do we need a whole theory? It all sounds so simple. There surely must be a simple way to decide what the outcome of any election is, a way that is straightforward, consistent, and fair.

For elections involving only two candidates or alternatives, there is indeed such a way, and we are all familiar with it: *majority wins.* (Of course, in case of a tie we need some rule for breaking it, but let's not worry about that now.) For elections involving three or more candidates or alternatives, *there is no consistent method by which a democratic society can make a choice that is always fair.* This remarkable fact was discovered in 1952 by the mathematical economist Kenneth Arrow and is known as **Arrow's impossibility theorem.**[1]

The primary purpose of this chapter is to clarify the full meaning and significance of Arrow's impossibility theorem. In so doing, we will discuss several commonly used methods for deciding the outcome of an election—how they work and what their implications are. We will also take a closer look at the notion of fairness in the context of voting theory.

Our discussion for the entire chapter is centered around the following deceptively simple, fictitious election.

The MAS Election. The Math Appreciation Society (MAS) is a student organization dedicated to spreading the radical idea that mathematics can be fun. The Tasmania State University chapter of MAS is holding an election to choose its president. There are four candidates: Alisha, Boris, Carmen, and Dave (*A, B, C,* and *D* for short). Each of the 37 members of the club is asked to submit a bal-

---

[1] In 1972, Arrow received the Nobel Prize in Economics for his many contributions to the mathematics of social choice theory.

Ballot	Ballot	Ballot	Ballot	Ballot	Ballot	Ballot	Ballot	Ballot	Ballot
1st A	1st B	1st A	1st C	1st B	1st C	1st A	1st D	1st A	1st A
2nd B	2nd D	2nd B	2nd B	2nd D	2nd B	2nd B	2nd C	2nd B	2nd B
3rd C	3rd C	3rd C	3rd D	3rd C	3rd D	3rd C	3rd B	3rd C	3rd C
4th D	4th A	4th D	4th A	4th A	4th A	4th D	4th A	4th D	4th D

Ballot	Ballot	Ballot	Ballot	Ballot	Ballot	Ballot	Ballot	Ballot	Ballot
1st C	1st A	1st C	1st D	1st C	1st A	1st D	1st D	1st C	1st C
2nd B	2nd B	2nd B	2nd C	2nd B	2nd B	2nd C	2nd C	2nd B	2nd B
3rd D	3rd C	3rd D	3rd B	3rd D	3rd C	3rd B	3rd B	3rd D	3rd D
4th A	4th D	4th A	4th A	4th A	4th D	4th A	4th A	4th A	4th A

Ballot	Ballot	Ballot	Ballot	Ballot	Ballot	Ballot	Ballot	Ballot	Ballot
1st A	1st B	1st C	1st C	1st D	1st A	1st D	1st C	1st A	1st D
2nd B	2nd D	2nd B	2nd B	2nd C	2nd B	2nd C	2nd B	2nd B	2nd C
3rd C	3rd C	3rd D	3rd D	3rd B	3rd C	3rd B	3rd D	3rd C	3rd B
4th D	4th A	4th A	4th A	4th A	4th D	4th A	4th A	4th D	4th A

Ballot	Ballot	Ballot	Ballot	Ballot	Ballot	Ballot
1st B	1st A	1st C	1st A	1st A	1st D	1st A
2nd D	2nd B	2nd D	2nd B	2nd B	2nd C	2nd B
3rd C	3rd C	3rd B	3rd C	3rd C	3rd B	3rd C
4th A	4th D	4th A	4th D	4th D	4th A	4th D

FIGURE 1-1

lot indicating his or her first, second, third, and fourth choices (ties are not allowed in individual ballots). The 37 ballots submitted are shown in Fig. 1-1. Who should be president? Why?

PREFERENCE BALLOTS AND PREFERENCE SCHEDULES

Ballots such as the ones shown in Fig. 1-1 in which the voter is asked to list all of the options in order of preference are called **preference ballots.**[2] A closer inspection of the 37 preference ballots in Fig. 1-1 shows that in many instances individual voters voted exactly the same way. In fact, the outcome of the election can be summarized in Fig. 1-2 and more conveniently in the table that follows, which is called a **preference schedule** for the election.

[2] To many of us, the first thing that comes into our minds when we think about voting is politics. Typically, in a political election we are asked to pick a single choice rather than to cast a preference ballot, so it is not surprising that, at first, the idea of voting by means of a preference ballot strikes many people as odd and removed from reality. This is far from being the case. First, in developing a theory, we must look at the subject in its broadest context. There are many real world scenarios outside the realm of political elections in which preferential voting is carried out (beauty pageants, sports awards, job-candidate selection, etc.), and in theory at least, they are all of equal importance. Second, even when we are asked only to pick our top choice, we almost always, consciously or unconsciously, end up ranking some or all of the other choices. (After all, if for whatever reason our first choice became unavailable, it is not unusual for us to know, almost as a reflex, who our new choice will be.)

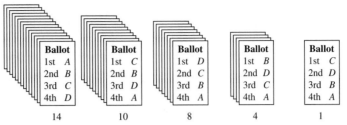

FIGURE 1-2

Number of voters	14	10	8	4	1
1st choice	A	C	D	B	C
2nd choice	B	B	C	D	D
3rd choice	C	D	B	C	B
4th choice	D	A	A	A	A

Some Basic Assumptions

Ballot
1st C
2nd B
3rd D
4th A

FIGURE 1-3

Before we take on the task of deciding who will be elected president of the Math Appreciation Society, we need to clarify some of the assumptions that we will make about preference ballots. The first of these assumptions is that the preferences of an *individual* are **transitive:** This means that if a voter prefers *X* to *Y* and *Y* to *Z*, then we can reasonably assume that the voter prefers *X* to *Z*. It follows therefore that a ballot such as the one shown in Fig. 1-3 tells us not only that this voter prefers *C* to *B*, *B* to *D*, and *D* to *A*, but also implicitly that the voter prefers *C* to *D*, *C* to *A*, and *B* to *A*.

A second important assumption we will make is that relative preferences are not affected by the elimination of one or more candidates. Consider, once again, a voter casting the preference ballot shown in Fig. 1-3 and let's suppose that, for whatever reason, candidate *B* drops out of the race. It is reasonable to assume that, as Fig. 1-4 shows, the relative order of preference for the other candidates remains unaffected: This voter still prefers *C* to *D* and *D* to *A*.

Ballot
1st C
~~2nd B~~
3rd D
4th A
⟶
Ballot
1st C
2nd D
3rd A

FIGURE 1-4

While there are occasional situations in which the above assumptions do not hold,[3] they are by and large exceptions. In any voting system in which voters

---

[3] See, for example, the discussion of how the 2000 Summer Olympics were awarded to Sydney in Appendix 2 at the end of this chapter.

act *rationally,* it is reasonable to expect that these assumptions will hold, and for the rest of this chapter we will take them for granted.

THE PLURALITY METHOD

A very well-known and seemingly reasonable way to decide an election is the **plurality method:** The candidate with the *most* first-place votes wins. For the MAS election, the results are

A: 14 first-place votes

B: 4 first-place votes

C: 11 first-place votes

D: 8 first-place votes.

The Math Appreciation Society News

ALISHA ELECTED PRESIDENT OF MAS!

Since Alisha has the most first-place votes, she is the winner!

Our fondness for the plurality method stems not only from its simplicity but also from the fact that it is a natural extension of the **majority rule**: In any democratic election between two candidates, the one with the majority (more than half) of the votes wins. Unfortunately, the majority rule cannot always be used when there are more than two candidates: In the MAS election with 37 voters, 19 first-place votes are needed for a majority, and none of the candidates received 19 first-place votes. Of course, if a candidate had received 19 or more first-place votes, that candidate would clearly have to be the winner under the plurality method since no other candidate could possibly receive as many. In the language of voting theory, we say that the plurality method *satisfies* the **majority criterion**.

> **Majority Criterion.** If there is a candidate or alternative that is the first choice of a majority of the voters, then that candidate or alternative should be the winner of the election.

On its surface, the majority criterion seems so logical and so much in accordance with our idea of democracy that we are inclined to take it for granted. Surprisingly, we will soon see that there are some well-known and widely used voting methods where a candidate could have a majority of the first-place votes and yet fail to win the election.

Problems with the Plurality Method

In spite of its frequent usage, the plurality method has several flaws and is generally considered a very poor method for choosing the winner of an election among more than two candidates. The principal weakness of the plurality method is that it fails to take into account preferences other than first choice. In fact, when we look at the MAS election preference schedule, we notice that a majority of the voters (23 to be exact) were not terribly crazy about Alisha being president and made her their last choice.

To underscore how the plurality method can sometimes produce some really bad results, consider the next example.

Example 1. An election is held among the 100 members of the Tasmania State University marching band to decide in which of five different bowl games they will march. The alternatives are the Rose Bowl (R), the Hula Bowl (H), the Cotton Bowl (C), the Orange Bowl (O), and the Sugar Bowl (S). The results of the election are shown in the following preference schedule:

Number of voters	49	48	3
1st choice	R	H	C
2nd choice	H	S	H
3rd choice	C	O	S
4th choice	O	C	O
5th choice	S	R	R

When the plurality method is used, the choice is the Rose Bowl (R), in spite of the fact that it is the *last* choice of a majority (51) of the band members, while the Hula Bowl (H) is the first choice of 48 voters (only 1 less than R) and also has 52 second-place votes. Under any reasonable interpretation, the Hula Bowl is much more representative of the band's wishes than the Rose Bowl. In fact, we can make the following persuasive argument on behalf of the Hula Bowl: When compared on a *one-to-one* basis with any other bowl, the Hula Bowl would always get a majority of the votes. Take, for example, a comparison between the Hula Bowl and the Rose Bowl: The Hula Bowl would get 51 votes (48 from the second column plus the 3 votes in the last column) versus 49 votes for the Rose Bowl. Likewise, a comparison between the Hula Bowl and the Cotton Bowl would result in 97 votes for the Hula Bowl (first and second columns) and 3 votes for the Cotton Bowl. Finally, we see that the Hula Bowl is preferred to both the Orange Bowl and the Sugar Bowl by all 100 voters.

We can now summarize the problem as follows: Although H wins in a one-to-one comparison between it and any other alternative, the plurality method fails to choose H as the winner. In the language of voting theory, we say that the plurality method *violates* a basic principle of fairness called the **Condorcet[4] criterion.**

[4] Named after Marie Jean Antoine Nicolas Caritat, Marquis de Condorcet (1743–1794). Condorcet was a French aristocrat, mathematician, philosopher, economist, and social scientist. As a member of a group of liberal thinkers (the *encyclopédistes*), his ideas were instrumental in leading the way to the French Revolution. Unfortunately, his independent ways eventually fell into disfavor and he died in prison.

> **Condorcet Criterion.** If there is a candidate or alternative that wins in a one-to-one comparison between it and any other alternative, then that candidate or alternative should be the winner of the election.

Before we go on, a word about how to interpret some of the terminology we have just introduced. When we say that the plurality method *violates the Condorcet criterion*, we mean that it is possible to come across situations in which there is a candidate that wins in a one-to-one comparison against every other candidate and yet that candidate is not the winner of the election under the plurality method. It is not necessary that this happen in every election—the criterion is violated once we know that the situation can arise (as we saw happen in the marching band example). Think about it this way: To violate the speeding laws we do not have to drive above the speed limit all the time—it is enough that we do it once in a while.

When there is a candidate that is preferred by the voters in each one-to-one comparison with the other candidates, we call such a candidate a **Condorcet candidate.** Using this wording, the Condorcet criterion has a nice ring to it: It simply says that whenever there is a Condorcet candidate, the Condorcet candidate should win the election. If it doesn't, the Condorcet criterion is violated.

We will return to the idea of comparing candidates on a one-to-one basis shortly. In the meantime, we conclude this section by discussing one last weakness of the plurality method known as **insincere** or **strategic voting.** Consider once again Example 1 and look at the last column of the preference schedule which represents the ballots of three band members—let's say Hekyll, Jekyll, and Hyde. Let's suppose, moreover, that all three have a pretty good idea of how everybody is going to vote and know that under the plurality method they will be stuck with going to the Rose Bowl (which they really don't want). Since they know that their first choice (the Cotton Bowl) has no chance of winning the election, their strategy is clear: Make the Hula Bowl the first choice on their ballots (even though it really isn't), thereby keeping the Rose Bowl from being the winner.

The ability of voters to manipulate the results of an election by insincere voting is a common but undesirable feature of many voting methods, and the plurality method is particularly susceptible to it. In presidential elections it has had a profound effect on the viability of third-party candidates: Often a voter may prefer such a candidate but thinks his vote may be wasted, so instead he votes (insincerely) for his preference between the two major-party candidates.

A classic example of this situation occurred during the 1992 presidential election. Many voters who were inclined to vote for Ross Perot opted to vote (insincerely) for either George Bush or Bill Clinton because of their fear of "wasting" their vote.

THE BORDA COUNT METHOD

In the **Borda count method**[5] each place on a ballot is assigned points. In an election with N candidates we give 1 point for last place, 2 points for second from last place, ..., and N points for first place. The points are tallied for each candidate separately, and the candidate with the highest total wins.

Let's use the Borda count method to choose the winner of the MAS election. Putting the point values in the preference schedule, we have

Number of voters	14	10	8	4	1
1st choice: 4 points	A: 56 pts	C: 40 pts	D: 32 pts	B: 16 pts	C: 4 pts
2nd choice: 3 points	B: 42 pts	B: 30 pts	C: 24 pts	D: 12 pts	D: 3 pts
3rd choice: 2 points	C: 28 pts	D: 20 pts	B: 16 pts	C: 8 pts	B: 2 pts
4th choice: 1 point	D: 14 pts	A: 10 pts	A: 8 pts	A: 4 pts	A: 1 pt

Consequently,

A gets $56 + 10 + 8 + 4 + 1 = 79$ points.

B gets $42 + 30 + 16 + 16 + 2 = 106$ points.

C gets $28 + 40 + 24 + 8 + 4 = 104$ points.

D gets $14 + 20 + 32 + 12 + 3 = 81$ points.

The Math Appreciation Society News

BORIS ELECTED PRESIDENT OF MAS!

The winner of the election under the Borda count method is Boris (B)! This is not a pleasant bit of news for Alisha (the winner under the plurality method).

	MATH APPRECIATION SOCIETY ELECTION RESULTS

Winner	Voting method
Alisha	Plurality
Boris	Borda count

Problems with the Borda Count Method

The most serious weakness of the Borda count method is that it clashes with what we have already agreed is a reasonable expectation in a democratic election: the majority criterion. Example 2 shows how under the Borda count

[5] This method is named after its proposer, the Frenchman Jean-Charles de Borda (1733–1799). Borda was a military man—a cavalry officer and naval captain—who wrote papers on such diverse subjects as mathematics, physics, the design of scientific instruments, and voting theory.

method, it is possible for a candidate to have a majority of the first-place votes and yet fail to win the election.

Example 2. Consider an election with 4 candidates (*A, B, C,* and *D*), 11 voters, and the preference schedule given in the following schedule.

Number of voters	6	2	3
1st: 4 points	*A*	*B*	*C*
2nd: 3 points	*B*	*C*	*D*
3rd: 2 points	*C*	*D*	*B*
4th: 1 point	*D*	*A*	*A*

Let's pick the winner using the Borda count method.

A gets $(6 \times 4) + (2 \times 1) + (3 \times 1) = 29$ points.

B gets $(6 \times 3) + (2 \times 4) + (3 \times 2) = 32$ points.

C gets $(6 \times 2) + (2 \times 3) + (3 \times 4) = 30$ points.

D gets $(6 \times 1) + (2 \times 2) + (3 \times 3) = 19$ points.

In this election *B* has the most points and wins under the Borda count method in spite of the fact that *A* has the majority of first-place votes (6 out of 11). In the language of voting theory, *the Borda count method violates the majority criterion.*

Another problem with the Borda count method is a direct consequence of the above: It also *violates the Condorcet criterion.* It is not hard to see that *A,* having a majority of first-place votes, would automatically win a one-to-one comparison against any other alternative (Exercise 49).

In spite of these weaknesses, the Borda count method is widely used and, in many cases, is a reasonably good method for deciding an election. In contrast to the plurality method, it uses *all* the information provided by the voters and produces as a winner the best compromise candidate.

In practice, the Borda count method is often used by assigning the points as follows: 0 points for last place, 1 point for second from last, 2 points for third from last, etc. It turns out (see Exercise 45) that tabulating the points this way does not in any way change the results of the election.

Sometimes elections are decided using simple variations of the Borda count method. In choosing the winner of the Heisman trophy (an annual award given to the "best collegiate football player" in the country), each of the voters (about 870 sportswriters and 50 former Heisman trophy winners) submits a ballot with a first, second, and third choice (3 points are awarded for first place, 2 points for second, and 1 point for third). The candidate with the highest total

The most coveted individual award in college football: The Heisman Memorial Trophy. (AP/Wide World Photos)

wins the award. This is just the Borda count method with many candidates but only three choices on each ballot.

THE PLURALITY-WITH-ELIMINATION METHOD

This is a sophisticated variation of the plurality method and is carried out in rounds using the following procedure:

- **Round 1.** Count the first-place votes for each candidate, just as you would in the plurality method. If a candidate has a majority of first-place votes, that candidate is automatically declared the winner. Otherwise, eliminate the candidate (or candidates if there is a tie) with the fewest number of first-place votes.

- **Round 2.** Cross out the name(s) of the candidates eliminated from the preference schedule and recount the first-place votes. If a candidate has a majority of first-place votes, declare that candidate the winner. Otherwise, eliminate the candidate with the fewest number of first-place votes.

■ **Rounds 3, 4, etc.**   Repeat the process, each time eliminating one or more candidates. Eventually, there will be a candidate with a majority of first-place votes. (In the worst of cases we might have to repeat the process until we end up with only two candidates.)

Let's apply the plurality-with-elimination method to the MAS election.

■ **Round 1.**

A: 14 first-place votes
B: 4 first-place votes
C: 11 first-place votes
D: 8 first-place votes

B has the fewest number of first-place votes and is eliminated.

Number of voters	14	10	8	4	1
1st choice	*A*	*C*	*D*	~~*B*~~	*C*
2nd choice	~~*B*~~	~~*B*~~	*C*	*D*	*D*
3rd choice	*C*	*D*	~~*B*~~	*C*	~~*B*~~
4th choice	*D*	*A*	*A*	*A*	*A*

The preference schedule (with *B* removed) now becomes

Number of voters	14	10	8	4	1
1st choice	*A*	*C*	*D*	*D*	*C*
2nd choice	*C*	*D*	*C*	*C*	*D*
3rd choice	*D*	*A*	*A*	*A*	*A*

■ **Round 2.**

A: 14 first-place votes
C: 11 first-place votes
D: 12 first-place votes

C is eliminated.

Number of voters	14	10	8	4	1
1st choice	*A*	~~*C*~~	*D*	*D*	~~*C*~~
2nd choice	~~*C*~~	*D*	~~*C*~~	~~*C*~~	*D*
3rd choice	*D*	*A*	*A*	*A*	*A*

The preference schedule (with *C* removed) now becomes

Number of voters	14	10	8	4	1
1st choice	*A*	*D*	*D*	*D*	*D*
2nd choice	*D*	*A*	*A*	*A*	*A*

■ **Round 3.**

A: 14 first-place votes

D: 23 first-place votes

The Math Appreciation Society News

DAVE ELECTED PRESIDENT OF MAS!

In this round Dave has a majority and is therefore the winner.

For the reader who likes straight answers to simple questions, it may be somewhat discomforting to realize that the question, Who is the real winner of the MAS election? may not make a lot of sense. The answer seems to depend on the voting method we choose.

MATH APPRECIATION SOCIETY ELECTION RESULTS

Winner	Voting method
Alisha	Plurality
Boris	Borda count
Dave	Plurality with elimination

We illustrate the plurality-with-elimination method with another example. This time, being a little more experienced, we take a shortcut, and instead of rewriting the preference schedule after each round, we just cross out the candidates as they are eliminated (and note that the highest candidate that is not crossed out in a column is the first-place candidate in that column).

Example 3. Consider an election with 5 candidates (*A, B, C, D,* and *E*), 24 voters, and the preference schedule

Number of voters	8	6	2	3	5
1st choice	*A*	*B*	*C*	*D*	*E*
2nd choice	*B*	*D*	*A*	*E*	*A*
3rd choice	*C*	*E*	*E*	*A*	*D*
4th choice	*D*	*C*	*B*	*C*	*B*
5th choice	*E*	*A*	*D*	*B*	*C*

We will determine the winner using the plurality-with-elimination method.

■ **Round 1.**

Candidate	1st-place votes
A	8
B	6
C	2 ⟶ *C* is eliminated
D	3
E	5

The preference schedule now looks like this:

Number of voters	8	6	2	3	5
1st choice	A	B	~~C~~	D	E
2nd choice	B	D	A	E	A
3rd choice	~~C~~	E	E	A	D
4th choice	D	~~C~~	B	~~C~~	B
5th choice	E	A	D	B	~~C~~

■ **Round 2.**

Candidate	1st-place votes
A	$8 + 2 = 10$
B	6
D	3 ⟶ *D* is eliminated
E	5

The preference schedule now is:

Number of voters	8	6	2	3	5
1st choice	A	B	~~C~~	~~D~~	E
2nd choice	B	~~D~~	A	E	A
3rd choice	~~C~~	E	E	A	~~D~~
4th choice	~~D~~	~~C~~	B	~~C~~	B
5th choice	E	A	~~D~~	B	~~C~~

■ **Round 3.**

Candidate	1st-place votes
A	$8 + 2 = 10$
B	6 ⟶ *B* is eliminated
E	$3 + 5 = 8$

The preference schedule once again:

Number of voters	8	6	2	3	5
1st choice	A	~~B~~	~~C~~	~~D~~	E
2nd choice	~~B~~	~~D~~	A	E	A
3rd choice	~~C~~	E	E	A	~~D~~
4th choice	~~D~~	~~C~~	~~B~~	~~C~~	~~B~~
5th choice	E	A	~~D~~	~~B~~	~~C~~

■ **Round 4.**

Candidate	1st-place votes
A	$8 + 2 = 10$
E	$6 + 3 + 5 = 14$

The winner of this election is E.

Plurality with elimination is most commonly used in elections in which there are few candidates (typically three or four). A simple variation of plurality with elimination commonly known as *plurality with a runoff* is often used in elections for local political office (city councils, county boards of supervisors, school boards, etc.). Plurality with a runoff works just like plurality with elimination except that *all* the candidates except the top two get eliminated in the first round (see Exercise 52). Another slight variation of the plurality-with-elimination method is used by the International Olympic Committee to select the cities in which the Olympic Games are held (see Appendix 2).

Like the Borda count method, plurality with elimination uses all the information provided by the voters' preferences (not just first choice), but unlike the Borda count method, it uses it sequentially rather than all at once. In one important respect, however, plurality with elimination is superior to the Borda count method: *It satisfies the majority criterion* (Exercise 47).

Problems with the Plurality-with-Elimination Method

The major flaw in the plurality-with-elimination method is quite subtle and is illustrated by the next example.

Example 4.   Consider an election with 3 candidates (A, B, and C), 29 voters, and the preference schedule

Number of voters	7	8	10	4
1st choice	A	B	C	A
2nd choice	B	C	A	C
3rd choice	C	A	B	B

Under plurality with elimination, B is eliminated in round 1, leaving us with the preference schedule

Number of voters	11	18
1st choice	A	C
2nd choice	C	A

It follows that the winner of this election is C.

Suppose now that due to some technicality the original election is declared void and a new election is held. All the voters vote exactly the same way as in the original election except for the four voters who voted A (1st choice), C (2nd choice), and B (3rd choice) (see the last column of the original preference schedule). They have a change of heart and decide to make C their first choice and A their second choice (they heard that C is going to lower their taxes). Since C won the original election and the only changes in the second election were to give C even more first-place votes, we would expect that C is still the winner. Just to be safe we will check it out.

Switching A and C in the last column makes that column the same as the third column, so the preference schedule for the second election is

Number of voters	7	8	14
1st choice	A	B	C
2nd choice	B	C	A
3rd choice	C	A	B

Under plurality with elimination, A is eliminated in round 1, leaving us with the preference schedule

Number of voters	15	14
1st choice	B	C
2nd choice	C	B

This means that B is now the winner of the election! To anyone with a sense of fairness this turn of events will seem absurd. How is it possible for candidate C, who had the original election won, to then lose the second election simply by virtue of having picked up some extra first-place votes? ▬

This example demonstrates that the plurality-with-elimination method violates a very important fairness criterion known as the **monotonicity criterion.** Informally, the monotonicity criterion says that a candidate that has an election won should not be penalized for getting additional votes. A more formal version of the monotonicity criterion follows.

> **Monotonicity Criterion.** If candidate or alternative X is the winner of an election and, in a reelection, all the voters who change their preferences do so in a way that is favorable only to X, then X should still be the winner of the election.

Example 4 demonstrates that plurality with elimination *violates the monotonicity criterion.* We leave it as an exercise for the reader to verify that plurality with elimination also *violates the Condorcet criterion* (Exercise 54).

THE METHOD OF PAIRWISE COMPARISONS

So far, all three voting methods we have discussed violate the Condorcet criterion. There are several voting methods that satisfy the Condorcet criterion (they are known generically as *Condorcet methods*), but we will discuss only one of the simplest ones: the method of pairwise comparisons.[6]

The **method of pairwise comparisons** is a little like a *round-robin tournament*; every candidate is matched on a one-to-one basis with every other candidate. Each of these one-to-one pairings is called a **pairwise comparison.** When comparing two candidates (say X and Y) one on one, a vote is assigned to X if that voter placed X higher than Y on the ballot. Likewise, if Y is higher than X, the vote goes to Y. When all the votes are tallied, the match (pairwise comparison) between X and Y can result in a win for X (1 point), a tie (1/2 point for each), or a loss for X (0 points). The candidate with the most points when all pairwise comparisons are tabulated is declared the winner. In case of a tie, the tie is broken by some previously agreed upon tie-breaking rule (see Example 6). (Of the various voting methods we discussed so far, this one is the most tie prone.)

For the MAS election, the method of pairwise comparisons works like this. First we compare A with B.

Number of voters	14	10	8	4	1
1st choice	Ⓐ←	C	D	B̲←	C
2nd choice	B̲←	B̲←	C	D	D
3rd choice	C	D	B̲←	C	B̲←
4th choice	D	Ⓐ←	Ⓐ←	Ⓐ←	Ⓐ←

---

[6] This method is sometimes known as *Copeland's method* and attributed to the late Prof. A. H. Copeland of the University of Michigan.

The 14 voters represented in the first column prefer A to B, but all the rest of the voters (23 of them) prefer B to A. Consequently, in the pairwise comparison between A and B, B is the winner. We summarize this result by

A versus B (14 to 23): B wins (B gets 1 point; A gets 0 points).

Our next pairwise comparison is between A and C.

Number of voters	14	10	8	4	1
1st choice	A	C	D	B	C
2nd choice	B	B	C	D	D
3rd choice	C	D	B	C	B
4th choice	D	A	A	A	A

Once again we can summarize the result by

A versus C (14 to 23): C wins (C gets 1 point).

In a like manner we compare A with D, B with C, B with D, and C with D. We leave it to the reader to check the following results:

A versus D (14 to 23): D wins (D gets 1 point)
B versus C (18 to 19): C wins (C gets 1 point)
B versus D (28 to 9): B wins (B gets 1 point)
C versus D (25 to 12): C wins (C gets 1 point).

The final scoreboard reads: A, 0 points; B, 2 points; C, 3 points, and D, 1 point. Since C has the most points, she is declared the winner by the method of pairwise comparisons.

The Math Appreciation Society News

CARMEN ELECTED PRESIDENT OF MAS!

MATH APPRECIATION SOCIETY ELECTION RESULTS

Winner	Voting method
Alisha	Plurality
Boris	Borda count
Dave	Plurality with elimination
Carmen	Pairwise comparisons

One election, four different methods, four different winners!

It is easy to see that the method of pairwise comparisons *satisfies the Condorcet criterion*. After all, if a candidate wins every match, then that candidate will automatically have the most points and win the election. It is not much harder to see that the method of pairwise comparisons *satisfies the majority criterion* (Exercise 48). Just a tad harder to see but still true is the fact that the method of pairwise comparisons *satisfies the monotonicity criterion* (Exercise 55).

Problems with the Method of Pairwise Comparisons

Since the method of pairwise comparisons satisfies all three of the criteria we have discussed so far, one may ask, Is there anything wrong with the method of pairwise comparisons? The answer, unfortunately, is yes. We illustrate one of the main problems with the method of pairwise comparisons in the next example.

Example 5. Consider an election with 5 candidates (*A, B, C, D,* and *E*), 22 voters, and the preference schedule

Number of voters	5	3	5	3	2	4
1st choice	A	A	C	D	D	B
2nd choice	B	D	E	C	C	E
3rd choice	C	B	D	B	B	A
4th choice	D	C	A	E	A	C
5th choice	E	E	B	A	E	D

We want to determine the winner of this election using the method of pairwise comparisons. We leave it to the reader to verify that the results of the ten possible pairwise comparisons are as follows:

A versus *B* (13 to 9): *A* wins

A versus *C* (12 to 10): *A* wins

A versus *D* (12 to 10): *A* wins

A versus *E* (10 to 12): *E* wins

B versus *C* (12 to 10): *B* wins

B versus *D* (9 to 13): *D* wins

B versus *E* (17 to 5): *B* wins

C versus *D* (14 to 8): *C* wins

C versus *E* (18 to 4): *C* wins

D versus *E* (13 to 9): *D* wins.

Since *A* has 3 points, *B* has 2 points, *C* has 2 points, *D* has 2 points, and *E* has 1 point, *A* is the winner of this election under the method of pairwise comparisons.

Now suppose that, for whatever reason, the votes have to be recounted, but before they are, candidates *B, C,* and *D* become discouraged and drop out of the race, leaving only candidates *A* and *E.* Eliminating candidates *B, C,* and *D* gives us the preference schedule

Number of voters	5	3	5	3	2	4
1st choice	A	A	~~C~~	~~D~~	~~D~~	~~B~~
2nd choice	~~B~~	~~D~~	E	~~C~~	~~C~~	E
3rd choice	~~C~~	~~B~~	~~D~~	~~B~~	~~B~~	A
4th choice	~~D~~	~~C~~	A	E	A	~~C~~
5th choice	E	E	~~B~~	A	E	~~D~~

After simplification, this schedule becomes:

Number of voters	10	12
1st choice	A	E
2nd choice	E	A

This means that the winner is now *E*! Under the method of pairwise comparisons the original winner was *A*, but when some of the candidates dropped out, the winner became *E*. This seems somewhat unfair: Why should a candidate be penalized for the actions of some of the other candidates?

In the language of voting theory we say that the method of pairwise comparisons violates the **independence of irrelevant alternatives criterion.**

> **Independence of Irrelevant Alternatives Criterion.** If candidate or alternative *X* is the winner of an election, and one or more of the other candidates or alternatives are removed and the ballots recounted, then *X* should still be the winner of the election.

Another problem with the method of pairwise comparisons is that often it is unable to produce a winner. Consider the following simple example.

Example 6

Number of voters	4	2	5
1st choice	A	B	C
2nd choice	B	C	A
3rd choice	C	A	B

Here *A* beats *B* (9 to 2), *B* beats *C* (6 to 5), and *C* beats *A* (7 to 4). This results in a three-way tie. How should we break the tie? There is no set way to break such a tie, and in practice, it is important to establish the rules as to how ties are to be broken ahead of time. Otherwise, consider what might happen. Candidate *C* could argue, not unreasonably, that the tie should be broken by counting first-place votes. In this case, candidate *C* would win. On the other hand, candidate *A* could make an equally persuasive argument that the tie should be broken by counting total points (Borda count). In this case *A* has 24 points and *C* has 23, so *A* would win. As the reader can see, it would have been smart to have thought about these things before the election. For a more detailed discussion of ties and how to break them, the reader is encouraged to look at Appendix 1 at the end of this chapter. ◼

The Number of Pairwise Comparisons

There is one final difficulty with the method of pairwise comparisons and it is a practical one: There seem to be a lot of pairwise comparisons to check out. In fact, how many pairwise comparisons do we really have to check? We saw in the MAS election that with 4 candidates we needed to check 6 pairwise comparisons, and in Example 5 we had 5 candidates and we checked 10 pairwise comparisons. To make it a little more challenging let's say that we have an election with 12 candidates. How many pairwise comparisons are possible? Let's try to systematically count the comparisons, making sure that we don't count any comparison twice.

- ◼ The first candidate must be compared with each of the other 11 candidates—*11 pairwise comparisons.*

- ◼ Compare the second candidate with each of the other candidates *except* the first one, since that comparison has already been made—*10 pairwise comparisons.*

- ◼ Compare the third candidate with each of the other candidates *except* the first and second candidates, since those comparisons have already been made—*9 pairwise comparisons.*

$$\vdots$$

- ◼ Compare the eleventh candidate with each of the other candidates *except* the first 10 candidates, since those comparisons have already been made. In other words, compare the eleventh candidate with the twelfth candidate—*1 pairwise comparison.*

We see that the total number of pairwise comparisons is

$$1 + 2 + 3 + 4 + 5 + 6 + 7 + 8 + 9 + 10 + 11 = 66.$$

How many pairwise comparisons are needed if there are 100 candi-

Well, using an argument similar to the preceding one, the total number of comparisons needed is

$$1 + 2 + 3 + 4 + \cdots + 99$$

and in a more generic vein we can say that if there are N candidates, the total number of comparisons is

$$1 + 2 + 3 + 4 + \cdots + (N - 1).$$

The next thing we are going to do is learn about a very useful mathematical formula. Let's go back to the case of 100 candidates. How much is $1 + 2 + 3 + \cdots + 99$? One could always add up these numbers, but that is not very imaginative, to say nothing of the fact that it is also a lot of work.

Let's try a different way to count the comparisons. Suppose that before a pairwise comparison between two candidates takes place, each candidate gives the other one his or her business card. Then clearly, each candidate would end up with the business card of every other candidate, and there would be a total of $99 \times 100 = 9900$ cards handed out (each of the 100 candidates would hand out 99 cards, 1 to each of the other candidates). But since each comparison resulted in 2 cards being handed out, the total number of comparisons must be half as many as the number of cards. Consequently,

$$1 + 2 + 3 + 4 + \cdots + 99 = \frac{99 \times 100}{2} = 4950.$$

Similar arguments show that if there are N candidates, the number of pairwise comparisons needed is

$$1 + 2 + 3 + \cdots + (N - 1) = \frac{(N - 1)N}{2}.$$

Because the number of comparisons grows quite fast in relation to the number of candidates, the method of pairwise comparisons is cumbersome and time-consuming and is seldom used in practice as a method for deciding elections. This is unfortunate, since in spite of its flaws, the method of pairwise comparisons can be a fairly good voting method.

RANKINGS

Quite often it is important not only to know who wins the election but also to know who comes in second, third, etc. Let's consider for example, the MAS election but suppose now that instead of electing just a president we need to elect a board of directors consisting of a president, a vice president, and a treasurer. The organization's constitution states that the winner of the election should become the president, the second-place candidate should become the vice president, and the third-place candidate should become the treasurer. Clearly, we need our voting method not only to tell us the winner but also who is second, third, etc.—in other words, a **ranking** of the candidates.

Extended Ranking Methods

Each of the four voting methods we discussed at the beginning of this chapter can be extended in a very natural way to produce a ranking of the candidates.

Let's start with the plurality method and see how we might extend it to produce a ranking of the four candidates in the MAS election. For the reader's convenience, here is the instant replay.

Number of voters	14	10	8	4	1
1st choice	A	C	D	B	C
2nd choice	B	B	C	D	D
3rd choice	C	D	B	C	B
4th choice	D	A	A	A	A

A: 14 first-place votes

B: 4 first-place votes

C: 11 first-place votes

D: 8 first-place votes.

We know that using the plurality method, A is the winner. Who should be second? The answer seems obvious: C has the second most first-place votes (11), so we declare C to come in second. By the same token, we declare D to come in third (8 votes) and B last. Summarizing, the extended plurality method gives us the following ranking:

Winner (president): A (14 first-place votes)

2nd place (vice president): C (11 first-place votes)

3rd place (treasurer): D (8 first-place votes)

Last: B (4 first-place votes).

Ranking the candidates using the *extended Borda count method* is equally obvious. In the MAS election, for example, the point totals under the Borda count method were A, 79 points; B, 106 points; C, 104 points; and D, 81 points. The resulting ranking, based on point totals would be

Winner (president): B (106 points)

2nd place (vice president): C (104 points)

3rd place (treasurer): D (81 points)

Last: A (79 points).

Ranking the candidates using *extended plurality with elimination* is only a bit more subtle: We rank them in reverse order of elimination (the first candidate eliminated is ranked last, the second candidate eliminated is ranked next to last, etc.).[7] If we applied this method to rank the candidates in the MAS election, the result would be

Winner (president): *D*

2nd place (vice president): *A* (eliminated last)

3rd place (treasurer): *C* (eliminated second)

Last: *B* (eliminated first).

Last, we can rank the candidates using the *extended method of pairwise comparisons* according to the number of points (recall that we count a tie as $\frac{1}{2}$ point). In the case of the MAS election, *C* had 3 points, *B* had 2 points, *D* had 1 point, and *A* had 0 points, so the results are

Winner (president): *C* (3 points)

2nd place (vice president): *B* (2 points)

3rd place (treasurer): *D* (1 point)

Last: *A* (0 points).

Recursive Ranking Methods

We will now discuss a different, somewhat more involved strategy for ranking the candidates which we will call the **recursive method.** When using this approach, the basic strategy is the same regardless of which voting method we choose—only the details are different.

Let's say we are going to use voting method *X* and a recursive strategy to get a ranking of the candidates after an election. We first use method *X* to find the winner of the election. So far, so good. We then cross out the name of the winner on the preference schedule and obtain a new, modified preference schedule with one less candidate on it. We apply method *X* once again to find the "winner" based on this new preference schedule and this candidate is ranked second. (This makes a certain amount of sense: What we're saying, is that after the winner is removed we run a brand new race and the best candidate in that race is the second-best candidate overall.) We repeat the process again (cross out the name of the last winner, calculate the new preference schedule, and apply method *X* to find the next winner who is then placed next in line in the ranking) until we have ranked as many of the candidates as we want.

We will illustrate the basic idea of recursive ranking with a couple of examples, both based on the MAS election.

---

[7] In cases where a candidate gets a majority of first-place votes before the ranking of all the candidates is complete, we continue the process of elimination to rank the remaining candidates.

Example 7. Suppose we want to rank the four candidates in the Math Appreciation Society election using the *recursive plurality method*. The preference schedule, once again is

Number of voters	14	10	8	4	1
1st choice	A	C	D	B	C
2nd choice	B	B	C	D	D
3rd choice	C	D	B	C	B
4th choice	D	A	A	A	A

■ **Step 1.** (Choose the winner using plurality.) We already know the winner is A with 14 first-place votes.

■ **Step 2.** (Choose second place.) First we cross out A from the original schedule—this gives us a "new" preference schedule to work with.

Number of voters	14	10	8	4	1
1st choice	A̶	C	D	B	C
2nd choice	B	B	C	D	D
3rd choice	C	D	B	C	B
4th choice	D	A̶	A̶	A̶	A̶

→

Number of voters	14	10	8	4	1
1st choice	B	C	D	B	C
2nd choice	C	B	C	D	D
3rd choice	D	D	B	C	B

In this schedule the winner using plurality is B, with 18 first-place votes. Thus, second place goes to B.

■ **Step 3.** (Choose third place.) We now cross out B from the preceding schedule. The resulting schedule is

Number of voters	14	10	8	4	1
1st choice	B̶	C	D	B̶	C
2nd choice	C	B̶	C	D	D
3rd choice	D	D	B̶	C	B̶

→

Number of voters	25	12
1st choice	C	D
2nd choice	D	C

Using plurality, the winner for this schedule is C with 25 first-place votes. This means that third place goes to C, and needless to say, last place goes to D.

We have the following ranking of the candidates:

Winner (president): A
2nd place (vice president): B
3rd place (treasurer): C
Last: D.

It is worth noting how different this ranking turned out to be when compared with the earlier ranking obtained using the extended plurality method. In fact, other than first place (which will always be the same), all the other positions turned out to be different.

Example 8. For our second example, we will apply the recursive plurality-with-elimination method to rank the candidates in the MAS election.

- ■ **Step 1.**   We apply the plurality-with-elimination method to the original preference schedule and get a winner: D. (We did all the busy work earlier.)
- ■ **Step 2.**   We now cross out the winner D on the preference schedule and get

Number of voters	14	10	8	4	1
1st choice	A	C	~~D~~	B	C
2nd choice	B	B	C	~~D~~	~~D~~
3rd choice	C	~~D~~	B	C	B
4th choice	~~D~~	A	A	A	A

This results in the revised schedule

Number of voters	14	10	8	4	1
1st choice	A	C	C	B	C
2nd choice	B	B	B	C	B
3rd choice	C	A	A	A	A

which is the same as

Number of voters	14	19	4
1st choice	A	C	B
2nd choice	B	B	C
3rd choice	C	A	A

Once again, we apply the plurality-with-elimination method to the revised schedule. B is eliminated first, and A second (we leave it to the reader to verify the details), leaving C as the winner. This means that second place in the original election goes to C!

- ■ **Step 3.**   We now cross out C on the last preference schedule, leaving the revised preference schedule

Number of voters	14	23
1st choice	A	B
2nd choice	B	A

The winner of this election under plurality with elimination is B. This means that third place goes to B (and of course it follows that A is last).

The final ranking under recursive plurality with elimination is

> Winner (president): D
> 2nd place (vice president): C
> 3rd place (treasurer): B
> Last: A.

Note that this results in a ranking different from the one produced by the extended plurality-with-elimination method.

While somewhat more complicated than the extended ranking methods, the recursive ranking methods are of both practical and theoretical importance. We encourage the reader to give them a try (Exercises 31 through 40).

CONCLUSION: FAIRNESS AND ARROW'S IMPOSSIBILITY THEOREM

When is a voting method fair? Throughout this chapter we have introduced several tests of fairness known as *fairness criteria*.[8] Let's review what they are.

- **Majority Criterion.**   If there is a candidate or alternative that is the first choice of a majority of the voters, then that candidate or alternative should be the winner of the election.

- **Condorcet Criterion.**   If there is a candidate or alternative that wins in a one-to-one comparison between it and any other alternative, then that candidate or alternative should be the winner of the election.

- **Monotonicity Criterion.**   If candidate or alternative X is the winner of an election and, in a reelection, all the voters who change their preferences do so in a way that is favorable only to X, then X should still be the winner of the election.

[8] Singular: criterion; plural: criteria.

■ **Independence of Irrelevant Alternatives Criterion.**   If candidate or alternative X is the winner of an election, and one or more of the other candidates or alternatives are removed and the ballots recounted, then X should still be the winner of the election.

Each one of the above criteria represents a reasonable expectation for fairness, and if a voting method does not satisfy all of them, it cannot be legitimately thought of as being completely fair. Is there a voting method that satisfies all four of the fairness criteria, if you will, a perfect or ideal voting method? Unfortunately, when choosing among three or more alternatives the answer is No! *No ideal voting method exists.*

At first glance, this fact seems a little surprising. Given the obvious importance of elections in a democracy and given the collective intelligence and imagination of mankind, how is it possible that no one has come up with a voting method that satisfies all of the fairness criteria? Up until the early 1950s this was a vexing question for scholars in social choice theory. Then, in 1952, a bombshell! Kenneth Arrow demonstrated the now famous **Arrow's impossibility theorem:** *When choosing among three or more alternatives, satisfying all of the fairness criteria is a mathematical impossibility.* No matter how hard we try, there can be no ideal voting method—when it comes to the mathematics of elections, fairness and democracy are inherently incompatible.

KEY CONCEPTS

Arrow's impossibility theorem	method of pairwise comparisons
Borda count method	monotonicity criterion
Condorcet candidate	plurality method
Condorcet criterion	plurality-with-elimination method
extended rankings methods	preference ballot
independence of irrelevant alternatives criterion	preference schedule
insincere (strategic) voting	rankings
majority criterion	recursive ranking methods
	transitive preferences

EXERCISES

Walking

1. The management of the XYZ Corporation has decided to treat their office staff to dinner. The choice of restaurants is The Atrium (*A*), Blair's Kitchen (*B*), The Country Cookery (*C*), and Dino's Steak House (*D*). Each of the 12 staff members is asked to submit a preference ballot listing his or her first, second, third, and fourth

<table>
<tr><td>

Ballot
1st A
2nd B
3rd C
4th D

</td><td>

Ballot
1st C
2nd D
3rd B
4th A

</td><td>

Ballot
1st B
2nd D
3rd C
4th A

</td><td>

Ballot
1st A
2nd B
3rd C
4th D

</td></tr>
</table>

Ballot	**Ballot**	**Ballot**	**Ballot**
1st C	1st C	1st A	1st C
2nd B	2nd B	2nd B	2nd D
3rd D	3rd D	3rd C	3rd B
4th A	4th A	4th D	4th A

Ballot	**Ballot**	**Ballot**	**Ballot**
1st A	1st A	1st C	1st A
2nd B	2nd B	2nd B	2nd B
3rd C	3rd C	3rd D	3rd C
4th D	4th D	4th A	4th D

choices among these restaurants. The resulting preference ballots are shown in the left margin.

(a) Write out the preference schedule for this election.
(b) Is there a majority winner?
(c) Find the winner of the election using the plurality method.
(d) Find the winner of the election using the Borda count method.

2. The Latin Club is holding an election to choose its president. There are three candidates, Arsenio, Beatrice, and Carlos (*A, B,* and *C* for short). The other 11 members of the club (the candidates are not allowed to vote) vote as shown below.

Voter	Sue	Bill	Tom	Pat	Tina	Mary	Alan	Chris	Paul	Kate	Ron
1st place	C	A	C	A	B	C	C	A	C	B	A
2nd place	A	C	B	B	A	B	A	C	A	A	B
3rd place	B	B	A	C	C	A	B	B	B	C	C

(a) Write out the preference schedule for this election.
(b) Is there a majority winner?
(c) Find the winner of the election using the plurality method.
(d) Find the winner of the election using the Borda count method.

Exercises 3 through 6 refer to the following: A math class is asked by the instructor to vote among four possible times for the final exam—A (December 15, 8:00 A.M.), B (December 20, 9:00 P.M.), C (December 21, 7:00 A.M.), and D (December 23, 11:00 A.M.). The class preference schedule is given below:

Number of voters	5	11	6	3	5	9	5	2	5
1st choice	A	A	A	A	B	B	B	C	D
2nd choice	B	B	C	D	A	A	D	B	B
3rd choice	C	D	B	B	C	D	A	A	A
4th choice	D	C	D	C	D	C	C	D	C

3. (a) How many students in the class voted?
   (b) Find the winner of the election using the plurality method.

4. Find the winner of the election using the Borda count method.

5. Find the winner of the election using the plurality-with-elimination method.

6. Find the winner of the election using the method of pairwise comparisons.

Exercises 7 through 13 refer to the election with 5 candidates (A, B, C, D, and E), 22 voters, and the preference schedule given below (see Example 5).

Number of voters	5	3	5	3	2	4
1st choice	A	A	C	D	D	B
2nd choice	B	D	E	C	C	E
3rd choice	C	B	D	B	B	A
4th choice	D	C	A	E	A	C
5th choice	E	E	B	A	E	D

7. Find the winner of the election using the Borda count method.

8. Find the winner of the election using
   (a) the plurality method
   (b) the plurality-with-elimination method.

9. Suppose that, for some unexplained reason, the votes in this election have to be recounted, but before this is done, candidate D drops out of the election.
   (a) Write out the preference schedule after D drops out.
   (b) Find the winner of the new election using the method of pairwise comparisons.

10. Find the ranking of the candidates in the original election using the extended plurality-with-elimination method.

11. Find the ranking of the candidates in the original election using the extended plurality method.

12. Find the ranking of the candidates in the original election using the extended pairwise comparisons method.

13. Find the ranking of the candidates in the original election using the extended Borda count method.

Exercises 14 through 20 refer to the election with 5 alternatives (R, H, C, O, and S), 100 voters, and the preference schedule given below (see Example 1).

Number of voters	49	48	3
1st choice	R	H	C
2nd choice	H	S	H
3rd choice	C	O	S
4th choice	O	C	O
5th choice	S	R	R

14. Find the winner of the election using the Borda count method.

15. Find the winner of the election using the plurality-with-elimination method.

16. Find the winner of the election using the method of pairwise comparisons.

17. Find the ranking of the alternatives using the extended plurality method.

18. Find the ranking of the alternatives using the extended plurality-with-elimination method.

19. Find the ranking of the alternatives using the extended Borda count method.

20. Find the ranking of the alternatives using the extended pairwise comparisons method.

Exercises 21 through 24 refer to an election with 5 candidates (A, B, C, D, and E), 24 voters, and the preference schedule given by the following table.

Number of voters	8	6	5	3	2
1st choice	B	D	D	C	E
2nd choice	A	B	B	A	A
3rd choice	C	A	E	B	D
4th choice	D	C	C	D	B
5th choice	E	E	A	E	C

21. **(a)** Find the winner of the election using the Borda count method.
   (b) Does this example illustrate a violation of the majority criterion? Why or why not?

22. **(a)** Find the winner of the election using the plurality-with-elimination method.
   (b) Does this example illustrate a violation of the Condorcet criterion? Why or why not?

23. Find the ranking of the candidates using the extended plurality method.

24. Find the ranking of the candidates using the extended Borda count method.

Exercises 25 through 27 refer to an election with 5 candidates (A, B, C, D, and E), 26 voters, and the preference schedule given below.

Number of voters	5	3	8	7	3
1st choice	C	D	E	B	A
2nd choice	B	A	B	A	D
3rd choice	D	C	A	C	C
4th choice	E	B	C	D	E
5th choice	A	E	D	E	B

25. **(a)** Find the winner of the election using the Borda count method.
   (b) Does this example illustrate a violation of the majority criterion? Why or why not?

26. **(a)** Find the winner of the election using the plurality-with-elimination method.
   (b) Does this example illustrate a violation of the Condorcet criterion? Why or why not?

27. Find the ranking of the candidates using the extended pairwise comparisons method.

Exercises 28 through 30 refer to an election with 4 candidates (A, B, C, and D), 27 voters, and the preference schedule given below.

Number of voters	5	7	5	4	3	3
1st choice	A	B	D	C	B	D
2nd choice	B	D	C	D	A	C
3rd choice	C	A	A	A	C	B
4th choice	D	C	B	B	D	A

28. **(a)** Find the winner of the election using the plurality method.
    (b) Does this example illustrate a violation of the Condorcet criterion? Why or why not?

29. **(a)** Find the winner of the election using the Borda count method.
    (b) Does this example illustrate a violation of the majority criterion? Why or why not?

30. Find the ranking of the candidates using the extended plurality-with-elimination method.

Exercises 31 and 32 refer to the MAS election. For the reader's convenience the preference schedule is given below.

Number of voters	14	10	8	4	1
1st choice	A	C	D	B	C
2nd choice	B	B	C	D	D
3rd choice	C	D	B	C	B
4th choice	D	A	A	A	A

31. Find the ranking of the candidates using the recursive Borda count method.

32. Find the ranking of the candidates using the recursive pairwise comparisons method.

Exercises 33 through 36 refer to the election with 5 candidates (A, B, C, D and E), 24 voters, and the preference schedule given below (see Example 3).

Number of voters	8	6	2	3	5
1st choice	A	B	C	D	E
2nd choice	B	D	A	E	A
3rd choice	C	E	E	A	D
4th choice	D	C	B	C	B
5th choice	E	A	D	B	C

33. Find the ranking of the candidates using the recursive plurality method.

34. Find the ranking of the candidates using the recursive plurality-with-elimination method.

35. Find the ranking of the candidates using the recursive pairwise comparisons method.

36. Find the ranking of the candidates using the recursive Borda count method.

Exercises 37 through 40 refer to the same election as Exercises 25 through 27.

37. Find the ranking of the candidates using the recursive Borda count method.

38. Find the ranking of the candidates using the recursive plurality method.

39. Find the ranking of the candidates using the recursive plurality-with-elimination method.

40. Find the ranking of the candidates using the recursive pairwise comparisons method.

Jogging

41. **(a)** Suppose that 50 players sign up for a round-robin tennis tournament (everyone plays everyone else). How many tennis matches must be scheduled?
    (b) If there are 50 people in a room and everyone kisses everyone else (on the cheek of course), how many kisses take place?

42. Show that the four voting methods we discussed in this chapter give the same winner when there are only two candidates, and in fact that the winner is just determined by straight majority.

43. **The mystery election problem.** You are given the following information about an election. There are 5 candidates and 21 voters. The preference schedule for the election has been lost—the only thing you know is that there were only *two* columns in the schedule. (As you can imagine, this is the key piece of information in the problem.)
    (a) Explain why there must be a majority winner in this election.
    (b) Explain why the argument given in (a) still works with any number of candidates and any odd number of voters.

44. An election is held among 4 candidates (*A, B, C,* and *D*) using the Borda count method. There are 11 voters. Suppose that after the ballots are in and the points are tallied, *B* gets 32 points, *C* gets 29 points, and *D* gets 18 points. How many points does *A* get, and why?

45. An election involving 5 candidates and 21 voters is held, and the results of the election are to be determined using the Borda count method. Unfortunately, the elections committee completely botches up the addition of the points. After a complaint is lodged by one of the candidates, the results of the election are recomputed using 4 points for a first-place vote, 3 points for a second-place vote, 2 points for a third-place vote, 1 point for a fourth-place vote, and 0 points for a fifth-place vote. Explain why computing the results this way gives the same election outcome as using the traditional Borda count method.

46. Explain why the plurality method satisfies the monotonicity criterion.

47. Explain why the plurality-with-elimination method satisfies the majority criterion.

48. Explain why the method of pairwise comparisons satisfies the majority criterion.

49. Explain why any method that violates the majority criterion must also violate the Condorcet criterion.

50. Give an example of an election decided under the Borda count method in which the Condorcet criterion is violated but the majority criterion is not.

51. The table below shows the three top-ranked college football teams at the end of the 1993 football season, according to the CNN/*USA Today* coaches poll.

Team	Points	Number of first-place votes
1. Florida State	1523	36
2. Notre Dame	1494	25
3. Nebraska	1447	1
⋮	⋮	⋮

This poll is based on the votes of 62 coaches, each one of whom ranks the top 25 teams. (The remaining 22 teams are not shown because they are irrelevant to this exercise.) A team gets 25 points for each first-place vote, 24 points for each second-place vote, 23 points for each third-place vote, etc.

(a) Based on the information given in the table, it is possible to conclude that all 62 coaches had Florida State, Notre Dame, and Nebraska in some order as their top three choices. Explain why this is true.

(b) Find the number of second- and third-place votes for each of the three teams.

52. Plurality with a runoff. This method is a simple variation of the plurality-with-elimination method. In this method, if a candidate has a majority of the first-place votes then that candidate wins the election; otherwise we eliminate all candidates except the two with the most first-place votes. The winner is then chosen between these two by recounting the votes in the usual way.

(a) Use the MAS election to show that plurality with a runoff can produce a different outcome than plurality with elimination.

(b) Give an example that shows that plurality with a runoff violates the monotonicity criterion.

(c) Give an example that shows that plurality with a runoff violates the Condorcet criterion.

Running

53. The Coombs method. This method is just like the plurality-with-elimination method except that in each round we eliminate the candidate with the *largest number of last-place votes* (instead of the one with the fewest number of first-place votes).

(a) Find the winner of the MAS election using the Coombs method.

(b) Give an example that shows that the Coombs method violates the Condorcet criterion.

(c) Give an example that shows that the Coombs method violates the monotonicity criterion.

54. Give an example (do not use one given in the book) of an election decided under the plurality-with-elimination method in which the Condorcet criterion is violated.

55. Show that the method of pairwise comparisons satisfies the monotonicity criterion.

56. Show that if in an election with an odd number of voters there is no Condorcet candidate, then any ranking of the candidates based on the extended pairwise comparisons method must result in at least two candidates that end up tied in the rankings.

57. The Pareto criterion. The following fairness criterion was proposed by the Italian economist Vilfredo Pareto (1848–1923): *If every voter prefers alternative X over alternative Y, then a voting method should not choose Y as the winner.* Show that all four voting methods discussed in the chapter satisfy the Pareto criterion. (A separate analysis is needed for each of the four methods.)

58. Suppose the following was proposed as a fairness criterion: *If a majority of the voters prefer alternative X to alternative Y, then the voting method should rank X above Y.* Give an example to show that all four of the extended voting methods discussed in the chapter can violate this criterion. (*Hint:* Consider an example with no Condorcet candidate.)

59. Consider the following fairness criterion: *If a majority of the voters prefer every* *alternative over alternative X, then a voting method should not choose alternative X as the winner.*

   (a) Give an example to show that the plurality method can violate this criterion.

   (b) Give an example to show that the plurality-with-elimination method can violate this criterion.

   (c) Explain why the method of pairwise comparisons satisfies this criterion.

   (d) Explain why the Borda count method satisfies this criterion.

60. The Condorcet loser criterion. *If there is an alternative that loses in a one-to-one comparison to each of the other alternatives, then that alternative should not be the winner of the election.* (This fairness criterion is a sort of mirror image of the regular Condorcet criterion.)

   (a) Give an example that shows that the plurality method can violate the Condorcet loser criterion.

   (b) Explain why the plurality-with-elimination method violates the Condorcet loser criterion.

   (c) Explain why the Borda count method satisfies the Condorcet loser criterion.

APPENDIX 1:
Breaking Ties

By and large, most of the examples given in the chapter were carefully chosen to avoid tied winners, but of course in the real world ties are sometimes inevitable.

In this appendix we will discuss very briefly the problem of how to break ties when they occur. Tie-breaking methods can raise some fairly complex issues, and it is not our purpose here to study such methods in great detail but rather to make the reader aware of the problem and give some inkling as to possible ways to deal with it.

For starters, consider the election with two candidates (A and B) and the preference schedule

Number of voters	10	10
1st choice	A	B
2nd choice	B	A

It should be clear that no reasonable voting method could choose A as the winner over B or choose B as the winner over A. If the roles of A and B were interchanged, the preference schedule would look exactly the same. Neither candidate is favored in any way over the other. A tie is inevitable. We might call this kind of tie an **essential tie**.

The election with three candidates (A, B, and C) and the preference schedule

Number of voters	7	7	7
1st choice	A	B	C
2nd choice	B	C	A
3rd choice	C	A	B

also represents an essential tie: No reasonable voting method can choose one candidate over the others. All three are interchangeable in the sense that any two candidates could switch names and we wouldn't be able to tell the difference (identical twins do it all the time!).

Essential ties cannot be broken using a rational tie-breaking procedure, and we must rely on some sort of outside intervention such as chance (flip a coin, draw straws, etc.), a third party (the president of the club, the chair of the committee, mom, etc.), or even some form of tradition (seniority, age, etc.).

Most ties are not essential ties, and they can often be broken in a more rational way: either by implementing some tie-breaking rule or by using a different voting method to break the tie. We illustrate some of these ideas with the following example:

Consider an election with 5 candidates (A, B, C, D, and E), 22 voters, and the preference schedule

Number of voters	5	3	5	3	2	4
1st choice	A	A	C	D	D	B
2nd choice	B	B	E	C	C	E
3rd choice	C	D	D	B	B	A
4th choice	D	C	A	E	A	C
5th choice	E	E	B	A	E	D

Using the method of pairwise comparisons, we have

      A versus B (13 to 9): A wins

      A versus C (12 to 10): A wins

      A versus D (12 to 10): A wins

      A versus E (10 to 12): E wins

      B versus C (12 to 10): B wins

      B versus D (12 to 10): B wins

      B versus E (17 to 5): B wins

      C versus D (14 to 8): C wins

      C versus E (18 to 4): C wins

      D versus E (13 to 9): D wins.

Since A has 3 points, B has 3 points, C has 2 points, D has 1 point, and E has 1 point, A and B tie for the winner.

We can break this tie in several different ways. For example,

1. Use the results of head-to-head competition between the winners. In the above example, since A beats B in a pairwise comparison, the tie would be broken in favor of A.

2. Use the *total point differentials*. For example, since A beats B 13 to 9, the point differential for A is $+4$, and since A lost to E 10 to 12, the point differential for A is -2. Computing the total point differentials for A gives $4 + 2 + 2 - 2 = 6$. Likewise, the total point differentials for B is $2 + 2 + 12 - 4 = 12$. In this case the point differentials favor B, so B would therefore be declared the winner.

3. Use first-place votes. In the example, A has 8 and B has 4. With this method, the winner would be A.

4. Use a Borda count between the winners. In the example, we have

$$\text{Borda count for } A: (5 \times 8) + (3 \times 4) + (2 \times 7) + (1 \times 3) = 69.$$

$$\text{Borda count for } B: (5 \times 4) + (4 \times 8) + (3 \times 5) + (1 \times 5) = 72.$$

With this method, the tie would be broken in favor of B.

By now we should not be at all surprised that different tie-breaking methods produce different winners and that there is no single *right* method for breaking ties. In retrospect, flipping a coin might not be such a bad idea!

APPENDIX 2: Realpolitik. Elections in the Real World

Olympic Venues. Selecting the city in which the next Olympic Games are to be held is a decision entrusted to the members of the International Olympic Committee. It is a decision that has tremendous economic and political impact on the cities involved, and it goes without saying that it always generates a fair amount of controversy. The actual voting method used to select the winner is a slight variation of the plurality-with-elimination method which we studied in the chapter. In this case, instead of indicating their preferences all at once, the voters make their preferences known one round at a time. We illustrate the procedures with the actual details of how Sydney, Australia was chosen to host the 2000 Summer Olympic Games.

On September 23, 1993, the 89 members of the International Olympic Committee met in Monte Carlo, Monaco, to vote on the selection of the site for the 2000 Summer Olympics. Five cities made bids: Beijing (China), Berlin (Germany), Istanbul (Turkey), Manchester (England), and Sydney (Australia). In each round, the delegates voted for just one city, and the city with the fewest votes was eliminated. The voting went as follows:

■ **Round 1.**

City	Beijing	Berlin	Istanbul	Manchester	Sydney
Votes	32	9	7	11	30

Istanbul is eliminated in round 1.

■ **Round 2.**

City	Beijing	Berlin	Manchester	Sydney
Votes	37	9	13	30

Berlin is eliminated in round 2.

■ **Round 3.**

City	Beijing	Manchester	Sydney	Abstentions
Votes	40	11	37	1

Manchester is eliminated in round 3.

■ **Round 4.**

City	Beijing	Sydney	Abstentions
Votes	43	45	1

Beijing is eliminated in round 4; Sydney gets the gold!

It is worth noting that, while in theory this method is equivalent to plurality with elimination, the fact that the voters are allowed to indicate their preferences in rounds rather than all at once allows for a certain irrational (or should we say fickle?) behavior on the part of the voters. After Berlin was eliminated in round 2, two of the voters who voted for Manchester changed their votes (Manchester went from 13 votes in round 2 to 11 votes in round 3.) What connection is there, one wonders, between the elimination of Berlin and Manchester's merit as an Olympic venue? One of the advantages of asking the voters to indicate their preferences all at once (by means of a preference ballot) is that this kind of irrational voting is eliminated.

There is one thing that is even more disturbing about the selection procedure: As we know, this voting method can violate the monotonicity criterion. A city may actually lose the election because of *too many* votes in an early round.

Sydney gets the 2000
Summer Olympics.
(AP/Wide World Photos)

Consider, for example, the following hypothetical scenario. There are three cities in the running for the next Olympic site: Antioch, Babylon, and Carthage. Two days before the election Antioch has the support of 33 delegates, Babylon has the support of 24 delegates, and Carthage has the support of 29 delegates. Let's assume that the second choice of all Antioch supporters is Babylon, but that the second choice of all Babylon supporters is Carthage. If the election had been held two days earlier, it would have gone like this:

■ **Round 1.**

City	Antioch	Babylon	Carthage
Votes	33	24→	29

Babylon is eliminated, and all of Babylon's votes are shifted to Carthage.

■ **Round 2.**

City	Antioch	Carthage
Votes	33	53

Carthage wins the election.

Over the next two days, the Carthage representatives press their advantage and continue their lobbying efforts trying to win even more delegates. Things go so well for them that 12 of the 33 Antioch supporters are persuaded to vote for Carthage. When the election is finally held, this is what happens.

■ **Round 1.**

City	Antioch	Babylon	Carthage
Votes	21→	24	41

Antioch is eliminated, and all of Antioch's votes are shifted to Babylon.

■ **Round 2.**

City	Babylon	Carthage
Votes	45	41

They are dancing in the streets of Babylon.

The Associated Press College Football Poll. In this election (which is held once a week during the college football season), the voters are a panel of approximately 60 national sportswriters, and the candidates are the top college football teams in the country. Each voter ranks their top 25 choices, with 25 points for first place, 24 points for second place, . . . , and 1 point for twenty-fifth place. The team with the highest number of total points is ranked first in the poll, the one with the next highest number of points second, etc. Essentially this method of ranking is the extended Borda count method with one subtle variation: The list of candidates is open-ended (any college football team in the country can in principle be considered a candidate). As we know, a problem with this voting method is that it can violate the majority criterion: A football team could have a majority of first-place votes and yet not be ranked first. In practice, the probability of this happening is very small.

The Academy Awards.   The Academy of Motion Picture Arts and Sciences gives its annual Academy Awards ("Oscars") for various achievements in connection with motion pictures (best picture, best director, best actress, etc.). The winner in each category is chosen by means of an election held among the eligible members of the Academy. The election process varies slightly from award to award and is quite complicated. For the sake of brevity we will describe the election process for best picture. (The process is almost identical for each of the major awards.) The election takes place in two stages: (1) the nomination stage in which the five top pictures are nominated, and (2) the final balloting for the winner.

We describe the second stage first because it is so simple: Once the five top pictures are nominated, each eligible member of the Academy is asked to vote for one picture, and the winner is chosen by simple plurality. Because the number of voters is large (somewhere between 4000 and 5000), ties are not likely to occur, but if they do, they are not broken. Thus, it is possible for two candidates to share an award.

The process for selecting the five nominations is considerably more complicated and is based on a voting method called **single transferable voting.** Each eligible member of the Academy is asked to submit a preference ballot with the names of their top five choices ranked from first to fifth. Based on the total number of valid ballots submitted, the minimum number of votes needed to get a nomination (called the **quota**) is established, and any picture with enough first-place votes to make the quota is automatically nominated.

The quota is always chosen to be a number that is over one-sixth (16.66%) but not more than one-fifth (20%) of the total number of valid ballots cast. (Setting the quota this way ensures that it is impossible for six or more pictures to get automatic nominations.) While in theory it is possible for five pictures to make the quota right off the bat and get an automatic nomination (in which case the nomination process is over), this has never happened in practice. In fact, what usually happens is that there are no pictures that make the quota automatically. Then, the picture with the fewest number of first-place votes (say X) is eliminated, and on all the ballots that originally had X as the first choice, X's name is crossed off the top and all the other pictures are moved up one spot. The ballots are then counted again. If there are still no pictures that make the quota, the process of elimination is repeated. Eventually, there will be one or more pictures that make the quota and are nominated.

The moment that one or more pictures are nominated there is a new twist: Nominated pictures "give back" to the other pictures still in the running (not nominated but not eliminated either) their "surplus" votes. This process of giving back votes (called a **transfer**) is best illustrated with an imaginary example. Suppose that the quota is 400 (a nice, round number) and at some point a picture (say Z) gets 500 first-place votes, enough to get itself nominated. The surplus for Z is $500 - 400 = 100$ votes, and these are votes that Z doesn't really need. For this reason the 100 surplus votes are taken away from Z and divided *fairly* among the second-place choices on the 500 ballots cast for Z.

The way this is done may seem a little bizarre, but it makes perfectly good sense. Since there are 100 surplus votes to be divided into 500 equal shares, each second-place vote on the 500 ballots cast for Z is worth $\frac{100}{500} = \frac{1}{5}$ vote. While one-fifth of a vote may not seem like much, enough of these fractional votes can make a difference and help some other picture or pictures make the quota. If that's the case, then once again the surplus or surpluses are transferred back to the remaining pictures following the procedure described above; otherwise, the process of elimination is started up again. Eventually, after several possible cycles of eliminations, transfers, eliminations, transfers, . . . , five pictures get enough votes to make the quota and be nominated, and the process is over.

The method of single transferable voting is not unique to the Academy Awards. It is used to elect officers in various professional societies as well as the members of the Irish Senate.

APPENDIX 3: Nonpreferential Voting

Arrow's impossibility theorem tells us that the search for an ideal voting method is hopeless. We should not take this to mean more than that. In particular, we should not draw the false conclusion that the study of voting methods is therefore a fruitless activity. Short of perfection, voting methods can be anywhere from very bad to very good, and the careful analysis of both new and old methods is still an active field of research among social scientists, economists, and mathematicians.

A relatively recent development in voting theory is the idea that voters should not be asked to order candidates according to preference (first choice is . . .; second choice is . . .; etc.), but rather to evaluate the candidates in absolute terms. Voting methods based on this type of balloting are known as **nonpreferential** (as opposed to **preferential**) voting methods. We will briefly mention one of the best known methods of this type: **approval voting.**

In approval voting, each voter votes for as many candidates as he or she wants. Each of these votes is simply a yes vote for the candidate, and it means that the voter approves of that candidate. The voters are not asked to list the candidates in order of preference. The candidate with the most approval votes wins the election.

The following table summarizes the results of a hypothetical election using approval voting.

VOTERS

Candidates	Sue	Bill	Tito	Prince	Tina	Van	Devon	Ike
A	Yes		Yes	Yes	Yes		Yes	Yes
B			Yes		Yes	Yes		
C	Yes				Yes	Yes	Yes	

The results of this election are as follows: winner, A (6 approval votes); second place, C (4 approval votes); last place, B (3 approval votes). Note that a voter can cast anywhere from no approval votes at all (such as Bill did above) to approval votes for all the candidates (such as Tina did above). It is somewhat ironical that the effect of Tina's vote is exactly the same as that of Bill's.

In the last few years a strong case has been made suggesting that approval voting is a considerable improvement over any of the more traditional preferential voting methods, and while it is not without flaws, it does indeed have several things going for it. We will mention three that are relevant to our discussion and which are important in the context of national politics.

First, approval voting is easy to understand and simple to implement in practice. Second, the approval voting process is unaffected by the number of candidates. In particular, if new candidates throw their hats into the ring, the voters can approve or disapprove of them without having to reconsider their votes for the original candidates. Last but not least, approval voting encourages voter turnout. The reason for this is psychological: Voters are more likely to vote when they feel they can make intelligent decisions, and unquestionably it is easier for a voter to give an intelligent answer to the question, Do you approve of this candidate—yes or no? than it is to the question, Which candidate is your first choice, second choice, etc.? The latter requires a much deeper knowledge of the candidates, and in today's complex political world it is a knowledge that very few voters have.

REFERENCES AND FURTHER READINGS

1. Arrow, Kenneth J., *Social Choice and Individual Values.* New York: John Wiley & Sons, Inc., 1963.

2. Black, Duncan, *The Theory of Committees and Elections.* New York: Cambridge University Press, 1968.

3. Brams, Steven J., and Peter C. Fishburn, *Approval Voting.* Boston: Birkhäuser, 1982.

4. Farquharson, Robin, *Theory of Voting.* New Haven, CN: Yale University Press, 1969.

5. Fishburn, Peter C., "A Comparative Analysis of Group Decision Methods," *Behavioral Science,* 16 (1971), 538–544.

6. Fishburn, Peter C., *The Theory of Social Choice.* Princeton, NJ: Princeton University Press, 1973.

7. Fishburn, Peter C., and Steven J. Brams, "Paradoxes of Preferential Voting," *Mathematics Magazine,* 56 (1983), 207–214.

8. Gardner, Martin, "Mathematical Games (From Counting Votes to Making Votes Count: The Mathematics of Elections)," *Scientific American,* 243 (October 1980), 16–26.

9. Niemi, Richard G., and William H. Riker, "The Choice of Voting Systems," *Scientific American,* 234 (June 1976), 21–27.

10. Saari, D. G., *The Geometry of Voting.* New York: Springer-Verlag, 1994.

11. Straffin, Philip D., Jr., *Topics in the Theory of Voting,* UMAP Expository Monograph. Boston: Birkhäuser, 1980.

CHAPTER 2

The Power Game

Weighted Voting Systems

The first principle of a civilized state is that power is legitimate only when it is under contract.
WALTER LIPPMANN*

In many voting situations, the *one person–one vote* principle is not justified. In a diverse society, it is in the very nature of things that individuals or groups are not always equal, and this needs to be recognized in the way that the voting is conducted by giving some voters more say than others. At a corporate shareholder's meeting, for example, each shareholder has as many votes as the number of shares he or she owns; in many committee situations the committee chairperson has tie-breaking powers not enjoyed by the other committee members; and even at home, when it's time to decide where to go on vacation, it always seems that mom and dad have more say than the kids.

In this chapter we will analyze voting systems in which the voters are not all equal. These are called **weighted voting systems** (the opinions of some voters have more weight than the opinions of others). To keep things simple we will assume that the voting will always be on two alternatives or candidates. Any vote involving only two alternatives can be thought of as a yes-no vote and is generally referred to as a **motion.**

We know from Chapter 1 (see Exercise 42) that when dealing with motions, we don't have to worry about the choice of voting method because all reasonable voting methods reduce to majority rule. The issue that will primarily concern us in this chapter is **power**: Who has it and how much of it do they really have? (Note that the issue of power is not very interesting in a one person–one vote situation. Clearly, if all voters are equal, then any voter has exactly the same amount of power as any other voter.) Understanding how power is distributed in a voting system is useful in several ways. In the first place it allows individuals to respond properly to a situation (if you are going to butter-up someone, you might as well butter-up those who have the most power). Most importantly, it allows institutions to rationally analyze power so that abuses may be prevented. (Does the Constitution give the vice president of the United States too much power by allowing him to cast the tie-breaking vote in the Senate? We'll see about that later.)

* Walter Lippmann, *The Public Philosophy* (Boston: Little, Brown and Co., 1955).

―――――――――――――――――――――――――――――――

WEIGHTED VOTING SYSTEMS

Before we pursue the trappings of power in more detail, we will introduce some terminology and discuss some examples of weighted voting systems.

Terminology

Every weighted voting system is characterized by three elements: the players, the weights of the players, and the quota. The **players** are just the voters themselves. (As much as possible we will adhere to the convention that the word "voter" refers to the one person–one vote situation, and the word "player" to the weighted voting situation.) We will use the letter N to represent the number of players and the symbols $P_1, P_2, \ldots, P_N$ to represent the names of the players—it is a little less personal but a lot more convenient than using Archie, Betty, Jughead, and Veronica. Each player controls a certain number of votes, and this number is called the player's **weight**. We will use the symbols $w_1, w_2, \ldots, w_N$ to represent the weights of $P_1, P_2, \ldots, P_N$, respectively. Finally, there is the **quota,** the minimum number of votes needed to pass a motion. We will use the letter q to denote the quota.

It is important to note that the quota q need not be restricted to be exactly a strict majority of the votes. There are many voting situations in which a strict majority of the votes is not enough to pass a motion—the rules may stipulate a different definition of what is needed for passing. When the U.S. Senate is attempting to override a presidential veto, for example, the rules state that two-thirds of the votes are needed. In other situations the rules may stipulate that 75% of the votes are needed or 83% (why not?) or even 100% (unanimous consent). In fact, any number can be a reasonable choice for the quota q, as long as it is more than half of the total number of votes but not more than the total number of votes. To put it more bluntly,

$$\frac{w_1 + w_2 + \cdots + w_N}{2} < q \leq w_1 + w_2 + \cdots + w_N.$$

Notation and Examples

A convenient way to describe a weighted voting system is

$$[q: w_1, w_2, \ldots, w_N].$$

The quota is always given first, followed by a colon and then the respective weights of the individual players. (It is customary to write the weights in nonincreasing order, and we will adhere to this convention throughout the chapter.)

―――――――――――――――――――――――――――――――

Example 1. [25: 8, 6, 5, 3, 3, 3, 2, 2, 1, 1, 1, 1].

This is a weighted voting system with 12 players $(P_1, P_2, \ldots, P_{12})$. P_1 has 8 votes, P_2 has 6 votes, P_3 has 5 votes, etc. The total number of votes is 36. The quota is 25.

Example 2. [7: 5, 4, 4, 2].

This weighted voting system doesn't make any sense because *the quota (7) is less than half of the total number of votes (15)*. If P_1 and P_4 voted yes and P_2 and P_3 voted no, both groups would win. This is a mathematical version of anarchy, and we cannot allow this to be a legal weighted voting system.

Example 3. [17: 5, 4 ,4, 2].

Here *the quota is too high.* In this weighted voting system no motion could ever pass. We can't allow this to happen either.

Example 4. [11: 4, 4, 4, 4, 4].

In this weighted voting system all 5 players are equal. To pass a motion at least 3 out of the 5 players are needed. Note that if the quota ($q = 11$) were changed to 12, the situation would still remain the same—at least 3 out of the 5 players would be needed. What we really have here is a *one person–one vote, strict majority situation,* and we can just as well describe it with the voting system [3: 1, 1, 1, 1, 1].

Example 5. [15: 5, 4, 3, 2, 1].

Here we have 5 players with a total of 15 votes. Since the quota is 15, the *only way a motion can pass is by unanimous consent of the players.* How does this voting system differ from the voting system [5: 1, 1, 1, 1, 1]? Well, this one also has 5 players, and the only way a motion can pass is by unanimous consent of the players. So for all practical purposes these two voting systems represent the same situation.

The surprising conclusion of Example 5 is that the weighted voting system [15: 5, 4, 3, 2, 1] represents a one person–one vote situation. This seems like a contradiction only if we think of a one person–one vote situation as implying that all players have an *equal number of votes rather than an equal say in the outcome of the election.* These two things are clearly not the same!

Power; More Terminology; More Examples

As Example 5 makes abundantly clear, a player's power cannot be measured by the numbers of votes he or she holds. This does not mean that there is no connection between votes and power. If player X and player Y have the same number of votes, then common sense tells us that they should have the same amount of power. On the other hand, if player X has four votes and player Y has two votes, we cannot conclude that player X has twice as much power as player Y. In fact just about anything is possible regarding the relative power of X and Y (of course, X can never be allowed to have *less* power than Y—that would be a gross violation of the idea of power!). (See Exercises 30 and 31.)

Before we formally describe the way in which power is defined mathematically (we will actually give two different definitions), let's consider a few more examples of weighted voting systems, now focusing in an intuitive way on the notion of power.

Example 6. [11: 12, 5, 4].

This is a situation in which a single player (P_1) controls enough votes to pass any measure single-handedly. Such a player has all the power, and not surprisingly, we call such a player a dictator.

Formally, we define a **dictator** as a player whose weight is bigger than or equal to the quota. Notice that whenever there is a dictator, all the other players have absolutely no power. A player without power is called a **dummy** (this is not a reflection on the player's intellect but rather on the fact that such a player has no say in the outcome of the election).

Example 7. [12: 11, 5, 4, 2].

Here we have a situation in which a player (P_1), while not a dictator, has enough votes to prevent any motion he or she doesn't like from passing. Even if the remaining players all band together, they can't force a motion against the will of P_1.

A player that is not a dictator but can single-handedly prevent any group of players from passing a motion is said to have **veto power**.

Example 8. [101: 99, 98, 3].

How is power distributed in this weighted voting system? At first glance it appears that P_1 and P_2 have lots of power while P_3 has very little power (if any). On closer inspection, however, we notice that it takes two of the players to pass a motion, and in fact, any two can do so. It seems appropriate, therefore, to claim that P_3, with a measly three votes, has as much power as either of the other two players. While hard to believe, this is in fact the case: The three players have equal power in this weighted voting system.

THE BANZHAF POWER INDEX

We are almost ready to formally introduce our first mathematical interpretation of power for weighted voting systems. This particular definition of power was suggested by John Banzhaf[1] in 1965.

[1] Banzhaf, who was a lawyer and not a mathematician, was mostly concerned with issues of equity and fairness in state and local systems of government.

Let's analyze the weighted voting system [101: 99, 98, 3] (Example 8) in a little more detail. Although this example itself is fairly simple, we will use it to introduce some important concepts.

Which groups of players could join forces to form a winning combination? Clearly, there are 4 such groups:

P_1 and P_2 (this group controls 197 votes)

P_1 and P_3 (this group controls 102 votes)

P_2 and P_3 (this group controls 101 votes, just enough to win)

P_1, P_2, and P_3 (this group controls all the votes).

From now on we will adhere to the standard language of voting theory and call any set of players that join forces to vote together a **coalition**. We use the word "coalition" in a rather generous way and will allow a single player to form a coalition alone. The total number of votes controlled by a coalition is called the **weight of the coalition**. Of course, some coalitions have enough votes to win and some don't. We will call the former **winning coalitions**, and the latter **losing coalitions**.

There is a particularly convenient way to describe coalitions mathematically using the language and notation of sets. For example, the coalition consisting of players P_1 and P_2 can be written as the set $\{P_1, P_2\}$; the coalition consisting of just player P_2 by itself can be written as the set $\{P_2\}$, and so on. Thus, for example 8, we can list all the possible coalitions as follows:

	Coalition	Coalition Weight	Winning or Losing?
a	$\{P_1\}$	99	Losing
b	$\{P_2\}$	98	Losing
c	$\{P_3\}$	3	Losing
d	$\{P_1, P_2\}$	197	Winning
e	$\{P_1, P_3\}$	102	Winning
f	$\{P_2, P_3\}$	101	Winning
g	$\{P_1, P_2, P_3\}$	200	Winning

▬ **TABLE 2-1**

If we now analyze the winning coalitions, we notice the following: In coalitions d, e, and f both players are needed for the win (if either player were to desert the coalition, the rest of the coalition would lose); in coalition g no single player is essential (if any single player were to desert the coalition, the rest of the coalition would still win).

A player whose desertion turns a winning coalition into a losing coalition obviously holds a certain amount of power, and we will call such a player a **crit-**

ical player for the coalition. Notice that a winning coalition often has more than one critical player, and occasionally a winning coalition has no critical players. Losing coalitions never have critical players.

The critical player concept is the basis for the definition of the **Banzhaf power index**. Banzhaf's key idea is that a player's power is proportional to the number of times that player is critical, so that the more often the player is critical, the more power he or she holds.

In our example each player is critical twice, so they all have equal power. Since there are three players, we can say that each player holds one-third of the power.

We can now formalize our approach for finding the Banzhaf power index of each player in an arbitrary weighted voting system with N players.

FINDING THE BANZHAF POWER INDEX OF PLAYER P

- ■ **Step 1.**   Make a list of all possible coalitions.
- ■ **Step 2.**   Determine which of them are winning coalitions.
- ■ **Step 3.**   In each winning coalition, determine which of the players are critical players.
- ■ **Step 4.**   Count the total number of times player P is critical and call this number B.
- ■ **Step 5.**   Count the total number of times all players are critical and call this number T.

The Banzhaf power index of player P is then given by the ratio B/T.

Example 9. Let's find the Banzhaf power index of each of the players in the weighted voting system [4: 3, 2, 1].

- ■ **Step 1.**   There are 7 possible coalitions. They are

$$\{P_1\}, \{P_2\}, \{P_3\}, \{P_1, P_2\}, \{P_1, P_3\}, \{P_2, P_3\}, \{P_1, P_2, P_3\}$$

- ■ **Step 2.**   The winning coalitions are

$$\{P_1, P_2\}, \{P_1, P_3\}, \text{ and } \{P_1, P_2, P_3\}.$$

- ■ **Step 3.**

Winning coalition	Critical players
$\{P_1, P_2\}$,	P_1 and P_2
$\{P_1, P_3\}$,	P_1 and P_3
$\{P_1, P_2, P_3\}$,	P_1 only

■ **Step 4.**

P_1 is critical three times.

P_2 is critical one time.

P_3 is critical one time.

■ **Step 5.** $T = 5$.

The Banzhaf power index of each of the players is

P_1: $\frac{3}{5}$; P_2: $\frac{1}{5}$; P_3: $\frac{1}{5}$

We will refer to the complete listing of the Banzhaf power indexes as the **Banzhaf power distribution** of the weighted voting system.

Some people prefer to deal with percentages rather than fractions. Percentagewise, the Banzhaf power distribution of the weighted voting system in Example 9 is

P_1: 60%; P_2: 20%; P_3: 20%.

Example 10. Let's consider the weighted voting system [6: 4, 3, 2, 1] and find its Banzhaf power distribution. Table 2-2 shows the 15 possible coalitions and the situation after steps 1 through 3 have been carried out. (The reader is encouraged to fill in the details.) The critical players in each coalition have been underlined.

Coalition	Weight	Winning or Losing?
$\{P_1\}$	4	Losing
$\{P_2\}$	3	Losing
$\{P_3\}$	2	Losing
$\{P_4\}$	1	Losing
$\{\underline{P_1}, \underline{P_2}\}$	7	Winning
$\{\underline{P_1}, \underline{P_3}\}$	6	Winning
$\{P_1, P_4\}$	5	Losing
$\{P_2, P_3\}$	5	Losing
$\{P_2, P_4\}$	4	Losing
$\{P_3, P_4\}$	3	Losing
$\{\underline{P_1}, P_2, P_3\}$	9	Winning
$\{\underline{P_1}, \underline{P_2}, \underline{P_4}\}$	8	Winning
$\{\underline{P_1}, \underline{P_3}, \underline{P_4}\}$	7	Winning
$\{\underline{P_2}, \underline{P_3}, \underline{P_4}\}$	6	Winning
$\{P_1, P_2, P_3, P_4\}$	10	Winning

■ TABLE 2-2   **The 15 Coalitions for Example 10 with Critical Players Underlined**

The Banzhaf power distribution is: P_1: $\frac{5}{12}$; P_2: $\frac{3}{12}$; P_3: $\frac{3}{12}$; P_4: $\frac{1}{12}$.

(Note that the power indexes add up to 1. This is not an accident! This fact provides a useful check on your calculations.)

How Many Coalitions?

Before we go on to the next example, let's take a brief detour and consider the following mathematical question: For a given number of players, how many different coalitions are possible? Here, our identification of coalitions with sets will come in particularly handy. Except for the empty subset { }, we know that every other subset of the set of players can be identified with a different coalition. This means that we can count the total number of coalitions by counting the number of subsets and subtracting one (that's because { } is not a coalition). So, how many subsets does a set have? The answer to this question can best be understood by looking at Table 2-3.

Set	Number of Subsets	Why?
$\{P_1, P_2\}$	4	{ }, $\{P_1\}$, $\{P_2\}$, $\{P_1, P_2\}$
$\{P_1, P_2, P_3\}$	8	{ }, $\{P_1\}$, $\{P_2\}$, $\{P_1, P_2\}$ $\{P_3\}$, $\{P_1, P_3\}$, $\{P_2, P_3\}$, $\{P_1, P_2, P_3\}$
$\{P_1, P_2, P_3, P_4\}$	16	{ }, $\{P_1\}$, $\{P_2\}$, $\{P_1, P_2\}$, $\{P_3\}$, $\{P_1, P_3\}$, $\{P_2, P_3\}$, $\{P_1, P_2, P_3\}$ $\{P_4\}$, $\{P_1, P_4, \}$, $\{P_2, P_4\}$, $\{P_1, P_2, P_4\}$, $\{P_3, P_4\}$, $\{P_1, P_3, P_4\}$, $\{P_2, P_3, P_4\}$, $\{P_1, P_2, P_3, P_4\}$
$\{P_1, P_2, P_3, P_4, P_5\}$	32	The 16 subsets inside the red box above plus the same 16 with P_5 added to each.

■ TABLE 2-3   **The Subsets of a Set**

What the preceding table shows us is that each time we add a new player we are doubling the number of subsets—the same subsets we had before we added the player plus an equal number consisting of each of the above but with the new player thrown in.

Since each time we add a new player we are doubling the number of subsets, we will find it convenient to think in terms of powers of 2. Table 2-4 summarizes what we have learned.

Example 11. Let's find the power distribution of the weighted voting system [15: 9, 6, 3, 3, 3].

We now know that with five players there are 31 ($2^5 - 1$) possible coalitions. Rather than plow straight ahead and list them all, why don't we try to fig-

Players	Number of Subsets	Number of Coalitions
$\{P_1, P_2\}$	$4 = 2^2$	3 $(4 - 1)$
$\{P_1, P_2, P_3\}$	$8 = 2^3$	7 $(8 - 1)$
$\{P_1, P_2, P_3, P_4\}$	$16 = 2^4$	15 $(2^4 - 1)$
$\{P_1, P_2, P_3, P_4, P_5\}$	$32 = 2^5$	31 $(2^5 - 1)$
$\vdots$	$\vdots$	$\vdots$
$\{P_1, P_2, ..., P_N\}$	2^N	$2^N - 1$

■ **TABLE 2-4**

ure out directly which are the winning coalitions. After all, they are the ones that count and there really are just a few of them. Table 2-5 shows the winning coalitions only, with the critical players in each coalition underlined. We leave it to the reader to verify the details which, while important, are not the main point of this example (the main point being that sometimes we can save a lot of work by directly zooming in on the winning coalitions).

Winning Coalitions

$\{\underline{P_1}, \underline{P_2}\}$ Only possible winning two-player coalition.

$\{\underline{P_1}, \underline{P_2}, P_3\}$
$\{\underline{P_1}, \underline{P_2}, P_4\}$
$\{\underline{P_1}, \underline{P_2}, P_5\}$ Winning three-player coalitions must contain P_1.
$\{\underline{P_1}, \underline{P_3}, \underline{P_4}\}$
$\{\underline{P_1}, \underline{P_3}, \underline{P_5}\}$
$\{\underline{P_1}, \underline{P_4}, \underline{P_5}\}$

$\{\underline{P_1}, P_2, P_3, P_4\}$
$\{\underline{P_1}, P_2, P_3, P_5\}$
$\{\underline{P_1}, P_2, P_4, P_5\}$ Any four-player coalition wins.
$\{\underline{P_1}, P_3, P_4, P_5\}$
$\{\underline{P_2}, \underline{P_3}, \underline{P_4}, \underline{P_5}\}$

$\{P_1, P_2, P_3, P_4, P_5\}$ The **grand coalition** (all players).

■ **TABLE 2-5** **Winning Coalitions for Example 11 with Critical Players Underlined**

The Banzhaf power distribution of the weighted voting system is

$$P_1: \tfrac{11}{25} = 44\%; \; P_2: \tfrac{5}{25} = 20\%; \; P_3: \tfrac{3}{25} = 12\%; \; P_4: \tfrac{3}{25} = 12\%; \; P_5: \tfrac{3}{25} = 12\%.$$

Example 12. A committee consists of four members, *A, B, C,* and *D.* In this committee each member has one vote, and a motion is carried by majority except that in case of a 2-2 tie, the coalition containing the chairperson (*A*) wins. What is the Banzhaf power distribution in this committee?

Although we don't have the player's weights to work with, we have all the necessary information to play the game. The winning coalitions are (1) any two-player coalition if it includes the chairperson, (2) any three-player coalition, and (3) the coalition containing all four players.

Table 2-6 shows the winning coalitions with the critical players underlined.

Winning Coalitions

{<u>A</u>, <u>B</u>}
{<u>A</u>, <u>C</u>}
{<u>A</u>, <u>D</u>}
{<u>A</u>, B, C}
{<u>A</u>, B, D}
{<u>A</u>, C, D}
{<u>B</u>, <u>C</u>, <u>D</u>}
{A, B, C, D}

◼◼ TABLE 2-6   **Winning Coalitions for
Example 12 with Critical Players Underlined**

The Banzhaf power distribution is

A: $\frac{6}{12} = 50\%$; B: $\frac{2}{12} \approx 16.67\%$; C: $\frac{2}{12} \approx 16.67\%$; D: $\frac{2}{12} \approx 16.67\%$.

What on the surface appeared to be a harmless tie-breaking rule gives the chairperson three times as much power as that of any of the other committee members. Knowing this ahead of time might make a difference in how one goes about choosing the chairperson.   ▭

Example 13. A committee consists of 5 members, the chairperson (Dr. *K*) and 4 other members of equal standing (*B, C, D,* and *E*). In this committee motions are carried by strict majority, but the chairperson never votes except to break a tie. How is power distributed in this voting system?

Once more, let's write down the winning coalitions with the critical players underlined.

$$
\begin{array}{l}
\text{Winning} \\
\text{coalitions} \\
\text{without the} \\
\text{chairman}
\end{array}
\left\{
\begin{array}{l}
\{\underline{B}, \underline{C}, \underline{D}\} \\
\{\underline{B}, \underline{C}, \underline{E}\} \\
\{\underline{B}, \underline{D}, \underline{E}\} \\
\{\underline{C}, \underline{D}, \underline{E}\} \\
\\
\{B, C, D, E\}
\end{array}
\right.
\qquad
\begin{array}{l}
\text{Winning} \\
\text{coalitions} \\
\text{with the} \\
\text{chairman}
\end{array}
\left\{
\begin{array}{l}
\{\underline{K}, \underline{B}, \underline{C}\} \\
\{\underline{K}, \underline{B}, \underline{D}\} \\
\{\underline{K}, \underline{B}, \underline{E}\} \\
\{\underline{K}, \underline{C}, \underline{D}\} \\
\{\underline{K}, \underline{C}, \underline{E}\} \\
\{\underline{K}, \underline{D}, \underline{E}\}
\end{array}
\right.
$$

▓▓ **TABLE 2-7   Winning Coalitions for Example 13 with Critical Players Underlined**

The Banzhaf power distribution in this committee is

$$
K: \tfrac{6}{30};\ B: \tfrac{6}{30};\ C: \tfrac{6}{30};\ D: \tfrac{6}{30};\ E: \tfrac{6}{30}.
$$

Surprise! All the members (including the chairperson) have the same amount of power.   ▭

The same situation described in Example 13 except on a larger scale exists in the U.S. Senate, where the vice president of the United States (what's his name?) votes only to break a tie. An analysis similar to the one in Example 13 would show that, assuming all 100 senators are voting, he has exactly the same amount of power as any other member of the Senate.

Applications of the Banzhaf Power Index

The Nassau County Board of Supervisors, New York.   John Banzhaf first introduced the Banzhaf power index in 1965 in an analysis of how power was distributed in the Board of Supervisors of Nassau County, New York. Although Banzhaf was a lawyer, it was his mathematical analysis of power in the Nassau County Board that provided the legal basis for a series of lawsuits[2] involving the mathematics of weighted voting systems and its implications regarding the equal protection (*one person–one vote*) guarantee of the Fourteenth Amendment.

Nassau County is divided into six different districts, and based on population figures, a total of 115 votes were allocated to each district as follows:

---

[2] For students of the law, here are the case references: *Graham v. Board of Supervisors* (1966); *Franklin v. Krause* (1974); *Bechtle v. Board of Supervisors* (1981); *League of Women Voters v. Board of Supervisors* (1983); and *Jackson v. Board of Supervisors* (1991, but still pending). All of the above lawsuits involved the Nassau County Board of Supervisors. Other important legal cases involving the Banzhaf power index are *Ianucci v. Board of Supervisors* (1967) and *Morris v. Board of Estimate* (U.S. Supreme Court, 1989).

District	Votes in 1964
Hempstead #1	31
Hempstead #2	31
Oyster Bay	28
North Hempstead	21
Long Beach	2
Glen Cove	2

The number of votes needed to pass a motion was 58, so that the Nassau County Board in effect functioned as the weighted voting system [58: 31, 31, 28, 21, 2, 2]. So far, so good, but what about the power of each of the districts? In his lawsuit, Banzhaf argued that in this instance, all the power in the County Board was concentrated in the hands of the top three players—Hempstead #1, Hempstead #2, and Oyster Bay. After a moment's reflection we can see why this is so: No winning coalition is possible without two of the top three players in it, and since any two of the top three already form a winning coalition, none of the last three players can ever be critical players. (We leave it to the reader to verify all the details—see Exercise 26.) The long and the short of it was that, as Banzhaf successfully argued, this County Board was in practice a three member board, with Hempstead #1, Hempstead #2, and Oyster Bay each having one third of the power, and North Hempstead, Glen Cove, and Long Beach having absolutely no power at all!

Based on Banzhaf's analysis, the number of votes allocated to each district was changed, and has indeed been changed several times since 1965. The status of the Nassau County Board as of 1994 is that it operates as the weighted voting system [65: 30, 28, 22, 15, 7, 6] (see Exercise 27).

The United Nations Security Council.   The main body responsible for maintaining the international peace and security of nations is the Security Council of the United Nations. At present the composition of the Security Council is as follows: There are 5 **permanent members** of the council (the United Kingdom, France, the People's Republic of China, Russia, and the United States), plus 10 additional **nonpermanent** slots filled by other countries on a rotating basis.[3] According to the voting rules of the Security Council, each of the permanent members has veto power, so that a resolution cannot pass unless each of them votes yes. In addition, at least 4 of the 10 nonpermanent members must also vote yes. It is a challenging (but not unreasonable) exercise (Exercise 46) to show that under these rules the U.N. Security Council can be formally described as the weighted voting system [39: 7, 7, 7, 7, 7, 1, 1, 1, 1, 1, 1, 1, 1, 1, 1]. Once the Security Council is described this way, it is possible to compute the Banzhaf power index of each country. While the calculations are

---

[3] When the original Security Council was set up by the League of Nations in 1945, the number of nonpermanent members was only 6. The number was increased to 10 in 1963.

The United Nations Security Council in session. (Yutaka Nagata/ United Nations)

not difficult, they go beyond the scope of this book so we will omit them. The long and the short of it is that the Banzhaf power index of each permanent member is $\frac{848}{5080} \approx 16.7\%$, while the Banzhaf power index of each nonpermanent member is $\frac{84}{5080} \approx 1.65\%$. In retrospect, one could ask: When the U.N. Security Council was originally set up, was it really intended that a permanent member have more than 10 times as much power as a nonpermanent member? The answer is most likely no, and in an ideal world one would expect that the voting rules of the Security Council would be reconsidered. Needless to say, this is not about to happen soon, as the 5 permanent members are not likely to voluntarily give up their power, and as things stand, the nonpermanent members can't do much about it!

The Electoral College. As we should all know, the president of the United States is chosen using an institution called the *electoral college*. In choosing the president, each state is allowed to cast a certain number of votes equal to the total number of members of Congress (senators plus representatives) from that state. The votes are cast by individuals called *electors* who are chosen to represent the citizens of their respective states. The general rule is that all the electors from a particular state vote the same way (for the presidential candidate who wins a plurality of the votes in that state). While there have been challenges to the constitutionality of this rule (known as the *unit* rule or *winner-take-all* rule), and in a few instances the rule has been violated by individual electors, it is standard procedure in the electoral college. We can summarize by saying that while in theory the electoral college is not a weighted voting system, in practice, if we assume that the unit rule is respected by all electors and the election is between two candidates (which it almost always is), the electoral college can be thought of as a very real example of a weighted voting system. In this system the players are the states (actually the 50 states plus the District of Columbia), and the weight of a state is the number of senators plus representatives from that state (the weight of the District of Columbia is set at 3). The quota is defined by a strict majority of the electoral vote. Since the 1964 presidential election, the total number of electoral votes has been set at 538 and the quota at 270. The appendix at the end of this chapter shows, among other things, the electoral votes for each state based on the 1990 census and the Banzhaf power index of each state. (The calculations for the Banzhaf power indexes require both sophisticated mathematical methods and a powerful computer.)

Perhaps the most significant thing to notice about the Banzhaf power indexes in the electoral college is the wide discrepancy between the power of the large states as opposed to that of the small states. (Contrast for example the Banzhaf power index of California at 11.14% with that of a small state such as Wyoming at 0.55%.) Understanding the impact of this wide spread in the power of the states is a fundamental element in the strategy of a presidential campaign. As a general rule, candidates allocate their resources (money, time, and personnel) to states in accordance to their power indexes and not in accordance to population. The net effect of this is that candidates usually spend a disproportionate amount of their campaign resources in the states with the greatest power indexes. A presidential candidate can win by the barest majority of the popular vote in just 12 states (the 12 most populous ones) and have enough electoral votes to win the election, a fact that is not lost on campaign managers and political strategists.

THE SHAPLEY-SHUBIK POWER INDEX

A different way of measuring power in weighted voting systems was proposed by Lloyd Shapley and Martin Shubik[4] in 1954. The key difference between their definition of power and Banzhaf's centers around the concept of a **sequential coalition.** According to Shapley and Shubik, coalitions are formed sequentially: Every coalition starts with a first player, who may then be joined by a second player, then a third, and so on. Thus, to an already complicated situation we are adding one more wrinkle—the question of the order in which the players joined the coalition.

Let's illustrate the difference with a simple example. According to Banzhaf, the coalition $\{P_1, P_2, P_3\}$ represents the fact that P_1, P_2, and P_3 have joined forces and will vote together. We don't care and don't even consider who joined the coalition first, second, or third. According to Shapley and Shubik the same three players can form *six* different coalitions: $\langle P_1, P_2, P_3 \rangle$ (this means P_1 started the coalition, P_2 joined in second, and P_3 third); $\langle P_1, P_3, P_2 \rangle$; $\langle P_2, P_1, P_3 \rangle$; $\langle P_2, P_3, P_1 \rangle$; $\langle P_3, P_1, P_2 \rangle$; $\langle P_3, P_2, P_1 \rangle$. Note the change in notation: From now on the notation $\langle\ \rangle$ will indicate that we are dealing with a sequential coalition; that is, we care about the order in which the players are listed.

---

[4] Lloyd S. Shapley was a mathematician at the Rand Corporation, and Martin Shubik an economist at Yale University.

Factorials

It is now time to consider another important mathematical question: For a given number of players N, how many sequential coalitions containing the N players are there? We have just seen that with three players there are six sequential coalitions: $\langle P_1, P_2, P_3 \rangle$, $\langle P_1, P_3, P_2 \rangle$, $\langle P_2, P_1, P_3 \rangle$, $\langle P_2, P_3, P_1 \rangle$, $\langle P_3, P_1, P_2 \rangle$, and $\langle P_3, P_2, P_1 \rangle$. What happens if we have four players? We could try to write down all of the sequential coalitions, a tedious and unimaginative task. Instead, let's argue as follows: To fill the first slot in a coalition we have 4 choices (any one of the four players); to fill the second slot we have 3 choices (any one of the players except the one in the first slot); to fill the third slot we have only 2 choices and to fill out the last slot we have only 1 choice. How do we combine these choices for the final product? We multiply! Thus, the total number of possible sequential coalitions with four players turns out to be $4 \times 3 \times 2 \times 1 = 24$.

One question remains to be answered: Why did we multiply these numbers? The answer lies in the following fundamental rule of mathematics called the **multiplication rule**: *If X can be done in m different ways and Y can be done in n different ways, then X and Y together can be done in m $\times$ n different ways.*

For example, if an ice cream shop offers 2 different types of cones and 3 different flavors of ice cream, then according to the multiplication rule there are $2 \times 3 = 6$ different cone/flavor combinations. Fig. 2-1 shows why this is so.

We will discuss the multiplication rule and its uses in greater detail in Chapter 15. Meanwhile, back to the issue at hand.

We now have a solid basis to draw some generalizations concerning the

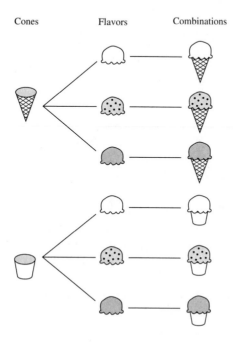

Cones Flavors Combinations

FIGURE 2-1

number of sequential coalitions. With 5 players, the number of sequential coalitions is $5 \times 4 \times 3 \times 2 \times 1 = 120$, and in a more general vein,

> The number of sequential coalitions with N players is
>
> $1 \times 2 \times 3 \times \cdots \times N.$

By the way, the number $1 \times 2 \times 3 \times \cdots \times N$ will show up several times in this book—in fact, it is a frequent flyer in the mathematical skies—and it is called the **factorial** of N and abbreviated $N!$. The factorial of 5, for example, is written $5!$ and equals 120 (because $1 \times 2 \times 3 \times 4 \times 5 = 120$), while $10! = 3,628,800$ (check it out!).

> For any positive integer N, the **factorial** of N is
>
> $N! = 1 \times 2 \times 3 \times \cdots \times N.$

Back to the Shapley-Shubik Power Index

Suppose that we have a weighted voting system with N players. We know from the preceding discussion that there is a total of $N!$ different sequential coalitions containing *all* the players. In each of these coalitions there is one player that tips the scales—the moment that player joins the coalition, the coalition changes from a losing to a winning coalition. We call such a player a **pivotal player** for the sequential coalition. The underlying principle of the Shapley-Shubik approach is that the pivotal player deserves special recognition. After all, the players who came before the pivotal player did not have it, and the players who came after are a bunch of Johnny-come-latelies. (Note that we can talk about "before" and "after" only because we are considering sequential coalitions.) According to Shapley and Shubik then, a player's power can be measured by the total number of times he is pivotal in relation to all other players.

The formal description of the procedure for finding the Shapley-Shubik power index of a player P is as follows:

> ### FINDING THE SHAPLEY-SHUBIK POWER INDEX OF PLAYER P
>
> ■ **Step 1.** Make a list of all sequential coalitions containing *all N* players. There are $N!$ of them.
> ■ **Step 2.** In each sequential coalition determine the pivotal player. There is one in each sequential coalition.

■ **Step 3.**   Count the total number of times P is pivotal and call this number S.

The Shapley-Shubik power index of P is then given by the ratio $S/N!$.

Listing the Shapley-Shubik power indexes of all the players gives the **Shapley-Shubik power distribution** of the weighted voting system.

Example 14. Let's consider, once again, the weighted voting system [4: 3, 2, 1]. This is the same weighted voting system we discussed in Example 9, but this time we will find its Shapley-Shubik power distribution.

■ **Step 1.**   There are $3! = 6$ sequential coalitions of the three players. They are

$\langle P_1, P_2, P_3 \rangle$, $\langle P_1, P_3, P_2 \rangle$, $\langle P_2, P_1, P_3 \rangle$, $\langle P_2, P_3, P_1 \rangle$, $\langle P_3, P_1, P_2 \rangle$, $\langle P_3, P_2, P_1 \rangle$.

■ **Step 2.**

Sequential coalition	Pivotal player
$\langle P_1, P_2, P_3 \rangle$	P_2
$\langle P_1, P_3, P_2 \rangle$	P_3
$\langle P_2, P_1, P_3 \rangle$	P_1
$\langle P_2, P_3, P_1 \rangle$	P_1
$\langle P_3, P_1, P_2 \rangle$	P_1
$\langle P_3, P_2, P_1 \rangle$	P_1

■ **Step 3.** P_1 is pivotal four times.
P_2 is pivotal one time.
P_3 is pivotal one time.

The Shapley-Shubik power distribution is

$P_1: \frac{4}{6} \approx 66.7\%$; $P_2: \frac{1}{6} \approx 16.7\%$; $P_3: \frac{1}{6} \approx 16.7\%$.

This power distribution is significantly different from the Banzhaf power distribution (P_1: 60%; P_2: 20%; P_3: 20%) obtained in Example 9.

Example 15.   Consider the weighted voting system [6: 4, 3, 2, 1].
This is the same weighted voting system that was discussed in Example 10. We will now find the Shapley-Shubik power index of each player.

There are 24 different sequential coalitions involving the 4 players. They are listed in Table 2-8 with the pivotal player underlined.

$\langle P_1, P_2, P_3, P_4 \rangle$	$\langle P_2, P_1, P_3, P_4 \rangle$	$\langle P_3, P_1, P_2, P_4 \rangle$	$\langle P_4, P_1, P_2, P_3 \rangle$
$\langle P_1, \overline{P_2}, P_4, P_3 \rangle$	$\langle P_2, \overline{P_1}, P_4, P_3 \rangle$	$\langle P_3, \overline{P_1}, P_4, P_2 \rangle$	$\langle P_4, P_1, \overline{P_3}, P_2 \rangle$
$\langle P_1, \overline{P_3}, P_2, P_4 \rangle$	$\langle P_2, \overline{P_3}, P_1, P_4 \rangle$	$\langle P_3, \overline{P_2}, P_1, P_4 \rangle$	$\langle P_4, P_2, \overline{P_1}, P_3 \rangle$
$\langle P_1, \overline{P_3}, P_4, P_2 \rangle$	$\langle P_2, P_3, \overline{P_4}, P_1 \rangle$	$\langle P_3, P_2, \overline{P_4}, P_1 \rangle$	$\langle P_4, P_2, \overline{P_3}, P_1 \rangle$
$\langle P_1, \overline{P_4}, P_2, P_3 \rangle$	$\langle P_2, P_4, \overline{P_1}, P_3 \rangle$	$\langle P_3, P_4, \overline{P_1}, P_2 \rangle$	$\langle P_4, P_3, \overline{P_1}, P_2 \rangle$
$\langle P_1, P_4, \overline{P_3}, P_2 \rangle$	$\langle P_2, P_4, \overline{P_3}, P_1 \rangle$	$\langle P_3, P_4, \overline{P_2}, P_1 \rangle$	$\langle P_4, P_3, \overline{P_2}, P_1 \rangle$

TABLE 2-8   The 24 Sequential Coalitions for Example 15 with Pivotal Players Underlined

The Shapley-Shubik power distribution is

$$P_1: \tfrac{10}{24};\ P_2: \tfrac{6}{24};\ P_3: \tfrac{6}{24};\ P_4: \tfrac{2}{24}.$$

Comparing with Example 10 we see that in this instance the Shapley-Shubik power distribution is exactly the same as the Banzhaf power distribution. (If nothing else, this example shows that it is not impossible for the two power distributions to produce the same answer. It doesn't happen often, but it does happen!)

Example 16. A committee consists of 4 members, *A*, *B*, *C*, and *D*. Each member of the committee has one vote except for the chairperson (*A*) who has veto power. A measure needs a minimum of two votes to pass (one of which must be the chairperson's). What is the Shapley-Shubik power distribution in this committee?

We know that there are 4! = 24 different sequential coalitions involving the four players, but we are not going to write them all down. Instead, let's analyze how each player can be pivotal. The chairperson (*A*) is pivotal in every sequential coalition except those in which she is the first player. *B* can be pivotal only if he is the second player in the sequential coalition and the chairperson is the first player. There are only two such sequential coalitions: $\langle A, B, C, D \rangle$, and $\langle A, B, D, C \rangle$. Thus, the Shapley-Shubik power index of *B* is $\tfrac{2}{24} \approx 8.33\%$. Exactly the same argument applies to *C* and *D*, and so each of them has Shapley-Shubik power index $\tfrac{2}{24} \approx 8.33\%$. The balance of the power is $\tfrac{18}{24} = 75\%$ and belongs to *A*. Thus the power distribution for this committee is

$$A: \tfrac{18}{24} = 75\%;\ B: \tfrac{2}{24} \approx 8.33\%;\ C: \tfrac{2}{24} \approx 8.33\%;\ D: \tfrac{2}{24} \approx 8.33\%.$$

As a final example involving the Shapley-Shubik power index, we choose one that is fairly complicated. Even those readers who have mastered this material shouldn't feel too bad if they have a harder time with this one.

Example 17. Let's find the Shapley-Shubik power distribution of the weighted voting system [15: 9, 6, 3, 3, 3]. This is the same weighted voting system we discussed in Example 11.

The number of sequential coalitions to consider is $5! = 120$. Maybe we should try to think of some shortcuts.

Let's start with player P_5 and try to figure out in how many sequential coalitions P_5 will be pivotal. If P_5 is the first player in the sequential coalition she can't be pivotal—that is clear. If P_5 is the second player in the coalition, even in the best of cases (when preceded by P_1), P_5 can't be pivotal. When P_5 is the third player in the coalition, then P_5 can be pivotal only when preceded by P_1 and P_3 or P_1 and P_4. (Think about it—if P_1 is not already there, there won't be enough votes; if P_1 and P_2 are both there, there will be too many votes!) Schematically, we can draw the following diagram of this situation

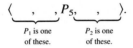

There is a total of eight such coalitions: $\langle P_1, P_4, P_5, P_2, P_3 \rangle$, $\langle P_4, P_1, P_5, P_2, P_3 \rangle$, . . . (we'll let the reader finish this list).

How can P_5 be pivotal when in the fourth position? A moment's reflection will show that this can happen only if P_1 does not precede P_5 in the coalition. In other words,

$$\underbrace{\langle \quad, \quad, \quad,}_{P_2, P_3, \text{ and } P_4} P_5, P_1 \rangle.$$

There is a total of 6 such coalitions. (It's not hard to list them.) Finally, P_5 can never be pivotal in the fifth (last) position.

We now know that P_5 is pivotal in a grand total of 14 coalitions, so her Shapley-Shubik power index is $\frac{14}{120}$.

How about P_4 and P_3? That's easy. Anything that is true of P_5 is true of P_4 and P_3, so each of them is also pivotal in 14 sequential coalitions. Let's do P_2 next, and let's try to let the pictures speak for themselves.

P_2 is pivotal in the second position:

$$\langle P_1, P_2, \underbrace{\quad, \quad, \quad}_{P_3, P_4, \text{ and } P_5} \rangle.$$

There is a total of 6 such coalitions. (There are 6 different ways in which P_3, P_4, and P_5 can be shuffled among themselves.)

P_2 is pivotal in the third position:

$$\langle \underbrace{\quad , \quad}_{\substack{P_1 \text{ is one} \\ \text{of these.}}} , P_2, \quad , \quad \rangle.$$

There is a total of 12 such coalitions. (Check it out!)

P_2 is pivotal in the fourth position:

$$\langle \underbrace{\quad , \quad , \quad}_{P_3, P_4, \text{ and } P_5} , P_2, P_1 \rangle.$$

There is a total of 6 such coalitions.

Since P_2 cannot be pivotal in either the first or last position, we are done. All in all, P_2 is pivotal in 24 coalitions, so that his Shapley-Shubik power index is $\frac{24}{120} = \frac{1}{5}$.

Finally there is P_1. By default, P_1 must be pivotal in each of the sequential coalitions that have not yet been counted. There are

$$120 - 14 - 14 - 14 - 24 = 54$$

of them. Thus, the Shapley-Shubik power index of P_1 is $\frac{54}{120}$.

Summarizing, the Shapley-Shubik power distribution is

$$P_1: \frac{54}{120} = 45\%; \ P_2: \frac{24}{120} = 20\%; \ P_3: \frac{14}{120} \approx 11.66\%; \ P_4: \frac{14}{120} \approx 11.66\%;$$
$$P_5: \frac{14}{120} \approx 11.66\%.$$

Comparing with Example 11, we get a slightly different distribution of power.

Applications of the Shapley-Shubik Power Index

The Electoral College Revisited. Calculating the Shapley-Shubik power index of the states in the electoral college is a task best left for a computer; but even with the world's fastest computer, calculating 51! sequential coalitions would take more than one hundred years. With a clever mathematical shortcut discovered by Irwin Mann and Lloyd Shapley in 1962, the calculations can be carried out by a personal computer in just a matter of hours. Applying this shortcut to the current electoral college gives the Shapley-Shubik power distribution shown in the last column of the appendix at the end of this chapter. If we compare the Banzhaf power indexes and the Shapley-Shubik power indexes, we notice that there is a very small difference between the two indexes. This is an example that shows that in some situations the Banzhaf and Shapley-Shubik power indexes give essentially the same answer. The next example illustrates a very different situation.

The United Nations Security Council Revisited. As we pointed out earlier in the chapter, the U.N. Security Council as presently structured can be formally described as the weighted voting system [39: 7, 7, 7, 7, 7, 1, 1, 1, 1, 1, 1, 1, 1, 1, 1] and the Banzhaf power index of the five permanent members (the ones with seven votes) is about 10 times larger than that of the nonpermanent members (16.7% against 1.65%). What if we use the Shapley-Shubik power index instead? It turns out (the calculations are a bit too complicated to describe in detail) that the Shapley-Shubik power index of each of the five permanent members is 19.6%, while that of each of the nonpermanent members is about 0.2%. According to the Shapley-Shubik power index, any one of the five permanent members has about 100 times more power than one of the nonpermanent members!

Here is a situation where the Banzhaf power index and the Shapley Shubik power index produce significantly different answers. (The Banzhaf power index of a nonpermanent member is about 9 times as much as its Shapley-Shubik power index.)

CONCLUSION

In any society, no matter how democratic, some individuals and groups have more power than others. This is simply a consequence of the fact that individuals and groups are not all equal. Diversity is the inherent reason why the concept of power exists.

Power itself comes in many different forms. We often hear expressions such as "In strength lies power" and "Money is power," and in our new information age, "Knowledge is power." In this chapter we discussed the notion of power as it applies to formal voting situations (weighted voting systems) and saw how mathematical methods allow us to measure the power of an individual or group by means of a *power index*. In particular, we looked at two different kinds of power indexes: the *Banzhaf power index* and the *Shapley-Shubik power index*.

The Banzhaf power index and the Shapley-Shubik power index provide two different ways to measure power, and while they occasionally agree (Example 15), they are often significantly different. Of the two, which one is closer to reality?

Unfortunately, there is no simple answer. Both of them are useful, and in some sense the choice is subjective. Perhaps the best way to evaluate them is to think of them as being based on a slightly different set of assumptions. The idea behind the Banzhaf interpretation of power is that *players are free to come and go,* negotiating their allegiance for power (somewhat like professional baseball players since the advent of free agency). Underlying the Shapley-Shubik interpretation of power is the assumption that when a player joins a coalition he or she is making a commitment to stay (as in an old-fashioned marriage). In the latter case a player's power is generated by his ability to be *in the right place at the right time.* In practice, the choice of which method to use in measuring power is based on which of the assumptions better fits the specifics of the situation. Mathematics, contrary to what we've often come to expect, does not give us the answer, just the tools that might help us make an informed decision.

KEY CONCEPTS

Banzhaf power distribution	**pivotal player**
Banzhaf power index	**players**
coalition	**quota**
coalition weight	**sequential coalition**
critical player	**Shapley-Shubik power**
dictator	**distribution**
dummy	**Shapley-shubik power index**
factorial	**veto power**
losing coalition	**weighted voting system**
motion	**weights**
multiplication rule	**winning coalition**

EXERCISES

Walking

1. Consider the weighted voting system [10: 6, 5, 4].
   (a) How many players are there?
   (b) What is the quota?
   (c) What is the weight of the coalition formed by P_1 and P_3?
   (d) Write down all winning coalitions.
   (e) Which players are critical in the coalition $\{P_1, P_2, P_3\}$?
   (f) Find the Banzhaf power distribution of this weighted voting system.

2. Consider the weighted voting system [4: 3, 2, 1, 1].
   (a) How many players are there?
   (b) What is the quota?
   (c) What is the weight of the coalition formed by P_1 and P_3?
   (d) Which players are critical in the coalition $\{P_2, P_3, P_4\}$?
   (e) Which players are critical in the coalition $\{P_1, P_3, P_4\}$?
   (f) Write down all winning coalitions.
   (g) Find the Banzhaf power distribution of this weighted voting system.

3. (a) Find the Banzhaf power distribution of the weighted voting system [6: 5, 3, 2].
   (b) Find the Banzhaf power distribution of the weighted voting system [8: 5, 3, 2].

4. (a) Find the Banzhaf power distribution of the weighted voting system [8: 5, 4, 3, 2, 1]. (If possible do it without writing down all coalitions—just the winning ones.)
   (b) Find the Banzhaf power distribution of the weighted voting system [10: 5, 4, 3, 2, 1]. (*Hint*: Note that the only change from [a] is in the quota and use this fact to your advantage.)

5. Find the Banzhaf power distribution of the weighted voting system [8: 5, 5, 3, 1, 1].

6. Find the Banzhaf power distribution of the weighted voting system [8: 5, 5, 2, 1, 1].

7. Consider the weighted voting system [10: 6, 5, 4].
   (a) Write down all the sequential coalitions involving all three players.
   (b) In each of the sequential coalitions in (a), underline the pivotal player.
   (c) Find the Shapley-Shubik power distribution of this weighted voting system.

8. Consider the weighted voting system [6: 5, 4, 1].
   (a) Write down all the sequential coalitions involving all three players.
   (b) In each of the sequential coalitions in (a), underline the pivotal player.
   (c) Find the Shapley-Shubik power distribution of this weighted voting system.

9. (a) Find the Shapley-Shubik power distribution of the weighted voting system [6: 5, 3, 2].
   (b) Find the Shapley-Shubik power distribution of the weighted voting system [8: 5, 3, 2].

10. Find the Shapley-Shubik power distribution of the weighted voting system [8: 4, 3, 2, 1].

11. Consider the weighted voting system [5: 3, 2, 1, 1].
    (a) Find the Banzhaf power distribution of this weighted voting system.
    (b) Find the Shapley-Shubik power distribution of this weighted voting system.

12. Consider the weighted voting system [60: 32, 29, 29, 29].
    (a) Find the Banzhaf power distribution of this weighted voting system.
    (b) Find the Shapley-Shubik power distribution of this weighted voting system.

13. In each of the following weighted voting systems, determine which players (i) are dictators; (ii) have veto power; (iii) are dummies:
    (a) [4: 2, 2, 1]
    (b) [10: 7, 3, 1]
    (c) [10: 9, 9, 1]

14. In each of the following weighted voting systems, determine which players (i) are dictators; (ii) have veto power; (iii) are dummies:
    (a) [95: 95, 40, 30, 20]
    (b) [125: 95, 40, 30, 20]
    (c) [10: 5, 5, 5, 2, 1, 1]

15. In each of the following weighted voting systems, determine which players (i) are dictators; (ii) have veto power; (iii) are dummies:
    (a) [20: 9, 8, 6, 3, 1]
    (b) [15: 15, 10, 3, 1]
    (c) [21: 13, 5, 2, 1]
    (d) [18: 12, 10, 3, 2]

16. In each of the following weighted voting systems, determine which players (i) are dictators; (ii) have veto power; (iii) are dummies:
    (a) [17: 17, 10, 4, 2]
    (b) [20: 10, 8, 7, 3, 1]
    (c) [21: 13, 10, 5, 2]
    (d) [19: 11, 5, 2, 1]

17. Consider the weighted voting system $[q: 12, 8, 5, 4, 2]$.
    (a) What is the smallest value that the quota q can take?
    (b) What is the largest value that the quota q can take?
    (c) How many coalitions are there for this weighted voting system?
    (d) How many sequential coalitions are there involving all the players?

18. Consider the weighted voting system $[q: 7, 3, 2, 1, 1]$.
    (a) What is the smallest value that the quota q can take?
    (b) What is the largest value that the quota q can take?
    (c) How many coalitions are there for this weighted voting system?
    (d) How many sequential coalitions are there involving all the players?

19. Consider the weighted voting system $[q: 5, 3, 1]$. Find the Shapley-Shubik power distribution of this weighted voting system when
    (a) $q = 5$
    (b) $q = 6$
    (c) $q = 7$
    (d) $q = 8$
    (e) $q = 9$.

20. Consider the weighted voting system $[q: 5, 3, 1]$. Find the Banzhaf power distribution of this weighted voting system when
    (a) $q = 5$
    (b) $q = 6$
    (c) $q = 7$
    (d) $q = 8$
    (e) $q = 9$.

21. This exercise is intended to help you develop a better understanding of factorials. If you have a fancy calculator with a factorial key, don't use it—use only the multiplication key!
    (a) Calculate 8!
    (b) Calculate 12!
    (c) Calculate 13! (*Hint*: Think of a shortcut using the answer for [b].)
    (d) Calculate 11! (*Hint*: Think of a shortcut using the answer for [b].)
    (e) $14x = 14!$ Solve for x. (*Hint*: You already calculated the answer.)

22. This exercise should be done *without* a calculator. The amount of arithmetic is minimal. Remember that it is always better to do cancellations first and multiplications later.

(a) Calculate $\frac{11!}{10!}$.

(b) Calculate $\frac{100!}{98!}$.

(c) Calculate $\frac{8!}{5!3!}$.

(d) Calculate $\frac{25!}{21!4!}$.

23. A business firm is owned by 4 partners, A, B, C, and D. When making group decisions, each partner has one vote and the majority rules except in the case of a 2-2 tie. Then, the coalition that contains D (the partner with the least seniority) *loses*. What is the Banzhaf power distribution in this partnership?

24. A business firm is owned by 4 partners, A, B, C, and D. When making group decisions, each partner has one vote and the majority rules except in the case of a 2-2 tie. Then, the coalition that contains D (the partner with the least seniority) *loses*. What is the Shapley-Shubik power distribution in this partnership?

25. A business firm is owned by 5 partners: A, B, C, D, and E. When making group decisions, each partner has one vote and majority rules except that A and B both have veto power and therefore *must* be in all winning coalitions. What is the Banzhaf power distribution in this partnership?

26. The 1964 Nassau County Board of Supervisors could be described by the weighted voting system [58: 31, 31, 28, 21, 2, 2].
    (a) Describe all the winning coalitions.
    (b) Find the Banzhaf power distribution for this weighted voting system.

Jogging

27. The 1994 Nassau County Board of Supervisors can be described by the weighted voting system [65: 30, 28, 22, 15, 7, 6]. Find the Banzhaf power distribution for this weighted voting system.

28. Consider the weighted voting system [21: 6, 5, 4, 3, 2, 1]. (Note that here the quota is the sum of *all* the weights of the players.)
    (a) How many coalitions are there?
    (b) Write down the winning coalitions only and underline the critical players.
    (c) Find the Banzhaf power index of each player.
    (d) Explain why in any weighted voting system $[q: w_1, w_2, \ldots, w_N]$, if $q = w_1 + \cdots + w_N$, then the Banzhaf power index of each player is $1/N$.

29. Consider the weighted voting system [21: 6, 5, 4, 3, 2, 1].
    (a) How many different sequential coalitions are there?
    (b) There is only one way in which a player can be pivotal in one of these sequential coalitions. Describe it.
    (c) In how many sequential coalitions is P_6 pivotal?
    (d) What is the Shapley-Shubik power index of P_6?
    (e) What are the Shapley-Shubik power indexes of the other players?
    (f) Explain why in any weighted voting system $[q: w_1, w_2, \ldots, w_N]$, if $q = w_1 + \cdots + w_N$, then the Shapley-Shubik power index of each player is $1/N$.

30. Give an example of a weighted voting system in which P_1 has twice as many votes as P_2 and
  (a) the Banzhaf power index of P_1 is greater than twice the Banzhaf power index of P_2.
  (b) the Banzhaf power index of P_1 is less than twice the Banzhaf power index of P_2.
  (c) the Banzhaf power index of P_1 is equal to twice the Banzhaf power index of P_2.
  (d) the Banzhaf power index of P_1 is equal to the Banzhaf power index of P_2.

31. Give an example of a weighted voting system in which P_1 has twice as many votes as P_2 and
  (a) the Shapley-Shubik power index of P_1 is greater than twice the Shapley-Shubik power index of P_2.
  (b) the Shapley-Shubik power index of P_1 is less than twice the Shapley-Shubik power index of P_2.
  (c) the Shapley-Shubik power index of P_1 is equal to twice the Shapley-Shubik power index of P_2.
  (d) the Shapley-Shubik power index of P_1 is equal to the Shapley-Shubik power index of P_2.

32. (a) Consider the weighted voting system [22: 10, 10, 10, 10, 1]. Are there any dummies? Explain your answer.
  (b) Without doing any work (but using your answer for [a]), find the Banzhaf and Shapley-Shubik power distributions of this weighted voting system.
  (c) Consider the weighted voting system [q: 10, 10, 10, 10, 1]. Find all the possible values of q for which player 5 is not a dummy.

33. (a) Verify that the weighted voting systems [12: 7, 4, 3, 2] and [24: 14, 8, 6, 4] result in exactly the same Banzhaf power distribution. (If you need to make calculations, do them for both systems side by side and look for patterns.)
  (b) Based on your work in (a), explain why the two proportional weighted voting systems [q: w_1, w_2, . . . , w_N] and [cq: cw_1, cw_2, . . . , cw_N] always have the same Banzhaf power distribution.

34. (a) Verify that the weighted voting systems [12: 7, 4, 3, 2] and [24: 14, 8, 6, 4] result in exactly the same Shapley-Shubik power distribution. (If you need to make calculations, do them for both systems side by side and look for patterns.)
  (b) Based on your work in (a), explain why the two proportional weighted voting systems [q: w_1, w_2, . . . , w_N] and [cq: cw_1, cw_2, . . . , cw_N] always have the same Shapley-Shubik power distribution.

35. A dummy is a dummy is a dummy. . . . This exercise shows that a player that is a dummy is a dummy regardless of which interpretation of power is used.
  (a) Show that if a player has a Banzhaf power index of 0 in a weighted voting system, then that player must also have a Shapley-Shubik power index of 0.
  (b) Show that if a player has a Shapley-Shubik power index of 0 in a weighted voting system, then that player must also have a Banzhaf power index of 0.

36. Consider the weighted voting system [q: 5, 4, 3, 2, 1].
  (a) For what values of q is there a dummy?
  (b) For what values of q do all players have the same power?

37. The weighted voting system [6: 4, 2, 2, 2, 1] represents a partnership between 5 people (P_1, P_2, P_3, P_4, and you!). You are the last player (the one with 1 vote), which in this case makes you a *dummy*! Not wanting to remain a dummy, you offer to buy one vote. Each of the other four partners is willing to sell you one of their votes, and they are all asking the same price. Which partner should you buy from in order to get as much power for your buck as possible? Use the Banzhaf power index for your calculations. Explain your answer.

38. The weighted voting system [27: 10, 8, 6, 4, 2] represents a partnership between 5 people (P_1, P_2, P_3, P_4, and P_5). You are P_5, the one with 2 votes. You want to increase your power in the partnership and are prepared to buy *one share* (one share = one vote) from any of the other partners. P_1, P_2, and P_3 are each willing to sell cheap ($1000 for one share), but P_4 is not being quite as cooperative—she wants $5000 for one share. Given that you still want to buy one share, who should you buy it from? Use the Banzhaf power index for your calculations. Explain your answer.

39. The weighted voting system [18: 10, 8, 6, 4, 2] represents a partnership between 5 people (P_1, P_2, P_3, P_4, and P_5). You are P_5, the one with 2 votes. You want to increase your power in the partnership and are prepared to buy shares (one share = one vote) from any of the other partners.

   (a) Suppose that each partner is willing to sell *one* share and they are all asking the same price. Assuming that you decide to buy only one share, which partner should you buy from? Use the Banzhaf power index for your calculations.

   (b) Suppose that each partner is willing to sell 2 shares and they are all asking the same price. Assuming that you decide to buy 2 shares from a single partner, which partner should you buy from? Use the Banzhaf power index for your calculations.

   (c) If you have the money and the cost per share is fixed, should you buy 1 share or 2 shares (from a single person)? Explain.

40. Sometimes in a weighted voting system, two or more players decide to **merge**—that is to say, to combine their votes and always vote the same way. (Notice that a merger is different from a coalition—coalitions are temporary, mergers are permanent.) For example, if in the weighted voting system [7: 5, 3, 1] P_2 and P_3 were to merge, the weighted voting system would then become [7: 5, 4]. In this exercise, we explore the effects of mergers on a player's power.

   (a) Consider the weighted voting system [4: 3, 2, 1]. In Example 9 we saw that P_2 and P_3 each have a Banzhaf power index of $\frac{1}{5}$. Suppose that P_2 and P_3 merge and become a single player P^*. What is the Banzhaf power index of P^*?

   (b) Consider the weighted voting system [5: 3, 2, 1]. Find first the Banzhaf power indexes of players P_2 and P_3, and then the Banzhaf power index of P^* (the merger of P_2 and P_3). Compare.

   (c) Same as (b) for [6: 3, 2, 1].

   (d) What are your conclusions from (a), (b), and (c)?

Running

41. **(a)** Give an example of a weighted voting system with 4 players and such that the Shapley-Shubik power index of P_1 is $\frac{3}{4}$.

   (b) Show that in any weighted voting system with 4 players, a player cannot have a Shapley-Shubik power index of more than $\frac{3}{4}$ unless he or she is a dictator.

   (c) Show that in any weighted voting system with N players, a player cannot have a Shapley-Shubik power index of more than $(N - 1)/N$ unless he or she is a dictator.

    (d) Give an example of a weighted voting system with N players and such that P_1 has a Shapley-Shubik power index of $(N-1)/N$.

42. (a) Give an example of a weighted voting system with 3 players and such that the Shapley-Shubik power index of P_3 is $\frac{1}{6}$.

    (b) Show that in any weighted voting system with 3 players, a player cannot have a Shapley-Shubik power index of less than $\frac{1}{6}$ unless he or she is a dummy.

43. (a) Give an example of a weighted voting system with 4 players and such that the Shapley-Shubik power index of P_4 is $\frac{1}{12}$.

    (b) Show that in any weighted voting system with 4 players, a player cannot have a Shapley-Shubik power index of less than $\frac{1}{12}$ unless he or she is a dummy.

44. (a) Give an example of a weighted voting system with N players having a player with veto power who has a Banzhaf power index of $1/N$.

    (b) Show that in any weighted voting system with N players, a player with veto power must have a Banzhaf power index of at least $1/N$.

45. (a) Give an example of a weighted voting system with N players having a player with veto power who has a Shapley-Shubik power index of $1/N$.

    (b) Show that in any weighted voting system with N players, a player with veto power must have a Shapley-Shubik power index of at least $1/N$.

46. The United Nations Security Council. The U.N. Security Council is made up of 15 countries. There are 5 permanent members (the People's Republic of China, France, Russia, the United Kingdom, and the United States) and 10 nonpermanent members. All 5 of the permanent members have veto power. A winning coalition must consist of the 5 permanent members plus at least 4 nonpermanent members. Explain why the Security Council can formally be described by the weighted voting system [39: 7, 7, 7, 7, 7, 1, 1, 1, 1, 1, 1, 1, 1, 1, 1].

47. The original United Nations Security Council. The original Security Council as set up in 1945 consisted of only 11 countries (5 permanent members plus 6 nonpermanent members). A winning coalition had to include all 5 permanent members plus at least 2 of the nonpermanent members. Give a formal description of the original Security Council as a weighted voting system.

48. The Fresno City Council. In Fresno, California, the city council consists of seven members (the mayor and six other council members). A motion can be passed by the mayor and at least three other council members, or by at least five of the six ordinary council members.

    (a) Describe the Fresno City Council as a weighted voting system.

    (b) Find the Shapley-Shubik power distribution for the Fresno City Council.

APPENDIX: Power Indexes in the Electoral College

State	Electoral Votes*	Banzhaf Power Index†	Shapley-Shubik Power Index
California	54	11.14%	10.81%
New York	33	6.20%	6.29%
Texas	32	6.00%	6.09%
Florida	25	4.63%	4.69%
Pennsylvania	23	4.25%	4.30%
Illinois	22	4.06%	4.11%
Ohio	21	3.87%	3.91%
Michigan	18	3.30%	3.33%
New Jersey	15	2.75%	2.76%
North Carolina	14	2.56%	2.57%
Georgia	13	2.38%	2.38%
Virginia	13	2.38%	2.38%
Indiana	12	2.19%	2.20%
Massachusetts	12	2.19%	2.20%
Missouri	11	2.01%	2.01%
Tennessee	11	2.01%	2.01%
Washington	11	2.01%	2.01%
Wisconsin	11	2.01%	2.01%
Maryland	10	1.82%	1.82%
Minnesota	10	1.82%	1.82%
Alabama	9	1.64%	1.64%
Louisiana	9	1.64%	1.64%
Arizona	8	1.46%	1.46%

* Number of seats in Congress (2 senators plus number of members in the House of Representatives).

† The Banzhaf power index was calculated with a computer program developed by Doug Shors.

State	Electoral Votes*	Banzhaf Power Index†	Shapley-Shubik Power Index
Colorado	8	1.46%	1.46%
Connecticut	8	1.46%	1.46%
Kentucky	8	1.46%	1.46%
Oklahoma	8	1.46%	1.46%
South Carolina	8	1.46%	1.46%
Iowa	7	1.28%	1.27%
Mississippi	7	1.28%	1.27%
Oregon	7	1.28%	1.27%
Arkansas	6	1.09%	1.09%
Kansas	6	1.09%	1.09%
Nebraska	5	0.91%	0.90%
New Mexico	5	0.91%	0.90%
Utah	5	0.91%	0.90%
West Virginia	5	0.91%	0.90%
Hawaii	4	0.73%	0.72%
Idaho	4	0.73%	0.72%
Maine	4	0.73%	0.72%
Nevada	4	0.73%	0.72%
New Hampshire	4	0.73%	0.72%
Rhode Island	4	0.73%	0.72%
Alaska	3	0.55%	0.54%
Delaware	3	0.55%	0.54%
Montana	3	0.55%	0.54%
North Dakota	3	0.55%	0.54%
South Dakota	3	0.55%	0.54%
Vermont	3	0.55%	0.54%
Wyoming	3	0.55%	0.54%
District of Columbia	3	0.55%	0.54%
Total	538	100%	100%

REFERENCES AND FURTHER READINGS

1. Banzhaf, John F., III, "Weighted Voting Doesn't Work," *Rutgers Law Review*, 19 (1965), 317–343.

2. Brams, Steven J., *Game Theory and Politics.* New York: Free Press, 1975, chap. 5.

3. Brams, Steven J., William F. Lucas, and Philip D. Straffin, *Political and Related Models.* New York: Springer-Verlag, 1983, chaps. 9 and 11.

4. Grofman, B., "Fair Apportionment and the Banzhaf Power Index," *American Mathematical Monthly,* 88 (1981), 1–5.

5. Imrie, Robert W., "The Impact of the Weighted Vote on Representation in Municipal Governing Bodies of New York State," *Annals of the New York Academy of Sciences,* 219 (November 1973), 192–199.

6. Lambert, John P., "Voting Games, Power Indices and Presidential Elections," *UMAP Journal,* 3 (1988), 213–267.

7. Merrill, Samuel, "Approximations to the Banzhaf Index of Voting Power," *American Mathematical Monthly,* 89 (1982), 108–110.

8. Riker, William H., "A Test of the Adequacy of the Power Index," *Behavioral Science,* 4 (1959), 120–131.

9. Riker, William H., and Peter G. Ordeshook, *An Introduction to Positive Political Theory.* Englewood Cliffs, NJ: Prentice-Hall, Inc., 1973, chap. 6.

10. Shubik, Martin, *Game Theory and Related Approaches to Social Behavior.* New York: John Wiley & Sons, Inc., 1964.

11. Straffin, Philip D., Jr., "Homogeneity, Independence and Power Indices," *Public Choice,* 30 (1977), 107–118.

12. Straffin, Philip D., "The Power of Voting Blocs: An Example," *Mathematics Magazine,* 50 (1977), 22–24.

13. Straffin, Philip D., Jr., *Topics in the Theory of Voting,* UMAP Expository Monograph. Boston: Birkhäuser, 1980, chap. 1.

The Slice Is Right

Fair Division

If we have fifty pieces of candy, how do we divide them fairly among five children?

PROBLEM IN FOURTH-GRADE SCHOOLBOOK

Sometime during our grammar school days, all of us have had to solve a problem like this—it is the classic application of whole-number division. It may come as a bit of a surprise, but we will resurrect the theme in this chapter. Why?

Implicit in the grammar school version of the problem is the assumption that the pieces of candy are all *identical.* But what if they aren't? What happens if we allow the pieces of candy to be different (say we have an assortment of Baby Ruths, Reese's Pieces, Snickers, Milky Ways, etc.) and therefore have potentially different values? Moreover, what if, as is usually the case, the children have different value systems (say Joey likes Reese's Pieces better than Baby Ruths and hates Snickers, while Tanya likes Snickers best, is allergic to Reese's Pieces, and on and on and on)? Under these circumstances the complexity of the problem seems daunting, and the task of dividing the pieces fairly among the children can seem hopeless. Actually, under the right set of circumstances we can accomplish this using relatively simple procedures called *fair division schemes.* Moreover—and this is almost magical—we can often divide the candy in such a way that each child is satisfied that he or she has received a fair share and still end up with one or more pieces of candy left over.

In this chapter we will introduce several fair division schemes and discuss how they work and the assumptions under which they make sense. Before we present specific descriptions of the various fair division schemes, there is a critical question that must be addressed: Is this topic really important? Who cares whether five little kids are happy with their shares of candy? A moment's reflection will show, however, that children and candy are but a metaphor for an important, and frequently occurring, problem: How can an object or a set of objects to be shared by a set of participants be divided among them in a way that ensures that each is satisfied that he or she has received a fair share of the total?

The settlement of an estate among heirs, the division of common property in a divorce proceeding, the subdivision of a parcel of land, and the apportionment of seats to states in the U.S. House of Representatives are all significant variations of a common theme: the problem of *fair division.*

FAIR DIVISION PROBLEMS AND FAIR DIVISION SCHEMES

In this section we will clarify the distinction between fair division problems and their possible methods of solution (fair division schemes).

The elements of every **fair division problem** are a set of N players (P_1, P_2, . . . , P_N) and a set of goods (which we will call S). The problem is to divide S into N shares (s_1, s_2, . . . , s_N) in such a way that each player gets a *fair share* of the total set S.

What is a fair share? By a **fair share** we mean any share that *in the opinion of the player receiving it* has a value that is at least $1/N$ of the total value of the set of goods S. For example, if we have 5 players, then any share that, in the opinion of player X, is worth $\frac{1}{5}$ (20%) or more of the total can be considered a fair share for player X.

Implicit in our description of a fair division problem is the assumption that each player has the ability to judge whether a share is fair or not. In fact we will assume much more and give each player the ability to assign exact fractional (or percent) values to each of the various shares (*I think that this piece is worth one-third of the total, this one is worth 40% of the total, . . .*). Although it is not always possible to precisely quantify our opinion about the worth of an object, it is nonetheless a practice that is very much part of our culture. We exercise it every time we make a bid at an auction, make an offer on a used car, or play Monopoly. Moreover, this is a skill that we acquire at an early age. (Any reader that doubts this should try playing Monopoly with a twelve year old!) Thus, the working assumption in this chapter will be that all players have the ability to precisely judge how they value things and to translate their judgments into numbers.

Any systematic procedure for solving a fair division problem is called a **fair division scheme**. A fair division scheme is essentially a set of rules that, when properly applied, *produces a fair division of the object or objects to be divided.* (In some ways, we can think of a fair division scheme as playing a role in solving a fair division problem that is very similar to the role played by a voting method in finding the winner of an election.)

We will expect any fair division scheme to satisfy the following conditions:

- The procedure is *decisive.* This means that if the rules are followed, the procedure is guaranteed to result in a fair division of the set S.

- The procedure is *internal* to the players. This means that the procedure does not require the intervention of an outside authority such as a judge, an arbitrator, a parent, etc.

■ The procedure assumes that the players have *no knowledge* about each others' value systems. This means that no player has information about the likes or dislikes of any of the other players.

■ The procedure assumes that the players are *rational*. This means that the players' values and strategies are assumed to be based on logic and not emotion.

A final word about fair division schemes. A fair division scheme does not necessarily guarantee that each player will receive a fair share. It is possible for a player to misplay the game (the most common cause for this is greed) and end up with an unfair share. What a fair division scheme does guarantee is that it is impossible for the remaining players or bad luck to conspire to deny any player his or her fair share.

Types of Fair Division Problems

Depending on the nature of the set of goods *S*, fair division problems can be classified into three types: continuous, discrete, and mixed.

In a **continuous** fair division problem the set *S* is divisible in infinitely many ways, and shares can be increased or decreased by arbitrarily small amounts. Typical examples of continuous fair division problems are dividing a parcel of land, a cake, a pizza, ice cream, or a bottle of wine. A large enough sum of money, while not continuous in the theoretical sense (pennies cannot be divided), can be considered for all practical purposes continuous (nobody argues over pennies any more).

A fair division problem is **discrete** when the set *S* is made up of objects that are indivisible (houses, cars, boats, jewelry, etc.). Most candy, while continuous in the theoretical sense, is for all practical purposes discrete (hard pieces of candy cannot easily be broken up without crumbling; soft pieces are hard to cut up without making a big mess), and for convenience throughout this chapter we will think of candy as indivisible and therefore discrete.

A **mixed** fair division problem is one in which some of the components are continuous and some are discrete. Dividing an estate consisting of a car, a house, and a parcel of land is a mixed division problem.

Depending on whether a fair division problem is continuous, discrete, or mixed, a different strategy is needed for finding a solution. In the rest of this chapter we will present several different schemes for solving continuous fair division problems, and two very different schemes for solving discrete fair division problems. We will not discuss schemes for solving mixed fair division problems as they can usually be solved by dividing the continuous and discrete parts separately.

We start with, what is undoubtedly, the best known of all fair division schemes: the divider-chooser method.

THE DIVIDER-CHOOSER METHOD

This classic scheme can be used anytime there is a continuous fair division problem involving two players. Most of us have unwittingly used it at some time or another, and it is commonly known as the *you cut–I choose* method. As this name suggests, one player divides the cake (we will use the word "cake" as a metaphor for any continuous set *S*) into two pieces, and the other player picks the piece he or she wants, leaving the other piece to the divider. When played honestly, this method guarantees that each player will get a share that he or she believes to be worth *at least* one-half of the total. The divider can guarantee this for herself in the mere act of dividing, and the chooser because of the simple fact that when anything is divided into two parts, one of the parts must be worth at least one-half or more of the total.

FIGURE 3-1

Example 1. On their first date, Bob and Rachel go to the county fair. With a $2.00 raffle ticket they win the chocolate-strawberry cake shown in Fig. 3-1. (Actually they had their hearts set on the red Corvette, but third prize is better than nothing.)

To Bob, chocolate and strawberry are equally yummy—he has no preference for one over the other. Thus, in Bob's eyes, the value of the cake is distributed evenly between the chocolate and strawberry parts (Fig. 3-2[a]). On the other hand, Rachel is allergic to chocolate (she gets terrific headaches when she eats it) so she won't eat any of the chocolate part. Thus, in Rachel's eyes the value of the cake is concentrated entirely in the strawberry half (Fig. 3-2[b]). Since this is their first date, neither one of them knows anything about the other's likes and dislikes.

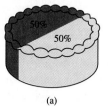
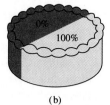

FIGURE 3-2 How Bob and Rachel eye the same cake.

(a)   (b)

Let's now see how Bob and Rachel might divide this cake using the divider-chooser method. Bob, being the gentleman that he is, volunteers to go first and be the divider. His cut is shown in Fig. 3-3[a]. There is no need to psychoanalyze the reasons for Bob's cut (granted, it is a little weird, but then, so is Bob). The important thing here is that mathematically speaking, it is a perfectly logical cut based on Bob's value system. It is now Rachel's time to choose, and her choice is obvious—she will pick

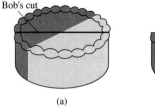

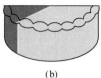

FIGURE 3-3 Bob's cut,
Rachel's pick.

(a)   (b)

berry and 25% chocolate (Fig. 3-3[b]). Notice that while Bob gets a share that is
(in his value system) worth exactly one-half of the total, Rachel ends up with a
share that (in her value system) is worth three-fourths of the total. ▬

It is clear from Example 1 that although the divider-chooser method satisfies our requirement for fair division (each player gets a share that in his or her
opinion is worth at least one-half of the total), there is a definite advantage in being the chooser. The simplest way to handle this problem is to randomly choose
who gets to be the divider and who gets to be the chooser (toss a coin, draw
straws, etc.).

The divider-chooser method for two players can be generalized for the
case of more than two players in several ways. We will present three of them in
this chapter: the lone divider method, the lone chooser method, and the last diminisher method. In the **lone divider method**, one of the players is the divider
and all the rest are choosers; in the **lone chooser method**, one of the players is
the chooser and all the rest are dividers; in the **last diminisher method**, each
player has a chance to be both a divider and a chooser.

Before we describe these methods we must mention another important assumption needed for all of these methods to work: *The value of the object S being divided does not diminish when the object is cut.* Thus, when cutting our theoretical cakes, there will be no crumbs.

THE LONE DIVIDER METHOD

For the sake of simplicity, we describe the method for the case of 3 players.
Here we have 1 divider and 2 choosers. The fairest way to decide which player
is the divider is by random selection (roll a die, draw straws, draw cards from a
deck, etc.). Let's say Diva is the divider, and Chooch and Cher are choosers.

- ■ **Move 1 (Divison).** The divider (Diva) cuts the cake into 3 slices (s_1, s_2,
  and s_3), each of which she judges to be a fair one-third of the cake.

- ■ **Move 2 (Declarations).** Each of the choosers declares independently
  (usually by writing it down on a slip of paper) which of the slices she be-

the bottom piece consisting of 75% straw-lieves to be a fair share (i.e., worth at least one-third) of the cake and therefore acceptable.

■ **Move 3 (Distribution).**   Who gets what? Depending on the declarations, several things can happen.

Case 1.   At least one of the choosers selects more than one slice as acceptable. In this case it is possible to distribute the three slices so that each player gets a slice that is acceptable.

Example 2.   Chooch: $\{s_1, s_2\}$ (which means that Chooch declares that s_1 and s_2 are both acceptable); Cher: $\{s_1\}$. A fair division of the cake can be obtained by the distribution: Chooch gets s_2, Cher gets s_1, and Diva gets s_3.   ▬

Example 3.   Chooch: $\{s_1, s_3\}$; Cher: $\{s_1, s_3\}$. In this case we give s_2 to Diva and distribute s_1 and s_3 between the choosers. Again, the fairest way to do it is to flip a coin and let the winning chooser be the first to choose. In any case everyone gets an acceptable slice.   ▬

Case 2.   Both choosers select just one slice as acceptable, but the slices are different. Once again we can distribute the slices so that each player gets a slice that is acceptable.

Example 4.   Chooch: $\{s_3\}$; Cher: $\{s_2\}$. Here a fair division is given by giving s_1 to Diva, s_3 to Chooch, and s_2 to Cher.   ▬

Case 3.   Both choosers select just one slice as acceptable, and it is the same slice. We now have a standoff as both choosers covet the same slice. One way out of the standoff is to get rid of the divider by giving her one of the other two slices (the fairest way to do it is to let the divider randomly pick between the two). After the divider has chosen her slice the remaining two slices are recombined into a single piece and divided between the two choosers using the divider-chooser method for two players. It all sounds complicated but it really isn't, and you can clearly see how the method handles this case in the next example.

Example 5.   Chooch: $\{s_1\}$; Cher: $\{s_1\}$. Here Diva gets to choose between s_2 and s_3. Let's say she end up with s_2. Now the two remaining slices s_1 and s_3 are combined into a single piece and divided between Chooch and Cher using the divider-chooser method for two players.

Why is this a fair division of the original cake? Clearly, Diva has no cause to complain—she gets s_2, a piece she believes to be a fair one-third of the cake. As far as Chooch and Cher are concerned, neither one of them considers s_2 to be worth one-third of the cake (otherwise they would have bid for it), which implies that both of them consider s_1 and s_3 combined to be worth more than two-thirds of the original cake. Dividing such a piece fairly between the two of them must result in shares that are acceptable to both. (In other words, if $x > \frac{2}{3}$ and $y \geq \frac{1}{2}x$, then we must have $y > \frac{1}{3}$.)

The lone divider method as described above for 3 players can be generalized for any number of players (see Exercises 8 through 15 and 50).

THE LONE CHOOSER METHOD

Once again, we start with a description of the method for the case of three players. Here we have one chooser and two dividers. As usual, we decide who is what by random selection. Let's say that Chuck is the chooser, and Dave and Dirk the dividers.

■ **Move 1 (First Division).**   Dave and Dirk cut the cake (Fig. 3-4[a]) into two slices using the divider-chooser method. Dave gets slice 1, and Dirk gets slice 2 (Fig. 3-4[b]). Each considers his slice worth at least one-half of the total.

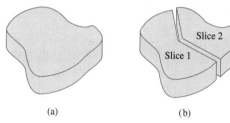

FIGURE 3-4   (a) The original cake. (b) First division. The cake is divided between Dave (slice 1) and Dirk (slice 2) using the divider-chooser method.

(a)   (b)

■ **Move 2 (Second Division).**   Dave divides slice 1 into three pieces that in his opinion have equal value. Likewise, Dirk divides slice 2 into three pieces that he considers to be of equal value (Fig. 3-5).

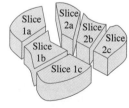

FIGURE 3-5   Second division. Slice 1 is divided into three pieces of equal value by Dave; slice 2 is divided into three pieces of equal value by Dirk.

■ **Move 3 (Selection).**   Chuck picks one of Dave's three pieces and one of Dirk's three pieces. These 2 pieces constitute Chuck's share. Dave gets to keep the remaining two pieces from slice 1, and Dirk gets to keep the remaining two pieces from slice 2 (Fig. 3-6).

Chuck picks slices 1a and 2c

Dave gets slices 1b and 1c.

Dirk gets slices 2a and 2b.

FIGURE 3-6 Selection.

Why is this a fair division of the cake? Since the first division was a fair division of the cake between two people, Dave's assessment of slice 1 is that it is worth at least one-half of the total. Since he got to keep two thirds of slice 1, his share must be worth at least two-thirds of one-half of the total, thereby making it a fair share ($\frac{2}{3} \times \frac{1}{2} = \frac{1}{3}$). The same argument applies to Dirk. As far as Chuck is concerned, slice 1 is worth some fraction of the total. We do not know what that fraction is, so we will call it F. We can now say that the value of slice 2 to Chuck is $1 - F$, since between them, slice 1 and slice 2 make up the whole cake. Since Chuck is the chooser and presumably picks slice 1a because he likes it best among slice 1a, slice 1b, and slice 1c, the value of slice 1a to Chuck must be at least one-third of F. Likewise, the value of slice 2c to Chuck must be at least one-third of $1 - F$. Since

$$\tfrac{1}{3}F + \tfrac{1}{3}(1 - F) = \tfrac{1}{3},$$

Chuck gets at least one-third, and therefore a fair share, of the cake.

The following example illustrates in detail how the lone chooser method works.

FIGURE 3-7

Example 6. David, Paul, and Katie want to divide an orange-pineapple cake using the lone chooser method. They draw straws and Katie gets to be the chooser, so David and Paul get to first divide the cake using the divider-chooser method. Since David drew the shortest straw, he gets to cut the cake.

The cake is a fancy cake costing $27.00 (David, Paul, and Katie are putting up $9.00 each) and is half orange and half pineapple as shown in Fig. 3-7.

Now, for their value systems:

■ David likes pineapple and orange equally, so to him, value equals size. In his eyes the cake looks like Fig. 3-8(a).

■ Paul likes orange but hates pineapple. In Paul's eyes the cake looks like Fig. 3-8(b).

■ Katie likes pineapple twice as much as she likes orange. In Katie's eyes the cake looks like Fig. 3-8(c).

FIGURE 3-8 How David, Paul, and Katie eye the same cake.

(a)

(b)

(c)

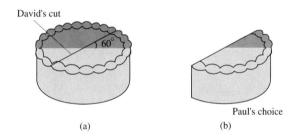

FIGURE 3-9 David's cut, Paul's choice.

(a)   (b)

■ **Move 1.** It's David's turn to cut. His cut is shown in Fig. 3-9(a). Of course, since Paul doesn't like pineapple, he will take the piece with the most orange as shown in Fig. 3-9(b).

The value of the two pieces in each player's eyes is shown in Fig. 3-10.

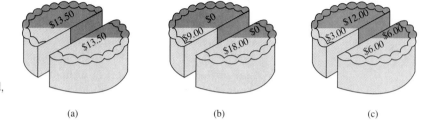

FIGURE 3-10 How David, Paul, and Katie view the two pieces.

(a)   (b)   (c)

■ **Move 2.** David divides his piece into three pieces that in his opinion are of equal value. Of course to David this only means the volumes are equal, so he cuts the pieces as shown in Fig. 3-11(a). Paul also divides his piece into three pieces that in his opinion are of equal value. Remember that Paul hates pineapple so his objective is to have one-third of the orange in each of his pieces, so he cuts the piece as shown in Fig. 3-11(b).

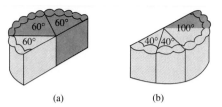

(a)   (b)

FIGURE 3-11 (a) David cuts "his" piece; (b) Paul cuts "his" piece.

FIGURE 3-12 The value of each of the six pieces in Katie's eyes.

■ **Move 3.** It's now Katie's turn to choose one piece from David's three pieces and one piece from Paul's three pieces. Fig. 3-12 shows the value of all the pieces to Katie. Obviously, Katie will choose from David's three pieces one of the $6.00 pieces and from Paul's three pieces the $8.00 piece.

Fair share	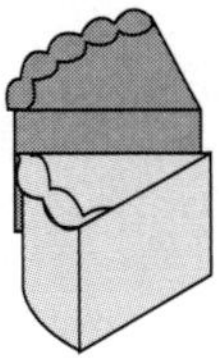	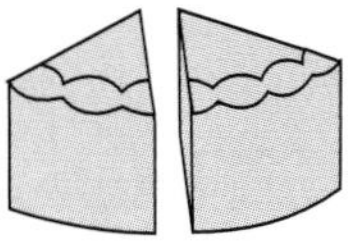	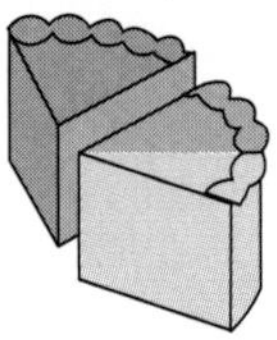
Recipient	David	Paul	Katie
Value in David's eyes	$9.00	$6.00	$12.00
Value in Paul's eyes	$9.00	$12.00	$6.00
Value in Katie's eyes	$9.00	$4.00	$14.00

Figure 3-13 The final division of the cake along with each player's assessment of it.

The final division of the cake and the value of the division in each of the three player's eyes is shown in Fig. 3-13.

Notice that each person has received a share that is worth at least $9.00 (one-third of the value of the cake) in his or her own eyes. Notice also that David feels that Katie got a more valuable share than he did. Remember—a fair division only guarantees that *each player will receive a fair share, not the best share.*

THE LAST DIMINISHER METHOD

We will describe the method for the case of five players. The method[1] can be generalized in a straightforward way to any number of players. Here there are no designated dividers and choosers, but before any action takes place the players are randomly assigned an order (P_1 first, P_2 second, . . . , P_5 last). The players will play in this order throughout the game, and the play takes place in rounds.

■ **Round 1.** P_1 cuts a slice from the cake that she believes to be a fair share (one-fifth) of the cake. This piece is P_1's staked claim (at least for the moment). P_1 must be careful to stake a claim that is neither too small (she may end up with that piece) nor too large (somebody else might end up with it). P_2 now has the right to pass or to play on P_1's claim. If he thinks that P_1's claim is a bad choice (worth less than one-fifth of the cake), he passes, remaining in contention for a fair share of the rest of the cake (worth in his opinion more than four-fifths of the cake). On the other hand if he thinks that P_1's claim is a good one (perhaps it appears to P_2 that P_1 might have been a little greedy), then he can make a claim on P_1's claim by staking out a subpiece of it. If P_2 opts to do this, then we call him a **diminisher**. Again, P_2 must be careful to make his claim just right—not

---

[1] To the best of our knowledge, this method was first described by the Polish mathematician Hugo Steinhaus in 1948 (see reference 6 at the end of this chapter). Steinhaus, however, credits the discovery of the method to two other Polish mathematicians: S. Banach and B. Knaster.

too small (he might end up with it) and not too large (someone else might end up with it). Also, P_2 cannot claim part of P_1's claim and also remain in contention for the rest of the cake (he cannot have his cake and eat it too!). If P_2 becomes a diminisher, then the difference between P_1's claim and P_2's claim is added to the remainder of the cake, and P_1 returns to the group of players contending for that remainder. P_1 should be pleased, since she is now in contention for a fair one-fourth of a piece that in her opinion is worth more than four-fifths of the whole. It is now P_3's turn to play. P_3 has a choice to pass and contend for the remainder of the cake or become a diminisher on the current claim (whether it is P_1's or P_2's). Similarly, P_4 and P_5 in turn have a chance to become diminishers on the current claim, or pass and remain in contention for the remainder. After all the players have had a chance to play, the player whose claim is current (the **last diminisher**) gets that piece and departs.

■ **Round 2.**   Reconstitute the cake, putting together all the slivers left over by the various cuts and subcuts. Start the process over with the remaining four players. At the end of the round the last diminisher gets his or her staked claim and departs.

■ **Round 3.**   There are three players left. Repeat the process with the remainder of the cake. The last diminisher gets his or her staked claim and departs.

■ **Round 4.**   At this point there are two players left and the leftover cake can be divided using the divider-chooser method.

Example 7. Five sailors are marooned on a deserted tropical island. Knowing how to make the best out of a good situation they decide to divide the island into fair shares and stay there forever. After some debate, they settle on the last diminisher method.

Paul Steele/ The Stock
Market

FIGURE 3-14 Round 1.

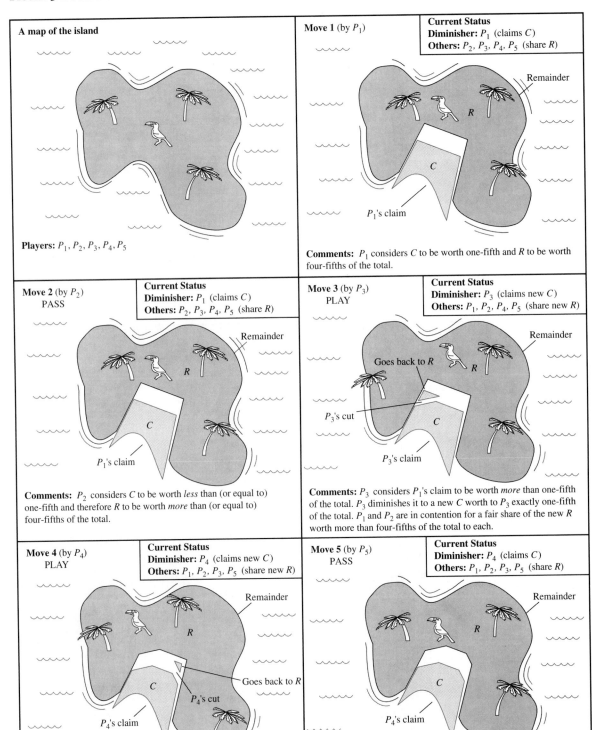

A map of the island

Players: P_1, P_2, P_3, P_4, P_5

Move 1 (by P_1)

Current Status
Diminisher: P_1 (claims C)
Others: P_2, P_3, P_4, P_5 (share R)

Remainder
R
C

P_1's claim

Comments: P_1 considers C to be worth one-fifth and R to be worth four-fifths of the total.

Move 2 (by P_2)
PASS

Current Status
Diminisher: P_1 (claims C)
Others: P_2, P_3, P_4, P_5 (share R)

Remainder
R
C

P_1's claim

Comments: P_2 considers C to be worth *less* than (or equal to) one-fifth and therefore R to be worth *more* than (or equal to) four-fifths of the total.

Move 3 (by P_3)
PLAY

Current Status
Diminisher: P_3 (claims new C)
Others: P_1, P_2, P_4, P_5 (share new R)

Goes back to R
Remainder
R
P_3's cut
C
P_3's claim

Comments: P_3 considers P_1's claim to be worth *more* than one-fifth of the total. P_3 diminishes it to a new C worth to P_3 exactly one-fifth of the total. P_1 and P_2 are in contention for a fair share of the new R worth more than four-fifths of the total to each.

Move 4 (by P_4)
PLAY

Current Status
Diminisher: P_4 (claims new C)
Others: P_1, P_2, P_3, P_5 (share new R)

Remainder
R
Goes back to R
C
P_4's cut
P_4's claim

Comments: P_4 considers P_3's claim to be worth *more* than one-fifth of the total. P_4 diminishes it to a new C worth to P_4 exactly one-fifth of the total. P_1, P_2, and P_3 are in contention for a fair share of the new R worth more than four-fifths of the total to each.

Move 5 (by P_5)
PASS

Current Status
Diminisher: P_4 (claims C)
Others: P_1, P_2, P_3, P_5 (share R)

Remainder
R
C
P_4's claim

Comments: P_5 considers C to be worth *less* than (or equal to) one-fifth and therefore R to be worth *more* than (or equal to) four-fifths of the total.

Round 1 is now over (all players have had a chance to pass or play). The final outcome of round 1 is to give the last diminisher (P_4) his or her claim C; the remainder R will be divided fairly among the four remaining players (P_1, P_2, P_3, and P_5).

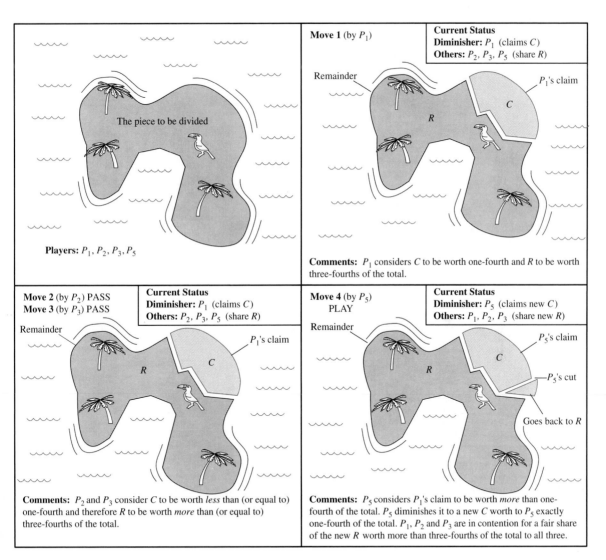

Players: P_1, P_2, P_3, P_5

The piece to be divided

Move 1 (by P_1)

Current Status
Diminisher: P_1 (claims C)
Others: P_2, P_3, P_5 (share R)

Remainder

R

P_1's claim

C

Comments: P_1 considers C to be worth one-fourth and R to be worth three-fourths of the total.

Move 2 (by P_2) PASS
Move 3 (by P_3) PASS

Current Status
Diminisher: P_1 (claims C)
Others: P_2, P_3, P_5 (share R)

Remainder

R

P_1's claim

C

Comments: P_2 and P_3 consider C to be worth *less* than (or equal to) one-fourth and therefore R to be worth *more* than (or equal to) three-fourths of the total.

Move 4 (by P_5)
PLAY

Current Status
Diminisher: P_5 (claims new C)
Others: P_1, P_2, P_3 (share new R)

Remainder

R

P_5's claim

C

P_5's cut

Goes back to R

Comments: P_5 considers P_1's claim to be worth *more* than one-fourth of the total. P_5 diminishes it to a new C worth to P_5 exactly one-fourth of the total. P_1, P_2 and P_3 are in contention for a fair share of the new R worth more than three-fourths of the total to all three.

FIGURE 3-15 Round 2

Round 2 is now over (all players have had a chance to pass or play). The final outcome of round 2 is to give the last diminisher (P_5) his or her claim C; the remainder R will be divided fairly among the remaining three players (P_1, P_2, and P_3).

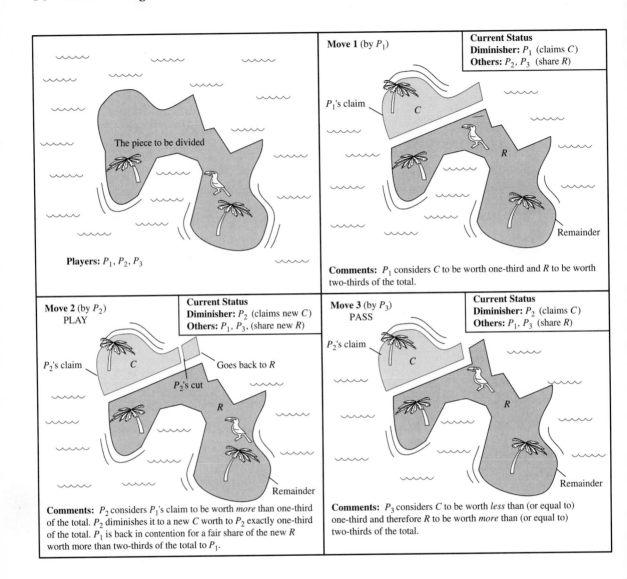

Move 1 (by P_1)

Current Status
Diminisher: P_1 (claims C)
Others: P_2, P_3 (share R)

P_1's claim

C

R

Remainder

Comments: P_1 considers C to be worth one-third and R to be worth two-thirds of the total.

The piece to be divided

Players: P_1, P_2, P_3

Move 2 (by P_2)
PLAY

Current Status
Diminisher: P_2 (claims new C)
Others: P_1, P_3, (share new R)

P_2's claim

C

Goes back to R

P_2's cut

R

Remainder

Comments: P_2 considers P_1's claim to be worth *more* than one-third of the total. P_2 diminishes it to a new C worth to P_2 exactly one-third of the total. P_1 is back in contention for a fair share of the new R worth more than two-thirds of the total to P_1.

Move 3 (by P_3)
PASS

Current Status
Diminisher: P_2 (claims C)
Others: P_1, P_3 (share R)

P_2's claim

C

R

Remainder

Comments: P_3 considers C to be worth *less* than (or equal to) one-third and therefore R to be worth *more* than (or equal to) two-thirds of the total.

FIGURE 3-16 Round 3.

Round 3 is now over (all players have had a chance to pass or play). The final outcome of round 3 is to give the last diminisher (P_2) his or her claim C; the remainder R will be divided fairly between the two remaining players (P_1 and P_3).

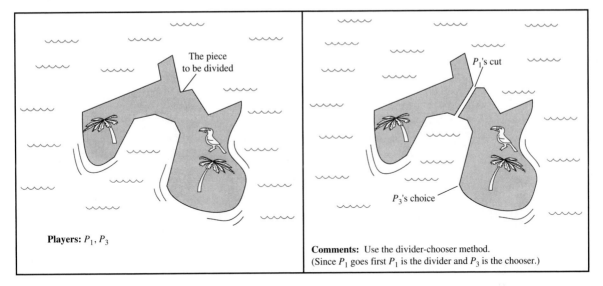

The piece
to be divided

P_1's cut

P_3's choice

Players: P_1, P_3

Comments: Use the divider-chooser method.
(Since P_1 goes first P_1 is the divider and P_3 is the chooser.)

FIGURE 3-17 Round 4.

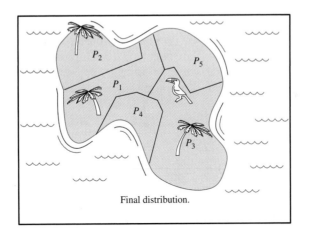

P_2

P_5

P_1

P_4

P_3

Final distribution.

FIGURE 3-18 The final di-
cision of the island.

In the next two sections we will discuss discrete fair division schemes. In
this case the set S consists of objects that are indivisible (houses, cars, paintings,
candy, etc.).

THE METHOD OF
SEALED BIDS

The best known of all discrete fair division schemes is the **method of sealed
bids** originally proposed by the Polish mathematician Hugo Steinhaus. The easiest way to illustrate how this method works is by means of an example.

Example 8. In her last will and testament, Grandma plays a little joke on her four grandchildren (Art, Betty, Carla, and Dave) by leaving three indivisible items—a house, a Rolls Royce, and a Picasso painting—to be divided equally among the four of them. After some discussion someone suggests that all three.

How can we divide this estate fairly among four heirs? A classic example of a discrete fair division problem. (Reprinted by permission of Rolls-Royce Motor Cars Inc.)

items be sold and the money divided equally among the four heirs. Unfortunately, each of the four heirs has a different opinion as to what the items are really worth (as shown in the following table), and they cannot therefore reach an agreement on what a reasonable selling price for each item ought to be. (By the way, we are assuming that all four grandchildren are honest people and that the figures shown in the table represent their true and sincere assessment of the dollar value of the items.)

	Art	Betty	Carla	Dave
House	220,000	250,000	211,000	198,000
Rolls Royce	40,000	30,000	47,000	52,000
Picasso	280,000	240,000	234,000	190,000

As a last recourse, they decide to consult a mathematician who suggests that it is possible to make everyone happy. His division of the estate (with explanation) goes as follows:

■ **Art.**   By his own estimation, the total value of the estate is $540,000. (This figure is obtained by adding the values that Art has assigned to each of the three items.) Since he is entitled to one-fourth of the estate, his fair share is worth one-fourth of that amount, $135,000. Since he was the highest bidder for the Picasso painting, he will get the painting, but since the painting is worth much more than what he is entitled to he must make up the difference by paying it back to the estate. In conclusion, Art gets the Picasso painting but pays back to the estate $145,000 ($280,000 − $135,000). Note that since Art was honest in his assessment of the value of each item, he is satisfied that this settlement is perfectly fair.

■ **Betty.** By her own estimation, the total value of the estate is $520,000 of which she is entitled to exactly one-fourth, which comes to $130,000. Being the highest bidder for the house, Betty gets it, but she must pay the estate the difference between her assessment of the value of the house ($250,000) and the fair share she is entitled to ($130,000). Thus, the settlement for Betty is that she gets the house but has to pay the estate $120,000.

■ **Carla.** Her assessment of the total value of the estate is $492,000, and her fair share is one-fourth of that or $123,000. Since Carla is not the highest bidder on any of the items, she doesn't get any of them. She is entitled, however, to get her fair share of $123,000 in cash.

■ **Dave.** Dave's assessment of the total value of the estate is $440,000, and his fair share is therefore $110,000. Since he is the highest bidder for the Rolls, he gets it for $52,000. He is still owed a balance of $58,000 (the difference between his fair share and his assessment of the value of the Rolls Royce) which is given to him in cash.

At this point each of the heirs has to be satisfied—after all, one way or another each has received a fair share. So much for fair shares. Now comes the real fun. If we add Art's and Betty's payments to the estate and subtract the payments made by the estate to Carla and Dave, we discover that something truly remarkable has happened: There is a surplus of $84,000 left over. The following table summarizes how this happened.

	Item Received	Cash
Art	Picasso painting	Pays $145,000 (+)
Betty	House	Pays $120,000 (+)
Carla	Nothing	Gets $123,000 (−)
Dave	Rolls Royce	Gets $58,000 (−)
		$84,000 surplus cash

Clearly, the only fair thing to do with this money is to offer it to the mathematician who came up with this brilliant scheme. Being a modest person, however, he declines and suggests instead that the money be divided equally among the four heirs. This is done, and each of the four heirs gets a bonus of $21,000.   ▬

The general procedure used in Example 8, which we will call the **method of sealed bids,** can be described as follows:

■ **Move 1 (Bidding).** Each player makes a sealed bid giving his or her honest assessment of the dollar value of each of the items in the estate. Each player's fair share is calculated by dividing the total of that player's bids by the number of players.

■ **Move 2 (Allocation).**   Each item goes to the highest bidder for that item. (In case of a tie, a predetermined tie-breaking procedure such as flipping a coin should be invoked.) Note that it is possible for a player to get more than one item—in fact even all of them. Each player puts in or takes out (in cash) from a common pot (the estate) the difference between his or her fair share and the total value of the items allocated to that player.

■ **Move 3 (Dividing the Surplus).**   After the original allocations are completed, there will almost always be a surplus of cash left in the estate. This surplus is divided equally among the players.

The method of sealed bids is the mathematical equivalent of pulling rabbits out of a hat. For the method to work, however, certain conditions must be satisfied.

1. Each player must have enough money to play the game. If a player is going to make honest bids on the items, he must be prepared to take some or all of them, which means that he may have to pay the estate certain sums of money. If the player does not have this money available, he is at a definite disadvantage in playing the game.

2. Each player must accept money (if it is a sufficiently large amount) as a substitute for any item. This means that no player can consider any of the items priceless (*I want Mom's diamond ring, and no amount of money in the world is going to make me change my mind!* is not an attitude conducive to a good resolution of the problem.)

3. Players must have no useful information about each other's value systems prior to the bidding. In particular it is critical that the bids be sealed and that no player sees another player's bids ahead of time. To illustrate why this is critical for a fair division, consider the following example: Player *A* thinks that the real value of item *X* is $1500 but before writing down this bid sees that player *B* has bid $2000 for item *X*. She can now bid as much as $1999.99 for *X* with the certain knowledge that she will not be stuck with *X* at that price, while at the same time inflating the total value of her own fair share.

The method of sealed bids takes a particularly simple form in the case of two players and one item. Consider the following example:

Example 9. Al and Betty are getting a divorce. The only common property of value is their house. Since the divorce is amicable and they are not particularly keen on going to court or hiring an attorney, they decide to divide the house using the method of sealed bids. The bids are

Al	$130,000
Betty	$142,000

Betty, being the highest bidder, gets the house but must pay the estate $71,000 (she is entitled to only half of the value of the house). Al's fair share is half of his bid, namely, $65,000. The surplus of $6000 is divided equally between Al and Betty, and the bottom line is that Betty gets the house but pays Al $68,000. Notice that this result is equivalent to assessing the value of the house as the value halfway between the two bids ($136,000) and splitting this value equally between the two parties, with the house going to the highest bidder and the cash to the other party (Exercise 40).

THE METHOD OF MARKERS

This is a discrete fair division scheme that does not require the players to put up any of their own money. In this sense it has a definite advantage over the method of sealed bids. On the other hand, unlike the method of sealed bids, this method cannot be used effectively unless there are many more items to be divided than there are players.

The basic idea in this method is that the items are lined up in a row (we call such a row an **array**) and that each player then breaks up the array of items into consecutive segments (as many segments as there are players). Each player must do this in such a way that the segments represent what in his or her opinion are fair (and therefore acceptable) shares of the entire set of items. Each player performs this task by laying down markers, each marker indicating the end of one segment and the beginning of another. The players lay down their markers independently and in such a way that no player can see the markers of any of the other players.

After all the players are through laying down their markers, the items are allocated among the players. We will describe how this is done with an example.

Example 10. Four players (P_1, P_2, P_3, and P_4) must divide the array of 20 items shown in Fig. 3-19. P_1 plays first and lays down her markers as indicated in Fig. 3-20. We interpret this to mean that P_1 considers ⬛ 🍪 🥫 a fair

FIGURE 3-19

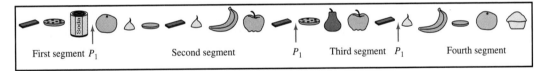

First segment P_1 Second segment P_1 Third segment P_1 Fourth segment

FIGURE 3-20 P_1's markers.

share, 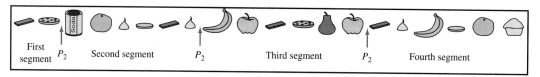 another fair share, and so forth.

Without seeing P_1's markers (P_1's markers can be recorded by a neutral person and then removed), P_2 now proceeds to play in a similar fashion. P_2 lays down his markers as shown in Fig. 3-21.

FIGURE 3-21 P_2's markers.

Similarly, P_3 and P_4 each get a turn to play by laying down their own markers, and they do so as shown in Figs. 3-22 and 3-23, respectively.

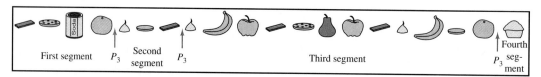

Figure 3-22 P_3's markers

FIGURE 3-23 P_4's markers.

We are now ready to distribute the items among the players. To do so we need to look at the entire picture, showing all the markers of all the players, as in Fig. 3-24.

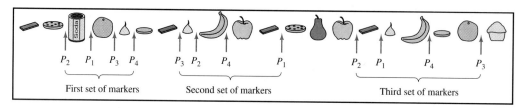

FIGURE 3-24

Moving from left to right, we assign the first segment to the player owning the first marker. That's P_2, who gets ⬤ ⬤. We know that these two items are in P_2's opinion a fair share. P_2 is now out of the game. All of P_2's markers can now be removed, and the new situation is described by Fig. 3-25.

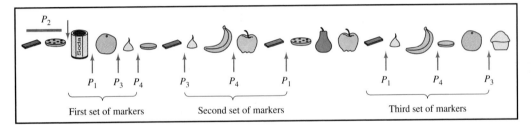

FIGURE 3-25

Once again we scan the picture from left to right until we find the first among the second set of markers. It belongs to P_3, and it is located to the right of the seventh item. We now look back and locate P_3's first marker (which is located to the right of the fourth item) and give P_3 ▵ ⬭ ▬ as shown in Fig. 3-26.

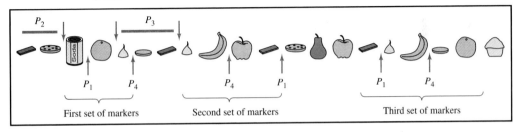

FIGURE 3-26

We repeat the process, locating the first among the third set of markers, which belongs to P_1 (remember P_2 and P_3 are out of the running now), and scanning backward to find P_1's previous (second) marker. The items between these markers are given to P_1 (Fig. 3-27).

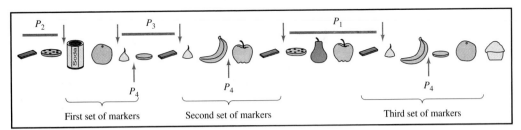

FIGURE 3-27

Finally, there is one last player left (P_4) who gets all the items from his last marker to the end of the array (Fig. 3-28).

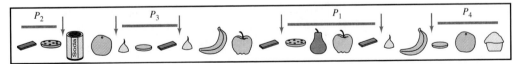

FIGURE 3-28

Much to our surprise, we see that after each player has received his or her fair share, there are a few items left over (Fig. 3-29). These leftover items can be distributed in various ways—given to charity, allocated to the players by drawing lots, or if there are enough of them, the method of markers can be used again.

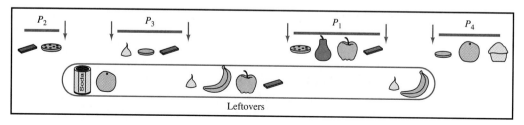

Leftovers

FIGURE 3-29

The general description of the **method of markers** is as follows: There are N players ($P_1, P_2, P_3, \ldots, P_N$) and M items arranged in an array. (Unlike the method of sealed bids, we must have $M > N$.)

- ■ **Move 1 (Bidding).** Each player independently and secretly divides the linear array into N consecutive segments any one of which the player will accept as a fair share of the total set of items.

- ■ **Move 2 (Allocation).** Assign to the player owning the first marker (going from left to right) in the first set of markers that player's first segment. Remove that player's markers. Assign to the player owning the first marker in the second set of markers that player's second segment. Remove that player's markers. Continue this process until each player has received a fair share. The last remaining player gets his or her last segment.

- ■ **Move 3 (Leftovers).** If there are leftover items, they can be allocated to the players by lottery, or if there are more items left than players, the method of markers can be used again.

In spite of its elegance and the nice touch added by the leftovers, the method of markers can be used only under some fairly restrictive conditions. In particular, the method assumes that every player is able to divide the array of items into segments in such a way that each of the segments has equal value (to that player). This precludes, for example, the existence of an item that has a disproportionately large value in relation to all the other items. Suppose that in Example 10, the 20 items consist of a gold watch and 19 pieces of candy (of assorted types). It would be impossible to distribute these items fairly among the four players, since no one except an insane candy lover could possibly divide them into four segments of equal value.

CONCLUSION

The problem of dividing an object or set of objects among the members of a group is a practical problem that comes up regularly in our daily lives. When the object is a pizza, a cake, or a bunch of candy, we don't always pay a great deal of attention to the issue of fairness, but when the object is an estate, a piece of land, or some other valuable asset, dividing things fairly becomes a critical issue.

On the surface, problems of fairness seem far removed from the realm of mathematics. We are more likely to think of ethics or law as being the proper fields for a discussion of this topic. It is surprising therefore, that when certain basic conditions are satisfied, mathematics can provide fair division methods that not only guarantee fairness but often turn out to actually do much better than that.

In this chapter we discussed several such methods, which we called *fair division schemes.* The choice of which is the best fair division scheme to use in a particular situation is not always clear, and in fact there are many situations in which a fair division is mathematically unattainable (we will discuss an important example of this in Chapter 4). At the same time, in a large number of everyday situations the fair division schemes we described in this chapter (or simple variations thereof) will work. Remember these methods the next time you must divide an inheritance, a piece of real estate, or even a fine old bottle of wine. They may serve you well.

KEY CONCEPTS

continuous fair division problem
discrete fair division problem
divider-chooser method
fair division problem
fair division scheme
fair share

last diminisher method
lone chooser method
lone divider method
method of markers
method of sealed bids

EXERCISES

Walking

1. Alex buys a chocolate-strawberry mousse cake for $12.00 (i). Alex values chocolate 3 times as much as he values strawberry.

(i)

(ii)

   (a) What is the value of the chocolate half of the cake to Alex?

   (b) What is the value of the strawberry half of the cake to Alex?

   (c) A piece of the cake is cut as shown in (ii). What is the value of the piece of cake to Alex?

2. Jody buys a chocolate-strawberry mousse cake for $9.00 (i). Jody values strawberry 4 times as much as she values chocolate.

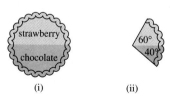

(i)   (ii)

   (a) What is the value of the chocolate half of the cake to Jody?

   (b) What is the value of the strawberry half of the cake to Jody?

   (c) A piece of the cake is cut as shown in (ii). What is the value of the piece of cake to Jody?

3. Kala buys a chocolate-strawberry-vanilla cake for $18.00 (i). Kala likes all 3 flavors but likes strawberry twice as much as vanilla and likes chocolate 3 times as much as vanilla.

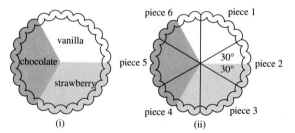

(i)   (ii)

   (a) What is the value of the chocolate part of the cake to Kala?

   (b) What is the value of the strawberry part of the cake to Kala?

   (c) What is the value of the vanilla part of the cake to Kala?

   (d) The cake is cut into 6 pieces equal in size (pieces 1 through 6) as shown in (ii). What is the value of each of the 6 pieces of cake to Kala?

4. Malia buys a chocolate-strawberry-vanilla cake for $14.00 (i). Malia likes all 3 flavors but likes strawberry twice as much as chocolate and likes chocolate twice as much as vanilla.

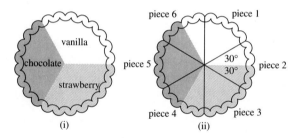

(i)   (ii)

   (a) What is the value of the chocolate part of the cake to Malia?

(b) What is the value of the strawberry part of the cake to Malia?

(c) What is the value of the vanilla part of the cake to Malia?

(d) The cake is cut into 6 pieces equal in size (pieces 1 through 6) as shown in (ii). What is the value of each of the 6 pieces of cake to Malia?

5. Two friends (David and Paul) decide to divide the pizza shown in the following figure using the divider-chooser method. David likes pepperoni, sausage, and mushrooms equally well but hates anchovies. Paul likes anchovies, sausage, and pepperoni equally well but hates mushrooms. Neither one knows anything about the other one's likes and dislikes (they are new friends).

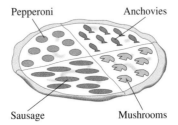

(a) Suppose that David is the divider. Which of the cuts (i) through (iv) show a division of the pizza into fair shares according to David?

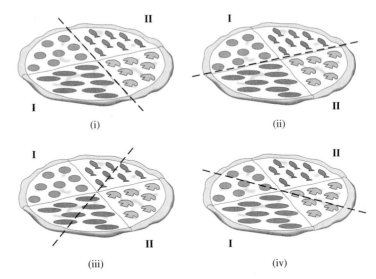

(b) For each of the cuts in (a), which piece should Paul choose?

6. Raul and Trudy want to divide a chocolate-strawberry mousse cake. Raul values chocolate 3 times as much as he values strawberry ($3S = C$); Trudy values strawberry 4 times as much as she values chocolate ($4C = S$).

(a) If Raul is the divider, which of the following cuts are consistent with Raul's value system?

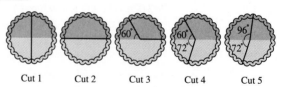

Cut 1   Cut 2   Cut 3   Cut 4   Cut 5

(b) For each of the cuts in (a), indicate which of the pieces is Trudy's best choice.

(c) Suppose Trudy is the divider. Draw three different cuts that are consistent with her value system.

(d) Explain why the following cut is consistent with any value system.

7. Three players want to divide a cake fairly using the lone divider method. The divider cuts the cake into 3 slices (s_1, s_2, s_3).

(a) If the chooser declarations are

Chooser 1: $\{s_2, s_3\}$

Chooser 2: $\{s_1, s_2\}$

describe a possible fair division of the cake.

(b) If the chooser declarations are

Chooser 1: $\{s_1, s_2, s_3\}$

Chooser 2: $\{s_3\}$

describe a possible fair division of the cake.

(c) If the chooser declarations are

Chooser 1: $\{s_1\}$

Chooser 2: $\{s_3\}$

describe a possible fair division of the cake.

(d) If the chooser declarations are

Chooser 1: $\{s_2\}$

Chooser 2: $\{s_2\}$

describe how to proceed to obtain a possible fair division of the cake.

8. Four players want to divide a cake fairly using the lone divider method. The divider cuts the cake into 4 slices (s_1, s_2, s_3, s_4), and the choosers make the following declarations:

Chooser 1: $\{s_2, s_3\}$

Chooser 2: $\{s_1, s_3\}$

Chooser 3: $\{s_1\}$.

(a) Describe a fair division of the cake.

(b) Explain why the answer in (a) is the only possible fair division of the cake.

9. Four players want to divide a cake fairly using the lone divider method. The divider cuts the cake into 4 slices (s_1, s_2, s_3, s_4), and the choosers make the following declarations:

Chooser 1: $\{s_2, s_4\}$

Chooser 2: $\{s_1, s_4\}$

Chooser 3: $\{s_1, s_2\}$.

   (a) Describe a fair division of the cake.
   (b) Describe a fair division of the cake different from the one given in (a).
   (c) Is it possible to find a fair division of the cake such that the divider doesn't get s_3? Explain your answer,

10. Four players want to divide a cake fairly using the lone divider method. The divider cuts the cake into 4 slices (s_1, s_2, s_3, s_4) and the choosers make the following declarations:

   Chooser 1: $\{s_3, s_4\}$
   Chooser 2: $\{s_3, s_4\}$
   Chooser 3: $\{s_3\}$.

   Describe how to proceed to obtain a possible fair division of the cake.

11. Five players want to divide a cake fairly using the lone divider method. The divider cuts the cake into 5 slices (s_1, s_2, s_3, s_4, s_5), and the choosers make the following declarations:

   Chooser 1: $\{s_3, s_4\}$
   Chooser 2: $\{s_3, s_4\}$
   Chooser 3: $\{s_2, s_3, s_4\}$
   Chooser 4: $\{s_2, s_3, s_5\}$.

   (a) Describe a fair division of the cake.
   (b) Describe a fair division of the cake different from the one given in (a).
   (c) Is it possible to find a fair division of the cake such that the divider doesn't get s_1? Explain your answer.

12. Five players want to divide a cake fairly using the lone divider method. The divider cuts the cake into 5 slices (s_1, s_2, s_3, s_4, s_5), and the choosers make the following declarations:

   Chooser 1: $\{s_3, s_5\}$
   Chooser 2: $\{s_1, s_4, s_5\}$
   Chooser 3: $\{s_1, s_3, s_4\}$
   Chooser 4: $\{s_3, s_4\}$.

   (a) Describe a fair division of the cake.
   (b) Describe a fair division of the cake different from the one given in (a).
   (c) Is it possible to find a fair division of the cake such that the divider doesn't get s_2? Explain your answer.

13. Six players want to divide a cake fairly using the lone divider method. The divider cuts the cake into 6 slices ($s_1, s_2, s_3, s_4, s_5, s_6$), and the choosers make the following declarations:

   Chooser 1: $\{s_2, s_3, s_5\}$
   Chooser 2: $\{s_1, s_5, s_6\}$
   Chooser 3: $\{s_3, s_5, s_6\}$
   Chooser 4: $\{s_1, s_6\}$
   Chooser 5: $\{s_6\}$.

   (a) Describe a fair division of the cake.
   (b) Explain why the answer in (a) is the only possible fair division of the cake.

14. Six players want to divide a cake fairly using the lone divider method. The divider cuts the cake into 6 slices ($s_1, s_2, s_3, s_4, s_5, s_6$), and the choosers make the following declarations:

Chooser 1: $\{s_1, s_3\}$
Chooser 2: $\{s_2\}$
Chooser 3: $\{s_4, s_5\}$
Chooser 4: $\{s_4, s_5\}$
Chooser 5: $\{s_2\}$.
Describe how to proceed to obtain a fair division of the cake.

15. Four players want to divide a parcel of land valued at $120,000 fairly using the lone divider method. The divider cuts the parcel into 4 slices (s_1, s_2, s_3, s_4) as shown in the figure.

The choosers value (in thousands of dollars) these pieces as follows:

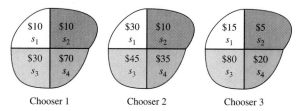

Chooser 1	Chooser 2	Chooser 3

(a) What should the chooser declarations be?
(b) Describe a possible fair division of the land.

16. Three players (X, Y, and Z) decide to divide a $12.00 vanilla-strawberry cake using the lone chooser method. The dollar amounts of the cake in each player's eyes are given in the following figure.

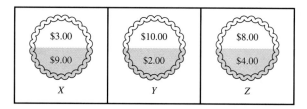

(a) Suppose that in the first division X cuts and Y chooses and that the player making the final selection is Z. Describe a possible fair division of the cake and for each player's final share give the dollar amounts of the share in the players' own eyes.
(b) Suppose that in the first division Z cuts and X chooses and that the player making the final selection is Y. Describe a possible fair division of the cake and for each player's final share give the dollar amounts of the share in the players' own eyes.
(c) Suppose that in the first division Y cuts and Z chooses and that the player making the final selection is X. Describe a possible fair division of the cake and for

each player's final share give the dollar amounts of the share in the players' own eyes.

17. Three players (*X, Y,* and *Z*) decide to divide the cake shown in the following figure using the lone chooser method.

    X likes chocolate and orange equally well, but hates strawberry and vanilla.
    Y likes chocolate and strawberry equally well, but hates orange and vanilla.
    Z likes chocolate and vanilla equally well, but hates orange and strawberry.
  (a) Suppose that in the first division *X* cuts and *Y* chooses and that the player making the final selection is *Z*. Describe a possible fair division of the cake and for each player's final share give the value (as a percentage of the total) of the share in the players' own eyes.
  (b) Suppose that in the first division *Z* cuts and *Y* chooses and that the player making the final selection is *Y*. Describe a possible fair division of the cake and for each player's final share give the value (as a percentage of the total) of the share in the players' own eyes.
  (c) Suppose that in the first division *Y* cuts and *Z* chooses and that the player making the final selection is *X*. Describe a possible fair division of the cake and for each player's final share give the value (as a percentage of the total) of the share in the players' own eyes.

18. A cake is to be divided among 4 people using the last diminisher method. The 4 people are randomly ordered and are called P_1, P_2, P_3, and P_4. In round 1, P_1 cuts a piece, P_2 and P_3 think it is fair and don't diminish it, and P_4 diminishes it.
  (a) Is it possible for P_2 to end up with that piece or just a part of that piece?
  (b) Who gets the piece at the end of round 1?
  (c) Who cuts the piece at the beginning of round 2?
  (d) Who is the last person who has an opportunity to diminish the piece in round 2?
  (e) How many rounds are required to divide the cake among the 4 people?

19. A cake is to be divided among 15 people using the last diminisher method. The 15 people are randomly ordered and are called P_1, P_2, P_3, . . . , P_{15}. In round 1, P_1 cuts a piece, and P_3, P_7, and P_{10} are the only diminishers. In round 2, the only diminisher is P_3, and in round 3 there are no diminishers.
  (a) Who gets the piece at the end of round 1?
  (b) Who cuts the piece at the beginning of round 2?
  (c) Who is the last person who has an opportunity to diminish the piece in round 2?
  (d) Who gets the piece at the end of round 2?
  (e) Who cuts the piece at the beginning of round 3?
  (f) Who is the last person who has an opportunity to diminish the piece in round 3?
  (g) Who gets the piece at the end of round 3?
  (h) How many rounds are required to divide the cake among the 15 people?

20. A cake is to be divided among 8 people using the last diminisher method. The 8 people are randomly ordered and are called $P_1, P_2, P_3, P_4, P_5, P_6, P_7, P_8$. In round 1, P_1 cuts a piece, and P_4, P_6, and P_8 are the only diminishers. In round 2, there are no diminishers.

   (a) Who gets the piece at the end of round 1? Why?

   (b) Who cuts the piece at the beginning of round 2?

   (c) Who is the last person who has an opportunity to diminish the piece in round 2? Why?

   (d) Who gets the piece at the end of round 2? Why?

   (e) Who cuts the piece at the beginning of round 3? Why?

   (f) Who is the last person who has an opportunity to diminish the piece in round 3?

   (g) How many rounds are required to divide the cake among the 8 people?

21. Three sisters (A, B, and C) wish to divide up 4 pieces of furniture they shared as children using the method of sealed bids. Their bids on each of the items are given in the following table.

	A	B	C
Dresser	$250	$310	$265
Desk	195	150	185
Vanity	175	215	235
Tapestry	510	490	475

Describe the outcome of this fair division problem.

22. Robert and Peter inherit their parents' old house and classic car. They decide to divide the 2 items using the method of sealed bids. Robert bids $29,235 on the car and $60,990 on the house. Peter bids $33,220 on the car and $65,300 on the house. Describe the outcome of this fair division problem.

23. Bob, Ann, and Jane wish to dissolve their partnership using the method of sealed bids. Bob bids $240,000 for the partnership, Ann bids $210,000, and Jane bids $270,000.

   (a) Who gets the business and for how much?

   (b) What do the other two people get?

24. Three heirs (A, B, and C) wish to divide up an estate consisting of a house, a small farm, and a painting, using the method of sealed bids. The heirs' bids on each of the items are given in the following table:

	A	B	C
House	$150,000	$145,000	$160,000
Farm	130,000	125,000	128,000
Painting	50,000	60,000	45,000

Describe the outcome of this fair division problem.

25. Three people (A, B, and C) wish to divide up 4 items using the method of sealed bids. Their bids on each of the items are given in the following table:

	A	B	C
Item 1	$20,000	$ 18,000	$ 16,000
Item 2	43,000	42,000	37,000
Item 3	4,000	2,000	1,000
Item 4	201,000	193,000	183,000

Describe the outcome of this fair division problem.

26. Three people (A, B, and C) wish to divide up 5 items using the method of sealed bids. Their bids on each of the items are given in the following table:

	A	B	C
Item 1	$14,000	$12,000	$22,000
Item 2	24,000	15,000	33,000
Item 3	18,000	24,000	14,000
Item 4	16,000	16,000	18,000
Item 5	16,000	18,000	20,000

(a) What does A end up with? (Does A pay anything?)
(b) What does B end up with? (Does B pay anything?)
(c) What does C end up with? (Does C pay anything?)

27. Five heirs (A, B, C, D, and E) wish to divide up an estate consisting of 6 items using the method of sealed bids. The heirs' bids on each of the items are given in the following table.

	A	B	C	D	E
Item 1	$352	$295	$400	$368	$324
Item 2	93	102	98	95	105
Item 3	461	449	510	501	476
Item 4	852	825	832	817	843
Item 5	512	501	505	515	491
Item 6	725	738	750	744	761

Describe the outcome of this fair division problem.

28. Three players (P_1, P_2, and P_3) agree to divide the 13 items shown by lining them up in order and using the method of markers. The players' bids are as indicated.

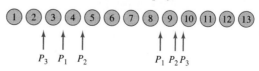

(a) Describe the allocation of items to each player.
(b) Which items are left over?

29. Three players (P_1, P_2, and P_3) agree to divide the 13 items shown by lining them up in order and using the method of markers. The players' bids are as indicated.

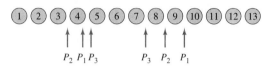

(a) Describe the allocation of items to each player.
(b) Which items are left over?

30. Two players (P_1 and P_2) agree to divide the 12 items shown by lining them up in order and using the method of markers. The players' bids are as indicated.

(a) Describe the allocation of items to each player.
(b) Which items are left over?

31. Three players (P_1, P_2, and P_3) agree to divide the 12 items shown by lining them up in order and using the method of markers. The players' bids are as indicated.

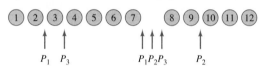

(a) Describe the allocation of items to each player.
(b) Which items are left over?

32. Three players (P_1, P_2, and P_3) agree to divide the twelve items shown by lining them up in order and using the method of markers. The players' bids are as indicated.

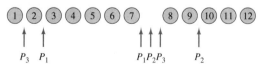

(a) Describe the allocation of items to each player.
(b) Which items are left over?

33. Five players (P_1, P_2, P_3, P_4, and P_5) agree to divide the 20 items shown by lining them up in order and using the method of markers. The players' bids are as indicated.

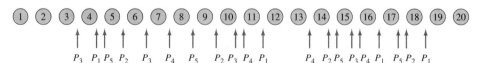

(a) Describe the allocation of items to each player.
(b) Which items are left over?

34. Four players (P_1, P_2, P_3, and P_4) agree to divide the 15 items shown below by lining them up in order and using the method of markers. The players' bids are as indicated.

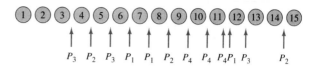

(a) Describe the allocation of items to each player.
(b) Which items are left over?

35. Four players (P_1, P_2, P_3, and P_4) agree to divide the 15 items shown by lining them up in order and using the method of markers. The players' bids are as indicated.

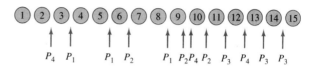

(a) Describe the allocation of items to each player.
(b) Which items are left over?

Jogging

36. Three players (P_1, P_2, and P_3) agree to divide the property shown using the last diminisher method. The order of the players is P_1, P_2, P_3. The first player to play, P_1, makes a claim C as shown in the figure.

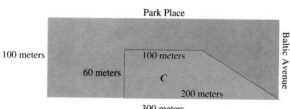

We know that both P_2's and P_3's value systems are the same and that they value the land uniformly.
(a) Give a geometric argument why P_2 and P_3 would both pass in round 1 and P_1 would end up with C.
(b) Describe a possible cut that the divider in round 2 might make.
(c) Suppose that after round 1 is over P_2 and P_3 discover that the city requires that the next cut be made parallel to Park Place. Describe a possible cut that the divider in round 2 might make in this case.
(d) Repeat (c) for a cut that must be made parallel to Baltic Avenue.

37. Three players (P_1, P_2, and P_3) agree to divide the property shown using the last diminisher method. The order of the players is P_1, P_2, P_3. The first player to play, P_1, makes a claim as shown in the figure.

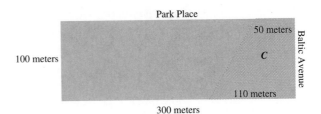

We know that both P_2's and P_3's value systems are the same and that they value the land uniformly.

(a) Give a geometric argument why P_2 and P_3 would both pass in round 1 and P_1 would end up with C.

(b) Describe a possible cut that the divider in round 2 might make.

(c) Suppose that after round 1 is over P_2 and P_3 discover that the city requires that the next cut be made parallel to Baltic Avenue. Describe a possible cut that the divider in round 2 might make in this case.

38. This exercise is based on Example 8. Suppose that in her will, Grandma stipulates that the estate should be divided among the 4 heirs as follows: Art, 25%; Betty, 35%; Carla, 30%; and Dave, 10%. Describe a variation of the method of sealed bids that will accomplish this. (Feel free to use a calculator.)

39. Say that N players $(P_1, P_2, \ldots, P_N)$ are heirs to an estate. According to the will, $P1$ is entitled to r_1% of the estate, P_2 is entitled to r_2%, etc. $(r_1 + r_2 + \ldots r_N = 100)$. Describe a general variation of the method of sealed bids that gives a fair division for this estate. (You should try this exercise only after you have finished Exercise 38.)

40. Two players (A and B) wish to dissolve their partnership using the method of sealed bids. A bids x and B bids y, where $x < y$.

(a) How much are A and B's original fair shares worth?

(b) How much is the surplus after the original allocations are made?

(c) When all is said and done, how much must B pay A for A's half of the partnership?

41. Three players (P_1, P_2, and P_3) agree to divide some candy using the method of markers. The candy consists of 3 Reese's Pieces (R), 6 caramels (C), and 6 mints (M). The players' value systems are as follows:

P_1 loves Reese's Pieces but does not like caramels or mints at all.

P_2 loves caramels and Reese's Pieces equally well (i.e., 1 Reese's Piece = 1 caramel) but does not like mints at all.

P_3 loves caramels and mints equally well (i.e., 1 caramel = 1 mint) but is allergic to Reese's Pieces.

The candy is lined up as shown.

(a) What bid would P_1 make to ensure that she gets her fair share (according to her value system)?

(b) What bid would P_2 make to ensure that he gets his fair share (according to his value system)?

(c) What bid would P_3 make to ensure that she gets her fair share (according to her value system)?

(d) Describe the allocations to each player.

(e) What items are left over?

42. Repeat Exercise 41 with the candy lined up as follows.

43. Suppose that two players (A and B) buy a chocolate-strawberry mousse cake with a caramel swirl and assorted frostings for $10.00. Since A contributes $7.00 and B only $3.00, they both agree that a fair division of the cake is one in which A gets a piece that is worth (in A's opinion) at least 70% of the cake and B gets a piece that is worth (in B's opinion) at least 30% of the cake. Describe a variation of the lone divider method that can be used in this situation. (*Hint:* Think of this problem as an ordinary fair division problem with many players.)

44. This problem is a variation of Exercise 43. Three players are involved (A, B, and C). A contributes $2.50, B contributes $3.50, and C contributes $4.00 toward the purchase of the cake. Describe a fair division scheme for this problem.

45. Consider the following variation of the divider-chooser method for 2 players: After the divider cuts the cake in 2 pieces, the chooser (who is unable to see either piece) picks his piece randomly by flipping a coin. The divider, of course, gets the other piece.

(a) Is this a fair division scheme according to our definition? Explain your answer.

(b) Who would you rather be—divider or chooser? Explain.

Running

46. **Alternative notion of a fair share.** To explain this exercise we will need some additional terminology. We will say that a player is *fairly happy* if she receives a share that is (in her opinion) worth at least $1/N$ of the total (what we have been calling a fair share). We will say that a player is *ecstatic* if she receives a share that is worth (in her opinion) at least as much as anyone else's—in other words, the player would not want to trade her share with anyone else. All the fair division schemes we presented in the chapter guarantee that all the players will be fairly happy but not that all the players will be ecstatic. Human nature being what it is, being fairly happy but not ecstatic might make some of the players unhappy. The purpose of this exercise is to discuss this alternative interpretation of fairness.

(a) Explain why the divider-chooser method for 2 players guarantees that each player will be ecstatic.

(b) Give an example of a fair division using the lone divider method with 3 players in which at least one of the players ends up with a piece that makes him happy but not ecstatic.

(c) Give an example of a fair division using the lone divider method with 3 players in which all 3 players are ecstatic.

47. Using the terminology of Exercise 46, describe a continuous fair division method for 3 players that guarantees that all three players will be ecstatic. (See references 7 and 8 at the end of the chapter.)

48. (a) Explain why after the original allocation is made in the method of sealed bids, the surplus produced must be either positive or zero.

   (b) Under what condition is the surplus zero?

49. The method of sealed bids for objects with negative value. Fair division schemes can sometimes be applied to objects having negative values, such as chores. The main purpose of this exercise is to have the student come up with a modified version of the method of sealed bids that can be used when dividing fairly a set of indivisible (discrete) objects with negative value (part c). First, let's clarify the exact meaning of a negative bid. When a player bids a negative amount (say −$10.00) on a chore, he is essentially judging what the *nuisance value* of the chore is. In this case, a bid of −$10.00 means that *if the job was entirely the player's responsibility he would be willing to pay $10.00 to get out of that responsibility* (and by the same token, it also means that *if the job was not in any way the player's responsibility, he would be willing to do the job for $10.00*).

   (a) Two players (*A* and *B*) are equally responsible for the completion of a single indivisible chore (say, for example, climbing a tree and sawing off a broken branch). Rather than flip a coin, they decide to try a modified version of the method of sealed bids. *A* bids −$10.00 and *B* bids −$5.00. Who do you think should do the job and how much should he be paid for it?

   (b) Three sisters (*A*, *B*, and *C*) are equally responsible for 4 indivisible chores. Their bids on each of the chores are given in the following table.

	A	B	C
Chore 1	−$250	−$310	−$265
Chore 2	−195	−150	−185
Chore 3	−175	−215	−235
Chore 4	−510	−490	−475

   Describe the outcome of this fair division problem. Who does which chore? Who pays money? Who gets paid? How much?

   (c) Describe a modified version of the method of sealed bids for an arbitrary number of players having to divide an arbitrary number of indivisible chores.

50. The purpose of this exercise is to extend the ideas of the lone divider method for 3 players to any number of players.

   (a) Describe how the lone divider method would work in the case of 4 players. (*Hint:* Consider several cases following the format for the case of 3 players. Also look at Exercises 8 through 10.)

   (b) Describe the lone divider method for the general case of *N* players.

REFERENCES AND FURTHER READINGS

1. Dubins, L. E., "Group Decision Devices," *American Mathematical Monthly,* 84 (1977), 350–356.

2. Fink, A. M., "A Note on the Fair Division Problem," *Mathematics Magazine,* 37 (1964), 341–342.

3. Gardner, Martin, *aha! Insight.* New York: W. H. Freeman, 1978, 124.

4. Kuhn, H. W., "On Games of Fair Division," *Essays in Mathematical Economics,* Martin Shubik, ed. Princeton, NJ: Princeton University Press, 1967, 29–37.

5. Steinhaus, Hugo, *Mathematical Snapshots.* New York: Oxford University Press, 1960, 65–71.

6. Steinhaus, Hugo, "The Problem of Fair Division," *Econometrica,* 16 (1948), 101–104.

7. Stromquist, Walter, "How to Cut a Cake Fairly," *American Mathematical Monthly,* 87 (1980), 640–644.

8. Woodall, D. R., "Dividing a Cake Fairly," *Journal of Mathematical Analysis and Applications,* 78 (1980), 233–247.

CHAPTER 4

Making the Rounds

Apportionment

If we have fifty pieces of candy, how do we divide them fairly among five children?

PROBLEM IN FOURTH-GRADE SCHOOLBOOK

Hard as it is to believe, we have yet to say the last word on this question. In this chapter we will discuss one more twist to the remarkable story of fair division (after which we really promise to lay the issue to rest!): What happens when the objects to be divided are all identical and indivisible but each participant is entitled to a different percentage of the total? The following two examples illustrate what we mean.

Example 1a (A home lesson in capitalism). Fifty identical pieces of candy (let's say for the sake of argument they are M&M's) are going to be divided among the 5 children in a household. Instead of dividing the candy equally among the 5 kids, mother has an inspired idea: Why not divide the candy in proportion to the amount of time each child spends helping with the chores around the house? Stopwatch in hand, mother boldly proceeds to carry out this ingenious exercise in capitalism. When the work is done and the final tally made, this is how things stand:

Child	Alan	Betty	Connie	Doug	Ellie	Total
Minutes worked	150	78	173	204	295	900

■ TABLE 4-1(a) **Division of Labor**

It is now time to divide the candy. According to the ground rules, Alan is entitled to $16\frac{2}{3}\%$ of the 50 pieces of candy ($\frac{150}{900} = 16\frac{2}{3}\%$), namely $8\frac{1}{3}$ pieces. Unfortunately, the individual pieces of candy are indivisible (ever try to cut an

111

M&M in thirds?). This means that Alan can't get his exact fair share: He can get 8 pieces (and get shorted) or he can get 9 pieces (and someone else will get shorted), but he can't get exactly what he earned! Of course, what goes for Alan also applies to each of the other children: Betty's exact fair share would be $4\frac{1}{3}$ pieces, Connie's $9\frac{11}{18}$ pieces, Doug's $11\frac{1}{3}$ pieces, and Ellie's $16\frac{7}{18}$ pieces (we leave it to the reader to verify these figures). Needless to say, none of these shares can be realized. Clearly, mother is going to need some help.   ▬

    If you are wondering what is so important about dividing a bunch of M&M's among a few children, stay tuned. . .

Example 1b (The Intergalactic Congress). It is the year 2525 and all the planets in the galaxy have finally signed a peace treaty. Five of the planets (Alanos, Betta, Conii, Dugos, and Ellisium) decide to join forces and form an Intergalactic Federation. The Federation will be ruled by an Intergalactic Congress consisting of 50 delegates, and each of the 5 planets will be entitled to a number of delegates that is proportional to its population. The population data for each of the planets is as follows:

Planet	Alanos	Betta	Conii	Dugos	Ellisium	Total
Population (in billions)	150	78	173	204	295	900

▬ Table 4-1(b)   **Intergalactic Federation: Population Figures for 2525 (Source: Intergalactic Census Bureau)**

How many delegates should Alanos get? How about Betta, Conii, Dugos and Ellisium?
    Once again, we can see that there is no perfectly fair way to divide the seats in the Intergalactic Congress among the 5 planets given the ground rules that each planet's share of seats should be proportional to its population. The very fact that the exact fair shares are fractional and yet we are required to give out shares that are whole numbers automatically implies that some planets will get more than their exact fair share while others will get less. Some element of unfairness is unavoidable.   ▬

APPORTIONMENT PROBLEMS

It is easy to see that be it M&M's or be it seats in the Intergalactic Congress, Examples 1a and 1b represent the same mathematical problem: How does one divide *fairly* a fixed number of *identical* and *indivisible* objects among a certain number of participants, each of which is entitled to a different fraction of the total? While the candy problem described in Example 1a is one many people can easily relate to (those of us who had brothers and sisters with whom we fought to get our due can do so for sure), it clearly represents a frivolous and unimportant situation. By the same token, Example 1b shows us that there is a serious and important side to the same problem. We will see a lot more of this latter side throughout the rest of this chapter.

Problems of the type described in Examples 1a and 1b are called **apportionment**[1] **problems**. Our goal in this chapter is to systematically study methods for deciding how to get around, as best we can, the inevitable element of unfairness lurking behind every apportionment problem. In other words, if, as we now know, there is no perfectly fair way to divide indivisible objects to participants that are entitled to fractional shares, then the next reasonable question we must tackle is, What is the best (least unfair) way to carry out the **apportionment**?

Over the years, mathematicians and others have designed many ingenious approaches to solving apportionment problems, and we will study several of the more interesting such **apportionment methods** in this chapter. Interestingly, the names associated with many of the apportionment methods we will study are names that one would expect to find in a history book, rather than a mathematics book: Alexander Hamilton, Thomas Jefferson, John Quincy Adams, and Daniel Webster. Why should statesmen and politicians care so much about a mathematics problem? The answer lies in Article 1, Section 2 of the Constitution of the United States:

> [Seats in the House of] *Representatives . . . shall be apportioned among the several States which may be included within this Union, according to their respective Numbers. . . . The actual enumeration shall be made . . . every . . . ten Years in such manner as they* [the House of Representatives] *shall by Law direct. The number of Representatives shall not exceed one for every thirty Thousand, but each State shall have at Least one Representative. . . .*

> ARTICLE 1, SECTION 2, CONSTITUTION OF THE UNITED STATES

From our perspective, the relevant facts implied in Article 1, Section 2, can be restated as follows:

- The seats in the House of Representatives shall be apportioned among the states on the basis of their respective populations.

- The Constitution does not spell out a specific method for apportioning these seats, only that each state's allotment of seats should (as much as possible) be proportional to its population (. . . *according to their respective Numbers* . . .). The actual details for implementing this goal are left to

---

[1] **ap·pôr´tion:** to divide and assign in due and proper proportion or according to some plan (Webster's New Twentieth Century Dictionary).

From the Granger
Collection

the House of Representatives (. . . *in such manner as they shall by Law direct* . . .).

■ The population figures on which the apportionments are based shall be determined every ten years. (This is done by a national census—the most recent census was taken in 1990.)

■ Each seat in the House of Representatives should represent at least 30,000 people; but no matter what its population, a state should always have at least one seat.

A LITTLE BIT OF U.S. HISTORY

Before we start a detailed discussion of apportionment methods, we will find it illuminating to briefly look at how Article 1, Section 2 played itself out historically. It is a real surprise to find that lurking behind an important aspect of American history one can find a purely mathematical problem.

It didn't take long after the drafting of the Constitution for the controversy over apportionment of the House of Representatives to surface. Following the census of 1790, two competing methods for apportioning the House were proposed—one by Alexander Hamilton, the other by Thomas Jefferson. After a heated debate, a bill apportioning the House of Representatives based on the method proposed by Hamilton was submitted to President Washington. After discussing the issue at length with his cabinet, Washington vetoed the bill (this was the first exercise of presidential veto power in U.S. history!). Unable to override the presidential veto and facing a damaging political stalemate, the House ended up adopting the method proposed by Jefferson.

Jefferson's method was used for five decades (until 1842) and then replaced by an apportionment method proposed by Daniel Webster. Webster's method was soon replaced by a method equivalent to the one originally proposed by Hamilton in 1790, which was then replaced again by Webster's method, which was eventually replaced (in 1941) by the current method of apportionment (called the Huntington-Hill method). Each change of method was preceded by considerable discussion and debate, often of a very nasty nature. There were decades in which the House was unable to pass an apportionment bill, making the very existence of the House unconstitutional.[2]

The latest apportionment controversy is of very recent vintage. In 1992, the state of Montana challenged the validity of the current apportionment method in the federal courts, with the case ending up in the Supreme Court. (Primarily, Montana was concerned about the fact that, based on the census of 1990, it would have to give up one of its two seats in the House to the state of Washington, and needless to say, they were not really happy about that.) After hearing the case, the Supreme Court ruled against Montana and upheld the validity of our current apportionment. Goodbye, Montana; Hello, Washington!

[2] For a little more detail, the reader is referred to Appendix 2 at the end of this chapter. A really detailed account of the history of the apportionment problem can be found in references 1 and 2.

George Washington and his Cabinet: lithograph, 1876, by Currier and Ives. (The Granger Collection)

THE MATHEMATICS OF APPORTIONMENT: BASIC CONCEPTS

We are now ready to take on the systematic study of apportionment methods. Much of our discussion for the rest of the chapter will be centered around the following simple, but important example.

Example 2 (Parador's Congress). Parador is a new republic located in Central America. It is made up of 6 states: Azucar, Bahia, Cafe, Diamante, Esmeralda, and Felicidad (A, B, C, D, E, and F for short). According to the recently adopted constitution of Parador, the 250 seats in its legislature must be divided among the 6 states according to their respective populations. Based on the most recent census, the populations of each state are:

State	A	B	C	D	E	F	Total
Population	1,646,000	6,936,000	154,000	2,091,000	685,000	988,000	12,500,000

■ TABLE 4-2   **Republic of Parador (Population Data by State)**

How many seats should each state get?

We will soon see that there is much more to this example than meets the eye, and that to get our feet wet we need to introduce some basic concepts.

We will start by calculating the number of people per seat in the legislature as a national average. Since the total population of Parador is 12,500,000 and the number of seats is 250, this average is given by 12,500,000/250 = 50,000. In general, *for any apportionment problem in which the total population of the country is P and the number of seats to be apportioned is M, the ratio P/M gives the number of people per seat in the legislature on a national basis.* We will call the number P/M the **standard divisor**. (Note that although our definition is given in the context of Example 2, the concept of standard divisor applies to any apportionment problem, regardless of the context. In our original candy example, the standard divisor is $P/M = 900/50 = 18$, and it represents the average number of minutes of work needed to earn a single M&M.)

Using the standard divisor, we can calculate the fraction of the total number of seats that each state would be entitled to if the seats were not indivisible. This number, which we will call the state's **standard quota**,[3] is obtained by dividing the state's population by the standard divisor. In other words,

---

[3] It is also known as the *exact quota*.

- Standard divisor = total population (*P*)/total number of seats (*M*)
- State's standard quota = state's population/standard divisor.

For Parador, we divide each state's population by 50,000 to get the standard quotas (as shown in Table 4-3).

State	A	B	C	D	E	F	Total
Population	1,646,000	6,936,000	154,000	2,091,000	685,000	988,000	12,500,000
Standard quota	32.92	138.72	3.08	41.82	13.70	19.76	250

■ TABLE 4-3   **Republic of Parador (Standard Quotas for Each State)**

The fact that the standard quota for state *A* is 32.92 simply means that if the seats in the legislature could be chopped up into fractional parts, then state *A* would get exactly 32.92 seats. Unfortunately this can't be done. The same thing goes for the rest of the states. (Notice that the sum of all the standard quotas should be the total number of seats to be divided, which we will usually denote by *M*. If it's close but not exactly *M*, it is due to roundoff errors.)

Associated with each state's standard quota are two whole numbers: the state's **lower quota** (the standard quota rounded down), and the state's **upper quota** (the standard quota rounded up). For example, state *A* (with a standard quota of 32.92) has a lower quota of 32 and an upper quota of 33. (In the unusual case that the state's quota is a whole number, then the lower and upper quotas are the same.)

Let's now return to the problem of how best to apportion the 250 seats in Parador's legislature based on the population figures given in Example 2. What seems to be the most sensible and logical approach is to use what we will call **conventional rounding**: Round the standard quotas the way we were taught to round numbers in school—up if the fractional part is over 0.5, down otherwise. Unfortunately, this approach does not always work. To see why this is the case, let's try it with Example 2.

State	Population	Standard quota	Rounded to
A	1,646,000	32.92	33
B	6,936,000	138.72	139
C	154,000	3.08	3
D	2,091,000	41.82	42
E	685,000	13.70	14
F	988,000	19.76	20
Total	12,500,000	250	251

■ TABLE 4-4

Oops! We have a slight problem here: We have just given out 251 seats in Parador's legislature, one more than what's available (kind of like overbooking an airplane flight!).

This is the major problem with conventional rounding of the standard quotas—we cannot count on the totals adding up to the right number of seats. Sometimes, as in this example, we end up giving out more seats than we are supposed to; other times we end up giving out less, and occasionally, by sheer luck, it can work out just right—but we simply can't count on it.

The fact that conventional rounding of the standard quotas does not work as an apportionment method is disappointing, but hardly a surprise at this point. After all, if this obvious and simple-minded approach worked all the time the whole issue of apportionment would be mathematically trivial (and there wouldn't be a reason for this chapter to exist!). Given this, we will be forced to consider more sophisticated approaches to apportionment. Our strategy for the rest of this chapter will be to look at several important (both historically and mathematically) apportionment methods and find out what is good and bad about each one (shades of Chapter 1?).

HAMILTON'S METHOD

While historically Hamilton's method did not come first, we will discuss it first because it is mathematically the simplest.[4]

HAMILTON'S METHOD

- **Step 1.** Calculate each state's standard quota. (Recall that the simplest way to do this is to first calculate the standard divisor and then divide each state's population by the standard divisor.)
- **Step 2.** Give to each state (for the time being) its *lower quota.*
- **Step 3.** If there are surplus seats (there usually are) give them (one at a time) to the states with the largest fractional parts until we run out of surplus seats.

Example 3 (Hamilton takes on Parador). Let's apply Hamilton's method to apportion Parador's legislature. Table 4-5 shows the details of the calculation. (The reader is reminded that the standard divisor in this example is 50,000.)

---

[4] Hamilton's method is also known as the *method of largest remainders* and sometimes as *Vinton's method.*

State	Population	Standard quota (Step 1)	Lower quota (Step 2)	Fractional part	Surplus seats (Step 3)	Final apportionment
A	1,646,000	32.92	32	.92	1 (1st)	33
B	6,936,000	138.72	138	.72	1 (4th)	139
C	154,000	3.08	3	.08		3
D	2,091,000	41.82	41	.82	1 (2nd)	42
E	685,000	13.70	13	.70		13
F	988,000	19.76	19	.76	1 (3rd)	20
Total	12,500,000	250	246	4.00	4	250

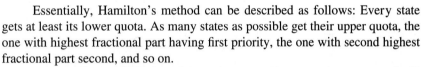

TABLE 4-5   Republic of Parador: Apportionment Based on Hamilton's Method

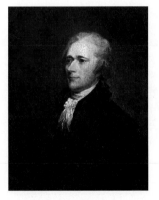

Alexander Hamilton

Essentially, Hamilton's method can be described as follows: Every state gets at least its lower quota. As many states as possible get their upper quota, the one with highest fractional part having first priority, the one with second highest fractional part second, and so on.

Is Hamilton's method a fair and reasonable apportionment method? Obviously, Hamilton thought it was, but we can already see in Example 3 that all is not quite as fair as we would hope. Consider the sad fate of state E next in line for a surplus seat with a hefty fractional part of 0.70 and yet getting none. Is state B (with a fractional part of 0.72) truly that much more deserving than state E in getting that last surplus seat? The answer is not quite clear. If we look at fractional parts in absolute terms the answer is yes (0.72 is more than 0.70). However, when we look at the fractional part as a percentage of the entire state's population, state E's 0.70 represents a much larger proportion of its quota (13.70) than state B's 0.72 is of its quota (138.72). One could argue that perhaps it would have been fairer to give the last surplus seat to state E, rather than B.[5]

While on the surface the rules of the game under Hamilton's method sound very fair, a little probing showed us that Hamilton's method appears to work to the advantage of the larger states and to the disadvantage of the smaller ones. Historically, this has been consistently the case, and there is some reason to suspect that Hamilton himself knew this.

THE QUOTA RULE   An appointment method that is always bound to allocate to each state either its lower quota or its upper quota is said to *satisfy* the **quota rule**. When an apportionment allocates to some state a number of seats that is either less than its

[5] The idea that relative rather than absolute fractional parts should determine the order in which the surplus seats are handed out is the basis for an apportionment method known as *Lowndes' method*. For details, the reader is referred to the explanation that precedes Exercise 36.

lower quota or more than its upper quota, then that apportionment is said to *violate quota,* and any apportionment method that allows this to happen, no matter how rarely, is said to *violate the quota rule.* Notice that we are drawing a subtle semantic distinction between *violating quota* and *violating the quota rule.* The latter is an expression that is applied to the *apportionment method* itself and it does not refer to any specific apportionment produced by it.

The legality of an apportionment that violates quota is clearly subject to question. (It would be hard to imagine that any unbiased person would consider as fair an apportionment which took a state with a standard quota of say 38.59 and allocated to it 40 seats.) Thus, violating the quota rule is a major flaw in an apportionment method. In view of this, one may wonder: Are there any serious apportionment methods that violate such a basic principle of fairness? The surprising answer is that indeed there are. In fact, several of the most important methods of apportionment we will study later on in this chapter (including the first method used in 1792 as well as the method currently in use) happen to violate the quota rule.

But that is then, and this is now. As far as Hamilton's method is concerned, it is clear that it has no problems with the quota rule. After all, when everything is said and done, states get either their upper quota (the lucky ones at the front of the line when surplus seats are given out) or their lower quota (the rest). This is one of the great things about Hamilton's method—it *satisfies the quota rule.* This may sound like faint praise right now, but we will later discover that there are not that many apportionment methods around that do this.

So far, Hamilton's method seems to have a lot going for it. It is easy to understand, it satisfies the quota rule, and except for what appears to be a natural favoritism towards large states over small ones, it seems reasonably fair. Why isn't Hamilton's method the answer to our prayers? As you may have guessed, we are about to discover the dark side of Hamilton's method.

THE ALABAMA PARADOX

The most serious (in fact, the fatal) flaw of Hamilton's method is commonly known as the **Alabama paradox**. In essence, the Alabama paradox occurs when an *increase in the total number of seats to be apportioned, in and of itself, forces a state to lose one of its seats.* The best way to understand what this means is to look carefully at the following example.

Example 4. A small country consists of 3 states: *A, B,* and *C.* Table 4-6(a) shows the apportionment under Hamilton's method when there are $M = 200$ seats to be apportioned. Table 4-6(b) shows the apportionment under Hamilton's method when there are $M = 201$ seats to be apportioned. The reader is encouraged to complete the details of those calculations. (Here is some help. The standard divisor when $M = 200$ is 100; the standard divisor when $M = 201$ is 99.5.)

State	Population	Standard quota when $M = 200$	Apportionment under Hamilton's method
A	940	9.4	10
B	9030	90.3	90
C	10,030	100.3	100
Total	20,000	200	200

■ TABLE 4-6(a)   **Apportionment under Hamilton's Method for $M = 200$**

State	Population	Standard quota when $M = 201$	Apportionment under Hamilton's method
A	940	9.45	9
B	9030	90.75	91
C	10,030	100.80	101
Total	20,000	201	201

■ TABLE 4-6(b)   **Apportionment under Hamilton's Method for $M = 201$**

When we use Hamilton's method to apportion the seats we can see that in the case of $M = 200$ seats A gets the only surplus seat and the final apportionment is A: 10 seats, B: 90 seats, and C: 100 seats.

What happens when the number of seats to be apportioned increases to $M = 201$? Now there are two surplus seats and they go to B and C, so that the final apportionment is A: 9 seats, B: 91 seats, and C: 101 seats.

■ **Conclusion:** State A loses a seat (its apportionment goes from 10 seats to 9 seats).

■ **Why?** There are more seats to go around!

This dreadfully unfair situation is what the Alabama paradox represents. Historically, the first serious occurrence of this problem came up in 1880, when it was noted that based on Hamilton's method, if the House of Representatives were to have 299 seats Alabama would get 8 seats, but if the House of Representatives were to have 300 seats Alabama would end up with 7 (see Table 4-7). This is how the name Alabama paradox came about.

State	Standard quota with $M = 299$	Apportionment with $M = 299$	Standard quota with $M = 300$	Apportionment with $M = 300$
Alabama	7.646	8	7.671	7
Texas	9.64	9	9.672	10
Illinois	18.64	18	18.702	19

■ TABLE 4-7   **Hamilton's Method and the Alabama Paradox, 1880**

Mathematically, the Alabama paradox is the result of some quirks of basic arithmetic. When we increase the number of seats to be apportioned, each state's standard quota goes up, but not in the same amount. Thus, the priority order for surplus seats used by Hamilton's method can get scrambled around, moving some states from the front of the priority order to the back, and vice versa. This can result in some state or states losing seats they already had. This is exactly what happened in Example 4 and in the Alabama fiasco of 1880.

MORE PROBLEMS WITH HAMILTON'S METHOD

The discovery of Alabama's paradox in 1880 turned out to be the kiss of death for Hamilton's method. Ironically, two other serious flaws of Hamilton's method were discovered later, when the method was no longer being used. We will briefly discuss these in this section, primarily because they are mathematically interesting, but also because they show that Hamilton's method has serious problems even if the Alabama paradox were not an issue.

The Population Paradox

Sometime around 1900, it was discovered that under Hamilton's method, *a state could lose seats simply because of its population having grown too much!* On its face this sounds totally crazy, but let's look at an example and we will see how this can actually happen.

Example 5. Remember the Intergalactic Federation of 2525 (Example 1b)? Here are the population figures once again:

Planet	Alanos	Betta	Conii	Dugos	Ellisium	Total
Population (in billions)	150	78	173	204	295	900

■ TABLE 4-8   **Intergalactic Federation: Population Figures for 2525. (Source: Intergalactic Census Bureau)**

Let's use Hamilton's method to apportion the 50 seats in the Intergalactic Congress. The total population for the 5 planets comes to 900 billion. If we divide this number by 50 we get a standard divisor of 18 billion. Using this standard divisor, we can get the standard quotas (column 2 of Table 4-9), and then carry out steps 2 and 3 of Hamilton's method as shown in columns 3 and 4 of Table 4-9 respectively. The final apportionment is shown in column 5. We call the reader's attention to two planets that play a key role in this story: Betta (4 delegates), and Ellisium (17 delegates).

Planet	Population (in billions)	Standard quota (Population ÷ 18)	Lower quota (Step 2)	Surplus seats (Step 3)	Final apportion-ment
Alanos	150	$8.\overline{3}$	8		8
Betta	78	$4.\overline{3}$	4		4
Conii	173	$9.6\overline{1}$	9	1	10
Dugos	204	$11.\overline{3}$	11		11
Ellisium	295	$16.3\overline{8}$	16	1	17
Total	900	50	48	2	50

■ TABLE 4-9   Intergalactic Federation: Apportionment of 2525 (Hamilton's Method)

Intergalactic Federation. Part II.   Ten years have gone by, and it is time to reapportion the Intergalactic Congress. Actually not much has changed (populationwise) within the Federation. Conii's population increased by 8 billion (Coniians are very prolific), and Ellisium's population increased by 1 billion. All other planets stayed exactly the same. (Table 4-10.)

Planet	Alanos	Betta	Conii	Dugos	Ellisium	Total
Population (in billions)	150	78	181	204	296	909

■ Table 4-10   Intergalactic Federation: Population Figures for 2535 (Source: Intergalactic Census Bureau)

Since the total population is now 909 billion and the number of delegates is still 50, the standard divisor we now get is 909/50 = 18.18. Table 4-11 shows the steps for Hamilton's method based on this new standard divisor. Once again, the final apportionment is shown in column 5. Do you notice something terribly wrong with this apportionment? Ellisium, whose population went up by 1 billion, is losing a delegate to Betta, whose population did not go up at all.

Planet	Population (in billions)	Standard quota (Population ÷ 18.18)	Lower quota (Step 2)	Surplus seats (Step 3)	Final apportion- ment
Alanos	150	8.25	8		8
Betta	78	4.29	4	1	5 ←!!!
Conii	181	9.96	9	1	10
Dugos	204	11.22	11		11
Ellisium	296	16.28	16		16 ←!!!
Total	909	50	48	2	50

■■■ TABLE 4-11   **Intergalactic Federation: Apportionment of 2535 (Hamilton's Method)**   ▬▬

The New States Paradox

This is, in essence, the **population paradox**: *State X has a population growth rate that is higher than that of state Y and yet, when the apportionment is recalculated based on the new population figures, state X loses a seat to state Y.*

Another paradox produced by Hamilton's method was discovered when Oklahoma became a state in 1907. Before Oklahoma joined the Union, the House of Representatives had 386 seats. Based on its population, Oklahoma was entitled to 5 seats, so the size of the House of Representatives was changed from 386 to 391. The obvious intent of adding the extra 5 seats to the House was to leave all the other state's apportionments unchanged. After all, it only seems reasonable that the other state's apportionments should not be affected by the addition of a new state if we also appropriately adjust the size of the House.

However, when the apportionments were recalculated under Hamilton's method using the same population figures but adding Oklahoma's population, and using 391 seats instead of 386, something truly bizarre took place: Maine's apportionment went up (from 3 to 4 seats) and New York's apportionment went down (from 38 to 37 seats). The mere addition of Oklahoma (with its fair share of seats) to the Union, would force New York to give a seat to Maine!

The perplexing fact that *the addition of a new state with its fair share of seats can, in and of itself, affect the apportionments of other states,* is called the **new states paradox**.

The following example gives a simple illustration of the new states paradox. For a change of pace, we will try something other than politics.

Example 6. Central School District has two high schools: North High has an enrollment of 1045 students; South High has an enrollment of 8955. The school district is allocated a counseling staff of 100 counselors, and these counselors are to be apportioned between the two schools using Hamilton's method. This results in an apportionment of 10 counselors to North High and 90 counselors to South High. The computation is summarized in Table 4-12.

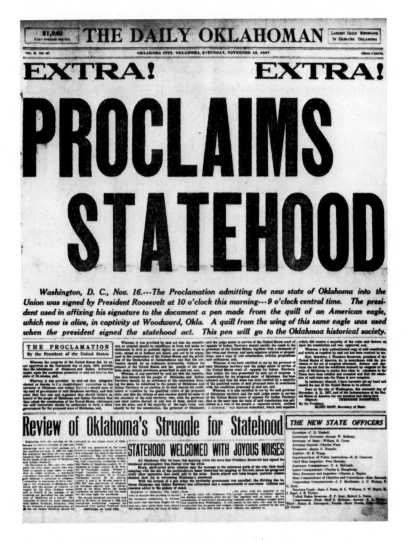

November 16, 1907: Oklahoma becomes the 46th state

School	Enrollment	Standard quota (Standard divisor = 100)	Apportionment
North High	1045	10.45	10
South High	8955	89.55	90
Total	10,000	100	100

■ **TABLE 4-12   Apportionment of Counselors to the Two High Schools Based on Hamilton's Method**

Suppose now that a new high school (New High) is added to the district. New High has an enrollment of 525 students so the district (using the same standard divisor of 100 students per counselor) decides to hire 5 new counselors and assign them to New High. After this is done, someone has the bright idea of having the entire apportionment recalculated (still using Hamilton's method). The surprising result is shown in Table 4-13.

School	Enrollment	Standard quota (Standard divisor = 100.238)	Apportionment
North High	1045	10.425	11 ←!!!
South High	8955	89.337	89 ←!!!
New High	525	5.238	5
Total	10,525	105	105

■■ TABLE 4-13   **Apportionment of Counselors to the Three High Schools Based on Hamilton's Method** ▬

JEFFERSON'S METHOD

We are now ready to take on the study of **Jefferson's method**,[6] an apportionment method of both historical and mathematical importance. Ironically, we will explain the idea behind Jefferson's method by taking one more look at Hamilton's method.

Recall that under Hamilton's method we start by dividing *every* state's population by a fixed number (the standard divisor). This gives us the quotas. Step 2 is then to round *every* state's quota down. Notice that up to this point Hamilton's method uses a uniform policy for all states—every state is treated in exactly the same way. If you are looking for fairness, this is obviously good! But now comes the bad part (step 3): We have some leftover seats which we need to distribute, but not enough for every state. Thus, we are forced to choose some states over others for preferential treatment. No matter how fair we try to be about it, there is no getting around the fact that some states get that extra seat, others don't. (One could say that to some extent, Hamilton's method follows the motto "to the victors go the spoils.") From the fairness point of view, this is the major weakness of Hamilton's method.

Now comes a new idea, based on the following thought: Wouldn't it be nice if we could eliminate step 3 in Hamilton's method? Or, to put it another way, Wouldn't it be nice if we could rig things up so that after dividing every state's population by the same number (step 1) and then rounding the resulting quotas down (step 2) we were to *end up with no surplus seats?*

[6] Jefferson's method is also known as the *method of greatest divisors,* and in Europe is known as the *method of d'Hondt.*

How could we work such magic? In theory, the answer is simple: We need to use a different **divisor** (instead of the standard divisor), so that we can come up with new **modified quotas** (no longer the standard quotas) that are just right (meaning that when we round all of them down we end up with a total that is the exact number of seats to be apportioned). In essence, we have just described **Jefferson's method**. Before we give a detailed description of the method, it's time to look at an example.

Example 7. Once again, we go back to the Parador example. (Recall that the standard divisor in this example is 50,000.)

Table 4-14 shows the calculations based on the standard divisor. We are already familiar with these calculations—they are exactly what we used in steps 1 and 2 of Hamilton's method.

Table 4-15 shows us similar calculations based on a **modified divisor** of $D = 49,500$. (Let's not worry for now about where the 49,500 came from—let's

State	Population	Standard quota (Population ÷ 50,000)	Lower quota (Step 2)
A	1,646,000	32.92	32
B	6,936,000	138.72	138
C	154,000	3.08	3
D	2,091,000	41.82	41
E	685,000	13.70	13
F	988,000	19.76	19
Total	12,500,000	250	246 ← 4 surplus seats (Too bad!)

■ TABLE 4-14   **Republic of Parador: Calculations using Standard Divisor**

State	Population	Modified quota (Population ÷ 49,500)	Modified lower quota
A	1,646,000	33.25	33
B	6,936,000	140.12	140
C	154,000	3.11	3
D	2,091,000	42.24	42
E	685,000	13.84	13
F	988,000	19.96	19
Total	12,500,000	252.52	250 ←No surplus seats (Eureka!)

■ TABLE 4-15   **Republic of Parador: Calculations using Modified Divisor $D = 49,500$**

just say it dropped out of the blue.) Note that using this smaller divisor, all the modified quotas are higher than the standard quotas, and the modified lower quotas add up to exactly the right total M. Thus, the last column of Table 4-15 shows an actual apportionment—the one produced by Jefferson's method.   ▬

We are now ready for a formal description of Jefferson's method.

JEFFERSON'S METHOD

■ **Step 1.**   Find a number D such that when each state's modified quota (state's population $\div D$) is rounded *downward*, the total is the exact number of seats to be apportioned.

■ **Step 2.**   Apportion to each state its modified lower quota.

Before continuing with our discussion of Jefferson's method, we will make official some of the terminology we have already used: We call the number D used in step 1 the **divisor**, and the result of dividing the state's population by D, the state's **modified quota** (based on the divisor D).

There is one important issue still hanging over our heads: How does one go about finding this "magic" divisor D that makes Jefferson's method work? For example, how did we come up with $D = 49,500$ in Example 7? The answer is educated trial and error. Let's start with the fact that the divisor we are looking for has to be a number smaller than the standard divisor (remember—we want the modified quotas to be bigger than the standard quotas, so we must divide by a smaller amount). So we pick a number D that we hope might work. We now carry out all the calculations asked for by Jefferson's method (divide the population by D; round the results downward; add up the total). If we are lucky, the total is exactly right and we are finished. Otherwise we change our guess (make it higher if the total is too high, lower if the total is too low) and try again. In most cases, it takes at most two or three guesses before we find a divisor D that works (usually there is more than one). Needless to say, a good calculator is essential.

Thomas Jefferson. (The White House Collection)

Let's go through the paces using Example 7. We know we are looking for a modified divisor that is less than 50,000. Let's start with a guess that is $D = 49,000$. The reader is encouraged to do the calculations (Exercise 34[a]). It turns out that this divisor doesn't work—it gives us a total of 252 which is too high. We need our guess to be a bit higher, so we try $D = 49,500$. Bingo! Note that the divisor $D = 49,450$ also works, and so do many others (Exercise 34[b]).

A constitutional footnote: When presented with Jefferson's method many people wonder, Can we get away with it? Is it legal to use divisors other than the standard divisor? The answer is yes. There is nothing in the Constitution that requires use of the standard divisor as the basis for apportionment. Remember that the standard divisor is the ratio of people per seat in the House of Representatives for the nation as a whole. Using the 1990 U.S. census, for example, we had a total U.S. apportionment population of $P = 248,102,973$ (for apportionment purposes the population of the District of Columbia is not included) and a House of Representatives of size $M = 435$ giving a standard divisor of approximately 570,352—give or take a limb or two. This means that for the *nation as a whole* there were about 570,352 people per seat in the House of Representatives. For individual states, however, the figures are different, because for each state the ratio of people per seat in the House of Representatives is based on the state's *actual apportionment* rather than the quota. Take for example California, with a population of 29,760,021 and 52 seats in the House of Representatives. If we divide these numbers out we get 572,308 Californians for every representative—a larger ratio than the national figure.

We can see from all of the above that there is no constitutional (or mathematical) requirement to use the standard divisor as a basis for apportionment. We should think of the standard divisor as an ideal goal rather than a requirement for apportionment. At the same time, fairness considerations require that we should not stray too far from this ideal. As we will see next, this can turn out to be a problem under Jefferson's method.

Problems with Jefferson's Method

Jefferson's method suffers from one major flaw: *It violates the quota rule.* If we go back to Example 7 and look at what Jefferson's method did for state B, we can see it: State B got 140 seats. So what, you say? Now look at its standard quota (138.72). According to the quota rule, the only fair apportionments for state B are either 138 seats or 139 seats. No matter how one cuts it, giving state B a windfall of 140 seats goes against a basic principle of fairness (remember that state B's gain has to be some other state's loss).

The kind of quota rule violation that took place in this example (where a state gets more than it should, in other words, more than its upper quota) is called a **violation of the upper quota**. The other possible violation of the quota rule is when a state gets less than its lower quota, which, of course, is called a **violation of the lower quota**. It turns out that Jefferson's method can only err in one direction—it can only violate the upper quota (Exercise 35[a]).

When Jefferson's method was adopted in 1791, it is doubtful that anyone realized (certainly neither Jefferson nor Washington did) that it suffered from such a major flaw. The first actual violation of the quota rule under Jefferson's method occurred in the apportionment of 1832, when New York, with a standard quota of 38.59 ended up with 40 seats. Needless to say, this horrified practically everyone except the New York delegation.

One of the strongest opponents of Jefferson's method was Daniel Webster:

> *The House is to consist of 240 members. Now, the precise portion of power, out of the whole mass presented by the number of 240, to which New York would be entitled according to her population, is 38.59; that is to say, she would be entitled to thirty-eight members, and would have a residuum or fraction; and even if a member were given her for that fraction, she would still have but thirty-nine. But the bill gives her forty . . . for what is such a fortieth member given? Not for her absolute numbers, for her absolute numbers do not entitle her to thirty-nine. Not for the sake of apportioning her members to her numbers as near as may be because thirty-nine is a nearer apportionment of members to numbers than forty. But it is given, say the advocates of the bill, because the process [Jefferson's method] which has been adopted gives it. The answer is, no such process is enjoined by the Constitution.*[7]

The apportionment of 1832 was to be the last apportionment of the House of Representatives based on Jefferson's method. It was clear that something new had to be tried, and the search for an apportionment method that did not violate the quota rule was on.

ADAMS' METHOD

At about the same time that Jefferson's method was falling into disrepute because of its violation of the quota rule, John Quincy Adams was proposing a method that was a mirror image of it. It was based on exactly the same idea but instead of going after the modified lower quotas, it went after the modified upper quotas.[8]

ADAMS' METHOD

- ■ **Step 1.** Find a divisor D such that when each state's modified quota (state's population $\div D$) is rounded *upward*, the total is the exact number of seats to be apportioned.
- ■ **Step 2.** Apportion to each state its modified upper quota.

---

[7] Daniel Webster, *The Writings and Speeches of Daniel Webster, Vol. VI* (Boston: Little, Brown and Company, 1903).

[8] Adams' method is also known as the *method of smallest divisors*.

John Quincy Adams. (The
Metropolitan Museum of
Art)

Undoubtedly, Adams thought that by doing this he could avoid the weakness of Jefferson's method—violating quota. He was only partly right.

Let's apply Adams' method to Parador's example.

Example 8. We will start by guessing a possible divisor D that we hope will work. We know that D will have to be bigger than 50,000 (we are trying to get modified quotas that are smaller than the standard quotas, so that, we hope, when we round them up, the total turns out to be 250). Remembering that 49,500 worked for Jefferson's method, a good guess might be $D = 50,500$.

Table 4-16 shows the calculations based upon $D = 50,500$.

State	Population	Modified quota (Population ÷ 50,500)	Modified upper quota	
A	1,646,000	32.59	33	
B	6,936,000	137.35	138	
C	154,000	3.05	4	
D	2,091,000	41.41	42	
E	685,000	13.56	14	
F	988,000	19.56	20	
Total	12,500,000	247.52	251	Too much!

■ **TABLE 4-16   Calculations for Adams' Method Based on $D = 50,500$**

131

The total is too high! We need to lower the modified quota just a little bit, so let's try a higher divisor, say $D = 50{,}700$. Table 4-17 shows the calculations based on $D = 50{,}700$.

State	Population	Modified quota (Population ÷ 50,700)	Modified upper quota
A	1,646,000	32.47	33
B	6,936,000	136.80	137
C	154,000	3.04	4
D	2,091,000	41.24	42
E	685,000	13.51	14
F	988,000	19.49	20
Total	12,500,000	246.55	250 That's it!

■ TABLE 4-17   **Calculations for Adams' Method Based on $D = 50{,}700$**

The last column of Table 4-17 shows the apportionment produced by Adams' method.

So, what's wrong with Adams' method? Once again, look at state B's apportionment of 137 seats. Since the standard quota for state B is 138.72, Adam's method robbed state B of its due by apportioning to it less than its lower quota.

Adams' method violates the quota rule in exactly the opposite direction of Jefferson's method—it violates *lower quota* (Exercise 35[b]). Thus, Adams' idea was only partly correct. This method is able to fix the tendency of Jefferson's method to be overly generous to some states but only by creating just as big of an injustice—the tendency to sometimes be overly stingy with some states.

WEBSTER'S METHOD

It is clear that both Jefferson's method and Adams' method share the same philosophy: Treat all states exactly the same way. (The only difference is that whereas Jefferson's method rounds all the quotas down, Adams' method rounds all the quotas up.) For a while this sounded like a good idea, but as we now know it has serious flaws. Besides, why should two quotas, say, of 52.1 and 3.9 be rounded both the same way? Is this just?

In 1832, Daniel Webster proposed a method based on a very basic idea. Let's round the quotas the way we round fractions in every other walk of life—up if the fractional part is more than 0.5, down otherwise.

Daniel Webster. (U.S.
Signal Corps photo)

We have tried this idea before, and it didn't work, but that was because we were married to the notion that we had to use the standard quotas. Now that we are more savvy, why not try it again using modified quotas chosen specifically so that when we are done rounding the conventional way (to the nearest integer) the total is exactly the number of seats to be apportioned?

WEBSTER'S METHOD[9]

- ■ **Step 1.**   Find a number D such that when each state's modified quota (state's population $\div$ D) is rounded the conventional way (to the nearest integer) the total is the exact number of seats to be apportioned.
- ■ **Step 2.**   Apportion to each state its modified quota rounded to the nearest integer.

Example 9.   Let's apportion Parador's legislature using Webster's method.

Our first decision is to make a guess at the divisor D. Should it be more than the standard divisor (50,000) or should it be less? Here we will use the standard quotas as a guideline. If we round off the standard quotas to the nearest integer (as we did at the beginning of this chapter), we get a total of 251. This number is too high, which tells us that we should guess a divisor D larger than the standard divisor. Let us try $D = 50,100$. Table 4-18 shows the calculations based on $D = 50,100$.

---

[9] Webster's method is sometimes known as the *Webster-Willcox method* as well as the *method of major fractions*.

State	Population	Modified quota (Population ÷ 50,100)	Rounded to	
A	1,646,000	32.85	33	
B	6,936,000	138.44	138	
C	154,000	3.07	3	
D	2,091,000	41.74	42	
E	685,000	13.67	14	
F	988,000	19.72	20	
Total	12,500,000	249.49	250	←It worked!

■ TABLE 4-18 **Calculations for Webster's Method Based on**
$D = 50,100$

The last column of Table 4-18 shows the actual apportionment produced by Webster's method.

Although Webster's method works in principle just like Jefferson's and Adams' methods, it is a little harder to use in practice. The divisor D we are looking for can be smaller than, equal to, or larger than the standard divisor. For guidelines as to how to go about making an educated guess, see Exercise 29.

There is something very gratifying about Webster's method. In some sense it resurrects and validates an appealingly simple idea that we tried and discarded at the beginning of the chapter: the notion that quotas should be rounded just like ordinary numbers. The reason it didn't work for us then is that we were basing it on the standard quotas—Webster's method makes the original idea work by modifying the quotas whenever necessary.

Our past experience has conditioned us to expect that before we are done with Webster's method we will have to endure some bad news. Actually, Webster's method turns out to be one of those bad news–good news situations. First the bad news: Webster's method also *violates the quota rule.* Our example (Example 9) does not show this (the reader is encouraged to verify that the apportionment given in Example 9 does not violate quota), but there are situations in which Webster's method produces an apportionment that violates quota. Now for the good news: The aforementioned situations are rare and somewhat contrived. In most practical apportionment situations, it is very unlikely that a violation of the quota rule would occur under Webster's method.

Thus, while in theory Webster's method suffers from the same "illness" that Jefferson's and Adams' methods do (violating the quota rule), in practice there is a big difference—Webster's method suffers from a very mild case of this illness, whereas the other two are seriously ill. In fact, Webster's method is considered a very good (but not perfect) apportionment method and in the opinion of some experts it is actually superior to the method that is currently being used to apportion the U.S. House of Representatives (see Appendix 1).

CONCLUSION: BALINSKI AND YOUNG'S IMPOSSIBILITY THEOREM

In this chapter we introduced four different *apportionment methods.* Table 4-19 summarizes the results of apportioning Parador's Congress (Example 2) under each of the four methods.

State	Population	Standard quota	Hamilton	Jefferson	Adams	Webster
A	1,646,000	32.92	33	33	33	33
B	6,936,000	138.72	139	140	137	138
C	154,000	3.08	3	3	4	3
D	2,091,000	41.82	42	42	42	42
E	685,000	13.70	13	13	14	14
F	988,000	19.76	20	19	20	20
Total	12,500,000	250	250	250	250	250

■ TABLE 4-19   **Parador's Congress: A Tale of Four Methods**

It is worthwhile to note that, here, each of the four methods produced a different apportionment. This clearly demonstrates that the methods are indeed all different. At the same time, we should warn the reader that it is not impossible for two different methods to produce identical apportionments (see Exercises 26, 27, 28, and 38).

Of the four methods we discussed, one is based on a strict adherence to the standard quotas (Hamilton's), whereas the other three (Jefferson's, Adams', and Webster's) are based on the philosophy that quotas can be conveniently modified by the appropriate choice of divisor. While some of the methods are clearly better than others, none of them is perfect: Each of the four methods either *violates the quota rule* or *produces paradoxes.* Table 4-20 summarizes the characteristics of the four methods.

For many years, the ultimate hope held by thoughtful students of the apportionment problem, both inside and outside Congress, was that mathematicians would eventually come up with an *ideal* apportionment method—one that

	Hamilton	Jefferson	Adams	Webster
Violates quota rule	no	yes	yes	yes
Alabama paradox	yes	no	no	no
Population paradox	yes	no	no	no
New states paradox	yes	no	no	no
Favoritism toward	large states	large states	small states	neutral

■ TABLE 4-20

never violates the quota rule, does not produce any paradoxes, and treats large and small states without favoritism. Even as far back as 1929, Representative Ernest Gibson of Vermont echoed these sentiments in Congress. "The apportionment of Representatives to the population is a mathematical problem. Then why not use a method that will stand the test [of fairness] under a correct mathematical formula?"[10]

Indeed, why not? The surprising answer was provided in 1980 by mathematicians Michel L. Balinski of the State University of New York at Stony Brook and H. Peyton Young of the University of Maryland who collaborated to produce a remarkable discovery known as **Balinski and Young's impossibility theorem**: *It is mathematically impossible for an apportionment method to be flawless. Any apportionment method that does not violate the quota rule must produce paradoxes and any apportionment method that does not produce paradoxes must violate the quota rule.*

Once again, an eerily familiar conclusion in a slightly different setting: Fairness and proportional representation are inherently incompatible.

KEY CONCEPTS

Adams' method	**lower quota**
Alabama paradox	**modified quota**
apportionment method	**new states paradox**
apportionment problem	**population paradox**
Balinski and Young's impossibility	**quota rule**
theorem	**standard divisor**
conventional rounding	**standard quota**
divisor	**upper quota**
Hamilton's method	**Webster's method**
Jefferson's method	

EXERCISES

Walking

Exercises 1 through 5 refer to a small country consisting of 4 states. There are 160 seats in the legislature and the populations of the states are shown as follows.

State	A	B	C	D
Population (in millions)	1.33	2.67	0.71	3.29

[10] *Congressional Record,* 70th Congress, 2nd Session, 70 (1929), p. 1500.

1. (a) Find the standard divisor.
   (b) Find each state's standard quota.
   (c) Find each state's apportionment under Hamilton's method.

2. (a) Using the divisor 49,400, find each state's modified quota.
   (b) Find each state's apportionment under Jefferson's method.

3. (a) Using the divisor 50,700, find each state's modified quota.
   (b) Find each state's apportionment under Adams' method.

4. (a) Using the divisor 50,650, find each state's modified quota.
   (b) Using the divisor 50,600, find each state's modified quota.
   (c) Explain why the divisor 50,650 works for Adams' method but the divisor 50,600 doesn't.
   (d) Find a divisor different from 50,700 and 50,650 that will also work for Adams' method.

5. (a) Find a divisor that works for Webster's method. (*Hint:* You should be able to do this without using a calculator!)
   (b) Find the apportionment for each state under Webster's method.

Exercises 6 through 9 refer to a bus company that operates 6 bus routes (A, B, C, D, E, and F) and 150 buses. The buses are apportioned among the routes on the basis of average number of daily passengers per route, which is given in the following table.

Route	A	B	C	D	E	F
Average number of daily passengers	12,550	38,623	19,781	31,112	33,280	14,654

6. (a) Find the standard divisor.
   (b) Find the standard quota for each bus route.
   (c) Apportion the buses among the routes using Hamilton's method.

7. Apportion the buses among the routes using Jefferson's method.

8. Apportion the buses among the routes using Adams' method.

9. Apportion the buses among the routes using Webster's method.

Exercises 10 through 14 refer to a clinic with a nursing staff consisting of 225 nurses working in four shifts: A (7:00 A.M. to 1:00 P.M.); B (1:00 P.M. to 7:00 P.M.); C (7:00 P.M. to 1:00 A.M.); and D (1:00 A.M. to 7:00 A.M.). The number of nurses apportioned to each shift is based on the average number of patients per shift, given in the following table.

Shift	A	B	C	D
Average number of patients	823	659	482	286

10. (a) Find the standard divisor. Explain what the standard divisor represents in this problem.
    (b) Find the standard quota for each shift.

11. Apportion the nurses to the shifts using Hamilton's method.

12. Apportion the nurses to the shifts using Jefferson's method. (*Hint:* Divisors don't have to be whole numbers.)

13. Apportion the nurses to the shifts using Adams' method. (*Hint:* Divisors don't have to be whole numbers.)

14. Apportion the nurses to the shifts using Webster's method. (*Hint:* Divisors don't have to be whole numbers.)

Exercises 15 through 19 refer to a small country consisting of five states. The total population of the country is 24.8 million. The standard quota of each state is given in the following table.

State	A	B	C	D	E
Standard quota	25.26	18.32	2.58	37.16	40.68

15. (a) Find the number of seats in the legislature.
    (b) Find the standard divisor.
    (c) Find the population of each state.

16. Find each state's apportionment using Hamilton's method.

17. Find each state's apportionment using Jefferson's method.

18. Find each state's apportionment using Webster's method.

19. Find each state's apportionment using Adams' method.

Exercises 20 through 23 refer to the following: Tasmania State University is made up of five different schools: Agriculture, Business, Education, Humanities, and Science. A total of 500 faculty positions must be apportioned based on the school's respective enrollments, shown below.

School	Agriculture	Business	Education	Humanities	Science	Total
Enrollment	2472	1314	2886	3405	4923	15,000

20. (a) Find the standard divisor. What does the standard divisor represent in this problem?
    (b) Find each school's standard quota.

21. Find the number of faculty members apportioned to each school using Hamilton's method.

22. Find the number of faculty members apportioned to each school using Jefferson's method.

23. Find the number of faculty members apportioned to each school using Webster's method.

24. A mother wishes to distribute 10 pieces of candy among her 3 children based on the number of minutes each child spends studying as shown in the following table.

Child	Bob	Peter	Ron
Minutes studied	703	243	54

   (a) Find each child's apportionment using Hamilton's method.
   (b) Suppose that just prior to actually handing over the candy, mom finds another piece of candy and includes it in the distribution. Find each child's apportionment using Hamilton's method and 11 pieces of candy.
   (c) Did anything paradoxical occur? (What's the name of this paradox?)

25. A mother wishes to distribute 11 pieces of candy among her 3 children based on the number of minutes each child spends studying as shown in the following table.

Child	Bob	Peter	Ron
Minutes studied	703	243	54

   (a) Find each child's apportionment using Hamilton's method. (Note that this is Exercise 24[b])
   (b) Suppose that before mom has time to sit down and do the actual calculations, the children decide to do a little more studying. Say Ron studies an additional 2 minutes, Peter an additional 12 minutes, and Bob an additional 86 minutes. Find each child's apportionment using Hamilton's method based on the new total time studied.
   (c) Did anything paradoxical occur? Explain.

Jogging

26. Make up an apportionment problem in which Hamilton's method and Jefferson's method result in exactly the same apportionment for each state. Your example should involve at least 3 states and the standard quotas should not be whole numbers.

27. Make up an appointment problem in which Hamilton's method and Adams' method result in exactly the same apportionment for each state. Your example should involve at least 3 states and the standard quotas should not be whole numbers.

28. Make up an apportionment problem in which Hamilton's method and Webster's method result in exactly the same apportionment for each state. Your example should involve at least 3 states and the standard quotas should not be whole numbers.

29. (a) Consider the following situation.

State	A	B	C	D	E
Standard quota	11.23	24.39	7.92	36.18	20.28

   You want to use Webster's method as your method of apportionment. Explain why you should look for a divisor that is *smaller* than the standard divisor.

(b) Consider the following situation.

State	A	B	C	D	E
Standard quota	11.73	24.89	7.92	35.68	19.78

You want to use Webster's method as your method of apportionment. Explain why you should look for a divisor that is *bigger* than the standard divisor.

(c) Under what conditions can you be assured that the standard divisor will work in Webster's method?

30. Consider an apportionment problem with only 2 states A and B. Suppose that state A has standard quota q_1 and state B has standard quota q_2, neither of which is a whole number. (Of course $q_1 + q_2 = M$ must be a whole number.) Let f_1 represent the fractional part of q_1 and f_2 the fractional part of q_2.

(a) Explain why one of the fractional parts is bigger than or equal to 0.5 and the other is smaller than or equal to 0.5.

(b) Assuming neither fractional part is equal to 0.5 explain why Hamilton's method and Webster's method must result in the same apportionment.

(c) Explain why in any apportionment problem involving only 2 states Hamilton's method can never produce the Alabama paradox or the population paradox.

(d) Explain why in the above situation Webster's method can never violate the quota rule.

31. The purpose of this example is to show that under rare circumstances using a modified divisor method may not work. A small country consists of 4 states with populations given as follows.

State	A	B	C	D
Population	500	1000	1500	2000

There are $M = 51$ seats in the House of Representatives.

(a) Find each state's apportionment using Jefferson's method.

(b) Attempt to apportion the seats using Adams' method with a modified divisor $D = 100$. What happens if you take $D < 100$? What happens if you take $D > 100$?

(c) Explain why Adams' method will not work for this example.

32. This exercise is based on actual data taken from the 1880 census. Here are some figures: In 1880, the population of Alabama was given at 1,262,505. With a House of Representatives consisting of $M = 300$ seats the standard quota for Alabama was 7.671.

(a) Find the 1880 census population for the United States (rounded to the nearest person).

(b) Given that the standard quota for Texas was 9.672, find the population of Texas (to the nearest person).

33. The following table shows the results of the 1790 census (the very first census of the United States taken after the Constitution was adopted).

State	Population
Connecticut	236,841
Delaware	55,540
Georgia	70,835
Kentucky	68,705
Maryland	278,514
Massachusetts	475,327
New Hampshire	141,822
New Jersey	179,570
New York	331,589
North Carolina	353,523
Pennsylvania	432,879
Rhode Island	68,446
South Carolina	206,236
Vermont	85,533
Virginia	630,560
Total	3,615,920

Based on the fact that the number of seats in the House of Representatives was set at $M = 105$:

(a) Find the apportionment that would have resulted under the original bill passed by Congress to use Hamilton's method.

(b) Find the apportionment that was actually used (remember that at the end it was based on Jefferson's method).

(c) Compare the answers in (a) and (b). Which state was the winner in the 1790 controversy between the two methods? Which state was the loser?

34. Suppose that you want to apportion Parador's Congress (Example 2) using Jefferson's method.

(a) Show that the divisor $D = 49,000$ does not work.

(b) Show that any divisor D between 49,401 and 49,542 (inclusive) works, and that there are no other whole number divisors that work.

35. (a) Explain why when Jefferson's method is used any violations of the quota rule must be violations of the upper quota.

(b) Explain why when Adams' method is used any violations of the quota rule must be violations of the lower quota.

(c) Use parts (a) and (b) to justify why in the case of an apportionment problem with just 2 states neither Jefferson's method nor Adams' method can possibly violate the quota rule.

*Exercises 36 and 37 refer to a variation of Hamilton's method known as **Lowndes' method**. (The method is also called the* modified Hamilton's method.) *The basic differ-*

ence between Hamilton's method and Lowndes' method is that after each state is assigned the lower quota, the surplus seats are handed out in order of relative fractional parts. (The relative fractional part of a number is the fractional part divided by the integer part. For example, the relative fractional part of 41.82 is 0.82/41 = 0.02 and the relative fractional part of 3.08 is 0.08/3 ≈ 0.027. Notice that while 41.82 would have priority over 3.08 under Hamilton's method, 3.08 has priority over 41.82 under Lowndes' method because 0.027 is greater than 0.02.)

36. (a) Find the apportionment of Parador's Congress (Example 2) under Lowndes' method.
    (b) Verify that the resulting apportionment is different from each of the apportionments shown in Table 4-19. In particular, list which states do better under Lowndes' method than under Hamilton's method.
    (c) Fill in the blank: Lowndes' method shows favoritism towards _____ (larger, smaller) states.

37. Consider an apportionment problem with only 2 states A and B. Suppose that state A has quota q_1 and state B has quota q_2, neither of which is a whole number. (Of course $q_1 + q_2 = M$ must be a whole number.) Let f_1 represent the fractional part of q_1 and f_2 the fractional part of q_2.
    (a) Find values q_1 and q_2 such that Lowndes' method and Hamilton's method result in the same apportionment.
    (b) Find values q_1 and q_2 such that Lowndes' method and Hamilton's method result in different apportionments.
    (c) Write an inequality involving q_1, q_2, f_1, and f_2 which would guarantee that Lowndes' method and Hamilton's method result in different apportionments.

Running

38. Make up an apportionment problem in which all 4 methods (Hamilton's, Jefferson's, Adams', and Webster's) result in exactly the same apportionment for each state. Your example should involve at least 4 states and the standard quotas should not be whole numbers.

Exercises 39 through 42 refer to the Huntington-Hill method described in Appendix 1. (These exercises should not be attempted without understanding Appendix 1.)

39. Use Huntington-Hill's method to find the apportionments of each state for a small country that consists of 5 states. The total population of the country is 24.8 million. The standard quotas of each state are as follows.

State	A	B	C	D	E
Standard quota	25.26	18.32	2.58	37.16	40.68

40. (a) Use the Huntington-Hill method to apportion Parador's Congress (Example 2).
    (b) Compare your answer in (a) with the apportionment produced by Webster's method. What's your conclusion?

41. A country consists of 6 states with populations as follows.

State	A	B	C	D	E	F	Total
Population	344,970	408,700	219,200	587,210	154,920	285,000	2,000,000

1a

1b

Imagine, if you will, having to visit each of the 48 state capitals in the continental United States, one after another, starting and ending in the same capital. In what order should you visit these cities so that the total length of the trip is as short as possible? The solution is shown on the top plate. The total length of the trip is approximately 12,000 air miles. This *optimal solution* was found using a sophisticated computer program which checked billions of possible combinations—an expensive proposition even under the best of circumstances. No algorithm is presently known that can efficiently solve this problem (commonly known as the *Traveling Salesman Problem*) when the number of cities becomes very large (see chapter 6). The bottom plate shows an *approximate solution* obtained using an extremely simple-minded procedure called the *nearest-neighbor algorithm*. The bad news is that the length of this trip is about 20% longer than the optimal solution (approx. 14,500 miles). The good news is that it can be found by anyone in a matter of minutes with just a pencil and a ruler.

2a

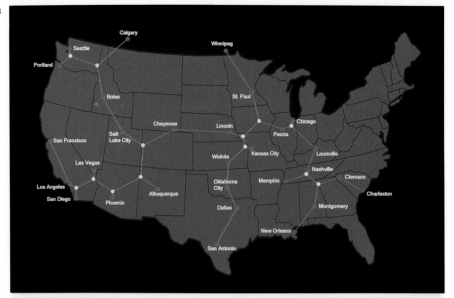

2b

A telephone company wants to connect the 29 cities shown on the map using the least possible amount of underground cable. The solution depends on whether junction points outside of the 29 cities are allowed. The top plate shows the shortest of all possible ways to connect the cities. The yellow dots are special junction points called *Steiner points.* This *optimal network,* which requires approximately 7400 miles of cable was found using a sophisticated computer algorithm developed by E. J. Cockayne and D. E. Hewgill of the University of Victoria. No algorithm is presently known that can efficiently solve this problem when the number of cities becomes large. The bottom plate shows the shortest network when no junction points other than the original cities are allowed. Although this network is about 3% longer than the optimal network (it requires approximately 7600 miles of cable) it can be found by hand in a matter of minutes using any one of several extremely simple algorithms (see chapter 7).

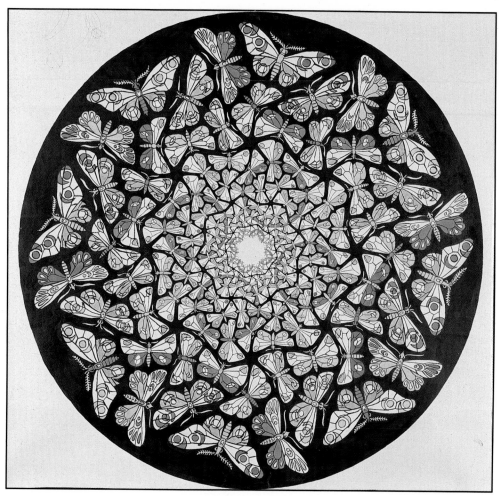

Butterflies by M. C. Escher. (© 1950 M. C. Escher/Cordon Art, Baarn, Holland.) This remarkable watercolor is rich in mathematical structure. In addition to black and white, Escher uses four other colors: red, yellow, blue and green. Each butterfly contains three of these colors (the body, the lower wings and the markings on the upper wings). There are 24 possible permutations of three colors and they are all present in equal proportions. Not only the butterflies but also the colored rings created by their arrangement exhibit *symmetry of scale* as they repeat themselves at smaller and smaller scales towards the center of the circle.

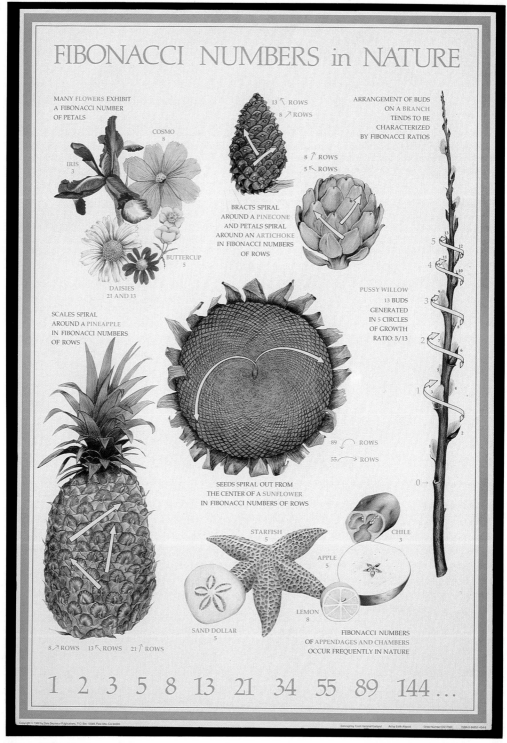

Fibonacci Numbers in Nature by Trudi Garland and Edith Allgood. (© 1988 by Dale Seymour Publications. Reprinted by permission.)

5a

5b   **5c**

A *logarithmic spiral* is exhibited by a cross section of the chambered nautilus shell (top) and some of its less glamorous cousins (bottom). Spirals of this type occur frequently in natural organisms whose growth is *gnomonic* (see chapter 9).

The Queen of Spades: an example of 180 degree rotational symmetry. (Turn the book upside down—the card still looks the same.)

6b

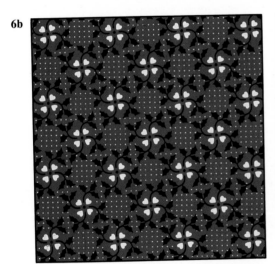

6c

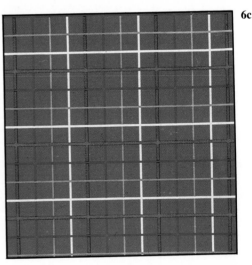

Although there are infinitely many different wallpaper designs, when classified according to their symmetries, wallpaper patterns can fall under one of only 17 different *pattern types* (see chapter 11). Two of these pattern types are illustrated by the "wallpaper patterns" shown above. To convince yourself that they are different turn the book upside down. One of them looks different now. Which one?

Although each queen by herself has 180 degree rotational symmetry, the pair of queens does not. What happens to the ladies' faces when you turn the book upside down?

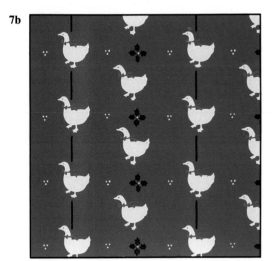

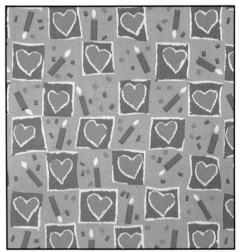

These are two very different looking wallpaper patterns which are nonetheless of the same pattern type: their only symmetries are translations (see chapter 11).

8a

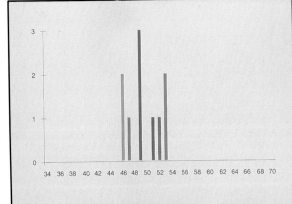

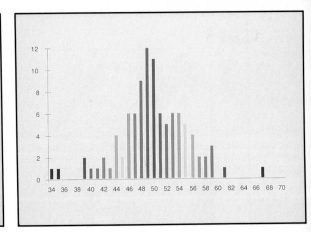

8c

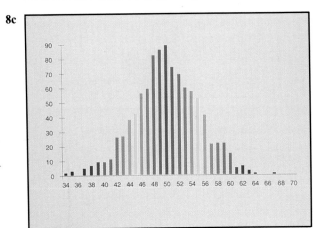

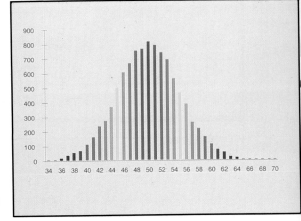

Imagine playing the following game: An honest coin is tossed 100 times in a row. If the number of heads falls between 45 and 55 you win $1, otherwise you lose $1. Can you predict whether you will win or lose at this game? While the results of playing this game once, twice or even ten times are unpredictable, a predictable pattern emerges if we play the game a very large number of times. The bar graphs show a computer simulation for the results of playing the game (from left to right) 10, 100, 1000, and 10,000 times respectively. The graph on the lower right shows the inescapable pattern that the outcomes of this game must follow: a bell shape called a *normal distribution* (for details see chapter 16).

There are 200 seats in the legislature.
(a) Find the apportionment under Webster's method.
(b) Find the apportionment under the Huntington-Hill method.
(c) Compare the divisors used in (a) and (b).
(d) Compare the apportionments found in (a) and (b).

42. A country consists of 6 states with populations as follows.

State	A	B	C	D	E	F	Total
Population	344,970	204,950	515,100	84,860	154,960	695,160	2,000,000

There are 200 seats in the legislature.
(a) Find the apportionment under Webster's method.
(b) Find the apportionment under the Huntington-Hill method.
(c) Compare the divisors used in (a) and (b).
(d) Compare the apportionments found in (a) and (b).

APPENDIX 1:
The Huntington-
Hill Method

The method currently used to apportion the U.S. House of Representatives is known as the **Huntington-Hill method**, and more commonly, as the **method of equal proportions**.

Let's start with some historical background. The method was discovered, sometime around 1911, by Joseph A. Hill, chief statistician of the Bureau of the Census, and Edward V. Huntington, professor of mechanics and mathematics at Harvard University. In 1929, the Huntington-Hill method was endorsed by a distinguished panel of mathematicians as the best apportionment method known. The panel, commissioned by the National Academy of Sciences at the formal request of the Speaker of the House, investigated many different apportionment methods and recommended the Huntington-Hill method as the best possible one.

On November 15, 1941, President Franklin D. Roosevelt signed "An Act to Provide for Apportioning Representatives in Congress among the Several States by the equal proportions method" (Public Law 291, H.R. 2665, 55 Stat 261). Under the same act, the size of the House of Representatives was fixed at $M = 435$. This act still stands today,[11] but political and legal challenges to it have come up periodically.

There are several ways to describe how the Huntington-Hill method works. For the purposes of explanation, the most convenient description of the method is by comparison to Webster's method. In fact, the two methods are almost identical. Just like Webster's method, we will find modified quotas and we will round some of them upwards and some of them downwards. The difference between the two methods is in the cutoff point for rounding up or down. Take for example, a state with a modified quota of 3.48. Under Webster's method, we know that we must round this quota downward because the cutoff

[11] The current apportionment of the House of Representatives based on the Huntington-Hill method can be found in the Appendix to Chapter 2 (each state's apportionment equals the number of electoral votes minus two).

point for rounding is 3.5. It may seem like a little bit of an overkill, but we can put it this way: The cutoff point for rounding quotas under Webster's method is exactly halfway between the lower quota and the upper quota.

$$\text{cutoff for Webster's method} = \frac{\text{modified lower quota} + \text{modified upper quota}}{2}$$

(Thus, for a state with a modified quota of 3.48, the cutoff point is 3.5.)

Now how about the cutoff point for rounding quotas under the Huntington-Hill method? It is given by the following formula:

$$\text{cutoff for the Huntington-Hill method} = \sqrt{\text{modified lower quota} \times \text{modified upper quota}}$$

Let's go back to the state with a modified quota of 3.48. Under the Huntington-Hill method, the cutoff for rounding this quota would be $\sqrt{3 \times 4} = \sqrt{12} = 3.464$. Since 3.48 is above this cutoff, this state's modified quota would have to be rounded upward. Thus, the Huntington-Hill method, unlike Webster's method, would give this state 4 seats.

HUNTINGTON-HILL ROUNDING RULES

For a quota that falls between n (lower quota) and $n + 1$ (upper quota) the Huntington-Hill cutoff point for rounding is $H = \sqrt{n \times (n + 1)}$. In other words, if the quota is above H, round upward, otherwise round downwards.[12]

HUNTINGTON-HILL METHOD

■ **Step 1.**   Find a number D such that when each state's modified quota (state's population $\div D$) is rounded according to the Huntington-Hill rules the total is the exact number of seats to be apportioned.

■ **Step 2.**   Apportion to each state its modified quota rounded using the Huntington-Hill rules.

Table 4-21 is a convenient table to have handy when working with the Huntington-Hill method.

We will conclude this section with a very simple example which shows that the Huntington-Hill method can produce an apportionment that differs from Webster's method.

---

[12] There is a handy mathematical name for the Huntington-Hill cutoffs. For any two positive numbers a and b, $\sqrt{a \times b}$ is called the *geometric mean* of a and b. Thus, we can describe each Huntington-Hill cutoff as the geometric mean between the modified lower and upper quotas.

Modified quota between	Cutoff point for rounding under Webster's method	Cutoff point for rounding under the Huntington-Hill method
1 and 2	1.5	$\sqrt{2} \approx 1.414$
2 and 3	2.5	$\sqrt{6} \approx 2.449$
3 and 4	3.5	$\sqrt{12} \approx 3.464$
4 and 5	4.5	$\sqrt{20} \approx 4.472$
5 and 6	5.5	$\sqrt{30} \approx 5.477$
6 and 7	6.5	$\sqrt{42} \approx 6.481$
7 and 8	7.5	$\sqrt{56} \approx 7.483$
8 and 9	8.5	$\sqrt{72} \approx 8.485$
9 and 10	9.5	$\sqrt{90} \approx 9.487$
10 and 11	10.5	$\sqrt{110} \approx 10.488$

■ TABLE 4-21

Example 10. A small country consists of 3 states. We want to apportion the 100 seats in its legislature to the three states according to the following population figures.

State	A	B	C	Total
Population	3480	46,010	50,510	100,000

■ TABLE 4-22

We will do it first using Webster's method and then using the Huntington-Hill method.

We start by computing the standard quotas. Since the standard divisor is 100,000/100 = 1000, this is really easy:

State	A	B	C	Total
Standard quota	3.48	46.01	50.51	100

■ TABLE 4-23

It so happens that rounding the standard quotas the conventional way gives a total of 100 so the standard quotas work for Webster's method (Table 4-24).

State	Population	Standard quota	Webster's apportionment
A	3480	3.48	3
B	46,010	46.01	46
C	50,510	50.51	51
Total	100,000	100	100

■ TABLE 4-24

Next, we notice that our old friend 3.48 has made an appearance. We know that under the Huntington-Hill method 3.48 is past the cutoff point of 3.464 so that it has to be rounded upwards (to 4). The other two standard quotas are not affected and would still be rounded as before (Table 4-25)

State	Population	Standard quota (Standard divisor = 1000)	Rounded under Huntington-Hill rules to
A	3480	3.48	4
B	46,010	46.01	46
C	50,510	50.51	51
Total	100,000	100	101

■ TABLE 4-25

Since the total comes to 101, these quotas don't work—we can see they are a little bit too high. But when we try a divisor just a tad bigger, ($D = 1001$) the totals do come out right (Table 4-26). The last column of Table 4-26 shows the way the 100 seats would be apportioned under the Huntington-Hill method. (Note that in this example the apportionment is different from the one produced by Webster's method.)

State	Population	Modified quota (Divisor = 1001)	Rounded under Huntington-Hill rules to
A	3480	3.477	4
B	46,010	45.964	46
C	50,510	50.460	50
Total	100,000	99.901	100

■ TABLE 4-26

APPENDIX 2:
A Brief History of
Apportionment in
the United States

■ **1787**
  ■ Constitutional Convention drafts the Constitution of the United States.
  ■ Small states vs. large states controversy dominates the Convention.
  ■ Article I, Section 2 gives Congress the authority to determine the method of apportion for the House of Representatives.

■ **1791**
  ■ Following the census of 1790 two methods of apportionment are proposed. Hamilton's method is supported by the Federalists; Jefferson's method is supported by the Republicans.
  ■ After considerable controversy and debate Congress approves a bill to apportion the House of Representatives using Hamilton's method with $M = 120$.

- ■ President Washington vetoes the bill (the first exercise of a presidential veto in U.S. history!).
- ■ Jefferson's method is adopted using $M = 105$ seats and a divisor $D = 33{,}000$. (Jefferson's method will remain in use until 1840.)

■ 1822

- ■ Rep. William Lowndes (South Carolina) proposes what we now call Lowndes' method. (See Exercise 36.) The proposal dies in Congress.

■ 1832

- ■ John Quincy Adams (former president and at this time a congressman from Massachusetts) proposes what we now call Adams' method. The proposal fails.
- ■ Senator Daniel Webster (Massachusetts) proposes what we now call Webster's method. His proposal also fails.
- ■ Jefferson's method prevails once again (with $M = 240$).

■ 1842

- ■ Webster's method is adopted with $M = 223$. This is one of the few times in U.S. history that M goes down. (Politicians are not inclined to legislate themselves out of work!)

■ 1852

- ■ Rep. Samuel Vinton (Ohio) proposes a bill adopting Hamilton's method as the *permanent* method of apportionment with $M = 233$ seats.
- ■ Congress approves the bill with the change $M = 234$. (For this particular value of M, Hamilton's method gives the same apportionment as Webster's method.)

■ 1872

- ■ $M = 283$ seats is chosen so that Hamilton's method and Webster's method result in the same apportionment.
- ■ For unexplained political reasons 9 additional seats are added to the House.
- ■ The final apportionment does not agree with either Hamilton or Webster's method. The whole thing is completely botched up by politics!

■ 1876

- ■ Rutherford B. Hayes becomes president of the United States based on the botched apportionment of 1872.
- ■ The electoral college vote is 185 votes for Hayes and 184 votes for Samuel J. Tilden. However, based on the correct apportionment required by law Tilden should have won the election.

■ 1880

- ■ The Alabama paradox surfaces as a serious flaw of Hamilton's method.

■ 1882

- ■ Congress begins to show a little frustration with the mathematical quirks of Hamilton's method.

> *I thought that mathematics was a divine science. I thought that mathematics was the only science that spoke to inspiration and was infallible in its utterances. I have been taught always that it demonstrated the*

The election of 1876. The botched apportionment of 1872 resulted in the election of Rutherford B. Hayes over Samuel Tilden. (Ohio Historical Society Library and the Granger Collection)

truth. I have been told . . . that mathematics, like the voice of Revelation, said when it spoke "thus saith the Lord." But here is a new system of mathematics [Hamilton's method] *that demonstrates the truth to be false.*

TEXAS CONGRESSMAN ROGER MILLS

■ Despite serious concerns about Hamilton's method an apportionment bill based on it with $M = 325$ seats is eventually approved. (This number is chosen so that Hamilton's method and Webster's method result in the same apportionment.)

■ **1901**

 ■ The Bureau of the Census submits to Congress tables showing apportionments based on Hamilton's method for all size Houses between $M = 350$ and $M = 400$.

 ■ For all values of M between 350 and 400 except one ($M = 357$) Colorado would get an apportionment of 3 seats. For $M = 357$ Colorado gets only 2 seats. (The Alabama paradox again!)

 ■ The chairman of the House Committee on Apportionment (Rep. Albert Hopkins of Illinois) does not care for the populist leanings of western states. The bill submitted by the Hopkins committee provides for —guess what?— $M = 357$ seats.

 ■ Congress is in an uproar! Hopkins' bill is defeated and Hamilton's method is finally abandoned for good.

 ■ Webster's method is adopted with $M = 386$.

■ **1907**

 ■ Oklahoma joins the Union. The new states paradox is discovered.

■ **1911**

 ■ Webster's method is readopted with $M = 433$. (A provision is made for Arizona and New Mexico to get one seat each if admitted into the Union.)

 ■ Joseph Hill (chief statistician of the Bureau of the Census) proposes a new method now known as the Huntington-Hill method or the method of equal proportions. (See Appendix 1.)

- **1921**
  - No reapportionment is done after the 1920 census (in direct violation of the Constitution).

- **1931**
  - Webster's method is used with $M = 435$.

- **1941**
  - Huntington-Hill method is adopted with $M = 435$. This remains (by law) as the permanent method of apportionment (until Congress votes to change it).

- **1992**
  - Montana challenges the constitutionality of the Huntington-Hill method in a lawsuit (*Montana v. U.S. Dept. of Commerce*). The Supreme Court upholds the Huntington-Hill method as constitutional.

REFERENCES AND FURTHER READINGS

1. Balinski, Michel L., and H. Peyton Young, "The Apportionment of Representation," *Fair Allocation: Proceedings of Symposia on Applied Mathematics,* 33 (1985) 1–29.

2. Balinski, Michel L., and H. Peyton Young, *Fair Representation: Meeting the Ideal of One Man, One Vote.* New Haven, CN: Yale University Press, 1982.

3. Balinski, Michel L., and H. Peyton Young, "The Quota Method of Apportionment," *American Mathematical Monthly,* 82 (1975), 701–730.

4. Eisner, Milton, *Methods of Congressional Apportionment,* COMAP Module #620.

5. Hoffman, Paul, *Archimedes' Revenge: The Joys and Perils of Mathematics.* New York: W. W. Norton & Co., 1988, chap. 13.

6. Huntington, E. V., "The Apportionment of Representatives in Congress," *Transactions of the American Mathematical Society,* 30 (1928), 85–110.

7. Huntington, E. V., "The Mathematical Theory of the Apportionment of Representatives," *Proceedings of the National Academy of Sciences,* USA 7 (1921), 123–127.

8. Meder, Albert E., Jr., *Legislative Apportionment.* Boston: Houghton Mifflin Co., 1966.

9. Saari, D. G., "Apportionment Methods and the House of Representatives," *American Mathematical Monthly,* 85 (1978), 792–802.

10. Schmeckebier, L. F., *Congressional Apportionment.* Washington, D.C.: The Brookings Institution, 1941.

11. Steen, Lynn A., "The Arithmetic of Apportionment," *Science News,* 121 (May 8, 1982), 317–318.

12. Webster, Daniel, *The Writings and Speeches of Daniel Webster, Vol. VI, National Edition.* Boston: Little, Brown, and Company, 1903.

Management Science

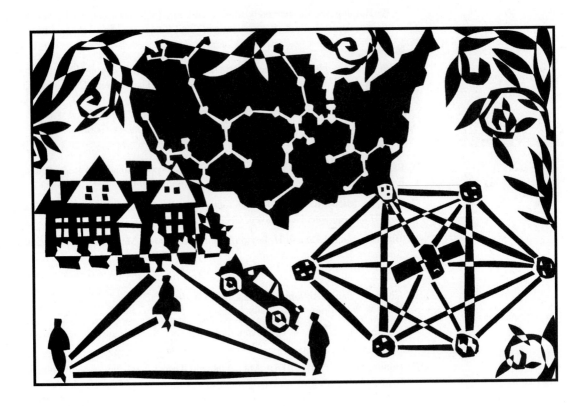

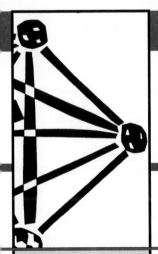

The Circuit Comes To Town

Euler Circuits

This chapter will take us through a guided tour along the charming streets of Cleansburg, Colorado—that famous ski resort and tourist mecca nestled in the foothills of the Sierra Serena mountains. Sounds dreamy? Well, don't get too excited. Our tour will be conducted in a garbage truck and our tour guides will be a couple of characters named Euler and Fleury.

Here is the situation in a nutshell: Cleansburg prides itself on having the cleanest streets anywhere ("you can eat off our streets" advertises the Chamber of Commerce) and, for as far back as anyone can remember, the Cleansburg Sanitation Department has provided *daily* garbage collection, as well as *daily* curb sweeping and snow removal (except for summer months) throughout the entire city. At the same time, just like everyone else, the Cleansburg Sanitation Department has been hit by big budgetary cutbacks, to the point that they are down to a single garbage truck, one curb sweeper, and two snowplows. Under these circumstances, is it still possible for them to continue to provide the level of service that the people of Cleansburg have grown accustomed to? If so, how? To help the Sanitation Department answer these questions, *you* have been hired as a consultant. Your specific job is to help design the most efficient routes for the garbage truck, curb sweeper, and snowplows as they crisscross the streets of Cleansburg (map in Fig. 5-1) on their appointed rounds.

ASSORTED ROUTING PROBLEMS

OK, you are wondering, What's up? If you are thinking that the little story about the Cleansburg Sanitation Department and *you* is not only corny but also preposterous, you are right on both counts. The fact is that to us, Cleansburg and its garbage collection problems will be just a backdrop. The true point of this chapter is to get us acquainted with an important class of mathematical problems called *routing problems* and to learn a little about the kinds of mathematical ideas needed to solve them.

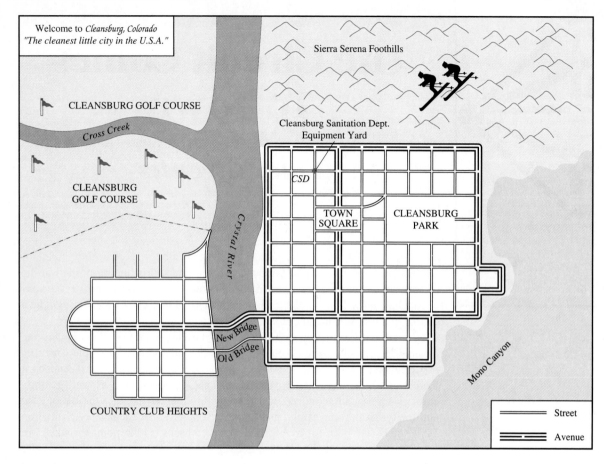

FIGURE 5-1 Street map for the city of Cleansburg.

What is a **routing problem**? To put it in the most general way, routing problems are concerned with finding the *most efficient* possible ways to route the delivery of goods or services to an assortment of destinations. The goods could be packages, mail, computer files, etc.; the services could be garbage collection, curb sweeping, snow removal, police protection, etc.; and the destinations could be cities in a country, houses along city blocks, computer terminals located across a university campus, etc.

While routing problems often sound very similar, subtle details can make a great difference in both the method and the nature of the solution. In this chapter we will concentrate on one particular category of routing problems called **Euler[1] circuit problems**, and in the next chapter we will concentrate on an entirely different class of routing problems (called Hamilton circuit problems, just in case you were interested).

[1] Pronounced "oiler."

The routing of the garbage truck, curb sweeper, and snowplows through the streets of Cleansburg are classic examples of Euler circuit problems. Other examples might involve routing water and electric meter readers, police patrols, ice cream trucks, etc. Whatever the case might be, the common thread in all Euler circuit problems is the need to traverse in an *efficient* way all the streets (or sidewalks, or lanes, etc.) within a designated locality—be it a town, a section of town, or a small neighborhood.

To clarify a few of the distinctions we have made, let's compare the case of an ice cream man, a mail carrier, and a snowplow operator, all assigned to carry on their particular duties in the same neighborhood. The ice cream man can accomplish his objective by making sure that he travels through each block of the neighborhood at least once; but once is not enough for the mail carrier, who must deliver mail on both sides of the street and therefore must travel through each block at least twice. And then there is the snowplow operator, who must travel through each block at least as many times as there are lanes on the road (a typical snowplow can only clear one lane at a time).

How does garbage collection fit in all of this? Well, there is actually more than one situation to consider. When it comes to garbage collecting, Cleansburg and New York City are not the same! In small towns with narrow streets and not much traffic, it is often the case that a garbage truck is able to collect the garbage on both sides of the street in a single pass through that street, especially if there are two garbage collectors in the truck. Of course this strategy would hardly be advisable on a busy street or in a large city—in these cases garbage is best collected in separate passes through each side of the street. To complicate matters further, there is the additional problem of one way streets, which may be rare in small towns but quite prevalent in larger cities. (To keep things simple, in this chapter we will make the assumption that all streets are two way streets.)

All of the preceding comments point to the fact that Euler circuit problems can come in many varieties, and as we work our way through the chapter, we will see how the setting of the problem will affect our formulation of it. Fortunately, once the problem is properly formulated, the basic mathematical theory needed to solve any Euler circuit problem is always the same. Our agenda for the chapter will be to develop the basic mathematical theory first and then apply it to illustrate how it can be used to solve assorted kinds of Euler circuit problems (including our original example for the city of Cleansburg).

Before we move on to the theoretical part of the chapter, let's look at another couple of simple examples of Euler circuit problems. They will help us set up the groundwork for the future.

The Königsberg Bridge Problem

There is a town in Russia called Kaliningrad (this one is for real!). In the 18th century it was a part of Prussia and was called Königsberg. In the 1700s the town was divided by a river (the Pregel), with four separate land areas (*A*, *D*, *L*, and *R*) joined by seven bridges as shown in Fig. 5-2.

For some unfathomable reason it became a great challenge to the locals to

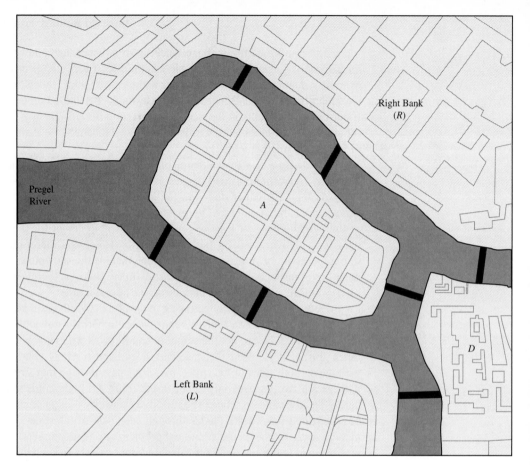

Figure 5-2 The seven bridges of Königsberg (circa 1735).

find a route that would allow them to walk around the town crossing each of the seven bridges once, never recrossing any of the bridges, and returning to the original starting point—a simple routing problem if there ever was one! While no one had been able to find such a route, people were not totally persuaded that the whole thing was impossible until one of the great mathematicians of all time, Leonhard Euler,[2] proved mathematically that such a route could not exist. We will discuss Euler's solution to the Königsberg bridge problem soon. But before we do, let's bring up a "different" problem.

A Children's Game   Can we trace the picture in Fig. 5-3(a), starting and ending at the same place, without lifting our pencil or retracing any lines?

Many of us played such games in our childhood (those were the good old days before video games). The pictures may have been different from Fig.

[2] Leonhard Euler (1707–1783) was one of the most brilliant and prolific mathematicians in history. It is estimated that, when finally compiled, his collected memoirs will fill nearly 100 volumes. In addition to mathematics, Euler loved children and had thirteen of his own. In the words of one biographer, "Euler was the Shakespeare of mathematics—universal, richly detailed and inexhaustible."

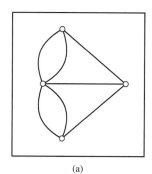 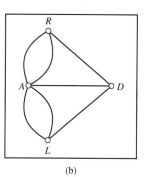

FIGURE 5-3 A children's game or the Königsberg bridge problem (or both?).

(a)   (b)

5-3(a), but the name of the game was the same. At this point we don't care too much about the answer (the willing reader is encouraged to play with it for a few minutes), but rather about the observation that this is the Königsberg bridge problem all over again. Although this is a critical observation, we will not explain it. Instead, we encourage each reader to personally reason out the connection. In the spirit of compromise we will give a hint without words in the form of Fig. 5-3(b).

GRAPHS

When Euler developed his solution to the Königsburg bridge problem, he actually did a lot more than that—he set the foundations for the study of an extremely important concept in mathematics: the concept of a *graph*.

For starters, let's just say that a **graph** is a picture consisting of dots (called **vertices**) and lines (called **edges**). The edges do not have to be necessarily straight lines (curvy, wavy, etc., are all acceptable), but they always have to connect two vertices. When an edge connects a vertex back with itself (which is also allowed), then it is called a **loop**.

The above description of a graph is not to be taken as a precise definition, but rather as an informal description that will help us get by for the time being. As with most concepts, the best way to get a feel for what a graph is, is to look at a few examples.

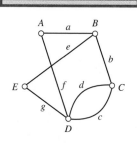

FIGURE 5-4

Example 1. Here is a graph with 5 vertices called *A, B, C, D,* and *E* and 7 edges called *a, b, c, d, e, f,* and *g.* (As much as possible, we will try to be consistent and use upper case letters for vertices and lower case letters for edges, but this is not mandatory.) A couple of comments about this graph: First, note that the point where edges *e* and *f* cross is not a vertex. Also note that there is no rule against having **multiple edges** connecting the same two vertices, as is the case with vertices *D* and *C.*

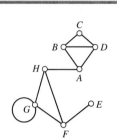

Example 2. This is a graph with 8 vertices (*A, B, C, D, E, F, G,* and *H*) and 11 unlabeled edges. (There is no rule that says that we have to give names to the edges.) We can still specify an edge by naming the two vertices that are its endpoints. For example, we can talk about the edge *AH* or the edge *BD* and so on. Note that there is a loop in this graph—it is the edge *GG.*

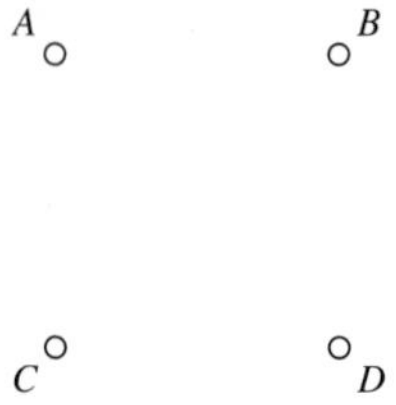

FIGURE 5-5

Example 3. Here is a graph with 4 vertices and no edges. While it does not make for a particularly interesting graph, there is nothing illegal about it. Graphs without any edges are permissible (graphs without vertices, however, are not!

A ○   ○ B

C ○   ○ D

FIGURE 5-6

Example 4.

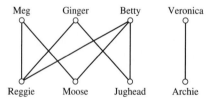

FIGURE 5-7   Reggie   Moose   Jughead   Archie

Here is a graph with 8 vertices and 8 edges. The vertices have funny names, but so what? Note that this graph is made up of two separate, disconnected pieces. Graphs of this type are called **disconnected,** and the individual "pieces" are called the **components** of the graph. (The very uninteresting graph in Fig. 5-6, for example, is disconnected and has four components.)

What might the graph in Fig. 5-7 represent? Let's suppose that Meg, Ginger, Betty, Veronica, Reggie, Moose, Jughead, and Archie are friends who go together to a party. A graph such as the one in Fig. 5-7 might be a way to describe who danced with whom at the party. We can learn at lot from such a picture (such as the fact that Veronica and Archie spent the evening dancing together), but most importantly, we should appreciate the fact that the picture provides such a crisp and convenient way to describe who danced with whom.

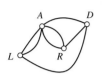

FIGURE 5-8

Example 5. Here is a graph with 4 vertices (*A, D, L,* and *R*) and 7 edges (*AR, AR, AD, AL, AL, DR,* and *DL*). A comparison with the graph for the Königsberg bridge problem (Fig. 5-3) shows that the two graphs have exactly the same vertices and edges—they are, in fact, the same graph even though they look very different. ▪

Example 5 illustrates an important point: A graph can be drawn in infinitely many different ways and *it is not the shape of the graph that matters, but rather the information telling us which vertices are connected to which other vertices.*

With all of the above examples under our belt, we might be ready for a definition of a graph. A graph is a *relational structure:* It tells us that we have a bunch of objects (the vertices) and that these objects are related to each other. The story of which objects are related to which other objects is given by the edges. That is the entire information carried by a graph—no more and no less.

It follows that any time we have a relationship between objects, whatever that relationship might be (love, hate, kinship, dance partner, etc.), we can describe such a relationship by means of a graph. To drive the point home, here is the last example for this section.

Example 6. For the last week of the baseball season (that's when the pennant race is hottest), the schedule for the National League East is

- Monday: Pittsburgh versus Montreal, New York versus Philadelphia, Chicago versus St. Louis

- Tuesday: Pittsburgh versus Montreal

- Wednesday: New York versus St. Louis, Philadelphia versus Chicago

- Thursday: Pittsburgh versus St. Louis, New York versus Montreal, Philadelphia versus Chicago

- Friday: Philadelphia versus Montreal, Chicago versus Pittsburgh

- Saturday: Philadelphia versus Pittsburgh, New York versus Chicago, Montreal versus St. Louis.

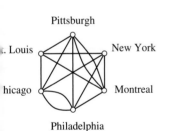

FIGURE 5-9

All the information about who plays whom is relational (each game "relates" two different teams) and can be conveniently represented by the graph shown in Fig. 5-9. (In this graph the position of the vertices has nothing to do with the geographic location of the cities. In describing who plays whom, the actual placement of the vertices is irrelevant.)

The main point of this example is the fact that the graph is in many ways a much more practical way to describe the schedule than the original listing. Say one wants to know if New York plays Pittsburgh the last week of the season. The answer is obvious if one looks at the graph, but much less so if one has to check through the listed schedule of games. ◼

Graph Concepts and Terminology

Every branch of mathematics has its own peculiar jargon, and the theory of graphs has more than its share. In this section we will introduce a few of the essential concepts and terms that we will need in the chapter.

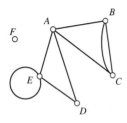

FIGURE 5-10

■ Two vertices are said to be **adjacent** if there is an edge joining them. In the graph shown in Fig. 5-10, vertices A and B are adjacent, but C and D aren't. Vertex E is adjacent to itself because there is a loop at E.

■ The **degree** of a vertex is the total number of edges at that vertex. (A loop contributes 2 toward the degree.) In the graph shown in Fig. 5-10, vertex A has degree 4 [we can write this as $\deg(A) = 4$], vertex B has degree 3 [$\deg(B) = 3$], $\deg(C) = 3$, $\deg(D) = 2$, $\deg(E) = 4$ (because of the loop), and $\deg(F) = 0$. (A vertex with degree 0 is called an **isolated** vertex.)

■ A **path** in a graph is a string of vertices, each one adjacent to the next one. Whereas a vertex can appear on the path several times, an edge can be part of a path *only once*. Here are some examples of paths taken from the graph in Fig. 5-10: A, B, C is a path; A, B, C, A, E, D is a longer path; D, E, E, A, C, B, C is also a path (E is adjacent to E because of the loop there, C and B are adjacent twice because of the double edge between them). A, C, D, E is *not* a path because C and D are not adjacent vertices; $A, B, C, A, D, E, A, C, B$ is *not* a path, because the edge joining A and C appears twice in the list.

■ If a path starts and ends at the same vertex, it is called a **circuit**. In the graph in Fig. 5-10, A, B, C, A is a circuit; so are D, E, A, C, B, A, D and B, C, B and E, E. On the other hand, A, B, C, B, A is not a circuit, since the same edge (AB) is used twice.

■ A graph is **connected** if any two of its vertices can be joined by a path. This essentially means that it is possible to travel from any vertex to any other vertex along consecutive edges of the graph. If a graph is not connected, it is said to be **disconnected.** A graph that is disconnected is made up of pieces that are by themselves connected. Such pieces are called the **components** of the graph. The graph in Fig. 5-11(a) is connected. The graphs in Figs. 5-11(b) and (c) are disconnected. The one in Fig. 5-11(b) has two components; the one in Fig. 5-11(c) has three.

■ Sometimes in a connected graph there is an edge such that if we were to erase it, the graph would become disconnected. For obvious reasons such an edge is called a **bridge** (burn a bridge behind you, and you'll never be

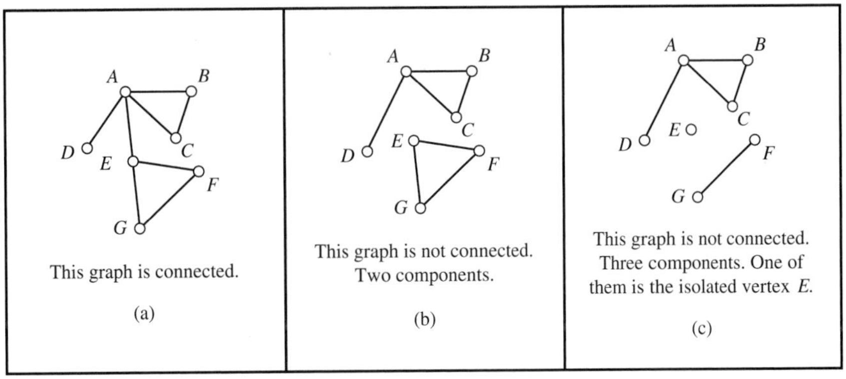

This graph is connected.

(a)

This graph is not connected. Two components.

(b)

This graph is not connected. Three components. One of them is the isolated vertex *E*.

(c)

FIGURE 5-11

able to get back to where you were). In Fig. 5-11(a), the edge *AE* is a bridge because when we remove it, the graph becomes disconnected (as in Fig. 5-11[b]). There is another bridge in Fig. 5-11(a). We leave it to the reader to find it.

■ When a path contains each and every edge of a connected graph exactly once, it is called an **Euler path**. For the graph in Fig. 5-11(a), the path *D, A, B, C, A, E, F, G, E* is an Euler path of the graph. The reader may verify this by tracing the path out and checking that each edge is traced exactly once. When an Euler path starts and ends at the same vertex it is called an **Euler circuit**.

There are three fundamental questions which we will learn how to answer in this chapter:

1. How can we tell if a graph has an Euler circuit or an Euler path?

2. If it does, how do we find it?

3. If it doesn't, how can we doctor up the graph so that it does?

Once we know how to answer these three basic questions we will have all the tools that we need to tackle just about any kind of Euler circuit problem, from the routing of garbage trucks in Cleansburg to the routing of police patrols in New York City.

EULER'S THEOREMS

Remember the Königsberg bridge problem? Finding a walk through the city that crosses each of the bridges once and returns to the starting place (Fig. 5-12[a]) is exactly the same problem as finding an Euler circuit in the graph shown in Fig. 5-12(b). Euler's answer to this problem was that it can't be done!

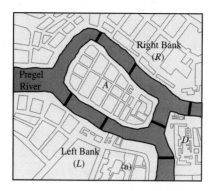 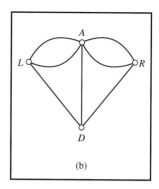

FIGURE 5-12

Why is such a walk impossible? Let's say for the sake of argument that the walk starts on the left bank L (it actually makes no difference where we pick the starting point). At least once along the way the walker must pass through the island A, and we suspect that the walker will have to visit A more than once. Let's count exactly how many times. The first visit to A will use up two bridges (one getting there and a different one getting out); the second visit to A will use up two other bridges; and the third visit to A will use up two more. Oops! There are only five bridges to get in and out of A, so two visits to A won't do it (there would be an unused bridge) and three visits are too many (we would have to re-cross one of the bridges). It follows that the walk is impossible! It's the odd number of bridges at A (or anywhere else) that causes the problem. The argument can be extended and made general in a very natural way. We present it without any further ado.[3]

Euler's Theorem 1.

If a graph has any vertices of odd degree, then it cannot have an Euler circuit.

If a graph is connected and every vertex has even degree, then it has at least one Euler circuit (usually more).

FIGURE 5-13

Note that a graph can have every vertex of even degree but be disconnected (as in Fig. 5-13), in which case it cannot have an Euler circuit.

What can we say about graphs with Euler paths? The same arguments apply for all the vertices (they must have even degree) except for the starting and ending vertices of the path which are different. The starting vertex requires one edge to get out and two more for each visit through that vertex, so it has odd degree. Likewise the ending vertex has odd degree (two edges for every visit plus one more to get to the vertex on the last leg of the trip). Thus, we have

---

[3] A translation of Euler's own account of how he developed these ideas is given in reference 4. Euler's original account, published in 1736, was the first paper in the theory of graphs. (Euler called it "the geometry of position".)

Euler's Theorem 2.

If a graph has more than two vertices of odd degree, then it cannot have an Euler path.

If a graph is connected and has exactly two vertices of odd degree, then it has at least one Euler path (usually more). Any such path must start at one of the odd-degree vertices and end at the other one.

Example 7. Fig. 5-14 shows some classic examples of graphs that children try to trace. From Euler's theorems, we know that Fig. 5-14(a) cannot be traced (it has neither an Euler circuit nor an Euler path), Fig. 5-14(b) can be traced but only if we start at *D* and end at *C* or vice versa (it has an Euler path but no Euler circuit), and Fig. 5-14(c) can be traced so that we start and end at the same place (it has an Euler circuit).

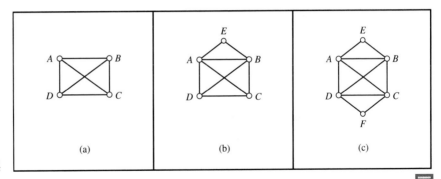

(a)   (b)   (c)

FIGURE 5-14

The careful reader may have noticed that there is an apparent gap in Euler theorems 1 and 2. The two theorems together cover the cases of graphs with zero vertices of odd degree (theorem 1), two vertices of odd degree (theorem 2), and more than two vertices of odd degree (theorem 2). What happens if a graph has just one vertex of odd degree? What then? Didn't Euler consider this possibility? It turns out that he did, but found that it is impossible for a graph to have *just one vertex of odd degree.*

The key observation that Euler made was that when *the degrees of all the vertices of a graph are added up, the total is exactly double the number of edges in the graph.* (Think about it: An edge *XY* contributes once to the degree of vertex *X* and once to the degree of vertex *Y*, so all in all that edge contributed twice to the sum of the degrees. If still unconvinced, the reader is referred to Exercise 33[a].) One of the important consequences of Euler's observation is that when the degrees of all the vertices of a graph are added up, *the total must be an even number,* which means therefore that it is impossible to have just one vertex of odd degree. In fact, we can push the logic one step further, and argue just as well that it is impossible for a graph to have 3, 5, 7, . . . vertices of odd degree. We summarize the preceding observations into a theorem.

Euler's Theorem 3.

The sum of the degrees of all the vertices of a graph is an even number (twice the number of edges, to be precise).

In every graph, the number of vertices of *odd* degree must be *even.*

FLEURY'S ALGORITHM

Euler's theorems are great because they give us an easy way to determine if a graph has an Euler circuit or an Euler path just by looking at the degrees of the vertices. Unfortunately they are of no help in finding the actual Euler circuit or path if there is one. Of course for graphs such as the ones shown in Figs. 5-14(b) and (c), we can find an Euler path (or circuit) by simple trial and error. In most real life applications, however, graphs can have hundreds and even thousands of vertices and edges, and for such graphs simple trial and error won't do.

Our next major accomplishment will be to learn a method for finding an actual Euler circuit (or Euler path) when there is one. The method comes packaged in the form of an **algorithm,** a set of mechanical rules that when followed are guaranteed to produce an answer. The fact that the rules that make up an algorithm are mechanical means that there is no thinking involved—that's why mindless but efficient things like computers are ideally suited to carrying out algorithms. For human beings, the difficulties in carrying out algorithms (once the rules are understood) are not intellectual but rather procedural. Accuracy and fastidious attention to detail are the important virtues when carrying out the instructions in an algorithm and being a brilliant mathematical thinker in and of itself is of little value here. We mention this as a piece of friendly advice because most of the practical things we will do in this part of the book will be based on the ability to correctly carry out algorithms—an ability that is acquired primarily through practice.

If every vertex of a connected graph has even degree, then Euler's theorem 1 assures us that the graph has an Euler circuit. We can find such an Euler circuit by using an algorithm known as **Fleury's algorithm**. Because this is our first encounter with a graph algorithm, we will describe it informally first, then work out a couple of examples, and give the formal description of the algorithm last.

The basic philosophy behind Fleury's algorithm is quite simple, and it can be summarized by an old piece of folk wisdom: *don't cross a bridge until you have to.* The only thing we have to be careful about is in our interpretation of the word *bridge.*

We know that when talking about graphs, a bridge is an edge whose removal disconnects the graph, and Fleury's algorithm specifically instructs us to travel along such edges only as a last resort. Simple enough, but there is a rub: The *graph* whose bridges we are supposed to avoid is not necessarily the original graph of the problem, but rather that part of the original graph which we haven't yet traversed. The point is this: Each time we traverse an edge, we are done with it! We will never cross it again, and as far as we are concerned, from that point on it is as if that edge never existed. Our concerns lie only on how we are going to get around the yet untraveled part of the graph. Thus, when we talk about bridges that we want to leave as a last resort, we are really referring to *bridges of the untraveled part of the graph.*

Since each time we traverse an edge the *untraveled part of the graph* changes (and consequently so do the bridges), Fleury's algorithm requires some careful bookkeeping. This does not make the algorithm difficult (in fact, its basic philosophy is quite simple), it just means that we must take extra pains in separating what we have already done from what we yet need to do.

While there are many different ways to accomplish this (and readers are certainly encouraged to invent their own) a fairly reliable way goes like this: We start with two separate copies of the graph, copy 1 for making decisions and copy 2 for record keeping. Every time we traverse another edge, we erase it from copy 1 but mark it (say in red) and label it with the appropriate number on copy 2. As we progress along our Euler circuit, copy 1 gets smaller and copy 2 gets redder. Copy 1 helps us decide where to go next; copy 2 helps us reconstruct our trip (just in case we are asked to demonstrate how we did it!). Let's try a couple of examples.

Example 8. The graph in Fig. 5-15 has an Euler circuit—we know this is so because every vertex has even degree. Let's use Fleury's algorithm to find an Euler circuit. This is a bit of an overkill (for this small a graph we can do it just as well by trial and error), but it will help us understand how the algorithm works.

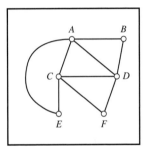

FIGURE 5-15

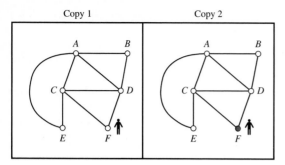

Start: We can pick any starting point
we want. Let's we start at F.

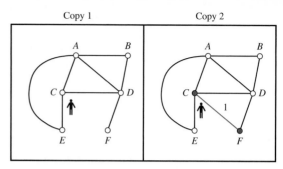

Step 1: Travel from F to C.
(Could have also gone from F to D.)

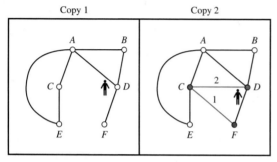

Step 2: Travel from C to D.
(Could have also gone to A or to E.)

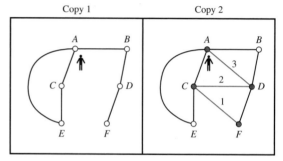

Step 3: Travel from D to A.
(Could have also gone to B but not to F—DF is a bridge!)

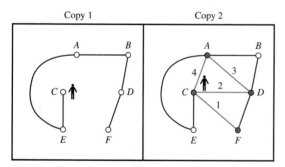

Step 4: Travel from A to C.
(Could have also gone to E but not to
B—AB is a bridge!)

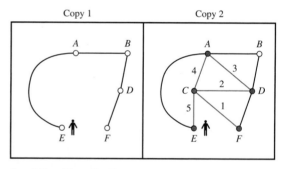

Step 5: Travel from C to E.
(There is no choice!). Since we are never coming back to C,
we can erase it from Copy 1 (keeps the picture uncluttered).

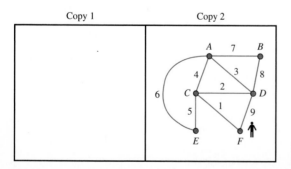

Step 6,7,8, and 9: Only one way to
go at each step.

Example 9. Let's apply Fleury's algorithm to the graph in Fig. 5-16. This graph is a little bigger than the one in our previous example, and it would be impractical to show each step of the algorithm with a separate picture as we did in Example 8. Thus, we ask for the reader's active involvement in working through this example. You should have in front of you two copies of the graph and follow the steps on those copies as we describe them in words. (Needless to say, pencil and eraser are strongly recommend for copy 1.)

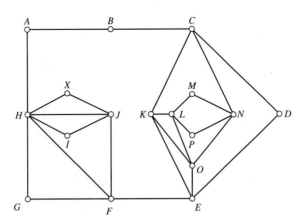

Figure 5-16

- ■ **Start:**   We pick an arbitrary starting vertex, say *X*.
- ■ **Step 1:**   From *X* we can go to either *J* or *H*. Since neither *XJ* nor *XH* is a bridge, we can pick either one. We pick *XJ*. (We now erase *XJ* on copy 1; mark it and label it with a 1 on copy 2.)
- ■ **Step 2:**   From *J* we can go to either *I*, *H*, or *F*. Any one of these choices is OK. We choose *JI*. (Erase *JI* from copy 1 and mark and label it on copy 2.)
- ■ **Step 3:**   There is only one way to go from *I* (to *H*). (Erase edge *IH* as well as vertex *I*—we won't be coming back to it—from copy 1 and mark and label *IH* on copy 2.)
- ■ **Step 4:**   From *H* we have several choices. One of them, *HX* is a bridge—it disconnects the vertex *X* from the rest of the graph. Any choice other than *HX* is OK. We choose *HG*. (Erase edge *HG* from copy 1; mark and label it on copy 2.)
- ■ **Step 5:**   There is only one way to go from *G* (to *F*). (Erase edge *GF* as well as vertex *G*—we won't be coming back to it—from copy 1 and mark and label *GF* on copy 2.)
- ■ **Step 6:**   From *F* we have several choices. All are OK since none are bridges. We choose *FE*. (To speed things up, from here on we will omit the "erase, mark, and label" stuff.)
- ■ **Step 7:**   No bridges at *E*. We choose *EO*.

- **Step 8:**   No bridges at *O*. We choose *OK*.
- **Step 9:**   No bridges at *K*. We choose *KC*.
- **Step 10:**   One of the choices at *C* (*CB*) is a bridge (we can't go that way!). We choose *CN*.
- **Step 11:**   No bridges at *N*. We choose *NP*.
- **Step 12:**   No choice here. Take *PL*. (Don't forget to erase vertex *P* also.)
- **Step 13:**   *LK* is a bridge. We choose *LO*.
- **Steps 14–23:**   No choices here. From *O* go to *N* then to *M* then to *L* then to *K* then to *E* then to *D* then to *C* then to *B* then to *A* then to *H*.
- **Step 24:**   We have some choices at *H* but *HX* is a bridge. We choose *HJ*.
- **Step 25–27:**   No more choices. From *J* go to *F* then to *H* and finally back to *X*. We are finished!

Here is the Euler circuit we found:

X, J, I, H, G, F, E, O, K, C, N, P, L, O, N, M, L, K, E, D, C, B, A, H, J, F, H, X.

Practice is the only way to really get the feel for how an algorithm works. Fleury's algorithm is no exception. The reader is strongly encouraged to try Exercises 27 through 30 at this juncture.

For the record, here is a formal description of the basic rules for Fleury's algorithm.

Fleury's Algorithm for Finding an Euler Circuit.

- First make sure that the graph is connected and all the vertices have even degree.

- Start at any vertex.

- Travel through an edge if
    - (a) it is not a bridge for the untraveled part, or
    - (b) there is no other alternative.

- Label the edges in the order in which you travel them.

- When you can't travel anymore, stop. (You are done!)

When a connected graph has exactly two vertices of odd degree, then we know that the graph does not have an Euler circuit, but it does have an Euler path. We can find such an Euler path by making one minor change in the rules for Fleury's Algorithm: *The starting point must be one of the vertices of odd degree.* Other than that, the rules are exactly the same, and when followed properly, it is guaranteed that the trip will end at *the other vertex of odd degree.* (See Exercises 31 and 32.)

EULERIZING GRAPHS

Fleury's algorithm gives us the means for finding efficient ways in which to traverse all the edges of a graph, provided the graph has either zero or two vertices of odd degree. But what happens when a graph has many (4, 6, 8, . . .) vertices of odd degree? We know from Euler theorems 1 and 2 that in this case there is no Euler circuit or Euler path, so that if we want to trace out all of the edges of the graph (of course without lifting our pen or pencil) we are going to have to retrace some of the edges. How can we do it so that we retrace as few edges as possible? This is the question we are going to tackle in this section.

Let's start with a simple example.

Example 10. Consider the graph in Fig. 5-17. Since it has eight vertices of odd degree (*B, C, E, F, H, I, K,* and *L*), the graph has no Euler circuit or Euler path. The graph in Fig. 5-18(a) is a close cousin to the original graph, but there is one big difference—all the vertices have even degree and therefore, this graph has an Euler circuit. Fig. 5-18(b) shows one such Euler circuit (just follow the numbers). The Euler circuit in Fig. 5-18(b) can be "reinterpreted" as a trip along the edges of our original graph as shown in Fig. 5-18(c). In this trip we are traveling along all of the edges of the graph, but we are retracing four of the edges (*BC, EF, HI,* and *KL*). While this is not an Euler circuit it shows us the most *efficient* trip (meaning a trip with the least amount of duplication) that covers all of the edges of the graph.

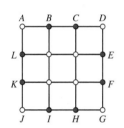

FIGURE 5-17 ● indicates vertices of odd degree.

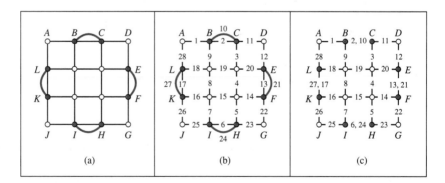

(a) (b) (c)

FIGURE 5-18

Before we go on to the next example let's introduce some convenient terminology. What is the connection between the graphs in Fig. 5-17 and 5-18(a)? A close look shows us that Fig. 5-18(a) is the result of applying the following straightforward process to the graph in Fig. 5-17: Add extra edges in such a way that the vertices of odd degree become vertices of even degree (in other words, neutralize the troublemakers). This process of changing a graph so that the vertices of odd degree are eliminated by adding additional edges is called **eulerizing** the graph. When eulerizing a graph, there is one thing that we must be careful about: The edges that we add *must be duplicates of edges that are already in the graph.* (Remember that the point of all of this is to cover the edges of the existing graph in the best possible way without creating any new edges.) Our next example, clarifies this point.

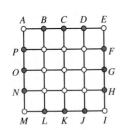

FIGURE 5-19 ● indicates vertices of odd degree.

Example 11. Consider the graph in Fig. 5-19. This graph has 12 vertices of odd degree, as shown in the figure. If we want to travel along all the edges of this graph and come back to our starting point we know we are going to have to double up on some of the edges. Which ones? The answer is provided by first eulerizing the graph, so let's discuss ways in which we can do this.

Figure 5-20(a) shows how *not to do it!* Adding the edges *DF* and *NL* is not allowed, since those edges were not in the original graph.

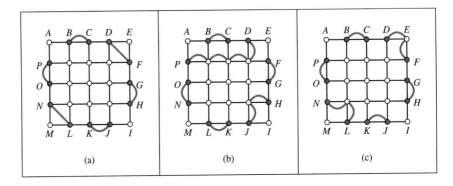

FIGURE 5-20

Figure 5-20(b) shows a legal, but wasteful eulerization of the original graph. It is legal because we have eliminated all the vertices of odd degree by adding edges that duplicate already existing edges, but it is wasteful because it is obvious that we could have accomplished the same thing by adding fewer duplicate edges.

Figure 5-20(c) shows an *optimal eulerization* of the original graph. It was obtained by adding only 8 duplicate edges. While there are other ways to do the same thing, there is no way to do it with less than 8 duplicate edges. This optimal eulerization gives us the blueprint for an optimal round trip along the edges of the original graph. We know that we are going to have to retrace exactly 8 of the edges, and in fact, we know exactly which ones. Figure 5-21 shows an actual example of an optimal trip, obtained using Fleury's algorithm on Fig. 5-20(c), and we can clearly see exactly which edges are being retraced.

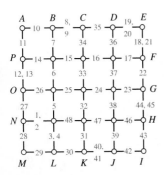

FIGURE 5-21

Example 12. Let's now try our hand at a slightly more challenging example. We want to trace the graph in Fig. 5-22 starting and ending at the same vertex (let's say X) and using as few duplicated edges as possible. Since the graph has quite a few vertices of odd degree, finding an optimal eulerization (i.e., one with the fewest number of duplicate edges) might take a little trial and error, and the reader is strongly encouraged to give it a try before looking at the answer (Fig. 5-23).

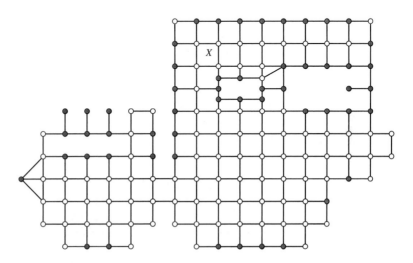

FIGURE 5-22 ● indicates vertices of odd degree.

We can now accomplish our main goal by finding an Euler circuit for the graph in Fig. 5-23. We can do this using Fleury's algorithm (painful and slow but guaranteed to work) or a little common sense and trial and error (possibly faster but riskier). One way or another, we encourage the reader to find such an Euler circuit.

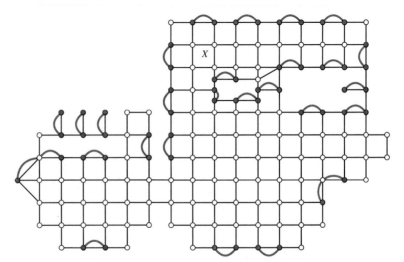

FIGURE 5-23

In some situations, we may want to trace a graph in the best possible way while starting in one place and ending in another. We know from Euler's theorem 2 that in this case, the starting and ending vertices must be of odd degree. Once we are given a designated starting and ending vertex both of odd degree (call them X and Y respectively), we can find an optimal tracing of the graph by finding an optimal **semi-eulerization**: leave X and Y as vertices of odd degree, and make all other vertices of odd degree even (exactly as before). Since the basic process is essentially the same, we leave it to the reader to try it out (see Exercises 41, 42, and 43).

GRAPH MODELS

It took some doing, but we finally have all the mathematical ingredients necessary for solving just about any real-world Euler circuit problem that might be thrown at us. It should be clear by now that solving an Euler circuit type problem boils down to finding an Euler circuit in an appropriately chosen graph. The question is, How do we *translate* the original real-world situation into the right graph—one that captures the essence of the problem we are trying to solve?

When a mathematical structure such as a graph is used to describe and study a real-world problem we call such a structure a mathematical **model** for the original problem. One of the most useful things that we will learn in this section, as well as in the next three chapters, is how to use a graph model as the basic analytical tool with which to study various important types of applied problems from the world of management science.

Let's consider a few simple examples.

Example 13. (The lazy ice cream man.) An ice cream man is hired to travel along the streets of the small neighborhood shown in Fig. 5-24(a), starting and ending at the ice cream parlor located at the corner of Baltic and Maine (P in Fig. 5-24[a]). It is a hot summer day and the air conditioning in the ice cream truck is broken, so he is anxious to get the job done with the least amount of waste.

Fortunately for him, he has read this chapter and realizes that if he is to use what he knows about Euler circuits, he first has to turn the little neighborhood into a graph. Since he only has to pass through a block once to sell ice cream, he can turn each neighborhood block into an edge and each corner into a vertex, as shown in Fig. 5-24(b). The original problem now boils down to finding an optimal trip along the edges of a graph (Fig. 5-25[a]), and as luck would have it, it is exactly the problem we solved in Example 11. Fig. 5-25(b) shows such an optimal trip. (It is exactly the trip shown in Fig. 5-21 but now shown in its natural habitat.)

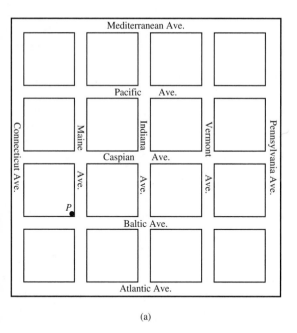

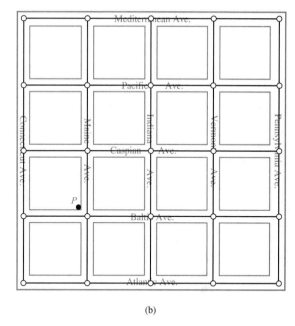

(a)   (b)

FIGURE 5-24

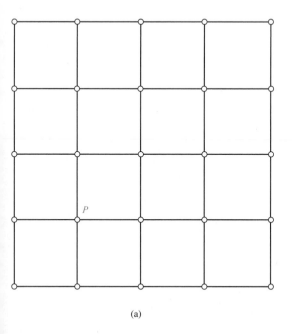

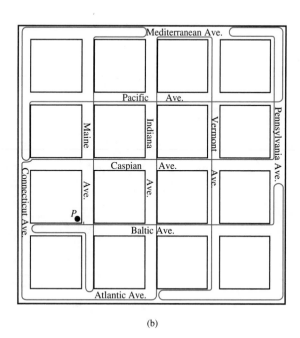

(a)   (b)

FIGURE 5-25

Example 14. (The efficient mail carrier.) Let's consider the routing of a mail carrier who must deliver mail in the same neighborhood as the ice cream man in Example 13. Her trip must start and end at the Post Office, located at the corner of Pacific and Vermont (Q in Fig. 5-26[a]). The main difference between her sit-

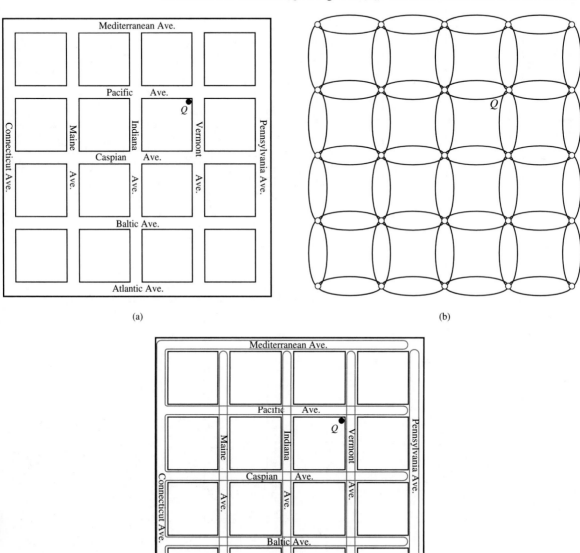

(a)

(b)

(c)

FIGURE 5-26

uation and that of the ice cream man in Example 13 is that the mail carrier must deliver mail on both sides of each street, which in practice means that she must travel twice through each block. This fact must be reflected in the graph that models the problem, and we accomplish this by putting two edges for every block, as shown in Fig. 5-26(b). Note that this graph does not have to be euler-ized—it already has all vertices of even degree. If we now apply Fleury's algo-rithm (using Q as the starting point) we can obtain an optimal route for the mail carrier. One such route is shown in Fig. 5-26(c).

In this example, the routing can be accomplished without any **deadhead** blocks (which is the name commonly given to those blocks added to the original graph in the process of eulerizing it).

Example 15 (Routing trucks for the Cleansburg Sanitation Department). This is our last example for this chapter, and it takes us back to our opening story and the problem of finding optimal routes for the Cleansburg Sanitation trucks.

Let's start with the problem of routing the garbage truck. You need to know that in Cleansburg, garbage trucks pick up garbage on both sides of the street as they pass through each block. Given this information, the appropriate graph model for the routing of the garbage truck requires one edge per city block. The appropriate graph is shown in Fig. 5-27(a). This is exactly the graph in Fig. 5-22, which we discussed in Example 12 (such good luck!). We already found an optimal eulerization for this graph (Fig. 5-23 reproduced here for the reader's convenience as Fig. 5-27[b]).

Fig. 5-27(b) shows which would be the deadhead blocks in an optimal route for the garbage truck. If we count them, we find 29 deadhead blocks, which is not too bad considering the total number of blocks involved. The last bit of mop-up work requires actually finding an optimal route, a job we delegate to the reader (see Example 12 and Exercise 39).

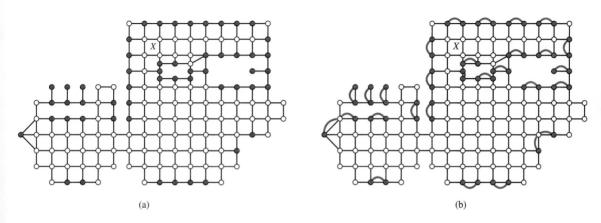

(a)   (b)

FIGURE 5-27

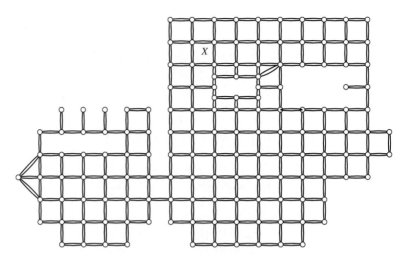

FIGURE 5-28 A graph model for sweeping curbs in Cleansburg.

How about a graph model for the curb sweeper? Here, the truck must pass through each block twice (once for each curb), so in our graph model we must put two edges for every block. The resulting graph is shown in Fig. 5-28. This graph has no vertices of odd degree, so all we have to do to find the optimal route is find an Euler circuit using X as the starting point. We leave this part to the truly ambitious reader. Likewise, a graph model for the snowplows would have one edge for every lane, so that, for example, a block in a four-lane avenue (see map, Fig. 5-1) would be represented by four edges.

CONCLUSION

In this chapter we learned several things, not the least of which is that there is probably a mathematical reason behind the sometimes curious manner in which the local garbage truck winds its way around the neighborhood. In fact, many cities nowadays systematically route their garbage trucks seeking maximum efficiency and using methods just like the ones we developed in this chapter. For a large city, the efficient routing of services (not only garbage collection but also street sweeping, snow removal, utility meter reading, mail delivery, etc.) can represent savings of millions of dollars.[4]

The efficient routing of garbage trucks is just one example of a broad category of problems called *routing problems*. The name of the game in routing problems is to find, among all the routes that accomplish a specified objective (pick up all the garbage, sweep every curb, inspect every utility meter, call on

---

[4] In New York City alone, for example, the routing of sanitation trucks using graph theory has resulted in savings estimated to be around 25 million dollars a year.

every customer, etc.), one that is as efficient as possible. In our somewhat simplified models we measured efficiency by the number of wasted (*deadhead*) blocks on the route, but in more complicated situations we might have to use other considerations such as total distance traveled or total money spent. Routing problems are tricky in this regard. While they all sound a lot alike and the broad objective is always to do things in the most efficient possible way, what this means in practice and how it is accomplished can differ greatly from problem to problem. This fact will become clear in the next chapter where we will explore a type of routing problem which marches to the tune of an altogether different drummer.

In this chapter we were introduced to the world of graphs. Graphs are to management science what squares, triangles, and circles are to traditional geometry: They provide an appropriate setting for analyzing (and hopefully solving) certain types of real-world problems. The study of graphs is in its own right an important branch of modern mathematics. Just like algebra and geometry, it has its own language and, more importantly, its own way of doing things. In this chapter we also had our first formal exposure to *algorithms* and learned about a specific algorithm called *Fleury's algorithm.* In the next three chapters we will get further glimpses of various parts of the world of graphs and discuss several new and surprising algorithms.

One final word about algorithms is in order: Carrying out an algorithm has often been compared to following a recipe or learning how to drive. In all these activities, intelligence is not a substitute for practice. Regardless of one's mathematical abilities, the only way to learn and understand an algorithm is to do it oneself as many times as is necessary.

KEY CONCEPTS

adjacent vertices	Euler path
algorithm	Euler's theorems
bridge	Fluery's algorithm
circuit	graph
components	graph model
connected graph	isolated vertex
deadhead block	loop
degree of a vertex	multiple edges
disconnected graph	path
edge	routing problems
Euler circuit	semi-eulerization
Euler circuit problem	vertex
eulerizing graphs	

EXERCISES

Walking

1. For each of the following graphs, list the vertices and edges.

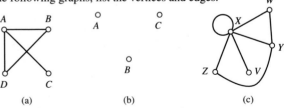

   (a)   (b)   (c)

2. For each of the following graphs, list the vertices and edges.

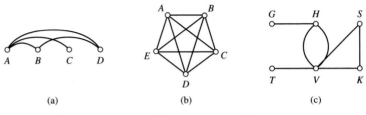

   (a)   (b)   (c)

3. For each of the following, draw two different pictures of the graph.
   (a) Vertices: A, B, C, D
       Edges: AB, BC, BD, CD
   (b) Vertices: K, R, S, T, W
       Edges: RS, RT, TT, TS, SW, WW, WS.

4. For each of the following, draw two different pictures of the graph.
   (a) Vertices: L, M, N, P
       Edges: LP, MM, PN, MN, PM
   (b) Vertices: A, B, C, D, E
       Edges: A is adjacent to C and E; B is adjacent to D and E; C is adjacent to A, D, and E; D is adjacent to B, C, and E; E is adjacent to A, B, C, and D.

5. For each of the following graphs, find the degree of each vertex.

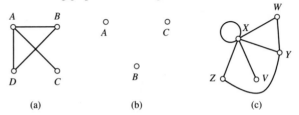

   (a)   (b)   (c)

6. For each of the following graphs, find the degree of each vertex.

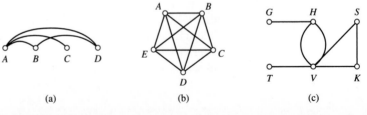

   (a)   (b)   (c)

7. **(a)** Explain why the following figures represent the same graph.

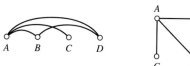 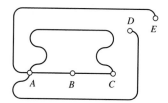

(b) Draw a third figure that represents the same graph.

8. (a) Explain why the following figures represent the same graph.

(b) Draw a third figure that represents the same graph.

9. (a) Draw a graph with four vertices and such that each vertex has degree 2.
   (b) Draw a graph with six vertices and such that each vertex has degree 3.

10. (a) Draw a graph with four vertices and such that each vertex has degree 1.
   (b) Draw a graph with eight vertices and such that each vertex has degree 3.

11. Draw a graph with 4 vertices, each of degree 3 and such that
   (a) there are no loops and no multiple edges
   (b) there are loops but no multiple edges
   (c) there are multiple edges but no loops
   (d) there are both multiple edges and loops.

12. Draw a connected graph with 5 vertices, each of degree 4 and such that
   (a) there are no loops and no multiple edges
   (b) there are loops but no multiple edges
   (c) there are multiple edges but no loops
   (d) there are both multiple edges and loops.

Exercises 13 through 16 refer to the graph in the margin:

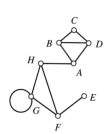

13. (a) Find a path from C to F passing through vertex B but not through vertex D.
   (b) Find a path from C to F passing through both vertex B and vertex D.
   (c) How many paths are there from C to A?
   (d) How many paths are there from H to F?
   (e) How many paths are there from C to F?

14. (a) Find a path from D to E passing through vertex G only once.
   (b) Find a path from D to E passing through vertex G twice.
   (c) How many paths are there from D to A?
   (d) How many paths are there from H to E?
   (e) How many paths are there from D to E?

15. (a) Find a circuit passing through vertex D.

   (b) How many circuits start and end at vertex *D*?
   (c) Which edges in the graph are bridges?

16. **(a)** Find a circuit passing through vertex *H*.
    (b) How many circuits start and end at vertex *H*?
    (c) Which edge can be added to this graph so that the resulting graph has no bridges?

17. **(a)** In the Königsberg bridge problem, which of the city bridges are bridges in the graph theory sense?
    (b) Give an example of a graph with 4 vertices in which every edge is a bridge.

18. **(a)** Give an example of a connected graph with 5 vertices and no bridges.
    (b) Give an example of a connected graph with 5 vertices in which every edge is a bridge.

19. For each of the following, determine if the graph has an Euler circuit, an Euler path, or neither of these. Explain your answer but do not find the actual path or circuit.

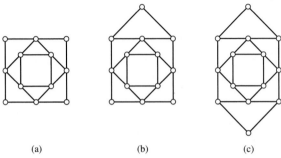

  (a)   (b)   (c)

20. For each of the following, determine if the graph has an Euler circuit, an Euler path, or neither of these. Explain your answer but do not find the actual path or circuit.

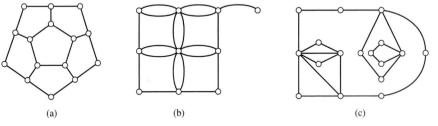

  (a)   (b)   (c)

21. For each of the following, determine if the graph has an Euler circuit, an Euler path, or neither of these. Explain your answer but do not find the actual path or circuit.

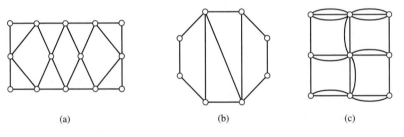

  (a)   (b)   (c)

22. Trace each of the following graphs without lifting your pencil or retracing any lines. (Indicate your answer by labeling the edges 1, 2, 3, etc., in the order in which they are traced.) You are not required to start and end in the same place. If it can't be done, explain why not.

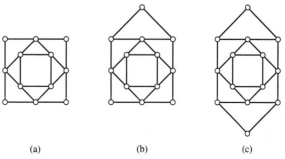

(a)   (b)   (c)

23. Trace each of the following graphs without lifting your pencil or retracing any lines. (Indicate your answer by labeling the edges 1, 2, 3, etc., in the order in which they are traced.) You are not required to start and end in the same place. If it can't be done, explain why not.

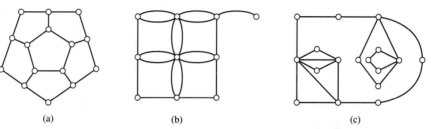

(a)   (b)   (c)

24. Trace each of the following graphs without lifting your pencil or retracing any lines. (Indicate your answer by labeling the edges 1, 2, 3, etc., in the order in which they are traced.) You are not required to start and end in the same place. If it can't be done, explain why not.

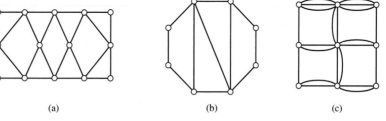

(a)   (b)   (c)

25. Find an optimal eulerization for each of the following graphs.

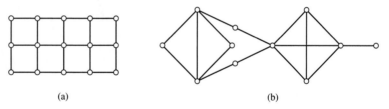

(a)   (b)

26. Find an optimal eulerization for each of the following graphs.

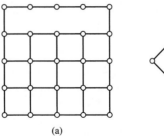

(a)

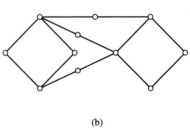

(b)

27. Find an Euler circuit in the following graph.

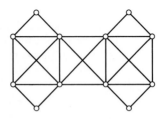

28. Find an Euler circuit in the following graph.

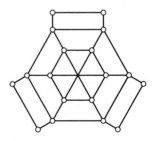

29. Use Fleury's algorithm to find an Euler circuit in the following graph.

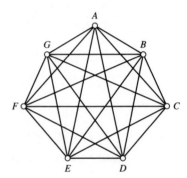

30. Use Fleury's algorithm to find an Euler circuit in the following graph.

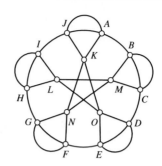

31. Find an Euler path in the following graph that starts at X and ends at Y.

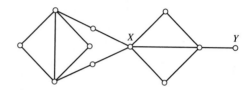

32. Find an Euler path in the following graph that starts at X and ends at Y.

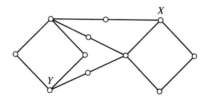

Jogging

33. (a) For each of the graphs in Exercises 1 and 2, complete the following table.

Graph	Number of Edges	Sum of the Degrees of all Vertices
Exercise 1(a)	4	$3 + 2 + 1 + 2 = 8$
Exercise 1(b)		
Exercise 1(c)		
Exercise 2(a)		
Exercise 2(b)		
Exercise 2(c)		

(b) Explain why in every graph the sum of the degrees of all the vertices equals twice the number of edges.

(c) Explain why every graph must have either zero or an even number of vertices of odd degree.

34. A garbage truck must pick up garbage along all the streets of the subdivision shown in the following figure (starting and ending at the garbage dump labeled *G*). All the streets are two-way streets, and garbage is picked up on both sides of the street simultaneously.

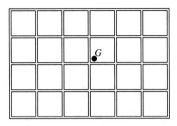

    (a) Draw an appropriate graph representing this problem.
    (b) Find an optimal eulerization of the graph in (a).
    (c) Find an optimal route for the garbage truck. Describe the route by labeling the edges 1, 2, 3, . . . , in the order in which they are traveled.

35. Consider the same subdivision as in Exercise 34, but this time assume the garbage is collected on each side of the street separately.
    (a) Draw an appropriate graph representing this problem.
    (b) Find an optimal route for the garbage truck. Describe the route by labeling the edges 1, 2, 3, . . . , in the order in which they are traveled.

36. Consider the following game: You must take a walk along the bridges of the city of Königsberg that starts and ends at the left bank (*L*) and crosses each bridge at least once. It costs $1 each time you cross a bridge. Describe the cheapest possible walk you can make.

Exercises 37 and 38 refer to a small town described by the following street map.

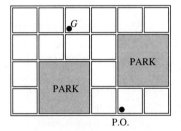

P.O.

37. A garbage truck must pick up garbage along the streets of the town (starting and ending at the garbage dump labeled *G*). All the streets are two-way streets, and garbage is picked up on both sides of the street simultaneously.
    (a) Draw an appropriate graph representing this problem.
    (b) Find an optimal eulerization of the graph in (a).
    (c) Find an optimal route for the garbage truck. Describe the route by labeling the edges 1, 2, 3, . . . , in the order in which they are traveled.

38. Suppose that we want to find an optimal route for a mail carrier to deliver mail along the streets of the town. In this case the mail carrier must walk along each block twice (once for each side of the street) except for blocks facing one of the parks (here only one pass is needed).

(a) Draw an appropriate graph representing this problem.

(b) Find an optimal eulerization of the graph in (a).

(c) Describe an optimal route for the mail carrier (the starting and ending point is the post office labeled P.O.).

39. Find an optimal route for the garbage collection problem in the city of Cleansburg. (Use the optimal eulerization of the city streets shown in Fig. 5-27; use Fleury's algorithm if you want to.)

40. The following diagram is of a hypothetical city with a river running through the middle of the city. There are three islands and seven bridges as shown in the figure. It is possible to take a walk and cross each bridge exactly once? If so, show how; if not, explain why not.

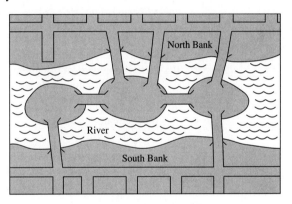

Exercises 41 through 43 refer to the problem of fixing up a graph so that it has an Euler path (with specified starting and ending vertices). This is done by semi-eulerizing the graph, i.e., leaving the starting and ending vertices of odd degree and making all the other vertices of odd degree even.

41. Suppose we want to trace the following graph with the least possible number of duplicate edges.

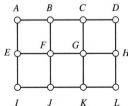

(a) Find an optimal semi-eulerization of the graph that starts at *E* and ends at *H*.

(b) Which edges of the graph will have to be retraced?

42. Suppose we want to trace the same graph as in Exercise 41 but now we want to start at *B* and end at *K*. Which edges of the graph will have to be retraced?

43. A policeman has to patrol along the streets of the subdivision represented by the following graph.

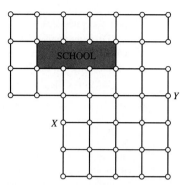

He wants to start his trip at the police station (located at *X*) and end the trip at his home (located at *Y*). He needs to cover each block of the subdivision at least once and at the same time he wants to duplicate the fewest possible number of blocks.

(a) How many blocks will he have to duplicate in an optimal trip through the subdivision?

(b) Describe an optimal trip through the subdivision. Label the edges 1, 2, 3, . . . , in the order the policeman would travel them.

44. (a) Give an example of a graph with 15 vertices and no multiple edges that has an Euler circuit.

(b) Give an example of a graph with 15 vertices and no multiple edges that has an Euler path but no Euler circuit.

(c) Give an example of a graph with 15 vertices and no multiple edges that has neither an Euler circuit nor an Euler path.

45. The following figure is the floorplan of an office complex.

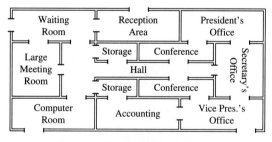

OFFICE COMPLEX

(a) Show that it is impossible to start outside the complex and walk through each door of the complex exactly once and end up outside.

(b) Show that it is possible to walk through every door of the complex exactly once (if you start and end at the right place).

(c) Show that by removing exactly one door it would be possible to start outside the complex, walk through each door of the complex exactly once, and end up outside.

Running

46. (a) Can a graph that has an Euler circuit have any bridges? If so, demonstrate it by showing an example. If not, explain why not.

   (b) Can a graph that has an Euler path have any bridges? If so, how many? Explain your answer.

47. Suppose G and H are two graphs that have no common vertices and such that each graph has an Euler circuit. Let J be a (single) graph consisting of the graphs G, H, and one additional edge joining one of the vertices of G to one of the vertices of H. Explain why the graph J has no Euler circuit but does have an Euler path.

48. Explain why in any graph in which the degree of each vertex is at least 2, there must be a circuit.

49. Suppose we have a graph with two or more vertices and without loops or multiple edges. Explain why the graph must have at least two vertices with the same degree.

50. (Open-ended question.) If we are given a list of N positive integers $d_1, d_2, \ldots, d_N$, is there always a graph having N vertices with exactly those degrees? You should consider separately the case when the sum of the integers $d_1, d_2, \ldots, d_N$ is odd (that one is easy) and the case when the sum of the integers is even (in this case the answer is yes, and your job is to describe how to construct the graph). In this second case, what happens if neither loops nor multiple edges are allowed? Explain and give examples to illustrate your explanations.

REFERENCES AND FURTHER READINGS

1. Beltrami, E., *Models for Public Systems Analysis.* New York: Academic Press, Inc., 1977.

2. Beltrami, E., and L. Bodin, "Networks and Vehicle Routing for Municipal Waste Collection," *Networks*, 4 (1973), 65–94.

3. Chartrand, Gary, *Graphs as Mathematical Models.* Belmont, CA: Wadsworth Publishing Co., Inc., 1977.

4. Euler, Leonhard, "The Königsberg Bridges," trans. James Newman, *Scientific American,* 189 (1953), 66–70.

5. Minieka, E., *Optimization Algorithms for Networks and Graphs.* New York: Marcel Dekker, Inc., 1978.

6. Newman, J., ed., *Mathematics—An Introduction to Its Spirit and Its Use.* New York: W. H. Freeman & Co., 1978.

7. Roberts, Fred S., "Graph Theory and Its Applications to Problems of Society," CBMS-NSF Monograph No. 29, Philadelphia: Society for Industrial and Applied Mathematics, 1978, chap. 8.

8. Tucker, A. C., "Perfect Graphs and an Application to Optimizing Municipal Services," *SIAM Review,* 15 (1973), 585–590.

9. Tucker, A. C., and L. Bodin, "A Model for Municipal Street-Sweeping Operations," in *Modules in Applied Mathematics,* Vol. 3, eds. W. Lucas, F. Roberts, and R. M. Thrall. New York: Springer-Verlag, 1983, 76–111.

CHAPTER 6

Hamilton Joins
the Circuit

The Traveling Salesman Problem

Two roads diverged in a wood, and I— / I took the one less traveled by / And that has made all the difference . . .

ROBERT FROST

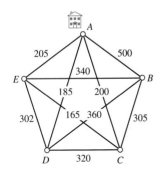

FIGURE 6-1

Sophie's Choices.   Sophie, a salesperson, must call on customers in five different cities (*A, B, C, D,* and *E*). Each edge of the graph shown in Fig. 6-1 shows the cost of travel (gas, insurance, wear and tear on the car, etc.)[1] between any two cities. The trip must start and end at Sophie's home town (*A*). What is the *cheapest* possible route Sophie can take (starting and ending at *A* and visiting each of the other cities exactly once)?

We will take up Sophie's problem (and variations thereof) in this chapter. Before we do so, we will introduce a little more graph terminology. Let's start with the observation that the graph in Fig. 6-1 is more than just a typical graph: In addition to vertices and edges, we have a number associated with each edge. Here the number represents the cost of traveling that particular edge—in another situation it could represent distance, time, the number of trees along the side of the road, or some other exotic variable. In general, when there are numbers associated with the edges of a graph, we call such numbers the **weights** of the edges[2] and we call such a graph a **weighted graph**. So Fig. 6-1 is a weighted graph, the weight of edge *AD* is 185, the weight of edge *BC* is 305, etc. Notice that there is no requirement for the length of the edges to be proportional to the weights. As with ordinary graphs, the lengths and shapes of the edges in a weighted graph are irrelevant.

When the weights of the edges represent variables other than cost (say, distance or time), then the wording of the problem can be suitably modified (find the shortest route or the fastest route). We will use the term **optimal route** generically to describe the cheapest, shortest, fastest, etc., whatever the case may be.

---

[1] In automobile travel, costs are generally taken to be directly proportional to distance (28 cents per mile, 21 cents per mile, etc.), but there are other variables (traffic, weather, road conditions, etc.) that can also affect costs.

[2] In scientific usage, the word "weight" is commonly used to generically represent variables such as cost, time, and distance—the kinds of things we usually want to minimize.

HAMILTON CIRCUITS

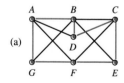

(a)

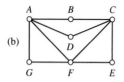

(b)

FIGURE 6-2

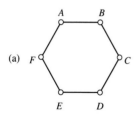

(a)

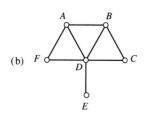

(b)

FIGURE 6-3

We are now in a position to describe a generic version of our original problem: Given a weighted graph, we want to find an optimal route that starts and ends at some specified vertex and passes through *every other vertex exactly once.*

A circuit that starts at a vertex of the graph, passes through every other vertex of the graph exactly once, and returns to the starting vertex is called a **Hamilton circuit.**[3]

The distinction between an Euler circuit and a Hamilton circuit may appear minor on the surface (just a word of difference—substitute "vertex" for "edge"). Mathematically speaking, however, there is a world of difference. Consider, for example, the two graphs in Fig. 6-2. The graph in Fig. 6-2(a) has no Euler circuits (lots of rotten vertices of odd degree to spoil our fun), but it has lots of Hamilton circuits (such as *A, B, D, C, E, F, G, A* and *A, D, C, E, B, G, F, A* and others that the reader can readily find—see Exercise 1). Contrast this situation with the one in Fig. 6-2(b) where we know there are Euler circuits (every vertex has even degree) and yet the graph has no Hamilton circuit (see Exercise 33).

While it would be nice to have a Hamilton's theorem (something like Euler's theorem for Hamilton circuits), there is unfortunately no such theorem. If we are given a graph and asked whether it has a Hamilton circuit or not, we have no sure-fire way to tell—unless of course we gut it out, try all possibilities, and either find one or give up. As a rule of thumb, the more "edge-rich" the graph, the more likely it is to have a Hamilton circuit. But that's just a rule of thumb. The graph in Fig. 6-3(a) has 6 vertices and 6 edges, and it has a Hamilton circuit (*A, B, C, D, E, F, A*). The graph in Fig. 6-3(b), on the other hand, has 6 vertices and 8 edges, but it doesn't have a Hamilton circuit.

There are situations in which it is possible to know for a fact that the graph has a Hamilton circuit. The following theorem exemplifies one such situation. We give it as a helpful fact, and for an explanation or proof the reader is encouraged to consult any standard book on graph theory.[4]

Dirac's Theorem.[5] Suppose that we have a connected graph with 3 or more vertices. If each vertex of the graph is adjacent to at least half of the remaining vertices, then the graph has a Hamilton circuit.

---

[3] Named after the great Irish mathematician and astronomer Sir William Rowan Hamilton (1805–1865). It is said that at the age of four Hamilton could read Latin, Greek, and Hebrew, as well as English. At the age of twenty-one he became a Professor of Astronomy at Trinity College in Dublin. Besides being a great scientist, Hamilton was an accomplished man of letters and a poet who counted Wordsworth and Coleridge among his closest friends.

[4] For example, references 2 and 8.

[5] This theorem was proved by the mathematician G. A. Dirac in 1952.

Example 1. The graph in Fig. 6-4 has 6 vertices each with degree 3. By Dirac's theorem it must have a Hamilton circuit. We leave it to the reader to find one (there are several!).

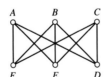

FIGURE 6-4

Example 2. G is a large graph (too large to draw). It is connected and has no loops or repeated edges. It has 20 vertices (V_1, V_2, V_3, . . . , V_{20}). The degrees of the vertices are $\deg(V_1) = 12$, $\deg(V_2) = 16$, $\deg(V_3) = 11$, $\deg(V_4) = 14$, $\deg(V_5) = 13$, $\deg(V_6) = 15$, $\deg(V_7) = 14$, $\deg(V_8) = 12$, $\deg(V_9) = 17$, $\deg(V_{10}) = 14$; all other vertices have degree 10. This graph has a Hamilton circuit (take it from Dirac—the man knew whereof he spoke!).

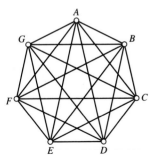

FIGURE 6-5

Example 3. (Complete Graphs).   The graph shown in Fig. 6-5 has 7 vertices, and every vertex is adjacent to every other vertex. A graph like this is called a **complete graph** (every pair of vertices is joined by exactly one edge). Fig. 6-6 shows complete graphs with 4, 5, and 6 vertices, respectively. Since in a complete graph each vertex is adjacent to all the others, if the graph has N vertices, each vertex has degree $N - 1$, and the total number of edges in the graph is $N(N - 1)/2$.

It is quite obvious that a complete graph has lots of Hamilton circuits. Let's consider, for example, the complete graph with 4 vertices shown in Fig. 6-6(a). It has the following Hamilton circuits:

1. *A, B, C, D, A*   **4.** *A, D, C, B, A*

2. *A, B, D, C, A*   **5.** *A, C, D, B, A*

3. *A, C, B, D, A*   **6.** *A, D, B, C, A*

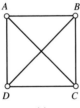 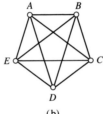 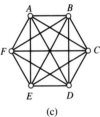

FIGURE 6-6   (a)   (b)   (c)

That's it! Notice that of the 6 Hamilton circuits listed above, the last 3 are repeats of the first 3 but read backward (circuit 4 is the same as circuit 1 but traveled backward, circuit 5 is circuit 2 traveled backward, and circuit 6 is circuit 3 traveled backward). Also note that there are many other ways to write down each of these circuits (*B, C, D, A, B* is the same circuit as *A, B, C, D, A* but read in a different order; so are *C, D, A, B, C* and *D, A, B, C, D*). ▤

The complete graph with 5 vertices (shown in Fig. 6-6[b]) has 24 Hamilton circuits. As before, of these 24, half are repeats of the other half but traveled backward. We leave it to the enterprising reader to write them down.

There is a convenient formula that gives the number of Hamilton circuits in any complete graph. It uses the factorial, a concept we first came across in Chapter 2. Recall that if N is any positive integer the number $N! = N \times (N-1) \times \cdots \times 3 \times 2 \times 1$ is called the **factorial** of N. The following table shows a few values of $N!$.

	Value
2!	$2 \times 1 = 2$
3!	$3 \times 2 \times 1 = 6$
4!	$4 \times 3 \times 2 \times 1 = 24$
5!	$5 \times 4 \times 3 \times 2 \times 1 = 120$
⋮	⋮
10!	$10 \times 9 \times 8 \times 7 \times 6 \times 5 \times 4 \times 3 \times 2 \times 1 = 3{,}628{,}800$

Here now is the formula for the number of Hamilton circuits in a complete graph:

> The complete graph with N vertices has $(N - 1)!$ Hamilton circuits. Of these, half are repeats of the other half but traveled backward.

THE TRAVELING SALESMAN PROBLEM

Our original problem for this chapter involved Sophie the traveling salesperson who wanted to find the cheapest trip that would allow her to visit each of the five cities shown in Fig. 6-1 once and return to the starting point (*A*). We can now rephrase the problem by saying we want to find an optimal Hamilton circuit in the complete weighted graph shown in Fig. 6-1.

The problem of finding an optimal Hamilton circuit in a complete weighted graph has many other important applications besides the example of a traveling salesperson. For reasons of tradition this type of problem has always been referred to as the **traveling salesman problem** regardless of the context in which it comes up. We will adhere to this tradition and refer to any problem in which we need to find an optimal Hamilton circuit in a complete weighted graph as a **TSP** (for traveling salesman problem). The following list gives a few examples of TSPs.

- ■ **Package Deliveries.** Companies such as United Parcel Service (UPS) and Federal Express deal with this situation daily. Each truck has packages to deliver to a list of destinations. The travel time between any two delivery locations is known or can be estimated (experienced drivers always know such things). The object is to deliver the packages to each of the delivery locations and return to the starting point in the least amount of time—clearly an example of a TSP. On a typical day, a UPS truck delivers packages to somewhere between 100 and 200 locations, so we are dealing with a TSP involving a graph with that many vertices.

- ■ **Fabricating Circuit Boards.** In the process of fabricating integrated circuit boards, tens of thousands of tiny holes must be drilled in each individual board. This is done by using a stationary laser beam and rotating the board around. Efficiency considerations require that the order in which the holes are drilled be such that the entire drilling sequence be completed in the least amount of time. This is an example of a TSP in which the vertices of the graph represent the holes on the circuit board and the weight of the edge connecting vertices X and Y represents the time needed to rotate the board from drilling position X to drilling position Y.

- ■ **Scheduling Jobs on a Machine.** In many industries there are machines that perform many different jobs. Think of the jobs as the vertices of the graph. After performing job X the machine needs to be set up to perform another job. The amount of time required to set up the machine to change from job X to job Y (or vice versa) is the weight of the edge connecting vertices X and Y. The problem is to schedule the machine to run through all the jobs in a cycle that requires the least total amount of time. This is another example of a TSP.

- ■ **Running Errands around Town.** If we have a lot of errands to do around town, organizing ourselves so that we can go from place to place following an optimal route and return home at the end of the day is an example of a TSP. For more details, the reader is encouraged to look at Exercise 29.

SOLVING THE TRAVELING SALESMAN PROBLEM

OK, so let's say that we have a real-life problem which when stripped down to its bare essence turns out to be a TSP. How do we solve it—in other words, how do we find an optimal Hamilton circuit? This problem turns out to be a lot more involved than one would expect. Let's start with the obvious approach: trial and error or, as it is more commonly known among scholars, the **brute force algorithm**.

Brute Force Algorithm for Solving TSPs

- List all possible Hamilton circuits for the weighted graph.

- For each Hamilton circuit, add up the weights of the edges in the circuit (this total is called the *weight of the circuit*).

- Of all the circuits, the one (or more) with the least weight is optimal and therefore a solution to the problem.

Example 4. Let's apply the brute force algorithm to solve Sophie's TSP which we introduced at the beginning of the chapter. For the reader's convenience we show the graph in Fig. 6-1 again (Fig. 6-7).

We already know that there are 24 possible Hamilton circuits and that 12 of them are duplicates of the other 12 but read backward (which allows us a shortcut in the computations). Since A is our starting point, we write down the Hamilton circuits as starting and ending at A. (If we had a different starting point, the circuits would still be exactly the same, but we would write them down differently. For example, if the starting point was D, then the circuit A, B, C, D, E, A would be written as D, E, A, B, C, D.) The computations (with the shortcut) are shown in Table 6-1.

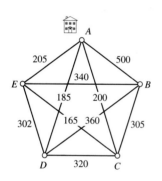

FIGURE 6-7

	Hamilton Circuit	Circuit Weight	Same Circuit Reversed
1	A, B, C, D, E, A	500 + 305 + 320 + 302 + 205 = 1632	A, E, D, C, B, A
2	A, B, C, E, D, A	500 + 305 + 165 + 302 + 185 = 1457	A, D, E, C, B, A
3	A, B, D, C, E, A	500 + 360 + 320 + 165 + 205 = 1550	A, E, C, D, B, A
4	A, B, D, E, C, A	500 + 360 + 302 + 165 + 200 = 1527	A, C, E, D, B, A
5	A, B, E, C, D, A	500 + 340 + 165 + 320 + 185 = 1510	A, D, C, E, B, A
6	A, B, E, D, C, A	500 + 340 + 302 + 320 + 200 = 1662	A, C, D, E, B, A
7	A, C, B, D, E, A	200 + 305 + 360 + 302 + 205 = 1372	A, E, D, B, C, A
8	A, C, B, E, D, A	200 + 305 + 340 + 302 + 185 = 1332	A, D, E, B, C, A
9	A, C, D, B, E, A	200 + 320 + 360 + 340 + 205 = 1425	A, E, B, D, C, A
10	A, C, E, B, D, A	200 + 165 + 340 + 360 + 185 = 1250	A, D, B, E, C, A
11	A, D, B, C, E, A	185 + 360 + 305 + 165 + 205 = 1220	A, E, C, B, D, A
12	A, D, C, B, E, A	185 + 320 + 305 + 340 + 205 = 1355	A, E, B, C, D, A

■ **TABLE 6-1**

If we now look down the circuit weight column, we see that circuit 11 (A, D, B, C, E, A) and its partner (A, E, C, B, D, A) have the smallest circuit weight. We have now found what we were after: The optimal route for Sophie is given by the Hamilton circuit A, D, B, C, E, A (or alternatively the same route reversed, A, E, C, B, D, A). The cost of the optimal route is $1220.   ▬

It is important to observe that when we talk about a solution to a TSP we mean a circuit and not a cost. In other words, the solution to Example 4 is not $1220. (It does poor Sophie little good to tell her that the cheapest possible route will cost her $1220 and not tell her what the route is.)

ANALYZING ALGORITHMS

So now we have a method of solution for any TSP—the brute force algorithm. It has only one minor flaw: It appears to be somewhat cumbersome. After all, it requires us to check the weight for each of the possible Hamilton circuits of the graph (or half of them if we use the shortcut). It is clear that this kind of checking can be done by a computer much better than by a human being because the brute force algorithm is essentially a mindless exercise in arithmetic with a little bookkeeping thrown in. Because computers are not only accurate but also extremely fast, we would expect that even if we throw a complicated graph at a computer, if the graph is within reason, the computer will quickly spit out an answer. (Let's arbitrarily decide that "within reason" means up to 1000 vertices in the weighted graph. For many applications that's not a lot.)

Table 6-2 (on the next page) shows some time calculations for solving TSPs using the brute force algorithm. The critical variable that affects the amount of time needed to find a solution is the number of vertices (N) in the graph. We show the computation time for two different imaginary computers. Computer 1 is a personal computer, and we will assume that it can run through 10,000 Hamilton circuits per second. Computer 2 is a large computer that can run through 1 million Hamilton circuits per second.

What Table 6-2 illustrates is an interesting phenomenon called the combinatorial explosion. Crudely put, this phenomenon says that any attempt to solve a problem such as a TSP by trial and error will sooner or later blow up in our faces. Even for a modest problem involving a weighted graph with 100 vertices, it would take computer 2 more than 10^{142} years to run through all possible Hamilton circuits. This is considerably longer than the age of the universe, currently estimated to be about 20 billion (i.e., 2×10^{10}) years. This also means that we cannot count on improved technology to bail us out. Let's say we use an imaginary computer that is 1 trillion times faster (that's 10^{12} times faster) than computer 2. It would take this technological marvel 10^{130} years (give or take an eon or two) to check all the Hamilton circuits in a graph with 100 vertices.

The upshot of all of this is that the brute force algorithm has very limited

Number of Vertices in Weighted Graph (N)	Number of Hamilton Circuits ($N-1$)!	Number of Circuits to Check Using Shortcut $\frac{1}{2}(N-1)!$	Time Needed by Computer 1 to Find a Solution	Time Needed by Computer 2 to Find a Solution
5	24	12	<1 second	<1 second
6	120	60	<1 second	<1 second
7	720	360	<1 second	<1 second
8	5,040	2,520	<1 second	<1 second
9	40,320	20,160	2 seconds	<1 second
10	362,880	181,440	18 seconds	<1 second
11	3,628,800	1,814,400	3 minutes	<2 seconds
12	39,916,800	19,958,400	33 minutes	<20 seconds
13	≈479 million	≈239.5 million	6 hr 39 min	≈4 minutes
14	≈6.2 billion	≈3.1 billion	≈86 hours	≈52 minutes
15	≈87 billion	≈43.6 billion	≈50 days	≈12 hr 6 min
16	≈1.3 trillion	≈654 billion	≈2 years	≈7 days 14 hr
17	≈20.9 trillion	≈10.5 trillion	≈33 years	≈121 days
18	≈356 trillion	≈178 trillion	≈564 years	≈5.64 years
19	≈6.4×10^{15}	≈3.2×10^{15}	≈10,100 years	≈101 years
20	≈1.2×10^{17}	≈6.1×10^{16}	≈193,000 years	≈1930 years
21	≈2.4×10^{18}	≈1.2×10^{18}	≈3.85 million years	≈38,500 years
⋮	⋮	⋮	⋮	⋮
100	≈10^{156}	≈5×10^{155}	≈1.5×10^{144} years	≈1.5×10^{142} years

■ TABLE 6-2

practical value. For a typical graph with more than just a handful of vertices, this algorithm is useless. This situation, of course, is not generally true for all algorithms. Fleury's algorithm, discussed in Chapter 5, is a perfectly useful algorithm. Even when the number of vertices and edges in the graph is large, the algorithm can produce a solution in a reasonable amount of time.

We will not go into a detailed discussion of what we mean by all these vague expressions such as "a reasonable amount of time" and "the number of vertices in the graph is large." This would lead us into an exciting but difficult area of mathematics and computer science that deals with measuring the performance of algorithms in a precise way.[6] For our purposes, suffice it to say that we will draw a simplistic but helpful distinction between algorithms: There are *good* algorithms (which we will call **efficient** algorithms), and there are *bad* algorithms (which we will call **inefficient** algorithms). We have already seen one of each type: Fleury's algorithm for finding an Euler circuit is an efficient algorithm; the brute force algorithm for solving a TSP is an inefficient algorithm.

The theoretical basis for the distinction between efficient and inefficient graph algorithms is the relationship between the size of the graph (number of

---

[6] An excellent introduction to this subject is given in references 4 and 7.

vertices and/or edges) and the number of steps needed to carry out the algorithm. When the number of steps needed to carry out the algorithm becomes disproportionately larger than the size of the graph, we have an inefficient algorithm. For example, in a complete weighted graph with 11 vertices, the brute force algorithm checks 10!/2 Hamilton circuits. If we double the number of vertices to 22, the brute force algorithm checks 21!/2 Hamilton circuits. The only problem is that 21!/2 is more than 100 billion times bigger than 10!/2 (Exercise 38). Thus, while the size of the problem doubled, the number of steps needed to solve it increased more than 100 billion times. This kind of explosion is characteristic of inefficient algorithms.

Approximate Algorithms

In a perfect world, our next step would be to present an algorithm that finds an optimal Hamilton circuit in any complete weighted graph and (unlike the brute force algorithm) is also an efficient algorithm. Unfortunately, no such algorithm is presently known. This can mean one of two things: (1) There is such an algorithm "out there," but no one has been clever enough to find it; or (2) such an algorithm is an impossibility (remember perfect voting methods?). So far, mathematicians have been unable to determine whether their inability to find an algorithm for solving TSPs that is both optimal and efficient is due to (1) or (2). In fact, there are reasons (that are too involved to explain here) that make resolving this dilemma one of the most important (and famous) unsolved problems in modern mathematics. Most experts are inclined to believe that an efficient algorithm that will give the optimal solution to any TSP is a mathematical impossibility. In fact, if such an algorithm were ever discovered, it would guarantee great fame and fortune to its discoverer. (For an excellent account of current progress on this famous unsolved mathematics problem see reference 5.)

In the meantime, we are faced with a quandary: In many industrial applications, it is important to solve a TSP for graphs that involve hundreds and even thousands of vertices, and to do so in real time (i.e., the boss wants an answer right away or, in the best of cases, by the end of the month). Since the brute force algorithm is out of the question and since no efficient algorithm that guarantees an optimal solution is known, the most reasonable strategy is one of partial retreat. We will give up on our requirement that the algorithm produce the very best solution and accept a solution that may not be optimal. In exchange, we ask for quick results. The name of the game nowadays in dealing with a large TSP, as well as with many other complex management science problems, is to find efficient *approximate algorithms.*

We will use the term **approximate algorithm** to describe any algorithm that produces solutions[7] that are, most of the time, reasonably close to the optimal solution. Sounds good, but what does *most of the time* mean? And how about *reasonably close*? Unfortunately, to properly answer these questions

---

[7] Note that in this context the word "solution" no longer means the "best answer," but simply "an answer."

would take us, once again, far beyond the scope of this book. We will have to accept the fact that in the area of analyzing algorithms we will be dealing with informal ideas rather than precise definitions.

Informally, there are two factors one needs to consider in evaluating an approximate algorithm: Its *worst case* performance and its *typical* performance. Very often these two things do not go hand in hand. What can we say about an approximate algorithm that is guaranteed to give solutions that are never off by more than 20% from the optimal solution. Would this be an acceptable approximate algorithm? How about a different algorithm that about 95% of the time gives solutions that are off by less than 2% but every once in a while gives solutions that can be off by 200% or more? Would it be an acceptable algorithm? Which is better, the former or the latter? (Who is a better child—one who is never an angel but never seriously misbehaves or one who is generally an angel but every once in a while throws a terrific temper tantrum?) These are all questions without clear-cut answers.

In the remaining sections of this chapter we will discuss two efficient approximate algorithms for solving TSPs: the *nearest neighbor algorithm* and the *cheapest link algorithm.*

THE NEAREST NEIGHBOR ALGORITHM

The **nearest neighbor algorithm** is almost completely described by its title. This is how the nearest neighbor algorithm would read (with embellishments) as a set of instructions to an imaginary traveling salesperson:

Nearest Neighbor Algorithm

- Start at home (make sure you've packed your toothbrush and wallet).

- Whenever you are in a city, pick the next city to visit from among the ones you haven't visited yet, and of all such cities pick the one that is closest to where you are (the nearest neighbor). In case of a tie, choose at random. Keep doing this until you've visited all the cities.

- From the last city go home.

Example 5. Let's apply the nearest neighbor algorithm to our original TSP. This is how it would go for Sophie: She starts at A (that's home). From A the cheapest trip is to D, so she goes there. From D the cheapest city to go to next (other than A) is E. From E the cheapest city to go to is C. From C the cheapest city to go to is B (she doesn't want to go back to E or A!). Once she is at B, she has visited all the cities, so the last leg of the trip is back to A. Thus, the nearest

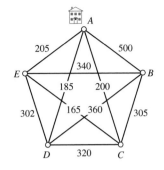

FIGURE 6-8

neighbor algorithm produces the Hamilton circuit *A, D, E, C, B, A* with a total cost of $1457.

If we compare this algorithm (and the answer it produced) with the brute force algorithm (and the answer it produced), we observe two things:

1. The Hamilton circuit produced by the nearest neighbor algorithm is more expensive than the optimal solution ($1457 versus $1220). (Given that we already know that the optimal solution costs $1220, we can calculate the **relative error:** $\frac{1457 - 1220}{1220} = \frac{237}{1220} \approx 0.194$, which tells us that the solution we found is off by about 19.4% from the optimal solution.)

2. As a reward for being willing to pay the penalty for a solution that is not optimal, we get an algorithm that is easily understood and quickly carried out.

To underscore this last point the reader is referred to Exercise 50 in which we need to route a circus on a tour through 21 cities. We know from the last column in Table 6-2 that even with the fastest computers the brute force algorithm would require thousands of years. With the nearest neighbor algorithm it takes the average student about 15 to 20 minutes to do it by hand and find an approximate solution.

Changing the Starting Vertex

There is a final twist to the nearest neighbor algorithm which we want to discuss next. It has to do with the choice of the starting vertex. Suppose that we choose *D* instead of *A* as the starting point in Example 5. We leave it as an exercise (Exercise 16) to verify that the solution produced by the nearest neighbor algorithm is *D, A, C, E, B, D* with a cost of $1250. This is a considerable improvement over the cost of our first solution based on the starting point *A*. If we choose *B* as the starting point for the nearest neighbor algorithm, we get a better solution yet: *B, C, E, A, D, B* with a cost of $1220 (Exercise 17).

The thoughtful reader might rightfully consider all of the above idle speculation—we can't choose *D* or *B* as a starting point because Sophie lives at *A* and must start her trip there. A different traveling salesperson living at *D* would (if using the nearest neighbor algorithm) be able to follow the better route *D, A, C, E, B, D*, and likewise a really lucky salesperson living at *B* could follow the route *B, C, E, A, D, B*, which at a cost of $1220 happens to be the cheapest possible. This gets Sophie thinking—Why can't I start at *A* and then follow one of these better routes? In other words, the circuit *D, A, C, E, B, D* can also be traveled in the sequence *A, C, E, B, D, A* (we just start at a different spot). Likewise, the circuit *B, C, E, A, D, B* can be traveled in the sequence *A, D, B, C, E, A*. A circuit doesn't change just because we start at a different spot!

The point of all of the above is that the nearest neighbor algorithm can be started at any vertex of the graph (not just the home vertex) and will produce a

Hamilton circuit. This same Hamilton circuit can then be rewritten with the home vertex as the starting point. This leads to an improved strategy which we will call the **repetitive nearest neighbor algorithm**.

Repetitive Nearest Neighbor Algorithm

- ■ **Step 1.** Pick an arbitrary vertex and apply the nearest neighbor algorithm with that vertex as the starting point.
- ■ **Step 2.** Repeat the process with each of the vertices of the graph.
- ■ **Step 3.** Of all the Hamilton circuits obtained in step 2, keep the best as a solution.
- ■ **Step 4.** Rewrite the solution obtained in step 3 using the home vertex as the starting point.

Let's illustrate the whole process with an example.

Example 6. It is the year 2020. A space probe launched from planet Earth is scheduled to go to each of the planetary moons shown in Fig. 6-9 to collect mineral samples and return to Earth with all the loot. The weighted graph shows the time (in years) needed to travel between any two of these planetary bodies.

We want to use the repetitive nearest neighbor algorithm to find a Hamilton circuit in the weighted graph shown in Fig. 6-9. For each possible starting vertex Table 6-3 shows the circuit obtained by using the nearest neighbor algorithm and the total weight of the circuit. We leave it to the reader to verify the details.

Of the circuits in Table 6-3, circuits 2 and 6 (which are the same circuit with different starting points) give us the best solution, with a total length of 12.3 years. Of course, we know that we can't start the trip at Phobos or Ganymede—the actual trip described by the solution is *E, H, T, P, C, G, E.* By

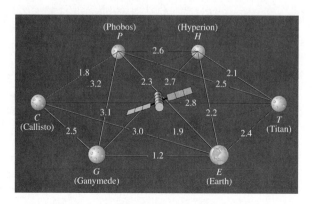

FIGURE 6-9

Circuit No.	Starting Point	Circuit Obtained by Nearest Neighbor Algorithm	Length of Trip (years)
1	Earth (*E*)	*E, G, T, H, P, C, E*	12.6
2	Phobos (*P*)	*P, C, G, E, H, T, P*	12.3
3	Hyperion (*H*)	*H, T, G, E, P, C, H*	12.5
4	Titan (*T*)	*T, G, E, H, P, C, T*	12.5
5	Callisto (*C*)	*C, P, E, G, T, H, C*	12.5
6	Ganymede (*G*)	*G, E, H, T, P, C, G*	12.3

◼ TABLE 6-3

sheer luck, this time the repetitive nearest neighbor algorithm gave us the optimal solution, although we can't except this to happen in general.

THE CHEAPEST LINK ALGORITHM

Our experience with the repetitive nearest neighbor algorithm has taught us that the order in which we build a Hamilton circuit and the order in which we actually travel it do not have to be the same. As a matter of fact, we can build a Hamilton circuit one edge at a time without even asking that the edges be consecutive. If we want to, we can first grab an edge here, then another edge over there, then another edge way yonder. This may look pretty disorganized at first, but there is no cause for concern as long as all the pieces come together at the end to form a Hamilton circuit. We often use this type of strategy in putting together a large jigsaw puzzle.

Our next algorithm, called the **cheapest link algorithm**, is based on this strategy. We look over the graph and grab the cheapest edge we can find, wherever it may be. We look again and grab the next cheapest. We keep doing this, each time grabbing the cheapest edge available but being careful that we don't close a circuit until the very end and that we don't have three edges coming out of the same vertex (either one of those two moves would make it impossible to put together a Hamilton circuit at the end).

Example 7. Before we give a more formal description of the cheapest link algorithm, let's apply it to Sophie's TSP.

The cheapest possible edge we can find anywhere in the graph in Fig. 6-10 is *EC* at \$165. We'll grab it! (For convenient bookkeeping let's indicate this by making it in red.) The next cheapest edge is *DA* at \$185. We'll take that one too! The next cheapest edge is *AC* with weight \$200. This one is also OK, so we mark it in red. So far, so good! Figure 6-11 shows how things stand at this point.

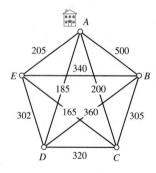

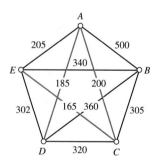

FIGURE 6-10   **FIGURE 6-11**

The next cheapest edge is *EA* at $205, but we can't use it because it would close the little circuit *E, A, C, E* and we can't have that! To remind ourselves that we have checked *EA* and can't use it, we can either erase it or cross it out. After doing that, the next cheapest edge available is *ED*, but we can't use it either because it creates the circuit *E, D, A, C, E*. We erase *ED* and try *BC*—the next cheapest available edge at $305. If we were to choose *BC* and mark it in red, we would have three different edges coming out of vertex *C* that would be part of our Hamilton circuit, so *BC* is also out. For exactly the same reason, we must also rule out the next cheapest edge, *DC*. The next possible choice is edge *EB*, and this one is OK so we mark it in red. Our last choice is edge *BD* which closes the Hamilton circuit. We can now read off our solution using any starting vertex we want. In this case our Hamilton circuit is *A, D, B, E, C, A* (or *A, C, E, B, D, A*) at a cost of $1250.

A formal description of the cheapest link algorithm can be given as follows:

Cheapest Link Algorithm

- ■ **Step 1.** Pick the edge with the smallest weight first (in case of a tie pick one at random). Mark the edge.
- ■ **Step 2.** Pick the next cheapest unmarked edge and mark it unless
    (a) it closes a smaller circuit or
    (b) it results in three marked edges coming out of a single vertex. If there are ties, break them arbitrarily.
- ■ **Steps 3, 4, etc.** Repeat step 2 until the Hamilton circuit is complete.

Example 8. If we use the cheapest link algorithm to find a Hamilton circuit for the space probe in Example 6 (as shown in Fig. 6-12), then the sequence of steps goes as follows:

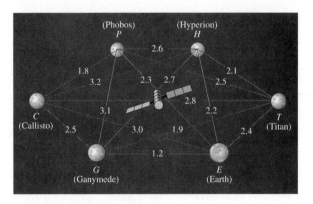

Figure 6-12

- **Step 1.** We pick (and mark) the edge *GE* (1.2 years). It is the cheapest edge in the graph.
- **Step 2.** Choose edge *CP* (1.8 years).
- **Step 3.** Choose edge *GT* (1.9 years).
- **Step 4.** Choose edge *HT* (2.1 years).
- **Step 5.** The next cheapest edge is *EH* (2.2 years), but it can't be used (it closes a circuit). We next choose *EP* (2.3 years).
- **Step 6.** There is only one way to close up the Hamilton circuit, and that's with edge *CH* (3.2 years).

The Hamilton circuit we get is *E, P, C, H, T, G, E* with total weight 12.5 years.

CONCLUSION

The nearest neighbor and cheapest link algorithms are two fairly simplistic approaches to the notoriously difficult problem of finding an optimal Hamilton circuit in a complete weighted graph (commonly known by the generic name of the *traveling salesman problem*). Both of them are based on what is aptly described as a *strategy of greed*: Go for whatever is cheapest at the moment without worrying about the long-term consequences. (Algorithms based on this kind of strategy are commonly referred to as *greedy algorithms*.) We are often reminded by well-meaning people that greed doesn't pay in the long run, and indeed it is possible to concoct examples in which the nearest neighbor or the cheapest link algorithm leads to just about the worst possible choice for a Hamilton circuit (see Exercises 43 and 44). What the nearest neighbor and cheapest link algorithms do have going for them is that they are easy to understand, can be carried out very efficiently, and in most typical problems give an approximate solution that is within a reasonable margin of error.

Many other more sophisticated algorithms for finding approximate solutions to the traveling salesman problem are known, and in most cases (unlike the nearest neighbor and cheapest link algorithms) they come with a performance guarantee (i.e., the approximate solution is guaranteed never to be off by more

How Big a TSP	Optimal Solution?	Approximate Solution within 3.5% of Optimal	Approximate Solution within 1% of Optimal	Approximate Solution within 0.75% of Optimal
About 3000 vertices	YES Computer time: Less than 1 week.	Computer time: A few seconds.	Computer time: A few minutes.	Computer time: Less than 1 hour.
About 100,000 vertices	NO	Computer time: A few minutes.	Computer time: Less than 2 days.	Computer time: About 7 months.
About 1 million vertices	NO	Computer time: Less than 3 hours.	Computer time: Hundreds of years.	Computer time: Millions of years.

TABLE 6-4:   A Sampler of the Current State of Affairs in Solving TSPs (All times based on the use of one super computer or several hundred small computers working in parallel.)

than a certain percent). One such algorithm is given in Exercise 49. At present, the best algorithms produce approximate solutions that are off by no more than 1% of the optimal solution for TSPs involving about 100,000 vertices and off by not more than 3.5% of the optimal solution for TSPs involving about 1 million vertices (see Table 6-4). Constant refinements of these algorithms and improvements in technology guarantee that such levels of performance are going to get even better.

In the meantime, the fundamental question remains unsolved: Is there an efficient algorithm for the traveling salesman problem guaranteed to always produce the optimal route or is such an algorithm a mathematical impossibility? This problem is still waiting for the next Euler to come along.

KEY CONCEPTS

approximate algorithms
brute force algorithm
cheapest link algorithm
complete graph
Dirac's theorem
efficient algorithms
factorial
Hamilton circuit
inefficient algorithms

nearest neighbor algorithm
optimal route
relative error
repetitive nearest neighbor
   algorithm
traveling salesman problem (TSP)
weighted graph
weights

EXERCISES

Walking

1. Two of the Hamilton circuits in the following graph are *A, B, D, C, E, F, G, A* and *A, D, C, E, B, G, F, A*. Find two more (not these circuits reversed). Use *A* as the starting point.

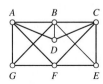

2. Find two Hamilton circuits in the following graph.

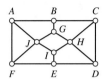

3. List all possible Hamilton circuits in the following graph.

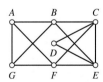

4. List all possible Hamilton circuits in the following graph.

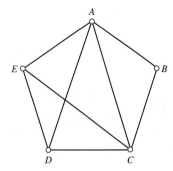

5. List all possible Hamilton circuits in the following graph starting
   (a) at vertex A
   (b) at vertex D.

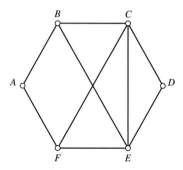

6. List all possible Hamilton circuits in the following graph starting
   (a) at vertex A
   (b) at vertex C.

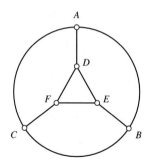

7.

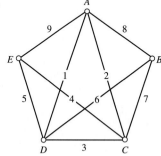

   (a) In the weighted graph above, find the weight of edge BD.
   (b) In the weighted graph above, find the weight of edge EC.
   (c) Find a Hamilton circuit in the graph and give its weight.
   (d) Find a different Hamilton circuit in the graph and give its weight.

8.

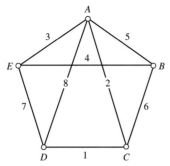

    (a) In the weighted graph above, find the weight of edge *AD*.
    (b) In the weighted graph above, find the weight of edge *AC*.
    (c) Find a Hamilton circuit in the graph and give its weight.
    (d) Find a different Hamilton circuit in the graph and give its weight.

9.

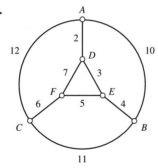

    (a) In the weighted graph above, find the weight of edge *BC*.
    (b) Find a Hamilton circuit in the graph and give its weight.
    (c) Find a different Hamilton circuit in the graph and give its weight.

10.

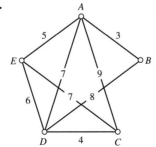

    (a) In the weighted graph above, find the weight of edge *AC*.
    (b) Find a Hamilton circuit in the graph and give its weight.
    (c) Find a different Hamilton circuit in the graph and give its weight.

11. (a) $5! = 120$. Use this fact to painlessly find $6!$ (No calculators please.)
    (b) Given that $10! = 3,628,800$, find $9!$ (No calculators please.)
    (c) How many different Hamilton circuits are there in a complete graph with 10 vertices?

12. (a) $7! = 5040$. Use this fact to painlessly find $8!$ (No calculators please.)
  (b) How many different Hamilton circuits are there in a complete graph with 9 vertices?

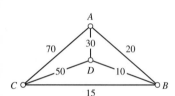

13. Consider the weighted graph shown in the margin.
  (a) Use the brute force algorithm to find an optimal Hamilton circuit.
  (b) Use the nearest neighbor algorithm with starting vertex A to find a Hamilton circuit.
  (c) Use the cheapest link algorithm to find a Hamilton circuit.
  (d) Compare the optimal solution obtained in (a) with the Hamilton circuits obtained in (b) and (c). Give the relative error in each of these solutions.

$$\text{Relative error} = \frac{(\text{cost of approximate solution}) - (\text{cost of optimal solution})}{\text{cost of optimal solution}}$$

14. Consider the weighted graph shown in the margin.
  (a) Use the brute force algorithm to find an optimal Hamilton circuit.
  (b) Use the nearest neighbor algorithm with starting vertex A to find a Hamilton circuit.
  (c) Use the cheapest link algorithm to find a Hamilton circuit.
  (d) Compare the optimal solution obtained in (a) with the Hamilton circuits obtained in (b) and (c). Give the relative error in each of these solutions.

$$\text{Relative error} = \frac{(\text{cost of approximate solution}) - (\text{cost of optimal solution})}{\text{cost of optimal solution}}$$

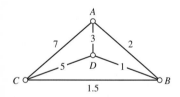

15. After a party at her house, Arlene has agreed to drive home Betty, Courtney, and Danette. The times (in minutes) to drive between her friends' homes are given in the graph shown in the margin.
  (a) Use the nearest neighbor algorithm with starting vertex A to find a route for Arlene to take.
  (b) Use the cheapest link algorithm to find a route for Arlene to take.
  (c) Find the route which gets Arlene home the quickest.

Exercises 16 and 17 refer to the following weighted graph.

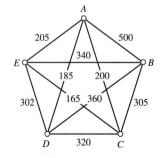

16. Apply the nearest neighbor algorithm with starting vertex D to find a Hamilton circuit in the graph.

17. Apply the nearest neighbor algorithm with starting vertex B to find a Hamilton circuit in the graph.

18. (a) Use the cheapest link algorithm to find a Hamilton circuit in the following weighted graph.

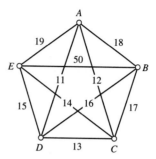

(b) Is there a cheaper Hamilton circuit? If so, find one. If not, explain why.

19. (a) Use the nearest neighbor algorithm to find a Hamilton circuit in the following weighted graph.

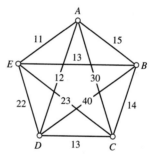

(b) Is there a cheaper Hamilton circuit? If so, find one. If not, explain why.

Exercises 20 and 21 refer to the following weighted graph.

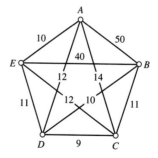

20. Use the repetitive nearest neighbor algorithm to find a Hamilton circuit in the graph.

21. (a) Use the cheapest link algorithm to find a Hamilton circuit in the graph.
(b) Is there a cheaper Hamilton circuit? If so, find one. If not, explain why.

Exercises 22 and 23 refer to the following weighted graph.

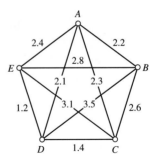

22. Use the cheapest link algorithm to find a Hamilton circuit in the graph.

23. Use the repetitive nearest neighbor algorithm to find a Hamilton circuit in the graph.

Exercises 24 through 26 refer to the following weighted graph.

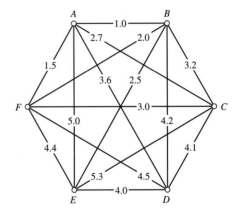

24. Use the cheapest link algorithm to find a Hamilton circuit.

25. Use the nearest neighbor algorithm starting at vertex *A* to find a Hamilton circuit.

26. Use the repetitive nearest neighbor algorithm to find a Hamilton circuit.

Exercises 27 and 28 refer to the following weighted graph.

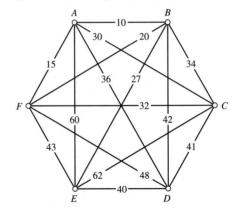

27. Use the cheapest link algorithm to find a Hamilton circuit.

28. Use the repetitive nearest neighbor algorithm to find a Hamilton circuit.

29. You have a busy day ahead of you. You must run the following errands (in no particular order): go to the post office, deposit a check at the bank, pick up some French bread at the deli, visit a friend at the hospital, and get a haircut at Karl's Beauty Salon. You must start and end at home. Each block on the map is exactly 1 mile.

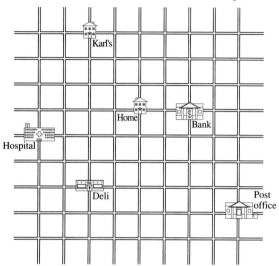

(a) Draw a weighted graph corresponding to this problem.

(b) Find the optimal (shortest) way to run all the errands. (Use any algorithm you think is appropriate.)

30. Rosa's Floral must deliver flowers to each of the five locations *A, B, C, D,* and *E* shown on the map. The trip must start and end at the flower shop, which is located at *X*. Each block on the map is exactly 1 mile.

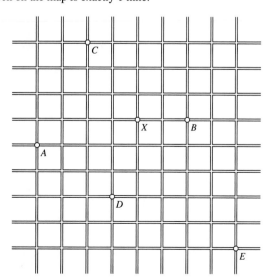

(a) Draw a weighted graph corresponding to this problem.

(b) Find the optimal (shortest) way to make all the deliveries. (Use any algorithm you think is appropriate.)

Jogging

31. Hamilton's puzzle.   Find a Hamilton circuit in the following graph. Indicate your solution by marking the circuit right on the graph.

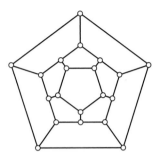

Historical footnote.   Hamilton made up and marketed a game which was a three-dimensional version of this exercise using a regular dodecahedron (see figure) in which the vertices were different cities. The purpose of the game was to find a "trip around the world" going from city to city along the edges of the dodecahedron without going back to any city (except for the return to the starting point). When a regular dodecahedron is flattened, we get the graph shown in this exercise.

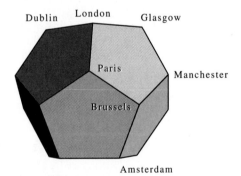

32. Find a Hamilton circuit in the following graph.

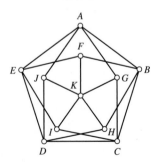

33. Explain why the following graph has no Hamilton circuit.

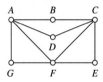

*Exercises 34 and 35 involve the concept of a **Hamilton path,** a path that passes through every vertex of the graph exactly once without necessarily returning to the starting vertex.*

34. Petersen graph. The following graph is called the Petersen graph. Find a Hamilton path in the Petersen graph.

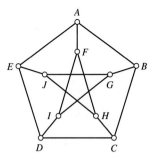

35. Find a Hamilton path in the following graph.

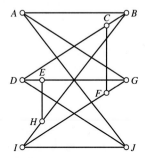

36. (a) Give an example of a graph with four vertices in which the same circuit can be both an Euler circuit and a Hamilton circuit.

(b) Give an example of a graph with N vertices in which the same circuit can be both a Euler circuit and a Hamilton circuit.

37. (a) Compare 2^3 with 3! Which is bigger?

(b) Compare 2^4 with 4! Which is bigger?

(c) Suppose N is more than 5. If you had a choice between 2^N or $N!$ dollars, which would you choose? Explain why the one you picked is the bigger number.

38. Explain why 21! is more than 100 billion times bigger than 10! (i.e., show that $21! > 10^{11} \times 10!$).

Exercises 39 and 40 refer to the following situation: A traveling salesperson's territory consists of the 11 cities shown on the following mileage chart. The salesperson must organize a round trip that starts and ends in Dallas (that's home) and will pass through each of the other 10 cities exactly once.

Mileage Chart

	Atlanta	Boston	Buffalo	Chicago	Columbus	Dallas	Denver	Houston	Kansas City	Louisville	Memphis
Atlanta	*	1037	859	674	533	795	1398	789	798	382	371
Boston	1037	*	446	963	735	1748	1949	1804	1391	941	1293
Buffalo	859	446	*	522	326	1346	1508	1460	966	532	899
Chicago	674	963	522	*	308	917	996	1067	499	292	530
Columbus	533	735	326	308	*	1028	1229	1137	656	209	576
Dallas	795	1748	1346	917	1028	*	781	243	489	819	452
Denver	1398	1949	1508	996	1229	781	*	1019	600	1120	1040
Houston	789	1804	1460	1067	1137	243	1019	*	710	928	561
Kansas City	798	1391	966	499	656	489	600	710	*	520	451
Louisville	382	941	532	292	209	819	1120	928	520	*	367
Memphis	371	1293	899	530	576	452	1040	561	451	367	*

39. Use the nearest neighbor algorithm with starting city Dallas to find a Hamilton circuit for the traveling salesperson. Try to apply the algorithm using the data directly from the chart.

40. Use the cheapest link algorithm to find a Hamilton circuit for the traveling salesperson. Try to apply the algorithm using the data directly from the chart.

Running

41. Complete bipartite graphs.   A complete bipartite graph is one in which the vertices can be divided into two sets A and B and each vertex in set A is adjacent to each of the vertices in set B. There are no other edges! (If there are m vertices in set A and n vertices in set B, the complete bipartite graph is written as $K_{m,n}$.)

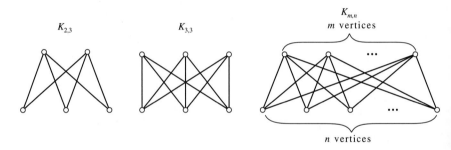

(a) Explain why when $m \neq n$, $K_{m,n}$ cannot have a Hamilton circuit.

(b) Explain why for $n > 1$, $K_{n,n}$ always has a Hamilton circuit.

42. Explain why the Petersen graph below (also see Exercise 34) does not have a Hamilton circuit.

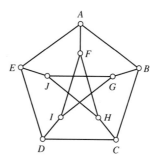

43. Make up an example of a complete weighted graph such that the Hamilton circuit produced by the nearest neighbor algorithm (you may pick the starting vertex) gives the worst possible choice of a circuit (in other words, one whose weight is bigger than any other).

44. Make up an example of a complete weighted graph such that the Hamilton circuit produced by the cheapest link algorithm gives the worst possible choice of a circuit.

45. Make up an example of a complete weighted graph such that the Hamilton circuit produced by the nearest neighbor algorithm (you can pick the starting vertex) has a relative percentage error of at least 100% (in other words, the weight of the Hamilton circuit produced by the nearest neighbor algorithm is at least twice as much as the weight of the optimal Hamilton circuit).

46. Make up an example of a complete weighted graph such that the Hamilton circuit produced by the cheapest link algorithm has a relative percentage error of at least 100% (in other words, the weight of the Hamilton circuit produced by the cheapest link algorithm is at least twice as great as the weight of the optimal Hamilton circuit).

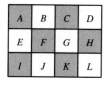

47. **The knight's tour.** A knight is on the upper left-hand corner of a 3 by 4 "chessboard" as shown in the figure.

(a) Draw a graph with the vertices representing the squares on the board and the edges representing the allowable chess moves of the knight (e.g., an edge joining vertices A and J means that the knight is allowed to move from square A to square J or vice versa in a single move).

(b) Find a Hamilton path starting at vertex A in the graph drawn in (a) and thus show how to move the knight so that it visits each square of the board exactly once starting at square A.

(c) Show that the graph drawn in (a) does not have a Hamilton circuit and consequently that it is impossible for the knight to move so that it visits each square of the board exactly once and then returns to its starting point.

48. Using the ideas of Exercise 47, show that it is possible for a knight to visit each square of a 6 by 6 "chessboard" exactly once and return to its starting point, and that this is true regardless of which square is used as the knight's starting point.

49. The nearest insertion algorithm. Here is a description of a different approximate algorithm for the traveling salesman problem. The basic idea is to start with a small subcircuit of the graph and enlarge it one vertex at a time until all the vertices are included and it is a full-fledged Hamilton circuit.

■ **Step 1.** Pick any vertex as a starting circuit (consisting of one vertex and zero edges). Mark it "red" ("red" is just a figure of speech for "any color.")

■ **Next step.** Suppose that at step k we have already built a red subcircuit with k vertices (call it C_k). We look for a black vertex in the graph that is as close as possible to some vertex of C_k. Let's call this black vertex B, and the vertex of C_k it is nearest to, R. We now create a new red circuit C_{k+1} which is the same as C_k except that B is inserted immediately after R in the sequence. Repeat until you have a Hamilton circuit.

(a) Verify that when the nearest insertion algorithm is applied to the original traveling salesman problem in this chapter (Sophie's) the following sequence of circuits is produced (we use A as our starting vertex):

■ C_1: A

■ C_2: A, D, A (D is the nearest vertex to C_1)

■ C_3: A, C, D, A (C is the nearest to A in C_2)

■ C_4: A, C, E, D, A (E is the nearest to C in C_3)

■ C_5: A, C, B, E, D, A (B is the nearest to C in C_4).

(b) Use the nearest insertion algorithm to find a Hamilton circuit for the graph in Exercise 22. Use A as the starting vertex.

(c) Use the nearest insertion algorithm to find a Hamilton circuit for the graph in Exercise 22. Use B as the starting vertex.

(d) Use the nearest insertion algorithm to find a Hamilton circuit for the graph in Exercise 22. Use C as the starting vertex.

50. (Open-ended question). The Great Kaliningrad Circus has been signed for an extended tour in the United States. The tour is scheduled to start and end in Miami, Florida, and visit 20 other cities in between. The cities and distances between the cities are shown in the accompanying mileage chart. The cost of transporting an entire circus the size of the Great Kaliningrad can be estimated to be about $1000 per mile, so finding a "good" Hamilton circuit for the 21 cities is clearly an important part of the organization of the tour. Your job is to do the best you can to come up with a reasonably good tour for the circus. You should not only describe the actual tour, but also explain what strategies you used to come up with it and why you think that your answer is a reasonable one. The tools at your disposal are everything you learned in this chapter (including Exercise 49), a limited amount of time, and your own ingenuity.

Mileage Chart

	Atlanta	Boston	Buffalo	Chicago	Columbus	Dallas	Denver	Houston	Kansas City	Louisville	Memphis	Miami	Minneapolis	Nashville	New York	Omaha	Pierre	Pittsburgh	Raleigh	St. Louis	Tulsa
Atlanta	*	1037	859	674	533	795	1398	789	798	382	371	655	1068	242	841	986	1361	687	372	541	772
Boston	1037	*	446	963	735	1748	1949	1804	1391	941	1293	1504	1368	1088	206	1412	1726	561	685	1141	1537
Buffalo	859	446	*	522	326	1346	1508	1460	966	532	899	1409	927	700	372	971	1285	216	605	716	1112
Chicago	674	963	522	*	308	917	996	1067	499	292	530	1329	405	446	802	459	763	452	784	289	683
Columbus	533	735	326	308	*	1028	1229	1137	656	209	576	1160	713	377	542	750	1071	182	491	406	802
Dallas	795	1748	1346	917	1028	*	781	243	489	819	452	1300	936	660	1552	644	943	1204	1166	630	257
Denver	1398	1949	1508	996	1229	781	*	1019	600	1120	1040	2037	841	1156	1771	537	518	1411	1661	857	681
Houston	789	1804	1460	1067	1137	243	1019	*	710	928	561	1190	1157	769	1608	865	1186	1313	1160	779	478
Kansas City	798	1391	966	499	656	489	600	710	*	520	451	1448	447	556	1198	201	592	838	1061	257	248
Louisville	382	941	532	292	209	819	1120	928	520	*	367	1037	697	168	748	687	1055	388	541	263	659
Memphis	371	1293	899	530	576	452	1040	561	451	367	*	997	826	208	1100	652	1043	752	728	285	401
Miami	655	1504	1409	1329	1160	1300	2037	1190	1448	1037	997	*	1723	897	1308	1641	2016	1200	819	1196	1398
Minneapolis	1068	1368	927	405	713	936	841	1157	447	697	826	1723	*	826	1207	357	394	857	1189	552	695
Nashville	242	1088	700	446	377	660	1156	769	556	168	208	897	826	*	892	744	1119	553	521	299	609
New York	841	206	372	802	542	1552	1771	1608	1198	748	1100	1308	1207	892	*	1251	1565	368	489	948	1344
Omaha	986	1412	971	459	750	644	537	865	201	687	652	1641	357	744	1251	*	391	895	1214	449	387
Pierre	1361	1726	1285	763	1071	943	518	1186	592	1055	1043	2016	394	1119	1565	391	*	1215	1547	824	760
Pittsburgh	687	561	216	452	182	1204	1411	1313	838	388	752	1200	857	553	368	895	1215	*	445	588	984
Raleigh	372	685	605	784	491	1166	1661	1160	1061	541	728	819	1189	521	489	1214	1547	445	*	804	1129
St. Louis	541	1141	716	289	406	630	857	779	257	263	285	1196	552	299	948	449	824	588	804	*	396
Tulsa	772	1537	1112	683	802	257	681	478	248	659	401	1398	695	609	1344	387	760	984	1129	396	*

**REFERENCES
AND FURTHER
READINGS**

1. Bellman, R., K. L. Cooke, and J. A. Lockett, *Algorithms, Graphs and Computers.* New York: Academic Press, Inc., 1970, chap. 8.

2. Chartrand, Gary, *Graphs as Mathematical Models.* Belmont, CA: Wadsworth, Publishing Co., Inc., 1977, chap. 3.

3. Knuth, Donald, "Mathematics and Computer Science: Coping with Finiteness," *Science,* 194 (December 1976), 1235–1242.

4. Kolata, Gina, "Analysis of Algorithms: Coping With Hard Problems," *Science,* 186 (November 1974), 520–521.

5. Kolata, Gina, "Math Problem, Long Baffling, Slowly Yields," *The New York Times,* March 12, 1991, B8–B10.

6. Lawler, E. L., J. K. Lenstra, A. H. G. Rinooy Kan, and D. B. Shmoys, *The Traveling Salesman Problem.* New York: John Wiley & Sons, Inc., 1985.

7. Lewis, H. R., and C. H. Papadimitriou, "The Efficiency of Algorithms," *Scientific American,* 238 (January 1978), 96–109.

8. Peterson, Ivars, *Islands of Truth.* New York: W. H. Freeman & Co., 1990, chap. 6.

9. Wilson, Robin, and John J. Watkins, *Graphs: An Introductory Approach.* New York: John Wiley & Sons, Inc., 1990.

10. Zimmer, Carl, "And one for the road," *Discover* (January 1993), 91, 92.

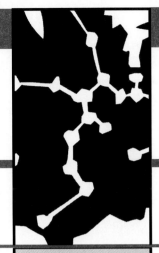

CHAPTER 7

Spanning Trees and Steiner Trees

Minimum Network Problems

I think that I shall never see / A billboard lovely as a tree / Indeed, unless the billboards fall / I'll never see a tree at all

OGDEN NASH

The Amazon Telephone Network Problem. Imagine this: After your big score as an efficiency expert with the Cleansburg Sanitation Department (see Chapter 5), you are taking a fabulously well paying job as a consultant to the Amazon Telephone and Telegraph Company (AT&T). One of the biggest challenges facing AT&T is the need to bring telephone service to some very exotic and remote places—small towns and villages buried deep in the Amazon jungle. For starters, AT&T needs your help with the following problem: The seven small villages shown in Fig. 7-1 have to be connected into a telephone network, with the connections made using underground fiber optic telephone lines. The graph in Fig. 7-2 shows the existing roads that connect these towns, with

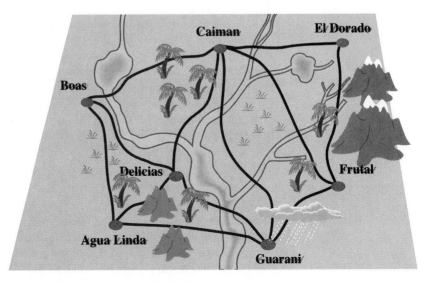

FIGURE 7-1 Reach out and tell someone? To the residents of these seven remote Amazonian villages, it's going to happen soon, thanks to the new telephone network about to be installed by AT&T (with your help).

217

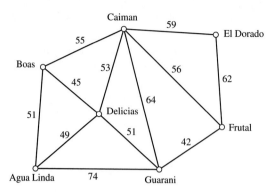

FIGURE 7-2 The cost (in millions of dollars) for each possible connection in the telephone network.

the weights of the edges showing the cost (in millions of dollars) of laying down the telephone lines along each road (laying down telephone lines anywhere else other than along an existing road would be prohibitively expensive in the jungle). AT&T wants to link all the cities so that a call can be made from any city to any other city either directly or by routing the call through one or more intermediate cities. At the same time, the company wants to spend as little money as possible. What is the *cheapest* way to link up the cities?

Problems just like the Amazon telephone network problem occur in many types of applications besides the construction of telephone networks: laying down tracks for a high-speed rail system, designing the layout of a computer chip, linking up a network of computers, designing an irrigation system, etc. The common thread in all these problems is the need to connect several locations (the vertices of a weighted graph) in such a way that it is always possible to get from any location to any other location and so that the total weight of the network is *minimal*. In connecting the various locations, no consideration is given to how roundabout or inconvenient the connections may be. The single-minded objective in determining the solution is to make the total weight of the network as small as possible. Problems of this type are known as **minimum network problems,** and in this chapter we will discuss two basic variations on this theme.

TREES

Let's start by asking (without getting into specifics), What should we expect the solution of the Amazon telephone network problem to be like?

First, we note that the solution to the problem is itself a graph that *lives inside* the original graph in Fig. 7-2. After all, Fig. 7-2 shows each of the potential connections between cities (the edges of the graph) and in finding the solution our job will be to select the right set of connections from among these. In the language of graph theory we say that the solution is a **subgraph** of the original graph. Moreover, since we are trying to build a telephone network that reaches each of the cities on the original graph, it is clear that the vertices of the solution subgraph must include each vertex of the original graph. In addition, the solution subgraph should have the following two characteristics:

1. It should be *connected*. This is an obvious consequence of the fact that our network must link every city with every other city.

2. It should *not contain any circuits* (i.e., nowhere in the solution subgraph should we be able to start at some vertex, travel along some edges, and return to the starting vertex). The reason for this is that in looking for an optimal (cheapest) subgraph we should never have more edges than are absolutely necessary to connect the various cities. Suppose, for example, that the solution subgraph contains a circuit like the one in Fig. 7-3. It is clear that in this case we could delete any one of the edges of the circuit and still keep the telephone calls going through the network.

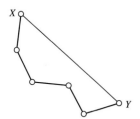

FIGURE 7-3 Laying telephone lines along the segment *XY* is a waste of money.

Any graph that is connected and has no circuits is called a **tree**. It is clear from our preceding discussion that the solution to the Amazon telephone network problem must be a tree.

Example 1. Figure 7-4 shows several examples of trees.

(a)

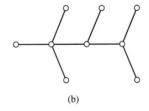

(b)

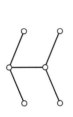

(c)

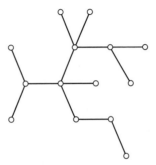

(d)

FIGURE 7-4

Example 2. Figure 7-5 shows examples of graphs that are *not* trees. Figures 7-5(a) and (b) have circuits; Fig. 7-5(c) has no circuits but is not connected; and Fig. 7-5(d) fails to be a tree on both accounts—it has circuits, and it is not connected.

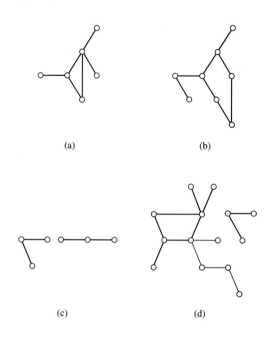

(a)   (b)

(c)   (d)

FIGURE 7-5

Properties of Trees

Trees are extremely important and useful structures. Not only do they help reduce carbon dioxide levels in the atmosphere, but they also show up in many mathematical applications, not the least of which are the minimum network problems we are discussing in this chapter. In what follows, we will briefly examine some of the main features of (mathematical) trees.

Let's start with the fact that in any connected graph there is always a path joining any vertex to any other vertex. What if there is more than one path joining a pair of vertices? If that is the case, we can be sure the graph is not a tree since two different paths joining the same two vertices must create a circuit as shown in Fig. 7-6.

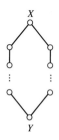

FIGURE 7-6 Two different paths joining vertices X and Y make a circuit.

Property 1. If a graph is a tree, there is one and only one path joining any two vertices. Conversely, if there is one and only one path joining any two vertices of a graph, the graph must be a tree.

One practical consequence of property 1 is that a tree is connected in a very precarious way: The removal of *any* edge of a tree will disconnect it (Exercise 30). We can restate this by saying that in a tree, every edge is a *bridge*.

> **Property 2.** In a tree, every edge is a bridge. Conversely, if every edge of a connected graph is a bridge, then the graph must be a tree.

From our preceding discussion, it seems intuitively obvious that a tree is "edge-poor." Having lots of edges is in some sense contrary to the nature of being a tree. At the same time, a tree must be connected, so a certain minimum number of edges is going to be necessary. A very important property of trees is that in a tree there is a very precise numerical relation between the number of edges and the number of vertices: The total number of edges is always 1 less than the number of vertices.

> **Property 3.** A tree with N vertices must have $N - 1$ edges.

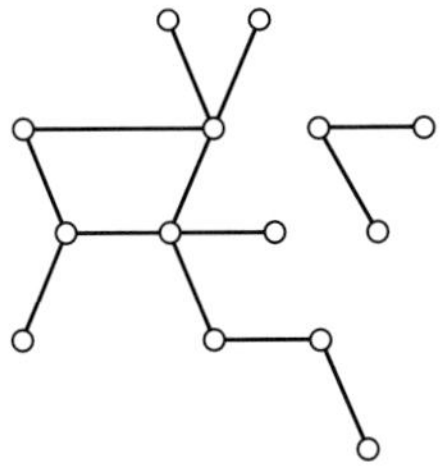

Property 3 also has a converse, but we must be a little careful. Can we say outright that if a graph has 1 less edge than it has vertices, then it must be a tree? The graph in Fig. 7-7 shows that this need not be the case—it has 14 vertices and 13 edges, and yet it is not a tree. The rub is that it is not connected. If we require the graph to be connected, then the converse of property 3 is indeed true.

FIGURE 7-7 A graph with 14 vertices and 13 edges that is not a tree.

> **Property 4.** A connected graph with N vertices and $N - 1$ edges must be a tree.

Example 3. This example illustrates some of the ideas discussed so far. Let's say that we have 5 vertices, and let's start putting edges on these vertices. At first, with 1, 2, or 3 edges (Figs. 7-8[a] through [c]), we just don't have enough edges to make the graph connected. When we get to 4 edges, we can, for the first time, make the graph connected. If we do so, we have a tree (Figs. 7-9[a] through [c]). As we add more edges (5, 6, etc.), the connected graph starts picking up circuits and stops being a tree (Figs. 7-10[a] through [c]).

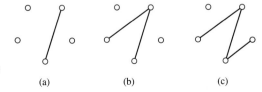

FIGURE 7-8 The graphs are disconnected. There are not enough edges.

(a)   (b)   (c)

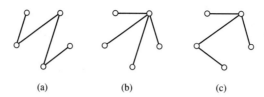

(a)   (b)   (c)

FIGURE 7-9 Just enough edges (four) to make a tree. Three different trees based on the same vertices.

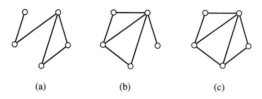

(a)   (b)   (c)

FIGURE 7-10 These graphs have circuits. They have too many edges to be trees.

MINIMUM SPANNING TREES

In Fig. 7-10 we have three examples of graphs that are connected but are not trees; they have more than the requisite number of edges. Within such a graph we can always find a tree reaching out to each of the vertices—somewhat like a skeleton holding up the rest of the body. We call such a tree a spanning tree of the original graph. Before we formally define a spanning tree, let's look at some examples.

Example 4. Consider the graph shown in Fig. 7-11(a). It is connected, it has 7 vertices and 9 edges, and we know it is not a tree. Inside such a graph we can find subgraphs like the one shown in Fig. 7-11(b) with the same 7 vertices but with only 6 of the edges. Such a subgraph is a tree that spans (reaches out to) all the vertices of the original graph and, as such, is called a **spanning tree** for the graph in Fig. 7-11(a). A graph may have more than one spanning tree: Fig. 7-11(c) shows a different spanning tree for the graph in 7-11(a). We leave it to the reader (Exercise 6) to find the other spanning trees for the graph in Fig. 7-11(a).

FIGURE 7-11

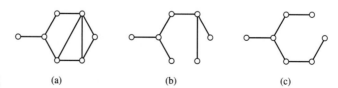

(a)   (b)   (c)

Any connected graph G has at least one spanning tree. If G is a tree, then it is its own spanning tree; otherwise, any tree contained within G and with the same vertices as G is a spanning tree. In general, the number of different spanning trees in a connected graph can be quite large. Consider the following example.

Example 5. The weighted graph in Fig. 7-12 has 8 different spanning trees, T_1, $T_2, \ldots, T_8$, as shown in Fig. 7-13.

FIGURE 7-12

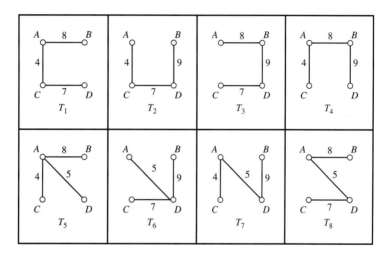

FIGURE 7-13

If we think of the graph in Fig. 7-12 as having edges which represent the possible connections between 4 cities (A, B, C, and D) and weights which represent the costs of installing these connections, then the 8 spanning trees shown in Fig. 7-13 will represent the 8 possible ways of connecting the cities in a treelike network. Of these, we can readily see (by just checking them one by one) that T_5 is the cheapest. Such a tree is called the *minimum spanning tree* of the original weighted graph.

With all this background, we should be ready now for a formal statement.

1. Suppose that G is an arbitrary connected graph. Then G has at least one (usually more) spanning tree. A **spanning tree** of G is a subgraph of G such that

    (a) Its vertices are exactly the vertices of G.

    (b) Its edges are some of the edges of G.

    (c) It is a tree.

2. Suppose that G is a connected weighted graph. Among all the spanning trees of G, there is one (maybe more) with the least total weight. Such a tree is called a **minimum spanning tree** (**MST**) of the weighted graph G.

Example 6. Here is an example to show that a weighted graph can have more than one minimum spanning tree.

The weighted graph in Fig. 7-14 has three different minimum spanning trees, each of which has a total weight of 4. They are shown in Fig. 7-15.

FIGURE 7-14

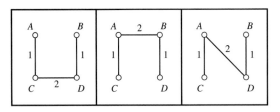

FIGURE 7-15

Example 7. The weighted graph in Fig. 7-16 is a tree, and therefore it has only one spanning tree—itself. Of course since it is the only one, it is a minimum spanning tree as well.

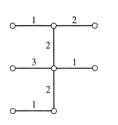

FIGURE 7-16

Using our new terminology, we can restate the Amazon telephone network problem at the beginning of the chapter (and any other problem like it) as one of finding a minimum spanning tree for the given graph. Not surprisingly, problems of this type are known as **minimum spanning tree problems**.

KRUSKAL'S ALGORITHM

We will now discuss a simple algorithm that will *always* find a minimum spanning tree for a connected weighted graph. This algorithm, known as **Kruskal's algorithm**,[1] is almost identical to the cheapest link algorithm in Chapter 6. It is nothing more than the result of combining greed with common sense.

---

[1] The algorithm is named after Joseph Kruskal, a mathematician working at the AT&T Bell Laboratories, who discovered it in 1956, although there is evidence that unbeknown to Kruskal the algorithm had already been discovered by several other mathematicians in Czechoslovakia, Poland, and France.

Kruskal's Algorithm.

- ■ **Step 1.**   Find the cheapest edge in the graph (if there is more than one, pick one at random). Mark it in red. (Needless to say, you can use any color.)
- ■ **Step 2.**   Find the cheapest unmarked (i.e., not red) edge in the graph that does not close a red circuit. (Remember: We don't want any circuits!) If there is more than one, pick one at random. Mark the edge red.
- ■ **Steps 3, 4, etc.**   Repeat Step 2 until the red edges reach out to every vertex of the graph. The marked edges form the desired minimum spanning tree.

Example 8.   Let's apply Kruskal's algorithm to the original Amazon telephone network problem.

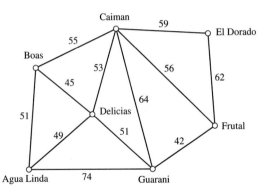

FIGURE 7-17   The cost (in millions of dollars) for each possible connection in the telephone network.

- ■ **Step 1.**   The cheapest connection is Guarani–Frutal, at $42 million. We mark the connection in red. (Note that this does not necessarily mean that these will be the first two towns actually connected. We are putting the network together on paper, and the rules require that we follow a certain sequence, but in practice we can build the connections in any order we want.)
- ■ **Step 2.**   The next cheapest connection is Boas–Delicias, at $45 million. We mark the connection in red.
- ■ **Step 3.**   The next cheapest connection is Agua Linda–Delicias, at $49 million. We mark the connection in red.
- ■ **Step 4.**   The next cheapest connection is a tie between Agua Linda–Boas and Delicias–Guarani, both at $51 million. Agua Linda–Boas, however, is a redundant connection, and we do not want to use it. (For bookkeeping purposes, the best thing to do is delete it.) Delicias–Guarani, on the other hand, is just fine, so we do mark that connection red.

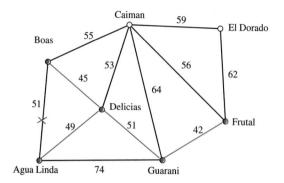

FIGURE 7-18 The network after step 4. Boas and Agua Linda can communicate with each other without the direct connection Agua Linda–Boas.

■ **Step 5.** The next cheapest connection is Caiman–Delicias, at $53 million. No problems here, so we mark the connection in red.

■ **Step 6.** The next cheapest connection is Boas–Caiman, at $55 million, but this is a redundant connection so we discard it. The next possible choice is Caiman–Frutal at $56 million but that is also a redundant connection (calls between Caiman and Frutal are already possible in our budding network), so we keep looking. The next possible choice is Caiman–El Dorado at $59 million, and this one is OK, so we mark the connection Caiman–El Dorado in red.

■ **Step ...** Wait a second—we are finished! (One way we can tell this is the case is by just looking at the red network we have built and verifying that it is a spanning tree. An even better way is by realizing that 6 edges—and therefore 6 steps—is exactly what it takes to build a tree on 7 vertices.)

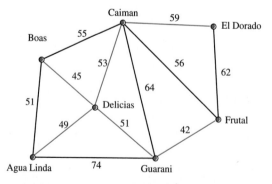

FIGURE 7-19 The minimum spanning tree (MST) for the Amazon telephone network problem. Junction points for the network are Delicias, Caiman, and Guarani.

The total cost of the telephone network we came up with is $299 million, and this is in fact, the optimal solution to the problem—there is no cheaper spanning tree! We call the reader's attention to one other fact that will become relevant later in the chapter: The network we have built has three **junction points**—a four-way junction at Delicias and two-way junctions at Caiman and Guarani.

Before concluding this section, let's look at a different application of minimum spanning trees, this time involving a problem in irrigation.

Example 9. The vertices of the graph shown in Fig. 7-20 are locations where we need to have water (for example, sprinkler heads). Vertex A is the main water source. The edges of the graph represent the places along which one could possibly lay irrigation pipes, and the weights represent the cost (in dollars) of each possible connection.

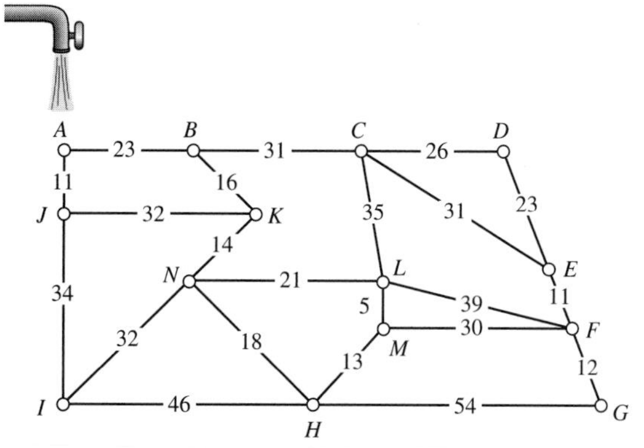

FIGURE 7-20

We want to lay down a network of pipes so that water from A can get to each of the sprinkler heads. We also want to do the job as cheaply as possible. Clearly, this is a minimum spanning tree problem. We will use Kruskal's algorithm to find the minimum spanning tree.

- ■ **Step 1.** Mark LM.
- ■ **Step 2.** Mark AJ.
- ■ **Step 3.** Mark EF.
- ■ **Step 4.** Mark FG.
- ■ **Step 5.** Mark HM.
- ■ **Step 6.** Mark NK.
- ■ **Step 7.** Mark BK.
- ■ **Step 8.** Mark HN.

■ **Step 9.**   *NL* is the next cheapest edge but it cannot be used because it closes a circuit, so we discard it. The next choice is *AB* or *DE*, either of which is OK. We mark *AB*. (We can now discard *JK* because it closes a circuit.)

■ **Step 10.**   We mark *DE* next.

■ **Step 11.**   Mark *CD*.

■ **Step 12.**   Discard *CE*. Mark *FM*.

■ **Step 13.**   Discard *BC, CL, FL,* and *GH*. Mark *IN*.

We are finished! The minimum spanning tree describing the cheapest irrigation system[2] (at a total cost of $234) is shown in Fig. 7-21.

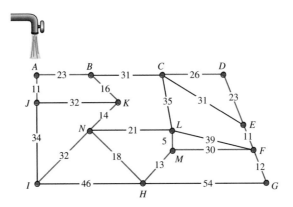

FIGURE 7-21

The reader has no doubt noticed that there is a great deal of similarity between Kruskal's algorithm and the cheapest link algorithm we discussed in Chapter 6. In fact, they are both based on an identical strategy: Be cheap (i.e., always make the most inexpensive choice possible), but follow the rules for putting together the particular structure you need (a spanning tree or a Hamilton circuit, respectively). Our previous experience with this strategy is that there usually is a price to pay—greed in the short term can produce bad results in the long term. Kruskal's algorithm is, however, a pleasant surprise. In this case the solutions we get with our greedy strategy are guaranteed to be optimal: There is no cheaper spanning tree than the one we get from Kruskal's algorithm, which means therefore that Kruskal's algorithm is an *optimal algorithm.*

With regard to its practicality, Kruskal's algorithm is an *efficient algorithm.* Just as with the cheapest link algorithm, it is not unreasonable to attempt a problem with hundreds of vertices by hand and one with hundreds of thousands of vertices by computer.

In short, the problem of finding minimum spanning trees is one of those rare situations where everything falls into place: We have an algorithm

---

[2] The reader is warned that there are important practical aspects of laying down irrigation pipe (such as water pressure) which we have neglected to take into account.

(Kruskal's) that is easy to understand and to carry out, and that is *optimal* and at the same time *efficient*—who could ask for better karma? Wouldn't it be great if things always went this well?

SHORTEST NETWORKS

Emboldened by our good fortune with minimum spanning trees, we will look once again at the problem of finding optimal networks for connecting a set of locations, but now under slightly different conditions: What if, in a manner of speaking, we don't have to *follow the road*? What if we can go from one location to another any way we want? To clarify the issue we are raising, let's look at a few simple examples.

Example 10 (The Australian telephone network problem). This is the story of 3 small fictional towns (Alcie Springs, Booker Creek, and Camoorea) located smack in the middle of the Australian outback. By sheer coincidence, the 3 towns happen to sit in such a way that they form an equilateral triangle, 500 miles a side, as shown in Fig. 7-22. Our problem, once again, is to lay out a telephone network that interconnects the 3 towns in the most inexpensive possible way. What is the shortest possible way to connect the towns?

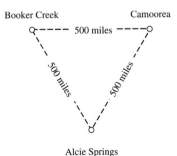

FIGURE 7-22 Three towns in the Australian outback.

While this example looks like just a simple version of the Amazon network problem, it is not. What makes this situation different is the nature of the terrain. This particular region of the Australian outback is mostly a flat and arid expanse of desert and, in contrast to the Amazon situation, there is little or no advantage in laying the telephone lines along roads. (In fact, let's assume that there are no roads to speak of connecting these three towns.) Because of the flat and homogeneous nature of the terrain, we can lay the telephone connections any way we want, and the cost per mile is always the same.

The question now is: Of all the possible networks that connect the 3 points, which one is the shortest? Let's start with a logical candidate—the minimum spanning tree connecting the 3 towns. Figure 7-23(a) shows a minimum spanning tree network, with a total length of 1000 miles.

It is not difficult to convince oneself that the MST is not the shortest possible answer. Look at Fig. 7-23(b). It shows a network that is definitely shorter than 1000 miles. It has a "T" junction at a new point we call *J*. A little high school geometry (Pythagorean theorem) and a calculator are sufficient to verify that it is approximately 433 miles from Alcie Springs to the junction *J*, and that the network therefore, is about 933 miles long (see Exercise 25).

Can we do even better? Why not? With a little extra thought and effort we might come up with the network shown in Fig. 7-23(c). Here, there is a "Y" junction at a new point called *S* located at the center of the triangle. This network is approximately 866 miles long (see Exercise 26). The most important feature of this network is that the 3 branches that come together at the junction point *S* do so by forming 3 *equal angles,* which forces each angle to be exactly 120 degrees. To the observant reader, it will not come as a tremendous surprise that this is the best we can do: Fig. 7-23(c) shows the *shortest possible network that one can have connecting the 3 vertices of the equilateral triangle.* (A good justification for this claim is that this is the only network that looks the same whether you are a resident of Alcie Springs or of Booker Creek or of Camoorea. For a more formal mathematical proof, see Exercise 45.)

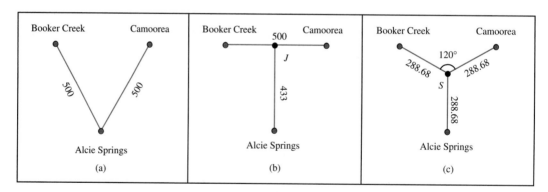

FIGURE 7-23 (a) A minimum spanning tree network with total length of 1000 miles. (b) A shorter network (total length of approximately 933 miles) with a T-junction point at *J*. (c) The shortest network (total length of approximately 866 miles) with a Y-junction point at *S*. The three branches of the "Y" meet at equal angles (120°).

Let's recap what we learned from Example 10: If we do not require the junction points of our network to be one of the original cities (in other words, if we are allowed to create new junction points in the network), then the minimum spanning tree may not be the best way to connect the cities. In Example 10, the best solution is a network with a new junction point in which 3 branches meet at 120° angles. If we compare the length of this solution (866 miles) with the MST solution (1000 miles) we can see that in this case, creating the new junction point produced a savings of 134 miles or 13.4%.

Before we go on to the next example, let's introduce some new terminology.

■ The shortest possible network connecting a set of points is called, not surprisingly, the **shortest network**.[3]

■ Any junction point in a network that is formed by 3 branches coming together at 120° angles is called a **Steiner point**[4] of the network.

Let's now move on to a more realistic example.

Example 11 (The Texas bullet train caper). There is a great deal of talk nowadays about "bullet" trains—high-speed trains traveling on specially designed magnetic tracks. One of the first bullet train systems being considered in the United States is in the state of Texas, where there are plans to interconnect the cities of Dallas, San Antonio, and Houston by means of a high-speed rail network. Figure 7-24 shows the exact geographic layout of the 3 cities, with the

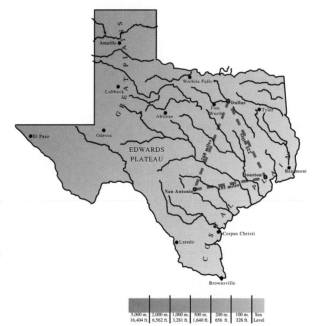

FIGURE 7-24 A topographical map of Texas. The San Antonio–Houston–Dallas triangle is in the central plains. The straight-line distances between the cities are San Antonio–Houston: 195 miles; San Antonio–Dallas: 260 miles; and Houston–Dallas: 225 miles.

[3] In theory, there could actually be more than one such network, so to be grammatically correct we should say *a shortest network* instead of *the shortest network*. In practice, the overwhelming majority of the time there is only one shortest network, in which case talking about *a shortest network* sounds a little peculiar. We think that in this case, common sense usage should override grammatical accuracy.

[4] Named after Jakob Steiner (1796–1863), a Swiss mathematician at the University of Berlin in the early nineteenth century.

straight-line distances between them. The most expensive part of this entire project would be the laying down of the new special tracks required by these trains. Given that the terrain in this part of Texas is relatively flat, we can pretty much assume that cost is proportional to distance, and if this is the case, then the cheapest layout for the tracks is given by the shortest network connecting the three cities. How do we find it?

Let's start with the minimum spanning tree for these 3 cities, which consists of the segments San Antonio–Houston and Houston–Dallas (Fig. 7-25[a]) and which has a total length of 420 miles. Based on what happened in Example 10, it is a good bet that we can shorten this network by introducing some junction points inside of the triangle Dallas–San Antonio–Houston. Should it be one junction point or several? Where should they be located? It turns out that all we need is just one junction point, and that it should be located so that the 3 branches of the network that meet at that point should do so at 120° angles—in other words, it should be a Steiner point of the network. Some basic (but not trivial) geometric reasoning can be used to show that the network obtained in this way and shown in Fig. 7-25(b) is the shortest network connecting the 3 cities. Its total length is approximately 390 miles, for a savings of 30 miles over

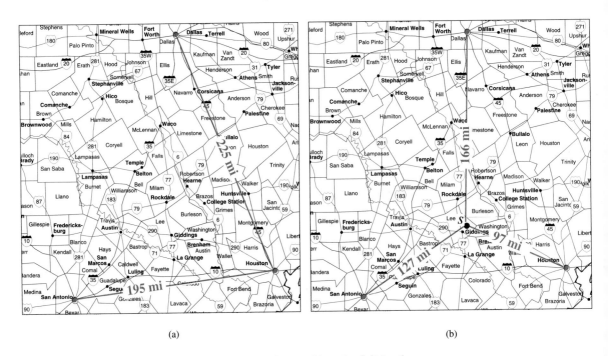

(a) (b)

FIGURE 7-25 (a) The minimum spanning tree, with a total length of 420 miles. The junction point is Houston. (b) The shortest network solution, with a total length of 390 miles. The junction point S is a Steiner point, located a few miles northeast of Giddings, Texas. The distances are: San Antonio–S: 127 miles; Houston–S: 97 miles; and Dallas–S: 166 miles.

the minimum spanning tree network. (While this may not seem like much, at several million dollars per mile of track, it does come out to some real money!) As a point of interest, the exact location of the junction point S can be pinpointed geometrically using a clever procedure called Torricelli's construction, which is described later in this section. In this particular example, the junction point S is located about 10 miles northeast of the town of Giddings, Texas.

Example 12. Suppose, once again, that we need to design the layout for a high-speed rail system, this time connecting the cities of Los Angeles, Las Vegas, and Salt Lake City. What is the shortest network connecting these three cities? Figure 7-26 shows the straight-line distances between the three cities.

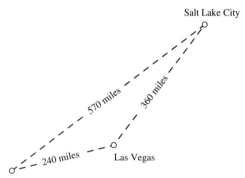

FIGURE 7-26

Our experience with Examples 10 and 11 tells us that to create the shortest network we should be looking for a junction point that is a Steiner point inside the triangle Los Angeles–Las Vegas–Salt Lake City. Unfortunately, when we try to find such a junction point we come up empty handed: There is no Steiner point inside the triangle! Why not? We can find the culprit in Las Vegas, or to be more precise, in the angle that the triangle forms at Las Vegas, which is approximately 155°. We know, from basic geometry, that for any point P inside of the triangle, the angle formed by joining P to Los Angeles and Salt Lake City has to be more than 155°. It follows that there can be no Steiner point inside of the triangle.

OK, but if there is no Steiner point inside the triangle, then how are we going to find the shortest network? The answer is provided by this wonderful bit of good news: *In this situation, the shortest network is the same as the minimum spanning tree.* Thus, for our example, the 600-mile network shown in Fig. 7-27(b), consisting of the segments Los Angeles–Las Vegas and Las Vegas–Salt Lake City, is indeed the shortest network.

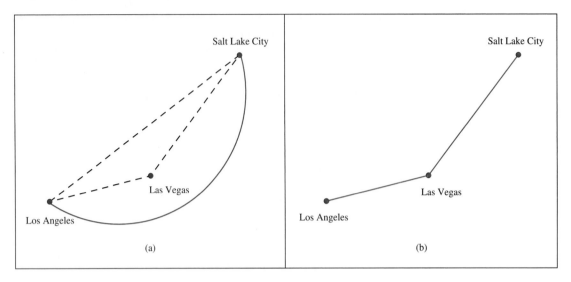

FIGURE 7-27 (a) The red arc shows the location of all the possible points that form a 120° angle when connected to Los Angeles and Salt Lake City. None of these points falls inside the triangle Los Angeles–Las Vegas–Salt Lake City. (b) The shortest network connecting 3 cities is the minimum spanning tree.

Let's summarize what we have learned from Examples 10 through 12 (not necessarily in order) about the shortest network connecting three points *A*, *B*, and *C*.

How to find the shortest network connecting three points:

1. When one of the angles of the triangle *ABC* is 120° or more, then the shortest network solution is the same as the minimum spanning tree solution (Fig. 7-28[a]), and finding such a network is a piece of cake (pick the two shortest sides of the triangle).

2. When all the angles of the triangle are less than 120°, then the shortest network solution and the minimum spanning tree solution part company. The shortest network solution is obtained by finding a Steiner junction point *S* and joining *S* to each of the vertices *A*, *B*, and *C* (Fig. 7-28[b]).

FIGURE 7-28 The shortest network connecting *A*, *B*, and *C*. (a) Case 1: an angle of 120° or more. The junction point is one of the vertices. (b) Case 2: all angles less than 120°. The junction point *S* is a Steiner point.

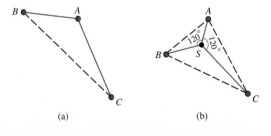

There is one more detail that we need to deal with involving case 2: how to pinpoint the location of the Steiner junction point *S*. We will now describe a simple, elegant geometric procedure that will allow us to do exactly that.

The Torricelli procedure[5] for finding a Steiner point inside a triangle:

Assume that the triangle *ABC* has all angles less than 120° and that *BC* is the longest side, as shown in Fig. 7-29.

- ■ **Step 1.** On the largest side of the triangle build an equilateral triangle *BCX* as shown in Fig. 7-29.
- ■ **Step 2.** Circumscribe a circle around the equilateral triangle *BCX*.
- ■ **Step 3.** Join *X* to *A*. The point where the line segment *XA* intersects the circle is the desired Steiner point *S*.

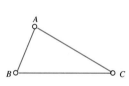

Start: The original triangle.

Step 1: On the longest side (*BC*), build the equilateral triange *BCX*.

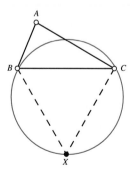

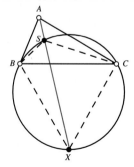

Step 2: Circumscribe a circle around the triangle *BCX*.

Step 3: Join *X* to *A*. The line segment *XA* intersects the circle at the Steiner point *S*.

FIGURE 7-29 Finding the Steiner point *S* inside a triangle.

---

[5] This procedure can be traced back to the Italian physicist-mathematician Evangelista Torricelli (1608–1647). Torricelli was a disciple of Galileo and is best remembered for discovering the barometer.

Why does the Torricelli procedure work? First, angle BXC is 60° (BXC is an equilateral triangle). It follows that angle BSC is 120° (opposite angles of a quadrilateral inscribed in a circle add up to 180°). Also, angles BSX and XSC are both 60° (they are equal to angles BCX and XBC, respectively). So, it follows that angles BSA and CSA are both 120° and that S is the desired Steiner point. As an additional point of interest, it turns out that the length of the entire network ($SA + SB + SC$) is exactly the same as the length of the segment AX.

Now that we have finally mastered the shortest network problem for three cities, let's look at some cases involving four cities.

Example 13. Four cities (A, B, C, and D) are to be connected into a telephone network. For starters, let's assume that the 4 cities form the vertices of a square 500 miles a side, as shown in Fig. 7-30(a). What does an optimal network connecting these four cities look like?

If we are *not allowed* to introduce any new junction points in the network, then the answer is a minimum spanning tree, such as the one shown in Fig. 7-30(b). The length of this network is 1500 miles, and it has junction points at A and B. On the other hand, if interior junction points are allowed, somewhat shorter networks are possible. One obvious improvement is the network shown in Fig. 7-30(c), having an "X" type of junction at O, the center of the square. The length of this network is approximately 1414 miles (see Exercise 37). An even more impressive improvement is shown by the network in Fig. 7-30(d), with 2 junction points S_1 and S_2 which happen to be Steiner points for this network. Using a little high school geometry (Exercise 38) we can find that the length of this network is approximately 1366 miles, a savings of about 134 miles (almost 9%) over the 1500 miles of the minimum spanning tree network. The best news of all is that the network in Fig. 7-30(d) is as short a network as one can find—no further improvements are possible. (There is, however, a second network that is also shortest—it is shown in Fig. 7-30[e].)

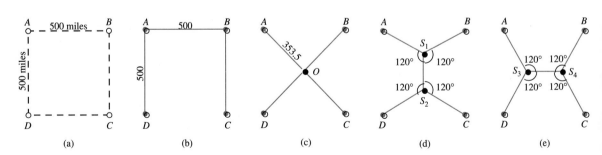

Figure 7-30 (a) Four cities located at the vertices of a square. (b) A minimum spanning tree network with a total length of 1500 miles. The only junction points are at A and B. (c) A shorter network, with a total length of approximately 1414 miles, obtained by placing a new junction point O at the center of the square. (d) A shortest network solution with a total length of approximately 1366 miles. The junction points S_1 and S_2 are both Steiner points. (e) A different shortest network solution, with Steiner points S_3 and S_4.

Example 14. Let's repeat what we did in Example 13, but this time assume that the 4 cities are located at the vertices of a rectangle, as shown in Fig. 7-31(a). By now, we have some experience on our side, so we can cut to the chase. We know that the minimum spanning tree solution is 1000 miles long (see Fig. 7-31[b]), and we have a hunch that to find the shortest network solution all we have to do is design a network with 2 interior junction points and make sure that these junction points are Steiner points. Taking a cue from Figs. 7-30(d) and (e) in Example 13 we can come up with 2 reasonable candidates, shown in Figs. 7-31(c) and (d). It is not difficult to convince oneself that just as in Example 13 these are the only 2 possible ways to create a network with 2 Steiner points. There is one thing, however, that is new: The network shown in Fig. 7-31(c) is approximately 993 miles (see Exercise 39[a]) while the network shown in Fig. 7-31(d) is approximately 919.6 miles (see Exercise 39[b]). At this point, our best candidate for the shortest network is given by the network shown in Fig. 7-31(d) (weighing in at a mere 919.6 miles). If there is any justice in this mathematical world, this particular network ought to be the shortest network. In fact, it is! (But don't jump to any conclusions about justice just yet!)

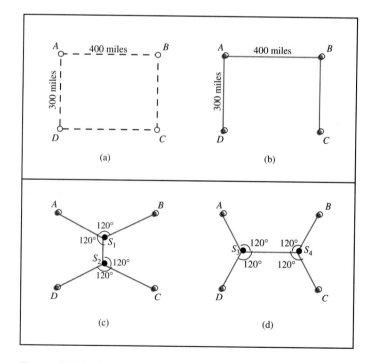

FIGURE 7-31 (a) Four cities located at the vertices of a rectangle. (b) A minimum spanning tree solution with a total length of 1000 miles. (c) A network with 2 interior Steiner points (S_1 and S_2). Total length is approximately 933 miles. (d) A different network with 2 interior Steiner points (S_3 and S_4). This is the shortest network, at approximately 919.6 miles.

Example 15. Let's now consider the case of 4 cities located at the vertices of a skinny trapezoid as shown in Fig. 7-32(a). The minimum spanning tree is shown in Fig. 7-32(b), and it is 600 miles long.

What about the shortest network? Once again, we are fairly certain that we should be looking for a network with a couple of interior junction points that are Steiner points. After a little trial and error, however, we realize that such a layout is impossible! The trapezoid is too skinny, or to put it in a slightly more formal way, the angles at A and B are more than 120°. Since no Steiner points can be placed inside the trapezoid, the shortest network, whatever it is, will have to be one without Steiner junction points.

Well, we say to ourselves, if not Steiner junction points, how about other kinds of interior junction points? How about X-junctions (Fig. 7-32[c]) or T-junctions (Fig. 7-32[d]), or other kinds of Y-junctions besides the ones required by Steiner points? As reasonable as this idea sounds, there is a remarkable thing that happens with shortest networks: *In any shortest network, every interior*

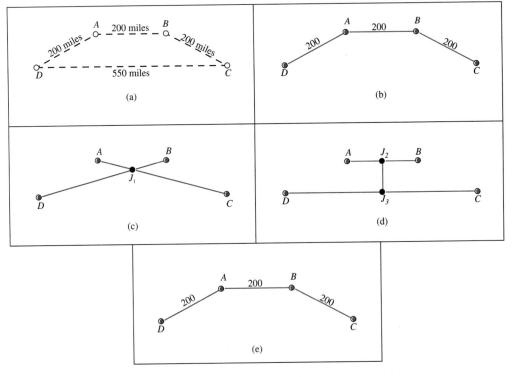

FIGURE 7-32 (a) Four cities located at the vertices of a trapezoid. (b) The minimum spanning tree for the 4 cities (total length is 600 miles). (c) A network with an X-junction at J_1 is unable to beat out the MST (total length is 774.6 miles). (d) A network with a couple of T-junction points at J_2 and J_3 is even worse (total length is 846.8 miles). (e) The shortest network and the MST are one and the same!

junction point has to be a Steiner point. (It's as if the shortest network is telling its interior junction points *turn yourself into a Steiner point or die.*) This is an important and powerful fact, and we will come back to it soon. Meanwhile, what does it tell us in this example? It tells us that the shortest network cannot have *any* interior junction points—remember, it's either Steiner or nothing! But we already know what is the best possible network without interior junction points—it is the minimum spanning tree! Conclusion: For the four cities shown in Fig. 7-32(a), *the shortest network and the minimum spanning tree happen to be one and the same* (Fig. 7-32[e]).

Example 16. For the last time, let's look at 4 cities *A, B, C,* and *D.* This time the 4 cities sit as shown in Fig. 7-33(a). The minimum spanning tree is shown in Fig. 7-33(b), and its length is 1000 miles. Based on what happened in Example 15 we have to consider this MST a serious candidate for the title of shortest network. At this point we know this much: Any network that is going to beat out

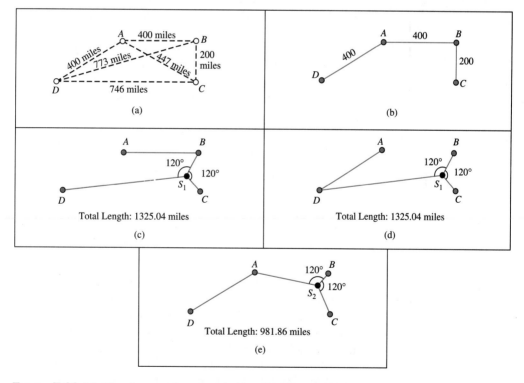

FIGURE 7-33 (a) The distances between 4 cities. (b) The minimum spanning tree (total length is 1000 miles). (c) and (d) Both of these networks have one Steiner point (S_1) connecting cities *B, C,* and *D.* Both networks have the same length (approximately 1325 miles) and are much longer than the MST. (e) A different challenger with one Steiner point (S_2) and length of approximately 982 miles beats out (just barely) the MST. It turns out to be the shortest network connecting the four cities.

the MST is going to have to have some interior junction points, and by golly, these junction points better be Steiner points. Because of the layout of the 4 cities, having 2 interior Steiner junction points is out of the question—it is a geometric impossibility (give it a try if you think otherwise!). On the other hand, there are 3 possible networks with a single interior Steiner junction point (Figs. 7-33[c], [d], and [e]). These are the only 3 possible challengers to the MST. Two of them (Figs. 7-33[c] and [d]) have a total length of 1325 miles, so they are not even close. On the other hand, the network shown in Fig. 7-33(e) has a total length of approximately 982 miles. Given that this is the best of all possible challengers, it must be the shortest network.

The main lesson to be learned from Examples 14, 15, and 16 is twofold: First, if a shortest network is not the MST, then it must have interior junction points, and these junction points must be Steiner points. (A useful bit of terminology: From now on, we will call a tree network in which every interior junction is a Steiner point a **Steiner tree**.) Second, just because a network is a Steiner tree, there is no guarantee that it is the shortest network. As Example 16 shows, there are many possible Steiner trees, and not all of them produce shortest networks.

What happens when the number of cities gets larger? How do we look for the shortest network? The answer is "very carefully." Unfortunately, we are on the same shaky grounds we were on in Chapter 6 (The Traveling Salesman Problem)—there is no efficient algorithm that is always guaranteed to produce the shortest network. At this point, the best one can do is to take advantage of the following rule, which we informally discovered as we plugged along the preceding examples.

The shortest network connecting a bunch of cities is either:

- the minimum spanning tree (no interior junction points), or

- a Steiner tree (some interior junction points).

This means that we can always find the shortest network by rummaging through all possible Steiner trees, finding the shortest one among them and comparing it with the MST. The shorter of these two has to be the shortest network. This sounds like a good idea, but once again, the problem is in the explosive growth of the number of possible Steiner trees. With as few as 10 cities, the possible number of Steiner trees we would have to rummage through is in the millions; with 20 cities, it's in the billions.

What's the alternative? Just as in Chapter 6, if we are willing to settle for less than the best deal (in other words, if we are willing to buy a network that is short but not necessarily the shortest) we can tackle the problem no matter how large the number of cities. To do this, we need some efficient approximate algorithms. Some excellent approximate algorithms for finding short networks are presently known, and interestingly enough, one of the best is a shockingly simple algorithm that we are well acquainted with: Kruskal's algorithm for finding a minimum spanning tree. It has been known for a long time that the minimum spanning tree is usually pretty close to the shortest network, and in 1990 mathematicians Frank Hwang of AT&T Bell Laboratories and Ding-Zhu Du of Princeton University nailed down the relationship between these two: *The gap between the MST and the shortest network is never more than 13.4% (and in the vast majority of cases it is actually less than 5%).*

We conclude this section with a last example. Our intent here is not to get bogged down with the details but rather to get a brief glimpse of the larger picture.

Example 17. The time: the year 2125. The place: the planet Romulak. Seven different groups of human explorers have settled into space colonies (A_1, A_2, A_3, A_4, A_5, A_6, and A_7) as shown in Fig. 7-34(a). The distances shown on the map are given in *romules*. (One *romule* is approximately equal to 3.14 miles). It is now time to connect the various colonies (so that the colonists can get around and trade with each other). To do this, we need to lay down a network of *romu-tubes* (giant suction tubes through which vehicles and cargo can travel at mind-boggling speeds). Romutubes are very expensive (about a billion *roteks* per *ro-mule*), so the name of the game is to look for the shortest network. The candidates are: the MST, shown in Fig. 7-34(b), plus all possible Steiner trees connecting the seven colonies. There are thousands of such Steiner trees, and just a few of them are shown in Figs. 7-34(c) through (g). A computer is needed to check through all Steiner trees, looking for a shortest one. After thousands of calculations, the answers comes back: The shortest network is the one shown in Fig. 7-34(e).

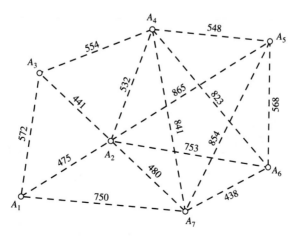

FIGURE 7-34(a) The 7 human colonies on the planet Romulak (circa 2125), with the distances between colonies (in romules).

FIGURE 7-34 (Continued)

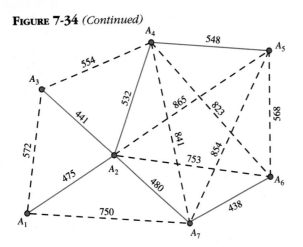

(b) The MST connecting the seven colonies. Total length: 2914 romules.

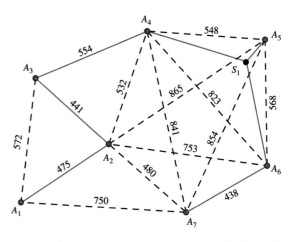

(c) A Steiner tree with 1 Steiner point. Total length: 2968 romules.

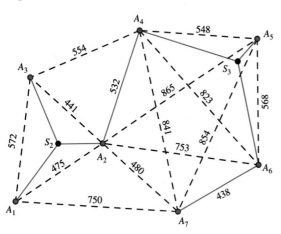

(d) A Steiner tree with 2 Steiner points. Total length: 2882.5 romules.

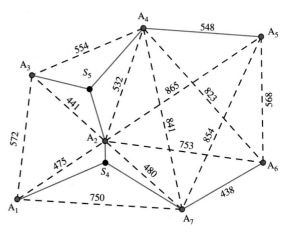

(e) Another Steiner tree with 2 Steiner points. Total length: 2809 romules.

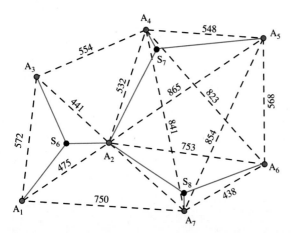

(f) A Steiner tree with 3 Steiner points. Total length: 2840 romules.

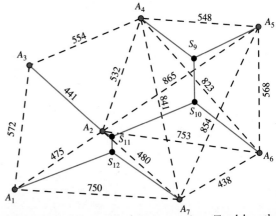

(g) A Steiner tree with 4 Steiner points. Total length: 3060 romules.

Did we save much by going through all the trouble? The MST is 2914 romules long, the shortest network is 2809 romules, for a savings of a mere 105 romules. You decide. ▬

CONCLUSION

In this chapter we discussed the problem of interconnecting a set of points in a network in an optimal way, where "optimal" usually means least expensive or shortest. In practice, the points represent geographical locations (cities, pumping stations, sprinkler heads, etc.), and the connections can be highways, pipelines, telephone lines, etc. Depending on the circumstances, we considered two different ways of doing this.

Version 1. In the first half of the chapter we were required to build the network in such a way that no junction points other than the original locations were allowed, in which case the optimal network is a *minimum spanning tree*. Finding a minimum spanning tree was the great success story of the chapter—Kruskal's algorithm provides the answer in an efficient and optimal way.

Version 2. In the second half of the chapter we considered what on the surface appeared to be a minor modification by removing the prohibition against new junction points. In this case the problem becomes one of finding the *shortest network* connecting the points. Here the situation gets considerably more complicated, but we do know a few things:

1. There are times when the minimum spanning tree and the shortest network are one and the same (as in Examples 12 and 15), but most of the time they are not. When they are not the same, the shortest network is obviously shorter than the minimum spanning tree, but by how much? It took mathematicians more than two decades to completely answer this question, but the answer is now known: The difference between the minimum spanning tree and the shortest network is in general, relatively small, and *under no circumstances can it ever be more than 13.4%.*

2. In a shortest network, any new junction points (we call them *interior* points) must have the form of a perfect Y-junction (we call such junction points *Steiner* points). Any tree network (i.e., without circuits) connecting the original points and such that all interior junction points are Steiner points is called a *Steiner tree*. It follows that *if a shortest network has interior junction points it must be a Steiner tree, and if it doesn't, then it must be the minimum spanning tree.*

3. While in general the number of minimum spanning trees for a given set of points is small (usually just one) and easy to find (use Kruskal's algorithm), the number of possible Steiner trees for the same set of points is usually very large and difficult to find.[6] With 7 points, for example, the

---

[6] There is a fun way to find some of the Steiner trees using soap film computers (see the Appendix "The Soap Bubble Solution" for details), but it is slippery to say the least.

possible number of Steiner trees is in the thousands, and with 10 points it's in the millions.

We may want to think of all the Steiner trees for a given set of points, together with the minimum spanning tree, as a haystack. The shortest network is the needle. We know that the needle is there, but rummaging through the entire haystack does not sound like a very good way to go about finding it. Unfortunately, except for a few relatively minor shortcuts, no better method is known. In more ways than one, the problem of finding the shortest network connecting a set of points is like the traveling salesman problem: No efficient algorithms are known, and it is very likely that none exist. On the other hand, excellent efficient but approximate algorithms have been discovered in recent years, and one of them happens to be our good friend, Kruskal's algorithm. With apologies for the mixed metaphor, our haystack has a silver lining!

KEY CONCEPTS

junction points
Kruskal's algorithm
minimum network problems
minimum spanning tree (MST)
minimum spanning tree problems
shortest network

spanning tree
Steiner points
Steiner tree
subgraph
tree

EXERCISES

Walking

1. For each of the following graphs determine whether the graph is a tree. If it is not a tree, give a reason why.

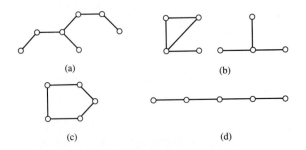

(a)

(b)

(c)

(d)

2. For each of the following graphs determine whether the graph is a tree. If it is not a tree, give a reason why.

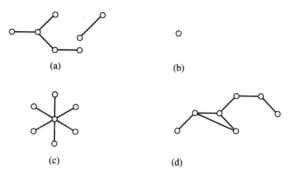

(a)   (b)

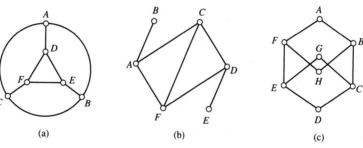

(c)   (d)

3. Find three different spanning trees for the following graph.

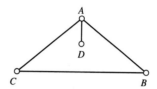

4. Find a spanning tree for each of the following graphs.

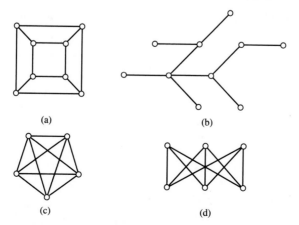

(a)   (b)   (c)

5. Find a spanning tree for each of the following graphs.

(a)   (b)

(c)   (d)

6. Find two different spanning trees for each of the following graphs.

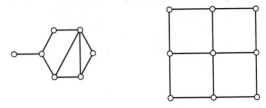

7. Find all the possible spanning trees of the following graph.

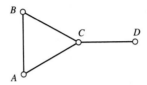

8. Find all the possible spanning trees of the following graph.

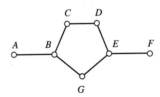

9. How many different spanning trees does each of the following graphs have?

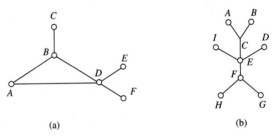

(a)   (b)

10. How many different spanning trees does each of the following graphs have?

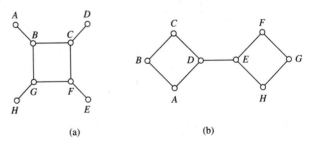

(a)   (b)

11. Use Kruskal's algorithm to find a minimum spanning tree of the following weighted graph.

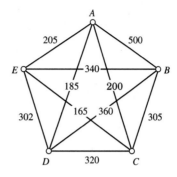

12. Use Kruskal's algorithm to find a minimum spanning tree of the following weighted graph.

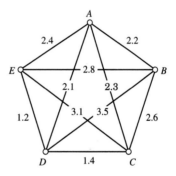

13. Use Kruskal's algorithm to find a minimum spanning tree of the following weighted graph.

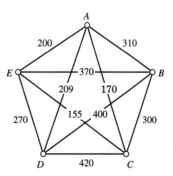

14. Use Kruskal's algorithm to find a minimum spanning tree of the following weighted graph.

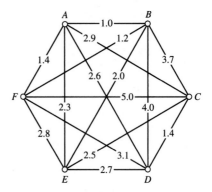

15. Use Kruskal's algorithm to find a minimum spanning tree of the following weighted graph.

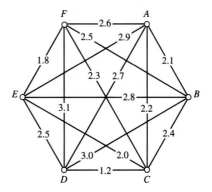

16. Use Kruskal's algorithm to find a minimum spanning tree of the following weighted graph.

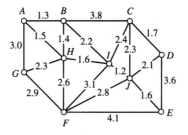

Exercises 17 and 18 refer to 5 cities (A, B, C, D, and E) located as shown in the following figure. In the figure, AC = AB, DC = DB, and angle CDB = 120°.

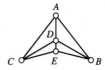

17. (a) Which is larger, *CD* + *DB* or *CE* + *ED* + *EB*? Explain.

(b) What is the shortest network connecting the cities *C, D,* and *B*? Explain.

(c) What is the shortest network connecting the cities *C, E,* and *B*? Explain.

18. (a) Which is larger, *CA + AB* or *DC + DA + DB*? Explain.

(b) Which is larger, *EC + EA + EB* or *DC + DA + DB*? Explain.

(c) What is the shortest network connecting the cities *A, B,* and *C*? Explain.

19. Find the length of the shortest network connecting the cities *A, B,* and *C* shown in the following figure.

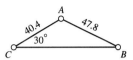

20. Find the length of the shortest network connecting the cities *A, B,* and *C* shown in the following figure.

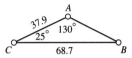

21. Find the length of the shortest network connecting the cities *A, B,* and *C* shown in the following figure.

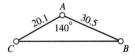

22. Find the length of the shortest network connecting the cities *A, B,* and *C* shown in the following figure.

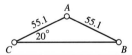

Exercises 23 through 26 refer to an equilateral triangle ABC with sides of length 500, altitude AJ, and Steiner point S as shown in the following figure.

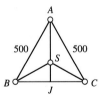

23. Show that triangles *ASB, BSC,* and *CSA* are congruent.

24. Show that *SA = SB = SC*. (*Hint*: Use Exercise 23.)

25. Use your knowledge of 30°–60°–90° triangles to find the lengths of

(a) *JA*

(b) *SA*.

26. Show that the length of the shortest network connecting A, B, and C is approximately 866. (*Hint*: Use Exercise 25[b].)

Jogging

27. Five cities (A, B, C, D, and E) are located as shown on the following figure. The 5 cities need to be connected by a railroad, and the cost of building the railroad system connecting any two cities is proportional to the distance between the cities. Find the length of the railroad network of minimum cost (assuming that no additional junction points can be added).

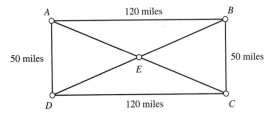

28. Four cities (A, B, C, and D), shown in the following figure, must be connected into a telephone network. The cost of laying down telephone cable connecting any two of the cities is given (in millions of dollars) by the weights of the edges in the graph. In addition, in any city that serves as a junction point of the network, expensive switching equipment must be installed. The cost of installing this equipment is given (also in millions of dollars) by the numbers inside the circles. Find the minimum cost telephone network connecting these 4 cities.

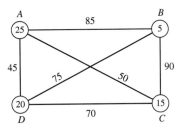

29. Explain why if G is a connected graph with N vertices, Kruskal's algorithm will require exactly N − 1 steps.

30. Explain why in a tree, every edge is a bridge. (Recall that a bridge is an edge whose removal disconnected the graph.)

31. **(a)** How many different spanning trees does the following graph have?

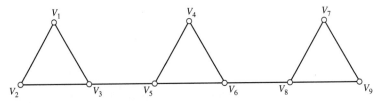

(b) How many different spanning trees does the following graph have?

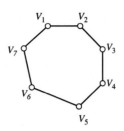

(c) How many different spanning trees does the following graph have?

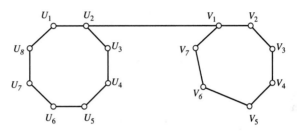

32. (a) Find all possible spanning trees of the following graph.

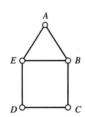

(b) How many different spanning trees does the following graph have?

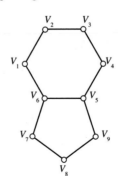

33. (a) Can you give an example of a tree with 4 vertices such that the degrees of the vertices are 2, 2, 3, and 3? If yes, do so. If not, explain why not.
   (b) If you have a tree with 4 vertices and you add up the degrees of all the vertices, what do you get?
   (c) If you have a tree with 5 vertices and you add up the degrees of all the vertices, what do you get?
   (d) If you have a tree with N vertices and you add up the degrees of all the vertices, what do you get?

34. (a) Give an example of a tree with 4 vertices such that the degrees of the vertices are 1, 1, 2, and 2.
   (b) Give an example of a tree with 6 vertices such that the degrees of the vertices are 1, 1, 2, 2, 2, and 2.
   (c) Give an example of a tree with N vertices such that the degrees of the vertices are 1, 1, 2, 2, 2, . . . , 2.

35. (a) Give an example of a tree with 4 vertices such that the degrees of the vertices are 1, 1, 1, and 3.
   (b) Give an example of a tree with 5 vertices such that the degrees of the vertices are 1, 1, 1, 1, 4.
   (c) Give an example of a tree with N vertices such that the degrees of the vertices are 1, 1, 1, . . . ,1, $N - 1$.

36. A highway system connecting 9 cities C_1, C_2, C_3, . . . , C_9 is to be built. Use Kruskal's algorithm to find a minimum spanning tree for this problem. The accompanying table shows the cost (in millions of dollars) of putting a highway between any 2 cities.

	C_1	C_2	C_3	C_4	C_5	C_6	C_7	C_8	C_9
C_1	*	1.3	3.4	6.6	2.6	3.5	5.7	1.1	3.8
C_2	1.3	*	2.4	7.9	1.7	2.3	7.0	2.4	3.9
C_3	3.4	2.4	*	9.9	3.4	1.0	9.1	4.4	6.5
C_4	6.6	7.9	9.9	*	8.2	9.7	0.9	5.5	4.9
C_5	2.6	1.7	3.4	8.2	*	4.8	7.4	3.7	3.5
C_6	3.5	2.3	1.0	9.7	4.8	*	8.9	4.4	5.8
C_7	5.7	7.0	9.1	0.9	7.4	8.9	*	4.7	3.9
C_8	1.1	2.4	4.4	5.5	3.7	4.4	4.7	*	2.8
C_9	3.8	3.9	6.5	4.9	3.5	5.8	3.9	2.8	*

37. Use 45°–45°–90° triangles to show that the length of the following network is $1000\sqrt{2} \approx 1414$.

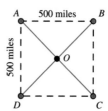

38. Use 30°–60°–90° triangles to show that the length of the following network is $500\sqrt{3} + 500 \approx 1366$.

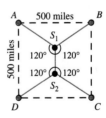

39. (a) Use 30°–60°–90° triangles to show that the length of the following network is $400\sqrt{3} + 300 \approx 993$.

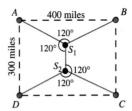

(b) Show that the length of the following network is $300\sqrt{3} + 400 \approx 919.6$.

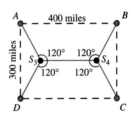

40. Consider triangle *ABC* with equilateral triangle *EFG* inside as shown in the following figure.

(a) Find angles *BFA, AEC,* and *CGB*.

(b) Explain why all the angles of triangle *ABC* are less than 120°.

(c) Explain why the Steiner point for triangle *ABC* lies inside triangle *EFG*.

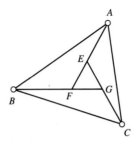

Running

41. (a) Show that in any tree with 2 or more vertices, there are at least 2 (maybe more) vertices of degree 1. (*Hint:* Use Exercise 33[d].)
   (b) Explain why in any tree with 3 or more vertices, it is impossible for all the vertices to have the same degree.

42. (a) Suppose that you are asked to find a minimum spanning tree for a weighted graph that must contain a given edge. Describe a modification of Kruskal's algorithm that accomplishes this.
   (b) Consider Exercise 36 again. Suppose that C_3 and C_4 are the two largest cities in the area, and the chamber of commerce insists that a section of highway directly connecting them must be built (or heads will roll). Find the minimum spanning tree that includes the section of highway between C_3 and C_4.

43. Prim's algorithm. The following algorithm for finding a minimum spanning tree is called Prim's algorithm.

■ **Step 0.** Pick any vertex as a starting vertex. (Call it S). Mark it red.

■ **Step 1.** Find the nearest neighbor of S (call it P_1). Mark both P_1 and the edge SP_1 in red.

■ **Step 2.** Find the nearest black neighbor to the red subgraph (i.e., the closest vertex to *any* red vertex). Mark it and the edge connecting the vertex to the red subgraph in red. Delete all black edges in the graph that connect red vertices.

Repeat Step 2 until all the vertices are marked red. The red subgraph is a minimum spanning tree.

Use Prim's algorithm to find a minimum spanning tree for the irrigation system problem given in Example 9.

44. Consider four points (A, B, C, and D) forming the vertices of a rectangle with length b and width a as shown in the following figure ($a < b$). Determine the conditions on a and b so that the length of the minimum spanning tree is less than the length of *one* of the (two) Steiner trees.

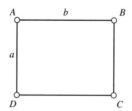

45. (a) Let P be an arbitrary point inside the equilateral triangle RQT as shown in the following figure. Suppose also that perpendiculars are drawn from P to the three sides as shown in the figure. Show that the sum of the lengths of PA, PB, and PC is equal to the length of the altitude of the triangle. [*Hint:* Compute the areas of triangle RPQ, triangle QPT, and triangle TPR. How does the sum of these three areas compare with the area of triangle RQT?]

(b) Let triangle *ABC* be an arbitrary triangle with all angles less than 120°, and let *S* be a Steiner point inside the triangle. Draw lines perpendicular to *SA, SB,* and *SC*. Explain why these lines intersect to form an equilateral triangle *RQT*.

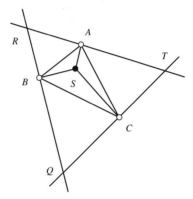

(c) Let *P* be any point other than *S* in triangle *ABC,* and draw perpendiculars from *P* to the three sides of triangle *RQT* as shown. Use (a) to conclude that *PA′* + *PB′* + *PC′* = *SA* + *SB* + *SC,* and yet *PA* ≥ *PA′*, *PB* ≥ *PB′*, and *PC* ≥ *PC′* (why?), and so *PA* + *PB* + *PC* ≥ *SA* + *SB* + *SC*.

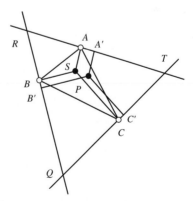

(d) Use (c) to conclude that no junction point *P* can give a shorter network than the Steiner point *S*.

46. Find the shortest network connecting the 12 vertices of the "Star of David" shown in the following figure.

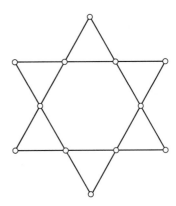

47. Eight cities (*A* through *H*) are located as in the following figure. Find the shortest network connecting the 8 cities. What is its length?

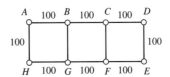

48. Find the shortest network connecting the 16 vertices of the 3 by 3 "checkerboard" shown in the following figure.

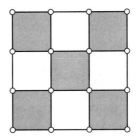

APPENDIX:
The Soap Bubble
Solution

Every child knows about the magic of soap bubbles: Take a wire or plastic ring, dip it into a soap-and-water solution, blow, and presto—beautiful iridescent geometric shapes magically materialize to delight and inspire our fantasy. Adults are not averse to a puff or two themselves.

What's special about soapy water that makes this happen? A very simplistic under-standing of the forces of nature that create soap bubbles will help us understand how these same forces can be used to find (imagine, of all things) Steiner trees connecting a given set of points.

Take a liquid (any liquid), and put it into a container. When the liquid is at rest, there are two categories of molecules: those that are on the surface and those that are below the surface. The molecules below the surface are surrounded on all sides by other molecules like themselves and are therefore in perfect balance—the forces of attraction between molecules all cancel each other out. The molecules on the surface, however, are only partly surrounded by other molecules and are therefore unbalanced. Because of this, an additional force called **surface tension** comes into play for these molecules. As a result of this surface tension, the surface layer of any liquid behaves exactly as if it were made of a very thin, elastic material. The amount of elasticity of this surface layer depends on the structure of the molecules in the liquid. Soap or detergent molecules are particularly well suited to create an extremely elastic surface layer. An ideal soap film solution can be obtained by mixing approximately one part of dishwashing detergent to ten parts of water and adding a small amount of glycerin (not nitro please!) to the solution.

The connection between the preceding brief lesson in soapy solutions and the material in this chapter is made through one of the fundamental principles of physics: A physical system will remain in a certain configuration only if it cannot *easily* change to another configuration that uses less energy. Because of its extreme elasticity, the surface layer of a soapy solution has no trouble changing its shape until it feels perfectly comfortable—i.e., at a position of relatively minimal energy. When the energy is proportional to the distance, minimal energy results in minimal distance—ergo Steiner trees.

Suppose that we have a set of points $(A_1, A_2, \ldots, A_N)$ for which we want to find a shortest network. We can find a Steiner tree that connects these points by means of an ingenious device which we will call a *soap bubble computer*. To begin with, we draw the points $A_1, A_2, \ldots, A_N$ to exact scale on a piece of paper. The scale should be such that points are not too close to each other (a minimum of 1 or 2 inches apart will do just fine). We now take two sheets of Plexiglas or Lucite and, using the paper map as a template, drill small holes on both sheets of Plexiglas at the exact locations of the points. Then we put thin metal or plastic pegs through the holes in such a way that the two sheets are held about an inch apart.

When we dip our device into a soap-and-water solution and pull it out, the soap bubble computer goes to work. The film layer that is formed between the plates connects the various pegs. For a while it will move seeking a configuration of minimal energy. Very shortly, it settles into a Steiner tree.

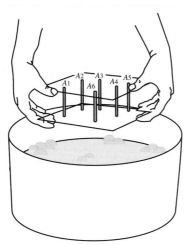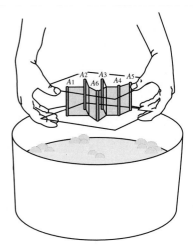

It is a bit of a disappointment that the Steiner tree we get is not necessarily the shortest network. The reasons for this are beyond the scope of our discussion. (The interested reader is referred to the excellent technical discussion of soap film computers in reference 1.) At the same time, we should be thankful for what nature has provided: a simple device that can compute in seconds what might take hours to do with pencil and paper.

REFERENCES AND FURTHER READINGS

1. Almgren, Fred J., Jr., and Jean E. Taylor, "The Geometry of Soap Films and Soap Bubbles," *Scientific American,* 235 (July 1976), 82–93.

2. Bern, M., and R. L. Graham, "The Shortest Network Problem," *Scientific American,* 260 (January 1989), 84–89.

3. Cheriton, D., and R. E. Tarjan, "Finding Minimum Spanning Trees," *SIAM Journal on Computing,* 5 (1976), 724–742.

4. Cockayne, E. J., and D. E. Hewgill, "Exact Computation of Steiner Minimal Trees in the Plane," *Information Processing Letters,* 22 (1986), 151–156.

5. Courant, R., and H. Robbins, *What is Mathematics?* New York: Oxford University Press, 1941.

6. Du, D.-Z., and F. K. Hwang, "The Steiner ratio conjecture of Gilbert and Pollak is true," Proceedings of the National Academy of Sciences, USA, 87 (December 1990), 9464–9466.

7. Gilbert, E. N., and H. O. Pollak, "Steiner Minimal Trees," *SIAM Journal of Applied Mathematics,* 16 (1968), 1–29.

8. Graham, R. L., and P. Hell, "On the History of the Minimum Spanning Tree Problem," *Annals of the History of Computing,* 7 (January 1985), 43–57.

9. Kruskal, J. B., Jr., "On the Shortest Spanning Subtree of a Graph and the Traveling Salesman Problem," *Proceedings of the American Mathematical Society,* 7 (1956), 48–50.

10. Ore, O., *Graphs and Their Uses.* New York: Random House, 1963.

11. Pierce, A. R., "Bibliography on Algorithms for Shortest Path, Shortest Spanning Tree and Related Circuit Routing Problems (1956–1974)," *Networks,* 5 (1975), 129–149.

CHAPTER 8

Directed Graphs and Critical Paths

Scheduling Problems

How long does it take to build a house? Here is a deceptively simple question that defies an easy answer. On the one hand, there is the obvious: the fact that the answer depends on the type and the size of the house, the size and skill of the work force, the type of tools and machinery used, etc. Less obvious, but equally important is another variable: someone's ability to organize and coordinate the timing of people, equipment, and work so that things get done in an efficient way. For better or for worse, this last issue boils down to a mathematical problem, just one in a large family of problems that fall under the purview of what is known as the *mathematical theory of scheduling*. Discussing some of the basics of this theory will be the theme of this chapter.

Let's get back to our original question. According to the Building Industry Association, a national association of home builders based in Washington, D.C., it takes 1092 man-hours to build the average American house. (Since we are not being very picky about details, we'll just call it 1100 hours.) One way to think about what the above statement means is this: Given one worker, and assuming he or she knows how to do every single job required for building a house, it would take that worker about 1100 hours of labor to finish this hypothetical average American house.

Let's now turn the question on its head: If we had 1100 workers, each of them qualified to do any construction type of work, could we get the same house built in one hour? We know that the answer is a resounding no, and that in fact, even if we put a million workers on the job, we still could not get the house built in one hour. There are some physical limitations to the speed with which a house can be built which are outside of the builder's control—the fact that there are some jobs that cannot be speeded up beyond a certain point regardless of how many workers one puts on that job and even more significantly, the fact that certain jobs can only be started after certain other jobs have been completed (roofing, for example, can only be started after framing has been completed).

OK, so given the fact that it is impossible to build the house in one hour, how fast can we build it if we had an unlimited number of workers at our dis-

259

posal and we only cared about speed? (To the best of our knowledge, the record is something a tad under 24 hours.) How fast could we build the house if we only had at our disposal 10 workers, all equally capable? What if we only had three workers? What if things are turned around on us and instead of a fixed number of workers we are given a certain time deadline—say we are allowed at most two weeks from start to finish? Should we contract for fifty workers? Could we get by with forty? What is the fewest number needed to meet the deadline? All of these are the kinds of questions scheduling theory is designed to deal with, and while they may sound simple, they are surprisingly difficult to answer, even in the most simplified sets of circumstances.

Scheduling problems analogous to the one of building a house occur in many other walks of life, and can range from the frivolous (preparing a dinner party) to the mundane (scheduling mechanics at a garage) to the deadly serious (scheduling computer programs in a "star wars" type of defense system). While each of these problems presents its own particular idiosyncrasies, and while it is rarely the case that two different real-world scheduling problems are exactly alike, it is nonetheless true that most scheduling problems share the same basic elements.

SCHEDULING: THE BASIC ELEMENTS

We will now introduce the principal characters in every scheduling story.

The Processors.   This is the name that we give to the "workers" that carry out the work. While the word *processor* sounds a little cold and inhuman, it does underscore an important point: Processors need not be human beings. In scheduling, a processor could just as well be a robot, or a computer, or even a cow milking machine. For the purposes of our discussion, we will represent the processors by P_1, P_2, P_3, . . . , P_N, where N represents the total number of processors.

The number of processors N can range from just 1, to the tens of thousands, but when N is 1, the whole question of scheduling is trivial and not very interesting. As far as we are concerned, real scheduling problems begin when N is 2 or more.

The Tasks.   In every complex project, there are individual pieces of work, often called "individual jobs" or "tasks." We will need to be a little more precise than that, however. We will define a *task* as an indivisible unit of work that (either by nature or by choice) cannot be broken up into smaller units. Moreover, and most importantly, in our definition of the term, *a task is always carried out by a single processor.*

To clarify the concept, let's consider a simple illustration. If a foreman assigns the wiring of a house to a single electrician, and it takes him six hours to do it, then we will consider wiring the house as a single, six-hour task. On the other hand, he may assign the job to two electricians, who together take, let's say, four hours. In this case, we will consider the job of wiring the house as two separate four-hour tasks. In general, we will use capital letters *A, B, C,* etc., to represent the tasks, although when convenient, we will also use appropriate abbreviations (clever things such as *WE* for "wiring the electrical system," *PL* for "plumbing," etc.).

At any given moment in time throughout the project, a task can be *in execution* (some processor is in the midst of carrying it out), *completed, ready* (can be started at that time), or *ineligible* (cannot be started at that time because certain other requirements have not yet been met).

The Processing Times.   Associated with every task there is a number, called the *processing time* of the task. It represents the amount of time, without interruption, required by a single processor to carry out the task. At this time, the alert reader would ask: The amount of time required by which processor? After all, it might take Archie four hours to carry out the same task that takes Betty only one hour. This surely complicates matters, so we will have to make an important concession to expediency. From now on, we will make the assumption that each processor can carry out every one of the tasks, and the processing time for a task does not depend on the processor chosen. To help things along even further, we will make a second assumption: *Once the task is started, the processor must execute it without interruptions.* A processor cannot stop in the middle of a task, be it to start another task or to take a break. With these assumptions, it now makes sense to talk about the processing time of each task, a single nonnegative number which we will attach (in parentheses) to the right of the task's name. Thus, when we see $A(5)$, we take this to mean that it takes 5 units of time (be it minutes, hours, or whatever) to execute the task called A, and that this is the case whether it is done by P_1, or P_2, or any other processor.

The Precedence Relations.   These are restrictions on the order in which the tasks can be executed. A typical precedence relation is of the form *task X precedes task Y,* and it means that task Y cannot be started until task X has been completed. Such a precedence relation can be conveniently abbreviated by writing $X \rightarrow Y$. Needless to say, precedence relations are the result of the fact that sometimes there are laws that govern the order in which things are done in the real world—we just don't put our shoes on before our socks! A single scheduling problem can have hundreds or even thousands of precedence relations, each of them adding another restriction on the scheduler's freedom. (As a general rule of thumb, the more precedence relations there are, the harder it is to create a good schedule.)

At the same time, we must remember that precedence relations between tasks are not necessarily the rule, and that in many cases there are no restrictions as to the order of execution between two tasks—some people put their shoes on before their shirt, others put their shirt on before their shoes, and occasionally, some of us have been known to put our shoes and shirt on at the same time. When a pair of tasks X and Y are not subject to any precedence requirements (neither $X \to Y$ nor $Y \to X$), we say that the tasks X and Y are **independent**. It is possible for a scheduling problem to be made up of all independent tasks (no precedence relations whatsoever to worry about). We will discuss this special situation in greater detail later in the chapter.

Two final comments about precedence relations. First, precedence relations are *transitive*: if $X \to Y$ and $Y \to Z$, then it must be true that $X \to Z$. In a sense, the last precedence relation is implied by the first two, and it is really unnecessary to specifically mention it. The second observation is that *we cannot have precedence relations forming a cycle!* As the reader can well imagine, if we have tasks X, Y, and Z such that $X \to Y$, $Y \to Z$, and $Z \to X$, then when it is time to schedule these three tasks, we will be in the same position as a dog chasing its tail. From here on, we will assume that all our scheduling problems are sensible, and that there are no cycles of precedence relations in them.

Processors, tasks, processing times, and precedence relations are the basic ingredients in every scheduling problem. They constitute, in a manner of speaking, the hand that is dealt to us. But how do we play such a hand? To get a small inkling of what's to come, let's look at the following (very) simple example.

Example 1. We have 2 processors (P_1 and P_2) to schedule the following 4 tasks (with the processing times in parentheses representing hours): $A(4)$, $B(5)$, $C(7)$, and $D(3)$. The only precedence relation is $A \to C$. Even in this very simple situation, many different schedules are possible. Figure 8-1 shows several different possibilities, each one illustrated by means of a timeline. Figure 8-1(a) shows a schedule that, to put it bluntly, is rather dumb. All the short tasks were assigned to one processor (P_1) and all the long tasks to the other processor (P_2)—obviously not a very clever strategy. Under this schedule, the **finishing time** (the time elapsed from the beginning of the project until the last task is completed) is 12 hours. Fig. 8-1(b) shows what looks like a much better schedule, but this particular schedule is in violation of the precedence relation $A \to C$. We just simply are not allowed to start task C on the third hour. In fact, this is true not only for this schedule, but for any other schedule we might think of. No matter how clever we are (and no matter how many processors we have at our disposal), *task C can never be started before the fourth hour.* Why not? Because of the requirement that $A(4)$ must be finished first. On the other hand, if we make P_2 sit idle for one hour, waiting for the green light so to speak, to start task C, we get a perfectly good schedule, shown in Fig. 8-1(c). The finishing time is 11 hours, and given the fact that we know that task $C(7)$ cannot be started until the fourth hour, this is as short as it is going to get. *No possible schedule can complete this project in less than 11 hours.* As far as finishing time is concerned, the schedule shown in Fig. 8-1(c) is *optimal.* In general, there is likely to be more than one optimal schedule, and Fig. 8-1(d) shows a different schedule with a finishing time of 11 hours.

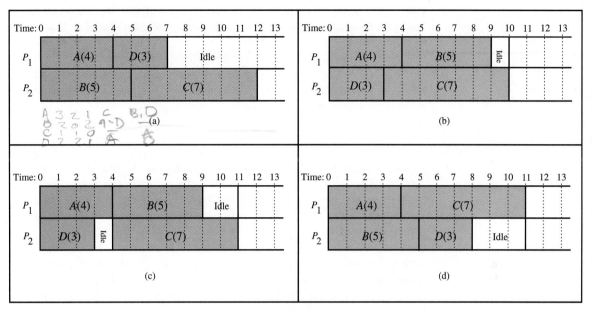

FIGURE 8-1 Some possible schedules for Example 1. (a) An *inefficient* schedule. Finishing time is 12 hours. (b) An illegal schedule. The precedence relation $A \rightarrow C$ is violated when C is started before A is completed. (c) A small adjustment in the preceding schedule makes things OK: Make P_2 sit idle for one hour before starting task C. Finishing time is 11 hours. (d) A different schedule with finishing time of 11 hours. Schedules (c) and (d) are both *optimal,* since it is impossible to finish the project in less than 11 hours.

As scheduling problems go, Example 1 was a fairly simple one. But even from such a simple example, we can draw some useful lessons. First, notice that even though we had only 4 tasks and 2 processors, we were able to create several different schedules, and the 4 we looked at were just a sampler—there are other possible schedules which we didn't bother to discuss. Imagine what would happen if we had hundreds of tasks and dozens of processors—the number of possible schedules to consider would boggle the mind. In looking for a good, or even optimal, schedule, we are going to need a systematic way to sort through the many possibilities—in other words, we are going to need some good *scheduling algorithms.*

The second useful thing we learned in Example 1 is that when it comes to the finishing time of a project, there is a bottom line—a minimum time barrier below which no schedule can reach, no matter how good an algorithm we use or how many processors we put to work. In Example 1, this minimum barrier was 11 hours, and as luck would have it, we found a schedule (two to be exact) with finishing time to match this minimum. Once we have this, we know we are home free—the schedule is optimal. While in general one isn't so lucky—the theoretical minimum cannot be accomplished by any schedule, no matter how good—knowing this minimum is particularly useful. One of the most important things we are going to learn in this chapter is how to calculate this theoretical minimum called the **critical time** of the project.

To set the stage for a formal discussion of scheduling algorithms, we will introduce our main example for the chapter. While couched in science fiction terms, the situation it describes is not entirely unrealistic.

Example 2 (Building the American dream home—in Mars). It is the year 2250. Due to your excellent job performance, your company has decided to transfer you to a new location: Scientific Base K1 on the planet Mars. As part of your job benefits, you will be provided with housing—your own individual deluxe model MHU (Martian Habitat Unit). The rub is that MHUs come in the form of prefabricated kits which are imported from Space Station USSS3 and have to be assembled together at the site, an elaborate and unpleasant job (see photo). Rather than doing the dirty work yourself, you decide to hire a couple of HUBRs (Habitat Unit Building Robots), affectionately known as "Hubies." Hubies can be rented by the hour at the local Rent-a-Robot outlet, but, needless to say, they are expensive. Given that you are footing the rental bill, your primary interest is to come up with a schedule that gets the MHU built as quickly as possible. That, in essence, is what this example is all about.

There are 15 separate tasks involved in building an MHU, and they are shown in Table 8-1, with the respective processing times given in Hubie-hours. In addition, the MHU manual shows 17 different precedence relations among these tasks, and they are listed in Table 8-2.

A completed Martian Habitat Unit (MHU) in the background, getting a last minute look over from one of the Hubies that built it (Photo Courtesy of NASA)

Task	Symbol (Processing time)
Assemble Pad	AP(7)
Assemble Flooring	AF(5)
Assemble Wall Units	AW(6)
Assemble Dome Frame	AD(8)
Install Floors	IF(5)
Install Interior Walls	IW(7)
Install Dome Frame	ID(5)
Plumbing	PL(4)
Install Atomic Power Plant	IP(4)
Install Pressurization Unit	PU(3)
Install Heating Units	HU(4)
Install Commode	IC(1)
Interior Finish Work	FW(6)
Pressurize Dome	PD(3)
Install Entertainment Unit (Virtual reality TV, computer, music box, communication port, etc.)	EU(2)

▨ **TABLE 8-1**

Precedence Relations
AP → IF
AF → IF
IF → IW
AW → IW
AD → ID
IW → ID
IF → PL
IW → IP
IP → PU
ID → PU
IP → HU
PL → IC
HU → IC
PU → EU
HU → EU
IC → FW
HU → PD

▨ **TABLE 8-2**

Before we proceed with the problem of how to best schedule the building of our MHU, we will take a brief but important detour. ▬

DIRECTED GRAPHS

There is a convenient way to summarize all the information presented in Tables 8-1 and 8-2, and it is shown in Fig. 8-2. In this figure, the tasks are represented by vertices and the precedence relations are represented by arrows pointing from one vertex to another, so that an arrow pointing from vertex X to vertex Y indicates that task X must be completed before task Y can be started.

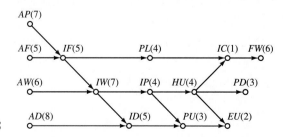

FIGURE 8-2

Notice that Fig. 8-2 looks a lot like an ordinary graph except that the edges point in a certain direction. A graph in which the edges have a direction associated with them is called a **directed graph**, and more commonly a **digraph**.

Just like graphs, digraphs are used to describe relationships between objects, but in this case the relationships do not necessarily flow both ways. When object X is related to object Y, we show this by having the arrow point from vertex X to vertex Y.

To distinguish digraphs from ordinary graphs, we use a slightly different terminology. Instead of "edge," we use the word **arc** to indicate the edge has a certain direction; instead of saying that vertices X and Y are adjacent, we say that X is **incident to** Y if the arc's arrowhead points toward Y ($X \rightarrow Y$), and that X is **incident from** Y if the arc's arrowhead points toward X ($X \leftarrow Y$); in addition to the degree of a vertex we speak about the indegree and the outdegree of a vertex (the **indegree** is the number of arrowheads pointing toward the vertex; the **outdegree** is the number of arrowheads coming out of the vertex). In a digraph, a **path** from vertex X to vertex Y is a directed sequence of arcs starting at X and ending at Y.

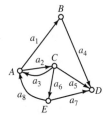

FIGURE 8-3

Consider the digraph in Fig. 8-3. This is a digraph with 5 vertices (A, B, C, D, and E) and eight arcs (a_1, a_2, a_3, a_4, a_5, a_6, a_7, and a_8). This digraph tells us, for example, that C is related to D but D is not related to C. On the other hand, the relationship between A and C is mutual. The indegrees and outdegrees of each vertex in Fig. 8-3 are shown in the following table.

Vertex	Indegree	Outdegree	Total Degree
A	2	2	4
B	1	1	2
C	1	3	4
D	3	0	3
E	1	2	3

Examples of paths from A to D are A, C, D and A, B, D and even A, C, A, B, D, but A, E, D is not (there is no arrow going from A to E).

There are many real life situations that can be represented by digraphs. Here are a few examples:

1. **Transportation.** Here the *vertices* might represent locations within a city, and the *arcs* might represent one-way streets.

2. **Communication.** Here the *vertices* might represent message centers or message sources, and the *arcs* the possible flows of information.

3. **Pipelines.** Here the *vertices* might represent pumping stations, and the *arcs* the direction of flow.

4. **Chain of command.** In a corporation or in the military we can use a digraph to describe the chain of command. The *vertices* are individuals, and an *arc* from *X* to *Y* indicates that *X* can give orders (is a superior) to *Y*.

5. **Asymmetric relationships.** Asymmetric relationships are those that are not always reciprocated. Being in love is a good example. We can describe asymmetric relationships by a digraph. In the case of being in love, the *vertices* are individuals, and we put an *(arc)* from *X* to *Y* if *X* is "in love" with *Y*. Sometimes there might also be an *arc* from *Y* to *X*, and sometimes (sigh) there might not.

**SCHEDULING:
THE MAIN TOOLS**

Let's return now to the problem of scheduling the construction of our Martian Habitat Unit. We now know that the main elements of the problem can be conveniently described by the directed graph shown in Fig. 8-4(a) (it's just Fig. 8-2 shown once again for the reader's convenience). Fig. 8-4(b) shows a slight modification of Fig. 8-4(a). Here we have added two fictitious tasks, START and END, with processing time 0. This is just a convenience that allows us to visualize the entire project as a flow that begins at START and concludes at END. By giving these fictitious tasks zero processing time we are not affecting the time calculations for the project. The digraph shown in Fig. 8-4(b) is called the **project digraph**.

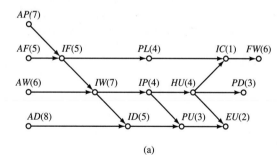

(a)

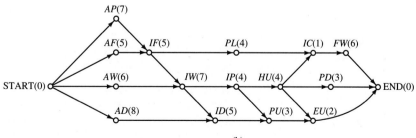

FIGURE 8-4

(b)

The Priority List Model

The project digraph is the basic tool that summarizes all the parameters of a scheduling problem, but there is nothing in the project digraph itself that specifically tells us how to create a schedule. To actually create a schedule, we are going to need something else, some set of instructions as to the order in which tasks should be executed. The basic tool that we will use to accomplish this is the **priority list**. A priority list is nothing more than a list of all the tasks in a particular order. The order is arbitrary (it doesn't have to make any special sense and it is unrelated to the precedence relations). Change the order of the tasks and you get a different priority list. Thus, there are as many priority lists as there are ways to order the tasks. (For 3 tasks there are 6 possible priority lists; for 4 tasks there are 24 priority lists; for 10 tasks there are more than three million priority lists; and for 100 tasks there are more priority lists than there are molecules in the universe. Clearly, a shortage of priority lists is not going to be our problem!) The alert reader may recognize a concept that we first saw in Chapter 2 and then again in Chapter 6—the concept of the **factorial** of a number.

For a project consisting of M tasks, the number of possible priority lists is

$$M! = 1 \times 2 \times 3 \times \ldots \times M.$$

How do we use the priority list to create a schedule? Think of the priority list as the order in which the scheduler would prefer to see the tasks executed. The only reason not to follow that exact order would be when some precedence relation prohibits one from doing so. Precedence relations override the priority list, but other than that we assign tasks to processors according to the priority list.

The ground rules for how we assign tasks to processors based on a priority list can be summarized as follows. At any given moment in time throughout the project, three different situations are possible:

1. *All processors are busy.* Nothing for us to do but wait.

2. *One of the processors is free.* We scan the priority list from left to right looking for the first *ready* task, which we assign to that processor.

   (Remember that for a task to be *ready*, all the tasks that are incident to it in the project digraph must have been completed.) If there are no ready tasks at that moment, the processor must sit idle until things change.

3. *More than one processor is free.* In this case the first ready task on the priority list is given to the first free processor, the second ready task is given to the second free processor, and so on. If there are more free processors than ready tasks, some of the latter processors will remain idle. Since the

processors are identical and (at least in theory) have no say in the matter, the choice of which free processor gets which ready task is totally arbitrary (we could just as well toss a coin). It simply makes things easier for us to have a consistent policy for choosing among the processors.

Before we take on our Martian example once again, it might be a good idea to illustrate how the priority list model for scheduling works with a couple of simple examples.

Example 3. Consider the project described by the project digraph shown in Fig. 8-5, and suppose that we are given the following priority list:

Priority List: $A(6), B(5), C(7), D(2), E(5)$.

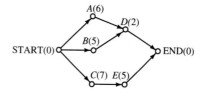

FIGURE 8-5

How would we create a schedule if we have 2 processors, P_1 and P_2? Here are all the steps, in full detail:

■ Time $T = 0$ (Start of project). $A(6)$, $B(5)$, and $C(7)$ are the only ready tasks. Following the priority list, we assign $A(6)$ to P_1 and $B(5)$ to P_2.

■ Time $T = 5$. P_1 is still busy with $A(6)$; P_2 has just completed $B(5)$. $C(7)$ is the only available ready task. We assign $C(7)$ to P_2.

■ Time $T = 6$. P_1 has just completed $A(6)$; P_2 is busy with $C(7)$. $D(2)$ has just become a ready task—its predecessors $A(6)$ and $B(5)$ have been completed. We assign $D(2)$ to P_1.

■ Time $T = 8$. P_1 has just completed $D(2)$; P_2 is still busy with $C(7)$. No ready tasks at this time, so P_1 is forced to sit idle.

■ Time $T = 12$. P_1 idle, P_2 has just completed $C(7)$. Both processors ready for work. $E(5)$ is the only ready task, so we assign $E(5)$ to P_1. P_2 sits idle.

■ Time $T = 17$. P_1 has just completed $E(5)$. Project is completed. Finishing time is 17 hours. The actual schedule is shown in Fig. 8-6.

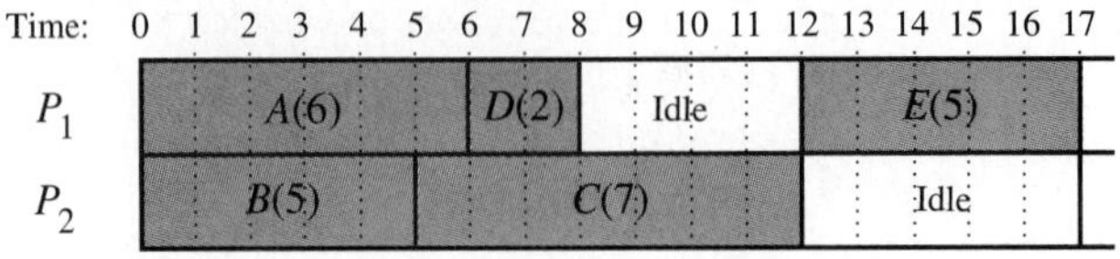

Time: 0 1 2 3 4 5 6 7 8 9 10 11 12 13 14 15 16 17

| P_1 | A(6) | D(2) | Idle | E(5) |
| P_2 | B(5) | C(7) | Idle |

FIGURE 8-6

Is the schedule above a good schedule? All we have to do is look at all the idle time and see that it is a very bad schedule. When it comes to finishing time, we would be hard put to come up with a worse schedule. Maybe we should try again!

Example 4. Let's schedule the project in Example 3 once again, but this time using a different priority list. (The project digraph is still shown in Fig. 8-5, and we still have 2 processors.) Our new priority list is:

Priority List: $E(5)$, $D(2)$, $C(7)$, $B(5)$, $A(6)$.

Here are the steps:

■ Time $T = 0$ (Start of project.) $C(7)$, $B(5)$, and $A(6)$ are the only ready tasks. Following the priority list, we assign $C(7)$ to P_1 and $B(5)$ to P_2.

■ Time $T = 5$. P_1 is still busy with $C(7)$; P_2 has just completed $B(5)$. $A(6)$ is the only available ready task. We assign $A(6)$ to P_2.

■ Time $T = 7$. P_1 has just completed $C(7)$; P_2 is busy with $A(6)$. $E(5)$ has just become a ready task, and we assign it to P_1.

■ Time $T = 11$. P_2 has just completed $A(6)$; P_1 is busy with $E(5)$. $D(2)$ has just become a ready task, and we assign it to P_2.

■ Time $T = 12$. P_1 has just completed $E(5)$; P_2 is busy with $D(2)$. No tasks left, so P_1 sits idle.

■ Time $T = 13$. P_2 has just completed the last task, $D(2)$. Project is completed. Finishing time is 13 hours. The actual schedule is shown in Fig. 8-7.

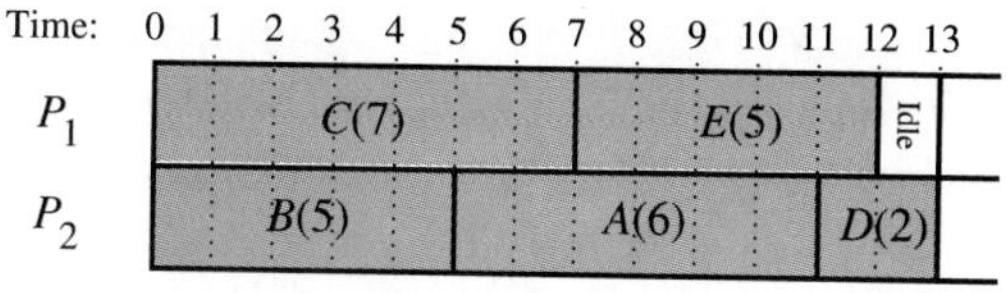

Time: 0   1   2   3   4   5   6   7   8   9   10   11   12   13

P_1 C(7)   E(5)   Idle

P_2 B(5)   A(6)   D(2)

FIGURE 8-7

Is this a good schedule? The answer is yes. Could we do even better (still using only two processors)? The answer is no. Why not? (See Exercise 21.) Could we do better if we used three processors? One would think the answer is a resounding Yes! Let's try it.

Example 5. This time we will schedule Example 4 (same project digraph, same priority list) but we will use 3 processors (P_1, P_2, and P_3). For the reader's convenience we show the priority list and project digraph again in Fig. 8-8.

Priority List: $E(5)$, $D(2)$, $C(7)$, $B(5)$, $A(6)$.

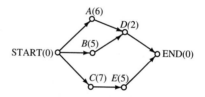

FIGURE 8-8

■ Time $T = 0$ (Start of project.) $C(7)$, $B(5)$, and $A(6)$ are the only ready tasks. We assign $C(7)$ to P_1, $B(5)$ to P_2, and $A(6)$ to P_3.

■ Time $T = 5$. P_1 is busy with $C(7)$; P_2 has just completed $B(5)$; and P_3 is busy with $A(6)$. There are no available ready tasks for P_2—$E(5)$ can't be started until $C(7)$ is done and $D(2)$ can't be started until $A(6)$ is done—so P_2 sits idle.

■ Time $T = 6$. P_3 has just completed $A(6)$; P_2 is idle; and P_1 is still busy with $C(7)$. $D(2)$ has just become a ready task. We assign $D(2)$ to P_2.

■ Time $T = 7$. P_1 has just completed $C(7)$ and $E(5)$ has just become a ready task, so without any further ado we assign it to P_1. There are no other tasks to assign.

■ Time $T = 8$. P_2 has just completed $D(2)$. There are no other tasks to assign, so P_2 sits idle.

■ Time $T = 12$. P_1 has just completed the last task, $E(5)$, so the project is completed. Finishing time is 12 hours. The actual schedule is shown in Fig. 8-9.

Time: 0 1 2 3 4 5 6 7 8 9 10 11 12

P_1	C(7) ... E(5)
P_2	B(5) ... Idle ... D(2) ... Idle
P_3	A(6) ... Idle

FIGURE 8-9

Examples 3, 4, and 5 are primarily intended to illustrate the way that the priority list model for scheduling works. It's fair to say that the basic idea is not difficult, but there is a lot of bookkeeping involved, and that becomes critical when the number of tasks is big. There are three things that we must keep track of at each stage of the schedule: Which tasks are *ready* for processing, which tasks are *being executed,* and which tasks have been *completed.* One simple way to do this is to keep updating the priority list using the following code: Ready tasks will be circled in red, tasks that are in execution will have a single red slash through them, and tasks that have been completed will have two slashes through them. Tasks that are not singled out at all are simply not ready.

We will show how to implement this strategy using the Martian Habitat Unit scheduling problem (Example 2).

Example 6. We are going to create a schedule for building a Martian Habitat Unit using 2 processors, which we will call *Hubie 1* and *Hubie 2.* For the reader's convenience, we show the project digraph again (Fig. 8-10). The schedule will be based on the following priority list:

Priority List: *AD*(8), *AW*(6), *AF*(5), *IF*(5), *AP*(7), *IW*(7), *ID*(5), *IP*(4), *PL*(4), *PU*(3), *HU*(4), *IC*(1), *PD*(3), *EU*(2), *FW*(6).

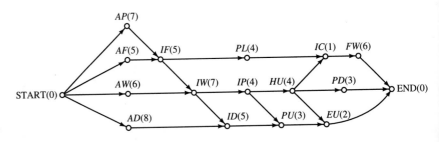

FIGURE 8-10

Priority List (Starting Status): *AD*, *AW*, *AF*, *IF*, *AP*, *IW*, *ID*, *IP*, *PL*, *PU*, *HU*, *IC*, *PD*, *EU*, *FW*.

■ **Time:** $T = 0$

Status of Processors: *Hubie 1* gets *AD*; *Hubie 2* gets *AW*.

Priority List (Updated Status): *AD*, *AW*, *AF*, *IF*, *AP*, *IW*, *ID*, *IP*, *PL*, *PU*, *HU*, *IC*, *PD*, *EU*, *FW*.

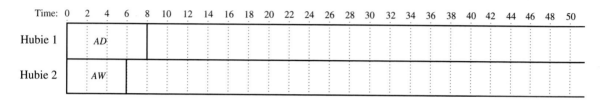

■ **Time:** $T = 6$

Status of Processors: *Hubie 1* busy (executing *AD*); *Hubie 2* gets *AF*.

Priority List (Updated Status): *AD*, *AW*, *AF*, *IF*, *AP*, *IW*, *ID*, *IP*, *PL*, *PU*, *HU*, *IC*, *PD*, *EU*, *FW*.

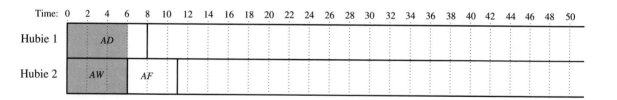

■ **Time:** $T = 8$

Status of Processors: *Hubie 1* gets *AP*; *Hubie 2* busy (executing *AF*).

Priority List (Updated Status): *AD*, *AW*, *AF*, *IF*, *AP*, *IW*, *ID*, *IP*, *PL*, *PU*, *HU*, *IC*, *PD*, *EU*, *FW*.

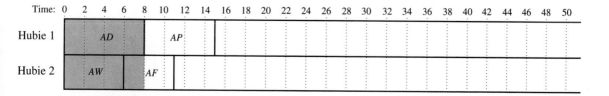

■ **Time:** $T = 11$

Status of Processors: *Hubie 1* busy (executing *AP*); *Hubie 2* idle (no ready tasks to take on).

Priority List (Updated Status): ~~AD~~, ~~AW~~, ~~AF~~, IF, ~~AP~~, IW, ID, IP, PL, PU, HU, IC, PD, EU, FW.

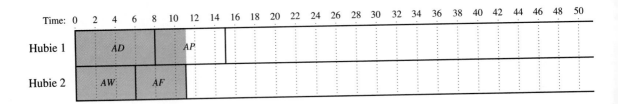

■ **Time:** $T = 15$

Status of Processors: *Hubie 1* just completed *AP*. *IF* becomes a ready task and is given to *Hubie 1*; *Hubie 2* still idle.

Priority List (Updated Status): ~~AD~~, ~~AW~~, ~~AF~~, ~~IF~~, ~~AP~~, IW, ID, IP, PL, PU, HU, IC, PD, EU, FW.

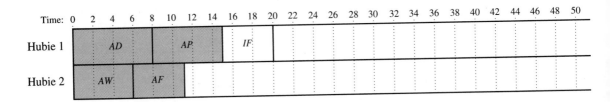

At this point we will let the reader take over and finish the schedule, hopefully following the above footsteps (see Exercise 22). Remember, the main point here is to learn how to keep track of the status of each task, and the only way to learn how to do it is with practice. (Besides, explaining the same thing over and over can get monotonous to both the explainer and the explainee.)

After a fair amount of work, we obtain the final schedule shown in Fig. 8-11. The finishing time for the project is 44 hours.

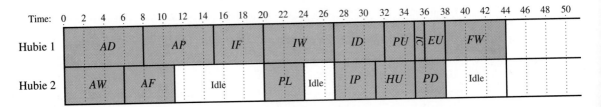

FIGURE 8-11

THE DECREASING TIME ALGORITHM

It is not difficult to convince oneself that the schedule obtained in Example 6 is not very good. What went wrong? Nothing, other than the fact that we used a priority list that had no rhyme or reason. If we want to improve our scheduling form, we will have to give a little more thought to our choice of priority list. An idea that seems intuitively obvious and that people informally use all the time is that longer jobs should be done first, shorter jobs should be done later. This is the basis for what would seem to be a somewhat more intelligent choice for the priority list: Put the tasks in decreasing order of processing times. (In case of a tie, break the tie randomly.) We will call this the **decreasing time list** and we will call the process of creating a schedule using the decreasing time list as the priority list the **decreasing time algorithm**.

Example 7. A decreasing time list for the 15 tasks required to build a Martian Habitat Unit is:

> **Decreasing Time List:** $AD(8)$, $AP(7)$, $IW(7)$, $AW(6)$, $FW(6)$, $AF(5)$, $IF(5)$, $ID(5)$, $IP(4)$, $PL(4)$, $HU(4)$, $PU(3)$, $PD(3)$, $EU(2)$, $IC(1)$.

Using the decreasing time list as our priority list we get the schedule shown in Fig. 8-12, with finishing time 42 hours. We leave the details to the reader (see Exercise 23).

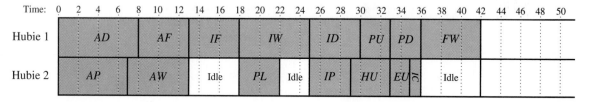

FIGURE 8-12

When looking at the finishing time under the decreasing time algorithm, one can't help but feel disappointed. This promising idea of doing the longer jobs first and the shorter jobs later turned out to be a little bit of a dud—at least in this example! What went wrong? If we work our way backward from the end, we can first see that we made a bad choice at time 33 hours. At this point there were three eligible tasks [$IC(1)$, $PD(3)$, and $EU(2)$] and two free processors. The decreasing time algorithm told us to choose the two longest tasks, $PD(3)$ and $EU(2)$. This was a very shortsighted strategy: If we had looked at what was down the road, we would have seen that $IC(1)$ is a much more critical task than the other two because we can't start $FW(6)$ until we finish IC. We see once again that looking at the immediate rather than the long-term consequences of our actions can result in bad decisions.

An even more blatant example of how the decreasing time algorithm can lead to bad choices in scheduling occurs at the very start: We failed to notice that it is critical to start *AP*(7) and *AF*(5) as early as possible. Until we finish *AP* and *AF*, we cannot start *IF*(5); and unless we finish *IF*, we cannot start *IW*(7); and until we finish *IW*, we cannot start *IP*(4) and *ID*(5); and so on down the line. The lesson to be learned here is that we might be better off if the priority we give to a task is based on the total length of all tasks that lie ahead of it. Simply put, the more the total amount of work lying ahead of a task, the sooner that task should be started.

THE CRITICAL PATH ALGORITHM

To formalize this idea we introduce the concept of a critical path from a vertex. Given a vertex *X* of a project digraph, of all the paths from vertex *X* to vertex END, the one that has the longest total sum of processing times is called the **critical path from *X*.**

Here are some examples using the MHU project diagraph:

Example 8. Let's suppose we want to find the critical path from the vertex *HU*. There are three paths from *HU* to END. They are (a) *HU, IC, FW,* END; (b) *HU, PD,* END; and (c) *HU, EU,* END. The sum of the processing times for the first path (a) is 4 + 1 + 6 = 11; the sum for the second path (b) is 4 + 3 = 7; and the sum for the third path (c) is 4 + 2 = 6. Of the three, the first path (a) has the largest sum, so it is the *critical path from HU.*

Example 9. If we use vertex *AD* as our starting point, there is only one path from *AD* to END—it is *AD, ID, PU, EU,* END. Since this is the only path, it is also the *critical path from AD.* Its total processing time is 18.

Example 10. There are quite a few paths from START to END, but after looking at the project digraph for a little while we can see that the one with the longest total processing time is START, *AP, IF, IW, IP, HU, IC, FW,* END with a total time of 34 hours. This is the *critical path from the vertex* START.

In Example 10 we obtain, more or less by trial and error, the critical path that starts with the vertex START. In any project digraph, this particular critical path is of fundamental importance, so much so that it is singled out with the special name of **the critical path of the project** or, more briefly, just **the critical path**. Thus, the critical path of the Martian Habitat Unit project is START, *AP, IF, IW, IP, HU, IC, FW,* END (which really means *AP, IF, IW, IP, HU, IC, FW* since START and END are fictitious tasks that were added for mere convenience). The total processing time for all the tasks in the critical path is called the **critical time** of the project. For the Martian Habitat Unit project, the critical time is 34 hours.

We will come back to the significance of the critical time and the project critical path soon. Meanwhile, let's discuss how to find critical paths from any vertex of a project digraph. After all, we can hardly be expected to find critical paths the way we did in Examples 8, 9, and 10 (by just "eyeballing" the picture) in examples where there are many vertices and many possible paths. We will now describe a simple procedure (which for lack of a better name we will call the **backflow algorithm**) which will allow us to find the length of the critical paths from each and every vertex of a project digraph. The basic idea is to start at END and move backward toward START. At each vertex we record the length of the critical path from that vertex (we will use square brackets for lengths of critical paths). When we move backward to a preceding vertex, we simply add the processing time of the vertex to the length of the longest critical path "ahead" of it.

Example 11. Let's concentrate on the last part of the MHU project digraph (Fig. 8-13) and use it to illustrate how the backflow algorithm works.

- ■ In step 1 we record the lengths of the critical paths from *FW*[6], *PD*[3], and *EU*[2].

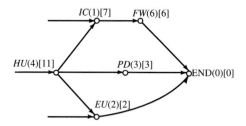

FIGURE 8-13

- ■ In step 2 we record the length of the critical path from *IC*[6 + 1 = 7].

- ■ In step 3 we record the length of the critical path from *HU*[7 + 4 = 11]. We note that the reason we got 11 is that of the three critical paths ahead of *HU* (at *IC, PD,* and *EU*), the *longest* one has length 7, which added to the 4 for *HU* gives us 11.

Continuing in this way, we get the critical paths from each vertex of the digraph as shown in Fig. 8-14. We leave it to the reader to verify the details.

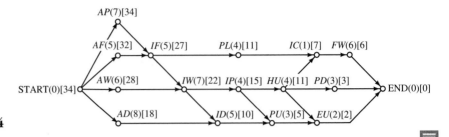

FIGURE 8-14

Why are the critical path and critical time of a project of special significance? There are two reasons: (1) The reader may recall that at the start of this chapter we discussed the fact that in every project, there is a theoretical time barrier for the finishing time of the project—a certain minimum amount of time below which the project can never be finished, regardless of how clever the scheduler is or how many processors are used. *This theoretical barrier turns out to be the project's critical time.* (2) If a project is going to be finished in the absolute minimum time (i.e., in the critical time), it is absolutely necessary that all the tasks in the critical path be done at the earliest possible time. Any delay in starting up one of the tasks in the critical path will necessarily delay the finishing time of the entire project. (By the way, this is why it is called the *critical path.*) Unfortunately, it is not always possible to schedule the tasks on the critical path one after the other, bang, bang, bang, without delay. For one thing, processors are not always free when we need them (and remember one of our rules—a processor cannot stop in the middle of one task, start a new task, and leave the other one for later). Another reason is the problem of uncompleted predecessor tasks. We cannot concern ourselves only with tasks along the critical path and disregard other tasks that might affect them through precedence relations. There is a whole web of interrelationships that we need to worry about.

It is possible, however, to use the concept of critical paths to generate very good (although not necessarily optimal) schedules. The idea is the same as the one we used with the decreasing time algorithm, but at a higher level of sophistication: Instead of using processing times to create the priority list, we will use critical path times. In other words, make a priority list in which the first task is the one with the longest critical path, the second task is the one with the second longest critical path, and so on. (As usual, break ties randomly.) The priority list obtained in this way is called the **critical path list**. The process of creating a schedule based on the critical path list is called the **critical path algorithm**.

Example 12. In Example 11 we saw how to calculate the length of the critical path from every vertex of the MHU project digraph. Arranging the tasks by critical path length we get

Critical Path List: *AP*[34], *AF*[32], *AW*[28], *IF*[27], *IW*[.
IP[15], *PL*[11], *HU*[11], *ID*[10], *IC*[7], *FW*[6], *PU*[5], *PD*[3],

Using the critical path list as the priority list we get the two-processor schedule shown in Fig. 8-15. The reader is encouraged to fill in the details (see Exercise 24). The project finishing time under this schedule is 36 hours, a big improvement over the finishing time of 42 hours produced by the decreasing time algorithm. Is this the best we can do?

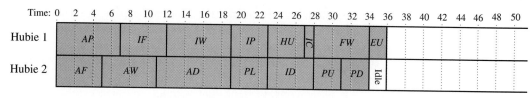

FIGURE 8-15 Schedule for the MHU project obtained using the critical path algorithm.

By and large, the critical path algorithm is an excellent method for scheduling a project (in most cases far superior to the decreasing time algorithm), but unfortunately it is not guaranteed to produce an optimal schedule. Even in the MHU project we can do better than the critical path algorithm, as illustrated by the optimal schedule shown in Fig. 8-16.

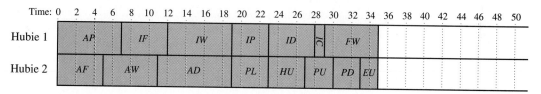

FIGURE 8-16 Optimal schedule for the MHU project using two processors.

Since none of the processors is idle at any time throughout the project, this schedule is optimal. That's the good news. The schedule was obtained basically by what could be best described as educated trial and error, rather than by some clever systematic strategy. That's the bad news. It turns out that scheduling problems have a lot in common with the traveling salesman problem (Chapter 6) and the shortest network problem (Chapter 7) in this regard—there are efficient algorithms that generally produce good schedules, but no efficient algorithm that is guaranteed to always produce an optimal schedule is presently known. Of the standard scheduling algorithms, the critical path algorithm is by far the best known and most commonly used. Other, more sophisticated algorithms have been developed in the last twenty years, and under specialized circumstances, they can outperform the critical path algorithm, but as an all-purpose algorithm for scheduling, the critical path algorithm is hard to beat.

SCHEDULING WITH INDEPENDENT TASKS

In this section we will briefly discuss what happens to scheduling problems in the special case when there are no precedence relations to worry about. This situation arises whenever we are scheduling tasks that are all independent—for example scheduling a group of typists in a steno pool to type a bunch of reports of various lengths.

It is tempting to think that without precedence relations hanging over one's head, scheduling becomes a simple problem, and one should be able to find optimal schedules without difficulty, but appearances are deceiving. *There are no optimal and efficient algorithms known for scheduling even when the tasks are all independent.*

While in a theoretical sense, we are not much better off in our ability to schedule independent tasks than we are scheduling tasks with precedence relations, from a purely practical point of view there are a few differences. For one thing, there is no getting around the fact that the nuts and bolts details of creating a schedule using a priority list become tremendously simplified when there are no precedence relations to mess with. In this case we just assign the tasks to the processors as they become free in exactly the order given by the priority list. Second, without precedence relations, the critical path time of a task equals its processing time which means that the *critical path list* and *decreasing time list* are exactly the same list and therefore the critical path algorithm is the same as the decreasing time algorithm. Before we go on, let's look at a couple of examples of scheduling with independent tasks.

Example 13. Suppose that we have the following 9 independent tasks: $A(70)$, $B(90)$, $C(100)$, $D(70)$, $E(80)$, $F(20)$, $G(20)$, $H(80)$, and $I(10)$ (all processing times are given in hours). We need to schedule these 9 tasks using three processors (P_1, P_2, and P_3). Let's do it first using the given priority list (A, B, C, D, E, F, G, H, I).

Priority List: $A(70)$, $B(90)$, $C(100)$, $D(70)$, $E(80)$, $F(20)$, $G(20)$, $H(80)$, $I(10)$.

In this case the resulting schedule is shown in Fig. 8-17. The fini[schedule is] 220 hours.

Time: 0 10 20 30 40 50 60 70 80 90 100 110 120 130 140 150 160 170 180 190 200 210 220 230 240 250

P_1	A(70)	D(70)	H(80)	
P_2	B(90)	E(80)	Idle	
P_3	C(100)	F(20)	G(20)	I(10)	Idle	

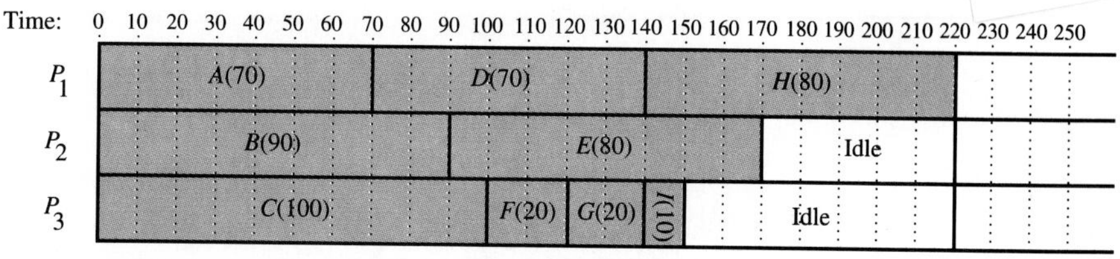

FIGURE 8-17

If we use the critical path algorithm, the priority list is the decreasing time list:

Decreasing Time List: C(100), B(90), E(80), H(80), A(70), D(70), F(20), G(20), I(10).

The resulting schedule is shown in Fig. 8-18 with a finishing time of 180 hours. Clearly, this schedule is optimal (all three processors are working for 180 hours).

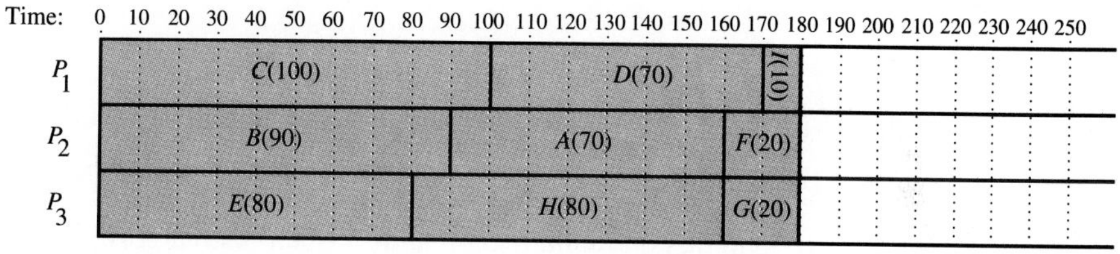

Time: 0 10 20 30 40 50 60 70 80 90 100 110 120 130 140 150 160 170 180 190 200 210 220 230 240 250

P_1	C(100)	D(70)	I(10)
P_2	B(90)	A(70)	F(20)
P_3	E(80)	H(80)	G(20)

FIGURE 8-18

In Example 13, the critical path algorithm gave us the optimal schedule, but unfortunately this need not always be the case. In some situations, the critical path algorithm can give us schedules that are a bit off. Example 14 illustrates one such situation.

Example 14. We want to schedule 7 independent tasks using three processors (P_1, P_2, and P_3). The tasks are as follows: A(50), B(30), C(40), D(30), E(50), F(30), and G(40).

If we use the critical path algorithm the priority list is $A, E, C, G, B, D, F,$ and we get the schedule shown in Fig. 8-19 with finishing time of 110 hours.

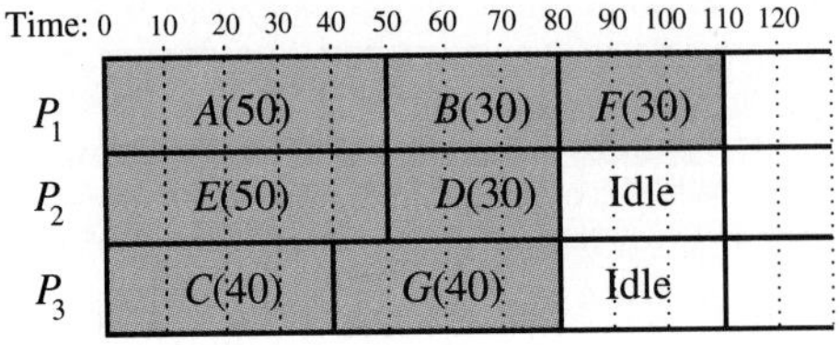

Time: 0 10 20 30 40 50 60 70 80 90 100 110 120

P_1	A(50)	B(30)	F(30)	
P_2	E(50)	D(30)	Idle	
P_3	C(40)	G(40)	Idle	

FIGURE 8-19

On the other hand, with just a minimal amount of effort, we can come up with an optimal schedule such as the one shown in Fig. 8-20.

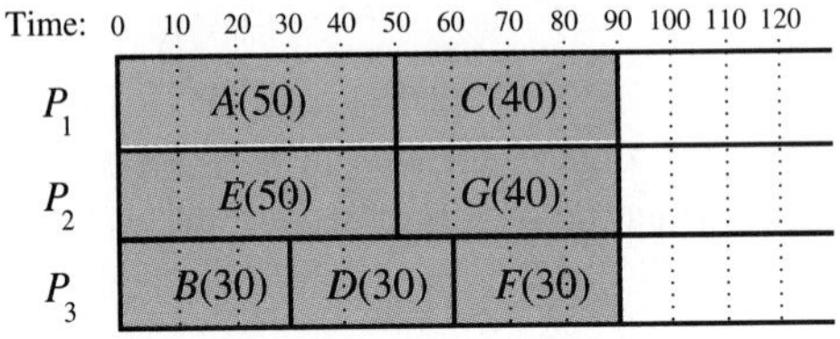

Time: 0 10 20 30 40 50 60 70 80 90 100 110 120

P_1	A(50)	C(40)	
P_2	E(50)	G(40)	
P_3	B(30)	D(30)	F(30)	

FIGURE 8-20

Given that the optimal finishing time is 90 hours we can precisely measure how "well" the critical path algorithm performed for us in terms of the relative percentage of error, which turns out to be $\frac{110 - 90}{90} = \frac{20}{90} \approx 0.2222 = 22.22\%$. Thus, in this particular example the critical path algorithm gave us a finishing time that is 22.22% over the optimal finishing time. ▪

In a sense, Example 14 is a snapshot of the critical path algorithm at its very worst. In 1969, American mathematician Ronald L. Graham of AT&T Bell Laboratories showed that for independent tasks, the critical path algorithm will always produce schedules with finishing times guaranteed to be within a certain percentage of the optimal finishing time. Moreover, Graham showed that when the tasks are independent the actual maximum error equals $(M - 1)/3M$ (where M is the number of processors) as shown in Table 8-3.

Number of Processors (M)	Max Error (CP) $\left(\dfrac{M-1}{3M}\right)$
2	$\dfrac{2-1}{3 \times 2} = \dfrac{1}{6} \approx 16.66\%$
3	$\dfrac{3-1}{3 \times 3} = \dfrac{2}{9} \approx 22.22\%$
4	$\dfrac{4-1}{3 \times 4} = \dfrac{3}{12} = 25\%$
5	$\dfrac{5-1}{3 \times 5} = \dfrac{4}{15} \approx 26.66\%$
$\vdots$	$\vdots$
100	$\dfrac{100-1}{3 \times 100} = \dfrac{99}{300} = 33\%$

■ **TABLE 8-3 Max Error (CP) represents the maximum possible error under the critical path algorithm when the tasks are independent**

(Note that with 3 processors, the percentage error in Example 14 of 22.22% is as high as it can be.) Graham's discovery essentially reassures us that when the tasks are independent we can't go too wrong using the critical path algorithm.

CONCLUSION

"Good order is the foundation of all things."
        EDMUND BURKE

"Waste neither time nor money but make the best use of both."
        BEN FRANKLIN

It is not much of an exaggeration to say that, in one form or another, the scheduling of human (and nonhuman) activity is one of the most pervasive and fundamental problems of modern life. At its most informal, it is part and parcel of the way we organize our everyday living (so much so that we are often scheduling things without realizing we are doing so). In its more formal incarnation, the systematic scheduling of a set of activities for the purposes of saving either time or money is a critical issue in industry, science, government, etc.

By now it should not surprise us that at their very core, scheduling problems are mathematical in nature, and that the mathematics of scheduling can be both simple and profound. By necessity, we focused on the simple side, but it is important to realize that there is a great deal more to scheduling than what we learned in this chapter.

In the chapter we discussed one of the more classic and important types of scheduling problems, commonly known as **machine scheduling problems**. In a machine scheduling problem we are given a set of *tasks,* a set of *precedence relations* among the tasks, and a set of identical *processors*. The objective is to schedule the tasks by properly assigning tasks to processors so that the *finishing time* for all the tasks is as small as possible.

To systematically tackle our scheduling problems we first developed a basic scheduling model called the *priority list model*. The priority list model gave us a general framework by means of which we can create, compare, and analyze schedules. Within the priority list model, many strategies can be followed (each strategy leading to the creation of a specific priority list). In the chapter we considered two basic strategies for creating schedules. The first was the *decreasing time algorithm,* a strategy that intuitively makes a lot of sense but which in practice often results in rather inefficient schedules. The second strategy, called the *critical path algorithm,* is generally a big improvement over the decreasing time algorithm, although it still falls short of the ideal of an optimal and at the same time efficient algorithm. The critical path algorithm is by far the best known and most widely used algorithm for scheduling in business and industry.

Although several other, more sophisticated strategies for machine scheduling have been discovered by mathematicians in the last thirty years, no scheduling algorithm is presently known that is efficient and always produces an optimal schedule, and the general feeling among the experts is that the chances that such an algorithm exists are slim to none.

KEY CONCEPTS

arc
backflow algorithm
critical path
critical path algorithm
critical path from a vertex
critical path list
critical time
decreasing time algorithm
decreasing time list
digraph
factorial
finishing time

incident (to and from)
indegree (outdegree)
independent tasks
machine scheduling problem
path
precedence relation
priority list
priority list model
processing time
processor
project digraph
task

EXERCISES

Walking

1. For each of the following digraphs, make and complete a table similar to the one shown here.

Vertex	Degree	Indegree	Outdegree	Vertex is incident to	Vertex is incident from
A					
B					
⋮					

(a)

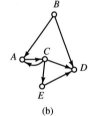

(b)

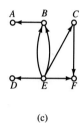

(c)

2. For each of the following digraphs list the vertices and arcs. (Use $\overrightarrow{XY}$ to represent an arc from X to Y.) Give the indegree and outdegree of each vertex and list the vertices incident to and incident from each vertex.

(a)

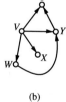

(b)

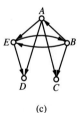

(c)

3. For each of the following, draw a picture of the digraph.
   (a) Vertices: A, B, C, D
   Arcs: A is incident to B and C; D is incident from A and B.
   (b) Vertices: A, B, C, D, E
   Arcs: A is incident to C and E; B is incident to D and E; C is incident from D and E; D is incident from C and E.

4. For each of the following, draw a picture of the digraph.
   (a) Vertices: A, B, C, D
   Arcs: A is incident to B, C, and D; C is incident from B and D.

(b) Vertices: *V, W, X, Y, Z*

Arcs: *X* is incident to *V, Z,* and *Y*; *W* is incident from *V, Y,* and *Z*; *Z* is incident to *Y* and incident from *W* and *V*.

5. A city has several one-way streets as well as two-way streets. The White Pine neighborhood is a rectangular area 6 blocks long and 2 blocks wide. Blocks alternate one-way, two-way, as shown in the following figure.

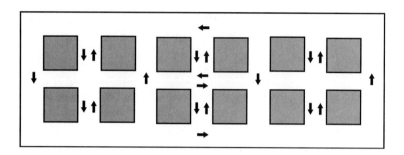

Draw a digraph that represents the traffic flow in this neighborhood.

6. A mathematics textbook for liberal studies students consists of 10 chapters. While many of the chapters are independent of the other chapters, some of the chapters require that previous chapters be covered first. The following diagram illustrates the dependence.

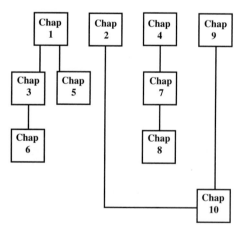

Draw a digraph that represents the dependence/independence relation among the chapters in the book.

7. Give an example of a directed graph with 4 vertices, no loops or multiple arcs, and
   (a) with each vertex having an indegree different than its outdegree
   (b) with one vertex of outdegree 3 and indegree 0, and the remaining three vertices each having indegree 2 and outdegree 1.

8. Give an example of a directed graph with 7 vertices, no loops or multip[le?]
vertices with indegree of 1 and outdegree of 1, 3 vertices of indegree 2 a[nd?]
2, and one vertex of indegree 3 and outdegree 3.

Exercises 9 through 13 refer to an apartment maintenance organization that refurbishes apartments before new tenants move in. The following tables show the tasks performed, the average time required for each task (measured in 15-minute increments), and the precedence relations between tasks.

Job	Symbol/Time	Precedence Relations
Bathrooms (clean)	$B(8)$	$L \rightarrow P$
Carpets (shampoo)	$C(4)$	$P \rightarrow K$
Filters (replace)	$F(1)$	$P \rightarrow B$
General cleaning	$G(8)$	$K \rightarrow G$
Kitchen (clean)	$K(12)$	$B \rightarrow G$
Lights (replace bulbs)	$L(1)$	$F \rightarrow G$
Paint	$P(32)$	$G \rightarrow W$
Smoke detectors (battery)	$S(1)$	$G \rightarrow S$
Windows (wash)	$W(4)$	$W \rightarrow C$
		$S \rightarrow C$

9. Using the priority list: *B, C, F, G, K, L, P, S, W*
   (a) make a schedule for refurbishing an apartment using a single worker
   (b) make a schedule for refurbishing an apartment using 2 workers.

10. Using the priority list: *W, C, G, S, K, B, L, P, F*
   (a) make a schedule for refurbishing an apartment using a single worker
   (b) make a schedule for refurbishing an apartment using 2 workers.

11. Explain what is illegal about the following schedule for refurbishing an apartment with one worker.

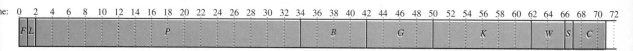

12. Explain what is illegal about the following schedule for refurbishing an apartment with 2 workers.

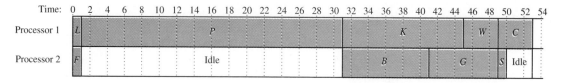

13. Explain what is illegal about the following schedule for refurbishing an apartment with 3 workers.

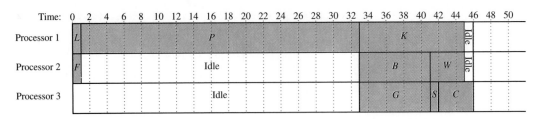

Exercises 14 through 16 refer to the problem of scheduling 7 tasks (A, B, C, D, E, F, and G) in accordance with the following project digraph.

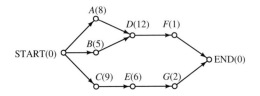

14. Using the priority list $G(2)$, $F(1)$, $E(6)$, $D(12)$, $C(9)$, $B(5)$, $A(8)$, schedule the project using 2 processors.

15. Use the decreasing time algorithm to schedule the project using 2 processors.

16. (a) Find the length of the critical path from each vertex.
    (b) What is the length of the critical path of the project digraph?
    (c) Use the critical path algorithm to schedule the project using two processors.
    (d) Explain why the schedule obtained in this case is optimal.

Exercises 17 through 20 refer to the problem of scheduling 11 tasks in accordance with the following project digraph.

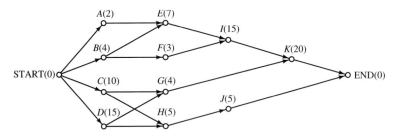

17. Use the decreasing time algorithm to schedule the project using 2 processors.

18. (a) Find the length of the critical path from each vertex.
    (b) What is the length of the critical path of the project digraph?
    (c) Use the critical path algorithm to schedule the project using 2 processors.
    (d) Explain why the schedule obtained in this case is not optimal.

19. Use the critical path algorithm to schedule the project using 3 processors.

20. Use the decreasing time algorithm to schedule the project using 3 processors.

21. Consider the schedule obtained in Example 4 in this chapter. Explain why this is an optimal schedule for 2 processors.

22. Find a schedule for building an MHU with 2 processors using the priority list: $AD(8)$, $AW(6)$, $AF(5)$, $IF(5)$, $AP(7)$, $IW(7)$, $ID(5)$, $IP(4)$, $PL(4)$, $PU(3)$, $HU(4)$, $IC(1)$, $PD(3)$, $EU(2)$, $FW(6)$. (See Example 6 in this chapter.)

23. Find a schedule for building an MHU with 2 processors using the decreasing time algorithm. (See Example 6 in this chapter.)

24. Find a schedule for building an MHU with 2 processors using the critical path algorithm. (See Example 12 in this chapter.)

Exercises 25 through 27 refer tto the following scheduling problem: There are ten computer programs that need to be executed. Three of the programs require 4 minutes each to complete, three more require 7 minutes each to complete, and four of the programs require 15 minutes each to complete. Moreover, none of the 15-minute programs can be started until all of the 4-minute programs have been completed.

25. Draw a project digraph for this scheduling problem.

26. Use the decreasing time algorithm to schedule the programs on
    (a) 2 computers
    (b) 3 computers
    (c) 4 computers.

27. Use the critical path algorithm to schedule the programs on
    (a) 2 computers
    (b) 3 computers
    (c) 4 computers.

Exercises 28 through 30 refer to the following scheduling problem: There are eight computer programs that need to be executed. One of the programs requires 10 minutes to complete, two programs require 7 minutes to complete, two more require 12 minutes each to complete, and three of the programs require 20 minutes each to complete. Moreover, none of the 20-minute programs can be started until both of the 7-minute programs have been completed, and the 10-minute program cannot be started until both of the 12-minute programs have been completed.

28. Draw a project digraph for this scheduling problem.

29. Use the decreasing time algorithm to schedule the programs on
    (a) 2 computers
    (b) 3 computers

30. Use the critical path algorithm to schedule the programs on
    (a) 2 computers
    (b) 3 computers

31. (a) Draw a project digraph for a project consisting of the 8 tasks described by the following table.

Task	Length of task	Tasks that must be completed before the task can start
A	5	C
B	5	C, D
C	5	
D	2	G
E	15	A, B
F	6	D, H
G	2	
H	2	G

(b) Use the critical path algorithm to schedule this project using 2 processors.

32. (a) Draw a project digraph for a project consisting of the 8 tasks described by the following table.

Task	Length of task	Tasks that must be completed before the task can start
A	3	
B	10	C, F, G
C	2	A
D	4	G
E	5	C
F	8	A, H
G	7	H
H	5	

(b) Use the critical path algorithm to schedule this project using 2 processors.

Jogging

33. Explain why in any digraph, the sum of all the indegrees must equal the sum of all the outdegrees.

34. Symmetric and asymmetric digraphs. A digraph is called **symmetric** if whenever there is an arc from vertex X to vertex Y there *is also* an arc from vertex Y to vertex X. A digraph is called **asymmetric** if whenever there is an arc from vertex X to vertex Y there *is not* an arc from vertex Y to vertex X. For each of the following, state whether the digraph is symmetric, asymmetric, or neither.

(a) A digraph representing the streets of a town in which all streets are one-way streets.

(b) A digraph representing the streets of a town in which all streets are two-way streets.

(c) A digraph representing the streets of a town in which there are both one-way and two-way streets.

(d) A digraph in which the vertices represent a bunch of men, and there is an arc from vertex X to vertex Y if X is a brother of Y.

(e) A digraph in which the vertices represent a bunch of men, and there is an arc from vertex X to vertex Y if X is the father of Y.

35. In 1961, T. C. Hu of the University of California showed that in any scheduling problem in which all the tasks have equal processing times and in which the original project digraph (without the START and END vertices) is a tree, the critical path algorithm will give an optimal schedule. Using this result, find an optimal schedule for the scheduling problem with the following project digraph using three processors. Assume each task takes 3 days. (Notice that we have omitted the START and END vertices.)

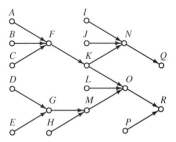

36. The following 9 tasks are all independent:
$A(4), B(4), C(5), D(6), E(7), F(4), G(5), H(6), I(7)$.
There are 4 processors available to carry out these tasks.
(a) Find a schedule using the critical path algorithm.
(b) Find an optimal schedule.

37. The following 7 tasks are all independent:
$A(4), B(3), C(2), D(8), E(5), F(3), G(5)$.
There are 3 processors available to carry out these tasks.
(a) Find a schedule using the critical path algorithm.
(b) Find an optimal schedule.

38. Use the critical path algorithm to schedule independent tasks of length 1, 1, 2, 2, 5, 7, 9, 13, 14, 16, 18, and 20 using 3 processors. Is this schedule optimal? Explain.

39. Consider the problem of scheduling 9 tasks in accordance with the following project digraph.

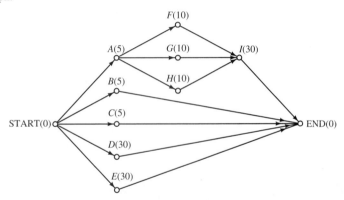

    (a) Use the critical path algorithm to schedule the project with 3 processors.

    (b) Find an optimal schedule for the project using 3 processors.

40. The faster processor paradox.

    (a) Use the critical path algorithm to schedule a project with 9 tasks using 2 processors according to the following project digraph.

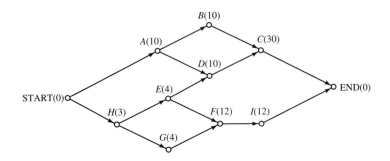

    (b) Now suppose that the processing times for each of the tasks are decreased by 1 (a faster model of processor is used), giving the following project digraph. Use the critical path algorithm to reschedule this project using 2 processors.

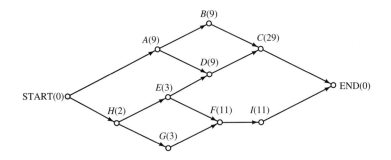

    (c) Compare the times obtained in (a) and (b). Explain how this can happen.

41. The more is less paradox.

    (a) Use the critical path algorithm to schedule a project with 8 tasks using 2 processors according to the following project digraph.

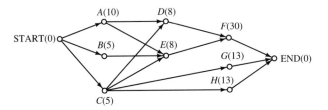

(b) Now suppose that the number of processors is increased to 3. Use the critical path algorithm to reschedule the project.

(c) Compare the finishing times obtained in (a) and (b). Explain how this can happen.

Running

Exercises 42 through 44 refer to the following fact: In 1966, Ronald L. Graham of AT&T Bell Laboratories showed that if T_{OPT} is the optimal finishing time for a given scheduling problem, then the finishing time T for any other schedule for the problem must satisfy the inequality $T \leq [2 - (1/M)]T_{OPT}$, where M is the number of processors. For example, with two processors $(M = 2)$ the finishing time T for any schedule must satisfy the inequality $T \leq \frac{3}{2}T_{OPT}$ so that no scheduling will result in a time longer than $1\frac{1}{2}$ times the optimal time.

42. Show that if $T_1 = 21$ hours and $T_2 = 12$ hours are finishing times of two different schedules for the same scheduling problem with 4 processors, then T_2 is the optimal finishing time for the scheduling problem and T_1 is the longest possible finishing time for the problem.

43. Suppose we have a scheduling problem with two processors and we came up with a schedule with finishing time $T_1 = 9$ hours. Explain why the optimal finishing time for this scheduling problem cannot be less than 6 hours.

44. Suppose we have a scheduling problem with 3 processors and we came up with two different schedules with finishing times $T_1 = 12$ hours and $T_2 = 15$ hours. Explain why the optimal finishing time for this scheduling problem has to be somewhere between 9 and 12 hours.

Exercises 45 and 46 refer to the following: It has been shown that if T_{OPT} is the optimal finishing time for a given scheduling problem in which all the tasks are independent, then the finishing time T for any schedule for the problem obtained by using the critical path algorithm must satisfy the inequality $T \leq (\frac{4}{3} - \frac{1}{3M})T_{OPT}$, where M is the number of processors. For example, with 4 processors $(M = 4)$ the finishing time T for any schedule obtained using the critical path algorithm must satisfy the inequality $T \leq \frac{5}{4} T_{OPT}$. (See Exercise 36 for an example where the equality holds.)

45. Give an example of a scheduling problem using 3 processors in which all the tasks are independent and such that the finishing time using the critical path algorithm is $\frac{11}{9}$ of the optimal finishing time.

46. Give an example of a scheduling problem using 5 processors in which all the tasks are independent and such that the finishing time using the critical path algorithm is $\frac{19}{15}$ of the optimal finishing time.

REFERENCES AND FURTHER READINGS

1. Baker, K. R., *Introduction to Sequencing and Scheduling*. New York: John Wiley & Sons, Inc., 1974.

2. Coffman, E. G., *Computer and Jobshop Scheduling Theory*. New York: John Wiley & Sons, Inc., 1976, chaps. 2 and 5.

3. Conway, R. W., W. L. Maxwell, and L. W. Miller, *Theory of Scheduling*. Reading, MA: Addison-Wesley Publishing Co., Inc., 1967.

4. Dieffenbach, R. M., "Combinatorial Scheduling," *Mathematics Teacher,* 83 (1990), 269–273.

5. Garey, M. R., R. L. Graham, and D. S. Johnson, "Performance Guarantees for Scheduling Algorithms," *Operations Research,* 26 (1978), 3–21.

6. Graham, R. L., "The Combinatorial Mathematics of Scheduling," *Scientific American,* 238 (1978), 124–132.

7. Graham, R. L., "Combinatorial Scheduling Theory," in *Mathematics Today,* ed., L. Steen. New York: Springer-Verlag, Inc., 1978, 183–211.

8. Graham, R. L., E. L. Lawler, J. K. Lenstra, and A. H. G. Rinnooy Kan, "Optimization and Approximation in Deterministic Sequencing and Scheduling: A Survey," *Annals of Discrete Mathematics,* 5 (1979), 287–326.

9. Hillier, F. S., and G. J. Lieberman, *Introduction to Operations Research* (3rd ed.). San Francisco: Holden-Day, Inc., 1980, chap. 6.

10. Roberts, Fred S., *Graph Theory and Its Applications to Problems of Society,* CBMS-NSF Monograph No. 29. Philadelphia: Society for Industrial and Applied Mathematics, 1978.

PART

III

Growth and
Symmetry

The Golden Ruler

Spiral Growth and Fibonacci Numbers

*To see a world in a grain of sand
And a heaven in a wild flower,
Hold infinity in the palm of your hand
And eternity in an hour.*

WILLIAM BLAKE

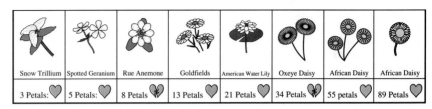

Snow Trillium	Spotted Geranium	Rue Anemone	Goldfields	American Water Lily	Oxeye Daisy	African Daisy	African Daisy
3 Petals:	5 Petals:	8 Petals	13 Petals	21 Petals	34 Petals	55 petals	89 Petals

FIGURE 9-1 . . . loves me . . . loves me not . . . loves me . . . loves me not . . . ?

The answer to this, one of the more celebrated questions posed by man (and woman) depends on whether the number of petals on the daisy is odd or even (odd means yes, even means a broken heart). But aren't the number of petals on a daisy random? Somewhat surprisingly, the answer is no! There are indeed rules that govern how many petals a daisy can have and knowing these rules can help us stack the love deck in our favor. Moreover, what is really intriguing is the fact that lurking behind these rules one can find some unusual and interesting mathematics.

Some of the same rules that govern the growth of petals on a daisy also govern the growth of other natural shapes: pineapples, pinecones, seashells, etc. All of these natural objects have one thing in common: They grow in a very special type of spiral pattern.

Spiral growth and its mathematical implications are the basic theme of this chapter. Along the way we will learn about some other mathematical concepts that are of interest in their own right: *Fibonacci numbers, the Golden Ratio,* and *gnomons.*

FIBONACCI NUMBERS

$$1 \quad 1 \quad 2 \quad 3 \quad 5 \quad 8 \quad 13 \quad 21 \quad 34 \quad 55 \quad 89 \quad \dots$$

FIGURE 9-2

Listed in Fig. 9-2 is a very remarkable group of numbers called the **Fibonacci numbers**. They are named after the Italian mathematician Leonardo of Pisa (better known by the nickname Fibonacci).[1] Notice that, except for the first three, the numbers in this list match up with the number of petals in the various flowers listed in Fig. 9-1. The remarkable frequency with which Fibonacci numbers occur among certain plants and flowers appears to be related to the equally remarkable mathematical properties of these numbers. We will discuss some of these now, and again in the next few sections.

First, the list of Fibonacci numbers is infinite. (This is what the three dots following the 89 indicate.) The list is also ordered, meaning that there is a first Fibonacci number (1), a second (1), a third (2), . . . , a seventh (13), . . . a tenth (55), an eleventh (89), etc. Needless to say, since there are infinitely many Fibonacci numbers, we can't write them all down. The entire infinite list of Fibonacci numbers is called the **Fibonacci sequence**.

Some logical questions immediately come to mind. What is the twelfth Fibonacci number? What is the one hundredth Fibonacci number? Is there a rule that would allow us to calculate the Nth Fibonacci number, for any positive integer N?

To answer these questions, we will introduce some convenient notation. We will write F_1 for the first Fibonacci number, F_2 for the second, . . . , F_{10} for the tenth, etc. Table 9-1 shows the actual values for the first eleven Fibonacci numbers once again.

| $F_1 = 1$ | $F_2 = 1$ | $F_3 = 2$ | $F_4 = 3$ | $F_5 = 5$ | $F_6 = 8$ | $F_7 = 13$ | $F_8 = 21$ | $F_9 = 34$ | $F_{10} = 55$ | $F_{11} = 89$ | $F_{12} = ?$ |

TABLE 9-1

We will use F_N to represent a Fibonacci number in a generic position N of the list. Because we are using numbers here to describe two different things (the Fibonacci number itself and the position in which it is located), there is some potential for confusion. For example, we must be able to distinguish between $F_8 + 1 = 21 + 1 = 22$ and $F_{8+1} = F_9 = 34$. Likewise, there is a big difference between $F_N - 1$ (subtract 1 from the Nth Fibonacci number) and F_{N-1} (the Fibonacci number immediately preceding the Nth Fibonacci number).

[1] Fibonacci (c. 1170–1250) was the son of a merchant and as a young man he traveled extensively with his father. Through his travels and studies in northern Africa, he became acquainted with the Arabic system of numeration and algebra, which he introduced to Christian Europe in his book *Liber Abaci* ("The Book of the Abacus"), published in 1202. Although he is best remembered for the discovery of Fibonacci numbers, they were only a minor part of his book and of his contributions to history.

Let's now go back to our original questions. First, how much is F_{12}? A quick glance at Table 9-1 suggests a pretty obvious pattern: Each Fibonacci number (from F_3 on) is the sum of the preceding two Fibonacci numbers. Thus, $F_{12} = 89 + 55 = 144$. Now, how about F_{100}? Presumably, this would be easy if we only knew F_{99} and F_{98}, which we don't. In fact, at this point we don't know $F_{99}, F_{98}, F_{97}, \ldots, F_{14}$, or F_{13}. On the other hand, it is clear that if we set our minds to it we could slowly but surely march up this Fibonacci ladder one rung at a time: $F_{13} = 144 + 89 = 233$, $F_{14} = 233 + 144 = 377$, etc. Let's cheat a little bit and say that we got $F_{97} = 83621143489848422977$ and $F_{98} = 135301852344706746049$ from a friend. We can now move up to

$$F_{99} = 135301852344706746049 + 83621143489848422977$$

$$= 218922995834555169026,$$

and finally,

$$F_{100} = 218922995834555169026 + 135301852344706746049$$

$$= 354224848179261915075.$$

Finally, let's tackle F_N. We know that this Fibonacci number, wherever it may be (as long as it is past F_2), is the sum of its two predecessors. In the notation we have developed,

$$\underbrace{F_N}_{\substack{\text{a generic}\\\text{Fibonacci}\\\text{number}}} = \underbrace{F_{N-1}}_{\substack{\text{the Fibonacci}\\\text{number right}\\\text{before it}}} + \underbrace{F_{N-2}}_{\substack{\text{the Fibonacci}\\\text{number two posi-}\\\text{tions before it}}} .$$

Of course the above rule cannot be applied to F_1 (which has no predecessors) or F_2 (which has only one predecessor), so to complete our description we add these two special cases by writing $F_1 = 1$ and $F_2 = 1$. The combination of facts

$$F_1 = 1, F_2 = 1, \text{ and } F_N = F_{N-1} + F_{N-2}$$

gives a complete description of everything we need to know to compute Fibonacci numbers. It is, in essence, their definition.

While this definition has an elegant simplicity to it, it does have one major drawback. We already saw that to calculate a Fibonacci number like F_{100} we would first have to calculate all of the preceding Fibonacci numbers ($\ldots, F_{96}, F_{97}, F_{98}, F_{99}$). Each one can be calculated by a simple addition; but the numbers get big fast, and the process, while simple, can be excruciatingly long and boring. Imagine, if you will, calculating $F_{10,000}$ this way. Just thinking about it is painful. Is there another way?

The following complicated-looking formula is known as **Binet's formula**.[2]

---

[2] The formula was actually published by Leonhard Euler in 1765. The Frenchman Jacques Binet rediscovered it in 1843 and got all the credit.

$$F_N = \frac{\left(\frac{1+\sqrt{5}}{2}\right)^N - \left(\frac{1-\sqrt{5}}{2}\right)^N}{\sqrt{5}}$$

In spite of its rather nasty appearance, there is one thing that makes this formula appealing—the fact that, in principle, it gives us a recipe for calculating any Fibonacci number without having to appeal to any of the Fibonacci numbers that come before it. Because of this, Binet's formula is said to give an *explicit* description of the Fibonacci numbers. Because it includes some "nasty" numbers like $\sqrt{5}$ and it involves exponents, doing hand calculations of Fibonacci numbers using Binet's formula is out of the question. A good calculator (with an exponent key[3]) is essential, and even then, large Fibonacci numbers can only be approximated. (That's because the calculator has to round off $\sqrt{5}$ to seven or eight decimal places.) In short, Binet's formula is primarily of theoretical interest.

THE EQUATION $x^2 = x + 1$

We will now discuss something that on its surface appears to be totally unrelated to Fibonacci numbers. In fact, on its surface, it appears to be one of those typical high school algebra questions: Solve the quadratic equation $x^2 = x + 1$. To do so, we first move all terms to the left-hand side to get $x^2 - x - 1 = 0$. We can now appeal to the famous *quadratic formula*[4] to get the solutions $\left(\frac{1+\sqrt{5}}{2}\right)$ and $\left(\frac{1-\sqrt{5}}{2}\right)$. To get an idea of the approximate size of these numbers, we need a calculator. For $\left(\frac{1+\sqrt{5}}{2}\right)$ we get 1.61803399 whereas for $\left(\frac{1-\sqrt{5}}{2}\right)$ we get -0.61803399.[5] Because $\sqrt{5}$ is an irrational number with an infinite, nonrepeating decimal expansion, both of these values are just approximations (albeit good ones) to the exact numbers $\left(\frac{1+\sqrt{5}}{2}\right)$ and $\left(\frac{1-\sqrt{5}}{2}\right)$. For most practical purposes rounding to three decimal places is sufficient, in which case we have $\left(\frac{1+\sqrt{5}}{2}\right) \approx 1.618$ and $\left(\frac{1-\sqrt{5}}{2}\right) \approx -0.618$. (The $\approx$ is there to remind us that things are not exactly equal.)

---

[3] On most calculators, the exponent key looks something like $\boxed{y^x}$. To calculate an exponent, say for example $(2.3)^7$, first enter 2.3, then enter $\boxed{y^x}$, and finally enter 7 followed by $\boxed{=}$.

[4] Just in case you forgot the quadratic formula: The solutions of $ax^2 + bx + c = 0$ are given by $x = \frac{-b \pm \sqrt{b^2 - 4ac}}{2a}$. For $x^2 - x - 1 = 0$ we get $x = \frac{1 \pm \sqrt{1+4}}{2}$. (For a brief review of the quadratic formula, see Exercises 17 through 20.)

[5] It is no coincidence that both numbers have exactly the same decimal expansion. This is clear once we observe that the two numbers must add up to 1. $\left(\frac{1+\sqrt{5}}{2}\right) + \left(\frac{1-\sqrt{5}}{2}\right) = 1$.

Notice that $\left(\dfrac{1 + \sqrt{5}}{2}\right)$ and $\left(\dfrac{1 - \sqrt{5}}{2}\right)$ happen to be the two expressions that appear (raised to the Nth power) in the numerator of Binet's formula for F_N. There are other interesting connections between the solutions of the equation $x^2 = x + 1$ and Fibonacci numbers. Let's concentrate, for a moment, on just the positive solution $\left(\dfrac{1 + \sqrt{5}}{2}\right)$. For convenience we will use the Greek letter Φ (phi) to represent $\left(\dfrac{1 + \sqrt{5}}{2}\right)$. The fact that Φ is a solution of the equation $x^2 = x + 1$ is synonymous with saying that $\Phi^2 = \Phi + 1$. Using this fact repeatedly, we can calculate other powers of Φ. For example, multiplying both sides of the equation $\Phi^2 = \Phi + 1$ by Φ we get $\Phi^3 = \Phi^2 + \Phi$. We can now substitute $\Phi + 1$ for Φ^2 and get $\Phi^3 = (\Phi + 1) + \Phi$. After all is said and done, we get $\Phi^3 = 2\Phi + 1$. To get Φ^4 we multiply both sides of our last equation ($\Phi^3 = 2\Phi + 1$) by Φ and get $\Phi^4 = 2\Phi^2 + \Phi$. Substituting $\Phi + 1$ for Φ^2 gives us $\Phi^4 = 2(\Phi + 1) + \Phi = 3\Phi + 2$. Continuing this way we get

$$\Phi^5 = 3\Phi^2 + 2\Phi = 3(\Phi + 1) + 2\Phi = 5\Phi + 3$$

$$\Phi^6 = 5\Phi^2 + 3\Phi = 5(\Phi + 1) + 3\Phi = 8\Phi + 5$$

$$\Phi^7 = 8\Phi^2 + 5\Phi = 8(\Phi + 1) + 5\Phi = 13\Phi + 8, \text{ and so on.}$$

We can now not only see that the Fibonacci numbers show up when we take the various powers of Φ but also possibly understand why it happens this way (see Exercise 37). The general rule that we have detected for the powers of Φ is

$$\Phi^N = F_N\Phi + F_{N-1}$$

THE GOLDEN RATIO

Let's go back, temporarily, to the Fibonacci numbers. Table 9-2 shows what happens when we divide consecutive Fibonacci numbers.

F_N	1	1	2	3	5	8	13	21	34	55	89	144	233
$\dfrac{F_N}{F_{N-1}}$	$\dfrac{1}{1}$	$\dfrac{2}{1}$	$\dfrac{3}{2}$	$\dfrac{5}{3}$	$\dfrac{8}{5}$	$\dfrac{13}{8}$	$\dfrac{21}{13}$	$\dfrac{34}{21}$	$\dfrac{55}{34}$	$\dfrac{89}{55}$	$\dfrac{144}{89}$	$\dfrac{233}{144}$	
In decimal form (rounded to 3 decimal places)	1.000	2.000	1.500	1.667	1.600	1.625	1.615	1.619	1.618	1.618	1.618	1.618	

TABLE 9-2   **The Ratio of Consecutive Fibonacci Numbers**

Table 9-2 shows that after some early hesitation, the ratio $\dfrac{F_N}{F_{N-1}}$ of consecutive Fibonacci numbers seems to "settle down" to a value of 1.618. It must be mentioned that none of these ratios are actually 1.618. If we go to seven decimal places, for example, $\frac{144}{89} \approx 1.6179775$ whereas $\frac{233}{144} \approx 1.6180556$. We leave it to the reader to experiment a little further with a calculator (see Exercises 5 and 6) and calculate a few more of the ratios $\dfrac{F_N}{F_{N-1}}$. It soon becomes clear that the further we go into the Fibonacci sequence, the closer the ratios $\dfrac{F_N}{F_{N-1}}$ get to some fixed number. This point can best be driven home by going to extremes. Let's calculate $\dfrac{F_{99}}{F_{98}}$ and $\dfrac{F_{100}}{F_{99}}$, both to 40 decimal places of accuracy. (Please understand that you need more than a calculator to do this!)

$$\frac{F_{99}}{F_{98}} = \frac{218922995834555169026}{135301852344706746049} \approx 1.6180339887498948482045868343656381177203$$

$$\frac{F_{100}}{F_{99}} = \frac{354224848179261915075}{218922995834555169026} \approx 1.6180339887498948482045868343656381177202$$

While these numbers are not identical, they match up in their first 39 decimal places, so that the difference between them is truly insignificant. Now, exactly what is the number that the ratios $\dfrac{F_N}{F_{N-1}}$ get closer and closer to? That's right, you guessed it—it is $\Phi = \left(\dfrac{1 + \sqrt{5}}{2}\right)$.

The number $\Phi = \left(\dfrac{1 + \sqrt{5}}{2}\right) \approx 1.618$ has turned up several times already in connection with Fibonacci numbers, and it will play a critical role for the rest of this chapter. Indeed, Φ is one of the more ubiquitous numbers in mathematics and is commonly known as **the golden ratio**. With the possible exception of π, there is no other number that plays a more important role in our physical world than the golden ratio. The ancient Greeks ascribed to it mystical characteristics and called it the *divine proportion*. The great astronomer Johannes Kepler wrote:

> *Geometry has two great treasures: one is the theorem of Pythagoras; the other, the . . . [golden] ratio. The first we may compare to a measure of gold; the second we may name a precious jewel.*

Many books have been written about the role of the golden ratio in geometry, art, architecture, music, and nature. The first known mathematical treatment of the golden ratio was *De Divina Proportione* (The Divine Proportion) written in 1509 by the Friar Luca Pacioli and illustrated by Leonardo da Vinci. Among the many modern books devoted to the golden ratio are references 1, 5, and 6.

What is it that sets the golden ratio Φ apart from other numbers? The answer is that Φ is blessed with a special and unique virtue: *It strikes a perfect balance between the large and the small.* Let's explain. Suppose we have a rod we want to break into two pieces (one large and one small) (Fig. 9-3) in some sort of perfect balance, that is, in such a way that the large piece is not too large and the small piece is not too small. In essence what this means is that the proportion between the large piece l and the small piece s should be the same as the proportion between the whole piece ($l + s$) and the large piece l.

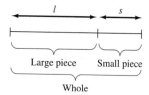

Large piece   Small piece

Whole

FIGURE 9-3

In other words, the large piece is to the small piece as the entire piece is to the large piece, i.e.,

$$\frac{l}{s} = \frac{l + s}{l}.$$

We will call this the **perfect balance equation**. To solve it, first rewrite it as

$$\frac{l}{s} = \frac{l}{l} + \frac{s}{l} = 1 + \frac{s}{l}.$$

If we now make the substitution $\frac{l}{s} = x$, the equation $\frac{l}{s} = 1 + \frac{s}{l}$ becomes

$$x = 1 + \frac{1}{x}.$$

Multiplying both sides of this last equation by x results in $x^2 = x + 1$, by now an old friend. Of the two solutions to $x^2 = x + 1$ one was a negative number $\left(\frac{1 - \sqrt{5}}{2} \approx -.618\right)$ and can be discarded (the ratio of the lengths of two rods cannot be a negative number!). This leaves one solution, and it is $\Phi = \frac{1 + \sqrt{5}}{2}$. Conclusion: The perfect balance equation between the large piece l and the small piece s is satisfied only when the ratio $\frac{l}{s}$ equals the golden ratio Φ.

The perfect balance between large and small represented by the golden ratio is particularly striking when it shows up in a rectangle. Suppose we have a rectangle with a long side (whether it be the base or the height is irrelevant) of length l and a short side of length s. If the ratio between long and short sides $\left(\frac{l}{s}\right)$ equals the golden ratio Φ, we call the rectangle a **golden rectangle**. The rectangles shown in Figs. 9-4 (a) and (b) are golden rectangles and those shown in Figs. 9-4 (c) and (d) are approximate golden rectangles.

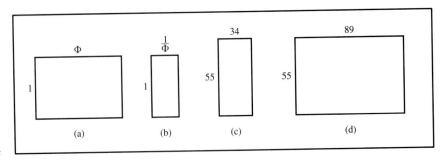

Φ

$\frac{1}{\Phi}$

34

89

1

1

55

55

(a)   (b)   (c)   (d)

FIGURE 9-4

The Golden Ratio in History

While examples of the use of the golden ratio in architecture can be traced as far back as biblical times (the ancient Pyramid of Cheops is said to have proportions based on Φ), the Greeks were the first to systematically incorporate the golden ratio into their art and architecture. The famous Greek sculptor Phidias consistently used the golden ratio to obtain the best proportions for his sculptures.[6] Another famous example of the way the golden ratio was used by the Greeks is in the proportions of the Parthenon, as illustrated in Fig. 9-5.

FIGURE 9-5 The face of the Parthenon fits almost exactly in a golden rectangle.

Horsemen. One of the friezes on the Parthenon attributed to the Greek sculptor Phidias (c. 440 B.C.). Phidias used the golden ratio to proportion not only the friezes but the Parthenon itself. (British Museum)

[6] In fact the choice of Φ to represent the golden ratio comes from Phidias' name.

The golden ratio was also extensively used by Renaissance artists, Leonardo da Vinci, Botticelli, and others. Detailed accounts of this can be found in reference 5.

In 1876 the famous German psychologist Gustav Fechner performed some experiments trying to establish whether certain proportions were more naturally pleasing to people than others. In one of his experiments he let subjects choose from an assortment of rectangles of various proportions the one they found the most aesthetically pleasing to the eye. Table 9-3 summarizes the results of Fechner's experiment. We can see that the golden rectangle was the overwhelming favorite and that over 75% of the subjects chose rectangles with ratios within 10% of the golden ratio.

$R = \dfrac{\text{longer side}}{\text{shorter side}}$	1	1.2	1.25	1.33	1.45	1.49	$\Phi \approx 1.618$	1.75	2	2.5
Percent of subjects preferring a rectangle with proportions R	3%	0.2%	2%	2.5%	7.7%	20.6%	35%	20%	7.5%	1.5%

■ TABLE 9-3   Fechner's Data

Fechner's experiments confirmed the intuitions of Greek sculptors and architects about the special aesthetic value of the golden ratio. Modern merchandisers have taken advantage of this by packaging products in boxes that are of aesthetically pleasing proportions. (Many cereal boxes, for example, have proportions close to the golden ratio. This is supposed to encourage "impulse" buying.)

From Greek temples to Renaissance art to cereal boxes, one number has stood out among all others in the search for beauty and balance—the golden ratio. Not surprisingly, nature itself discovered this number long before humans did.

GNOMONS

So far we have discussed some exotic mathematical concepts such as Fibonacci numbers and the golden ratio Φ, but not a word about the theme of the chapter—spiral growth in nature. The connection between these ideas is made through an even more exotic piece of mathematics, the concept of a gnomon.

As far as we can tell, gnomons were first discussed by Aristotle.[7] In geometry, a **gnomon** to a figure A is another figure which, when suitably attached to A results in a new figure A' which is similar (in the geometric sense) to A.

[7] Aristotle and the Pythagorean school of Greek geometry were fascinated by gnomons and ascribed mystical properties to them. (One must remember that there was a thin line separating religion and geometry in ancient Greece.)

Before we go into gnomons let's have a very quick review of the concept of geometric **similarity**. We know from high school geometry that two objects are similar if one is a scaled version of the other. Thus, two objects are similar when they represent the same picture at different scales. When we take a photo to the photo lab and have it blown up we are in fact dealing with the notion of similarity. Likewise, a slide projector takes a slide and blows it up onto a screen—once again we are dealing with similarity.

Here is a list of some very basic facts from elementary geometry that we will use in this section.

■ Two triangles are similar if their sides are proportional. Alternatively, two triangles are similar if the sizes of their respective angles are the same.

■ Two squares are always similar.

■ Two rectangles are similar if their sides are proportional, that is,

$$\frac{\text{long side 1}}{\text{short side 1}} = \frac{\text{long side 2}}{\text{short side 2}}.$$

■ Two circles are always similar.

■ Two circular rings are similar if their inner and outer radii are proportional, that is,

$$\frac{\text{outer radius 1}}{\text{inner radius 1}} = \frac{\text{outer radius 2}}{\text{inner radius 2}}.$$

We are now ready to take on gnomons.

Example 1. Consider square A in Fig. 9-6(a). The L-shaped figure in Fig. 9-6(b) is a gnomon to square A, because when suitably attached to square A (Fig. 9-6[c]), it results in a figure similar to A—the square A'.

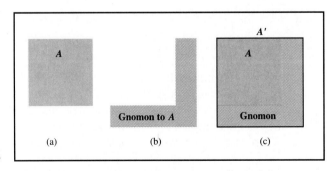

(a) (b) (c)

FIGURE 9-6

Note that the wording is *not* reversible: The square A is not a gnomon to the L-shaped figure since there is no way to combine the two to form a similar L-shaped figure.

Example 2. The O-ring in Fig. 9-7(b) (with inner radius r and outer radius R) is a gnomon to the circle of radius r (Fig. 9-7[a]), since when we attach the O-ring to the original circle A, we get a new circle A' (and as we know, two circles are always similar).

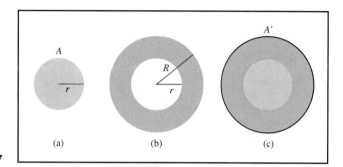

(a)   (b)   (c)

FIGURE 9-7

Example 3. Suppose A is an O-ring with outer radius r (Fig. 9-8[a]). Consider an O-ring with inner radius r and outer radius R (Fig. 9-8[b]).

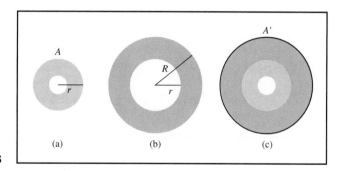

(a)   (b)   (c)

FIGURE 9-8

Is this latter O-ring a gnomon to A? The answer is no, because while we do get another O-ring A' when we attach the two O-rings together (Fig. 9-8[c]), A' is not similar to A. What would a gnomon to A look like? (For an answer, see Exercise 34.)

Example 4. Suppose that we have an arbitrary rectangle A of height h and base b (Fig. 9-9[a]). The L-shaped object shown in Fig. 9-9(b) is a gnomon to rectangle A as long as the ratios b/h and y/x are equal. In this case, they can be combined to form a rectangle A' similar to A (see Exercise 39). An elegant geometric way to build the L-shaped gnomon to the rectangle is by drawing the line through the diagonal of the starting rectangle A—it must also be the diagonal through the corner of the L.

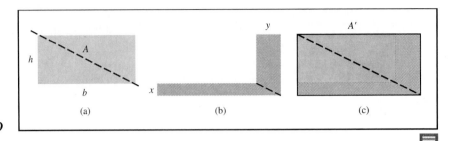

FIGURE 9-9

Example 5. Say *ABC* is an arbitrary triangle with interior angles of α degrees, β degrees, and γ degrees, and suppose α is larger than β (Fig. 9-10[a]). We can now construct a triangle *ABD* such that side *AD* is a continuation of side *AC* and the interior angle at *B* is α − β degrees (as in Fig. 9-10[b]). The triangle *ABD* is a gnomon to the original triangle *ABC* because the resulting triangle *BDC* (Fig. 9-10[c]) has angles the same size as those of the original triangle and is therefore similar to it.

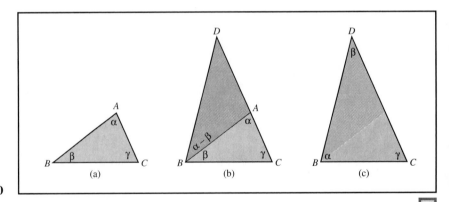

FIGURE 9-10

Example 5 illustrates a particularly nice situation much appreciated by Pythagoras and his followers:[8] gnomons for triangles that are themselves triangles. A particularly interesting case of this construction occurs when α = 72°, β = 36°, and γ = 72° (see Exercise 40). In this case both the original triangle and the gnomon are isosceles triangles.

Example 6. Here is another question that fascinated Greek geometers: Can a figure be its own gnomon? The answer is yes. Consider the rectangle *A* with sides *l* and *s* as shown in Fig. 9-11(a). For *A* to have itself as a gnomon, it must be the case that the rectangle *A′* of sides *l* and 2*s* (Fig. 9-11[c]) is similar to *A*. (Here the longer side has length 2*s*, and the shorter side has length *l*.) This means

---
[8] The original idea is attributed to Hero of Alexandria, a disciple of Pythagoras.

$$\frac{l}{s} = \frac{2s}{l} \qquad \text{or} \qquad l^2 = 2s^2; \qquad \text{i.e., } l = s\sqrt{2}.$$

In short, if the longer side of a rectangle is $\sqrt{2}$ (≈ 1.414) times the shorter side, the rectangle is its own gnomon.

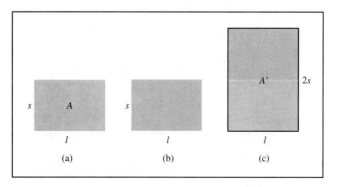

FIGURE 9-11

Example 7. In this example we raise a slightly more difficult but important question: Under what circumstances does a rectangle have a square gnomon? Consider again the rectangle A with sides l and s (Fig. 9-12[a]). The only possible hope we could have for a square gnomon is by using a square of side l, as in Fig. 9-12(b). (A square of side s could never work. The resulting rectangle would have a short side of length s and a long side of length $l + s$ and couldn't possibly be similar to A.) If the square is a gnomon the resulting rectangle A' (Fig. 9-12[c]) is similar to A, which means $\dfrac{l}{s} = \dfrac{l+s}{l}$. This is none other than the perfect balance equation which we saw earlier, and which we now know can only be satisfied by the proportions for a golden rectangle. Conclusion: *A rectangle can have a square gnomon if and only if it is a golden rectangle.*

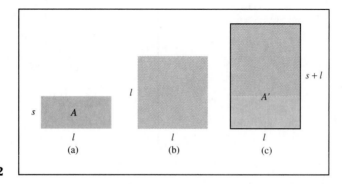

FIGURE 9-12

OK, but what makes this fact so interesting? Consider the following process for shape building: Start with a 1 by 1 square (marked ❶ in Fig. 9-13). Tack on to it another 1 by 1 square (marked ❷ in Fig. 9-13). Squares ❶ and ❷ together form a 2 by 1 rectangle, as shown in Fig. 9-13(a). Call this the "second generation" shape. For the third generation, tack on a 2 by 2 square ❸ as shown in Fig. 9-13(b). The "new shape" (❶, ❷, and ❸ together) is a 2 by 3 rectangle. Next tack on to it a 3 by 3 square ❹ as shown in Fig. 9-13(c). Then tack on a 5 by 5 square ❺ as shown in Fig. 9-13(d).

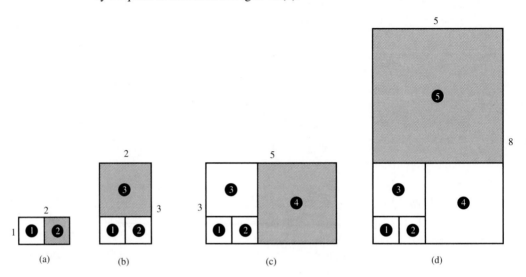

(a)   (b)   (c)   (d)

Figure 9-13

As we continue building shapes this way, we can see that at each generation we have a rectangle whose sides are consecutive Fibonacci numbers. After several generations, for example, we will have a 34 by 55 rectangle. At this point the rectangle is, for all practical purposes, a golden rectangle, and every subsequent generation will essentially produce golden rectangles (Fig. 9-14). The object we are building is now growing in a self-similar way: Each generation is similar to the preceding one.

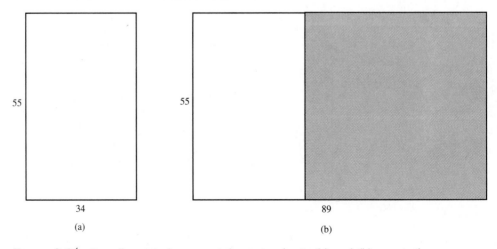

(a)   (b)

Figure 9-14   For all practical purposes the rectangles in (a) and (b) are similar (both are approximately golden rectangles).

GNOMONIC GROWTH

We started this chapter with a question about daisies which we now reconsider in a slightly more general form: What are the rules that govern the geometric arrangement of leaves on a stem or of petals on a flower?[9]

The starting point for our discussion is the observation that there is a surprisingly large and varied group of natural objects that have a special affinity for Fibonacci numbers. Figure 9-15 illustrates some typical examples of this. While there is no generally accepted explanation for this affinity, there is one common element in all the natural objects shown in Fig. 9-15: the nature of their growth.

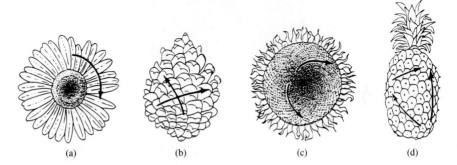

| (a) | (b) | (c) | (d) |

FIGURE 9-15 (a) Daisies: Numbers of petals are most frequently 8, 13, 21, 34, 55, or 89. (b) Pinecones: bracts spiral in sets of 8 and 13 rows. (c) Sunflowers: Florets spiral out from the center in sets of 55 rows and 89 rows. (d) Pineapples: Scales spiral in sets of 8, 13, and 21 rows.

Natural organisms grow in essentially two different ways. The more common type of growth (and the one we are most familiar with) is the growth exhibited by humans, animals, and many plants. This can be called *all-around growth*, in which all living parts of the organism grow simultaneously (although not necessarily at the same rate). One of the characteristics of this type of growth is that there is no obvious way to distinguish between the newer and the older parts of the organism. In fact, the distinction between new and old parts does not make much sense. The historical record (so to speak) of the organism's growth is lost. By the time the child becomes an adult, no identifiable traces of the child (as an organism) remain—that's why we need photographs!

Contrast this with the kind of growth exemplified by the shell of the chambered nautilus, a ram's horn, the trunk of a redwood tree, or the inflorescence of a daisy. This is a kind of growth that we may informally call *growth at one end* or *asymmetric growth*. In this type of growth the organism has a part added to it (either by its own or outside forces) in such a way that the old organism together with the added part form the new organism. At any stage of the growth process we can see not only the present but the organism's entire past. All the previous stages of its growth are the building blocks that make up the present structure.

---

[9] The study of the arrangement and distribution of parastiches (leaves, stalks, petals, bracts, etc.) on plants has a technical name: *phyllotaxis*.

A sunflower loves Fibonacci numbers: 34 petals on the outside; 55 and 89 rows of florets on the inside. (Wendell/Photo Researchers)

A second important fact is that in most instances organisms that grow in this fashion do so in a way that preserves their overall shape; in other words, they remain similar to themselves. This is where gnomons come into the picture: Regardless of how the new growth comes about, its shape is a gnomon of the entire organism. We will call this kind of growth process **gnomonic growth**.

We will illustrate the process of gnomonic growth with some examples.

Example 8. We know from Example 2 that the gnomon to a circle is an O-ring with an inner radius equal to the radius of the circle. We can thus have circular gnomonic growth (Fig. 9-16). Rings added one layer at a time to a starting circular structure preserve the circular shape throughout the structure's growth. When carried to three dimensions, this is a good model for the way the trunk of a redwood tree grows.

FIGURE 9-16 Circular gnomonic growth.

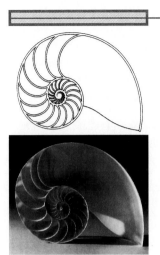

FIGURE 9-17 A cross section of the chambered nautilus shell.

Example 9. Figure 9-17 shows a cross section of a well-known seashell—that of the chambered nautilus. The spiral-shaped shell of the chambered nautilus is a wonder of natural architecture and a classic example of nature's ability to combine function with beauty. The nautilus builds its shell in stages, each stage consisting of the addition of a chamber to the already existing shell. At every stage of its growth, the shape of the chambered nautilus shell remains the same—the beautiful and distinctive spiral shown in Fig. 9-17 as well as in the accompanying photograph. In essence, we can interpret this as a classic case of gnomonic growth—each new chamber that is added to the shell is a gnomon to the entire shell. The gnomonic growth of the shell proceeds in essence as follows: Starting with the shell of a baby chambered nautilus (which is a tiny spiral similar in all respects to the adult spiral shape), the animal builds a chamber (by producing a special secretion around its body that calcifies and hardens). The resulting, slightly enlarged spiral shell is similar to the original one. The process then repeats itself in a *recursive* way: A new chamber is added (it is a gnomon to the shell that is similar but slightly larger than the first one), resulting in a new enlarged spiral. This process continues ad infinitum or until the animal reaches maturity.

Example 10. This is a fictitious example. Imagine, if you will, a family of small extraterrestrial beings that wants to build itself a structure in which to live. Let's say, for the sake of argument, that the perfect floor plan for this family is an isosceles triangle (Fig. 9-18[a]). As the family grows, the original triangle becomes too small. They need to move into a new shelter just like the original one only bigger. They can do this by adding a triangle that is a gnomon to the original triangle (see Example 5). This results in the structure shown in Fig. 9-18(b). As the process of growth continues, we get something like the structure shown in Fig. 9-18(c). The successive stages of growth are labeled ❶, ❷, ❸, ❹, and ❺. This means that by the fifth generation the structure is the entire triangle shown in Fig. 9-18(c).

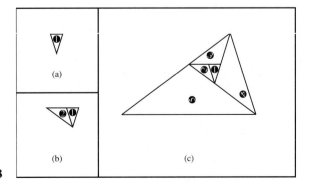

(a)

(b)

(c)

FIGURE 9-18

If we track the vertex V (Fig. 9-19[a]) through the various generations of gnomonic growth (V_1, V_2, V_3, V_4, V_5, . . .) we see a very special kind of spiral shape (Fig. 9-19[c]) called a **logarithmic spiral**. The logarithmic spiral is characteristic of gnomonic growth in nature and can be seen in many seashells, animal horns, etc. The most familiar and beautiful example of the logarithmic spiral is the one traced by the outer edge of the chambered nautilus shell (Fig. 9-17).

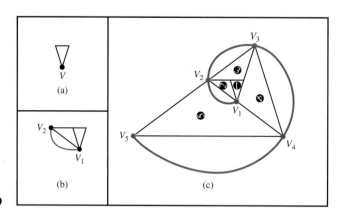

Figure 9-19

Example 11. Figure 9-20 shows a set of squares of various sizes each one having a quarter circle inscribed in it. The dimensions of the squares are the Fibonacci numbers, 1, 1, 2, 3, 5, 8, 13, 21, ...

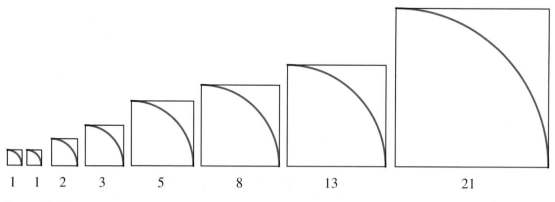

Figure 9-20

Let's play nature's game and think of these squares as building blocks for a shape. We will build the shape one step at a time, always attaching the next square in the sequence to what we had built before and making sure of two things: (1) the total shape always is contained in a rectangle whose dimensions are consecutive Fibonacci numbers (we did this earlier); and (2) the quarter cir-

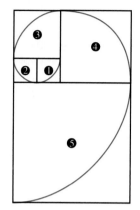

FIGURE 9-21 Early stages of growth. Rectangles are not yet golden; growth is not yet gnomonic.

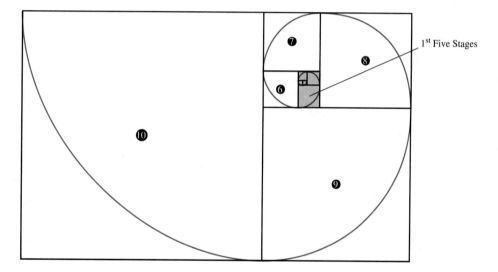

1ˢᵗ Five Stages

FIGURE 9-22 A few generations later rectangles start getting close to golden rectangles, the growth becomes gnomonic, and the spiral becomes logarithmic.

cles match up, so that each new one picks up where the other one left off. Fig. 9-21 shows the first five stages of the construction. Up to this point the growth is not gnomonic. Figure 9-22 shows the first ten stages (sized down so that the entire picture can be seen).

At this point, we can say that the rectangles have settled into golden rectangleship. This implies that each new square is basically a gnomon to the whole structure and the spiral has started taking the distinctive shape of a logarithmic spiral. We can continue building the spiral for as long as we want to—twenty, thirty, or even one hundred steps are not uncommon in nature. To us, size is a problem, so we leave the rest to the reader's imagination, but in case help is needed, we recommend looking at the picture of the chambered nautilus once again.

CONCLUSION

Fibonacci numbers, gnomons, the golden ratio, logarithmic spirals—these are the mathematical elements nature uses to produce some of its complex, beautiful forms, among them the daisy, the sunflower, and the chambered nautilus. This chapter was not an attempt to explain exactly how and why these relationships between nature and mathematics occur (to the best of our knowledge there are several theories but no confirmed facts) but rather to point out that a relationship does indeed exist.

A technical account of some of the competing theories that attempt to explain the surprising connections between mathematics and spiral growth can be found in references 3, 7, 9, and 10.

KEY CONCEPTS

Binet's formula
Fibonacci number
Fibonacci sequence
gnomon
gnomonic growth

golden ratio
golden rectangle
logarithmic spiral
perfect balance equation
similarity

EXERCISES

Walking

In Exercises 1 through 14, F_N refers to the Nth Fibonacci number.

1. Find F_{15}, F_{16}, F_{17}, and F_{18}.

2. Find F_{19}, F_{20}, F_{21}, F_{22}, and F_{23}.

3. Given that $F_{36} = 14930352$ and $F_{37} = 24157817$,
   (a) find F_{38}
   (b) find F_{35}.

4. Given that $F_{31} = 1346269$ and $F_{33} = 3524578$,
   (a) find F_{32}
   (b) find F_{34}.

5. Use your calculator to find $\dfrac{F_{14}}{F_{13}}$, $\dfrac{F_{16}}{F_{15}}$, and $\dfrac{F_{18}}{F_{17}}$. What are your conclusions?

6. Use your calculator to find $\dfrac{F_{20}}{F_{19}}$, $\dfrac{F_{21}}{F_{20}}$, $\dfrac{F_{22}}{F_{21}}$, and $\dfrac{F_{23}}{F_{22}}$. What are your conclusions?

7. Find integers N and M (other than $N = 1$, $M = 1$) satisfying the equation $F_N = M^2$.

8. Find integers N and M (other than $N = 1$, $M = 1$) satisfying the equation $F_N = F_M^3$.

9. Write each of the following integers as the sum of *distinct* Fibonacci numbers.
   (a) 47

    (b) 48

    (c) 207

    (d) 210

10. Write each of the following integers as the sum of *distinct* Fibonacci numbers.

    (a) 52

    (b) 53

    (c) 107

    (d) 112

11. Fact: $(F_1 + F_2 + F_3 + \cdots + F_N) + 1 = F_{N+2}$. Verify this fact for:

    (a) $N = 4$

    (b) $N = 5$

    (c) $N = 10$

    (d) $N = 11$.

12. Fact: *If we make a list of any ten consecutive Fibonacci numbers, the sum of all these numbers divided by 11 is always equal to the seventh number in the list.*

    (a) Using F_N as the first Fibonacci number in the list, write the above fact as a mathematical equation.

    (b) Verify this fact for $N = 5$.

    (c) Verify this fact for $N = 6$.

13. Fact: *If we make a list of any four consecutive Fibonacci numbers, twice the third one minus the fourth one is always equal to the first one.*

    (a) Using F_N as the first Fibonacci number in the list, write the above fact as a mathematical equation.

    (b) Verify this fact for $N = 1$.

    (c) Verify this fact for $N = 4$.

    (d) Verify this fact for $N = 8$.

14. Fact: *If we make a list of any four consecutive Fibonacci numbers, the first one times the fourth one is always equal to the third one squared minus the second one squared.*

    (a) Using F_N as the first Fibonacci number in the list, write the above fact as a mathematical equation.

    (b) Verify this fact for $N = 1$.

    (c) Verify this fact for $N = 4$.

    (d) Verify this fact for $N = 8$.

Exercises 15 and 16 require the use of a calculator with an exponent key. (On most calculators, the exponent key looks something like $\boxed{y^x}$ *. To calculate an exponent, say for example* $(2.3)^7$*, enter first 2.3, then enter* $\boxed{y^x}$ *, and finally enter 7 followed by* $\boxed{=}$ *.*

15. Calculate

    (a) $\left(\dfrac{1 + \sqrt{5}}{2}\right)^{10}$

    (b) $\left(\dfrac{1 - \sqrt{5}}{2}\right)^{10}$

    (c) [Answer (a) − Answer (b)]/$\sqrt{5}$

16. Calculate

(a) $\left|\dfrac{1 + \sqrt{5}}{2}\right|^{25}$

(b) $\left|\dfrac{1 - \sqrt{5}}{2}\right|^{25}$

(c) [Answer (a) − Answer (b)]/$\sqrt{5}$

Exercises 17 through 20 are intended for readers who need a brief review of the quadratic formula. (For a more extensive review readers are encouraged to look at any intermediate algebra textbook.) Any quadratic equation can be solved by first putting it in the form $ax^2 + bx + c = 0$ and then using the **quadratic formula** $x = \dfrac{-b \pm \sqrt{b^2 - 4ac}}{2a}$. *A calculator with a square root key is often needed to carry out this calculation.*

17. Use the quadratic formula to find the two solutions of $x^2 = 2x + 1$. Use a calculator to approximate the solutions to three decimal places.

18. Use the quadratic formula to find the two solutions of $x^2 = 3x + 2$. Use a calculator to approximate the solutions to three decimal places.

19. Use the quadratic formula to find the positive solution of $3x^2 = 5x + 8$. Use a calculator to approximate the solution to five decimal places.

20. Use the quadratic formula to find the two solutions of $5x^2 = 8x + 13$. Use a calculator to approximate the solutions to five decimal places.

Exercises 21 through 24 require the use of the quadratic formula together with the **Pythagorean theorem**. *(In any right triangle with legs of lengths a and b, and hypotenuse of length c, $a^2 + b^2 = c^2$.)*

21. The right triangle shown in the figure has hypotenuse of length x and legs of length $\sqrt{x}$ and 1. Solve for x.

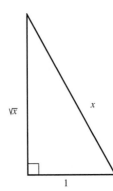

22. The right triangle shown in the figure has hypotenuse of length $2x$ and legs of length $\sqrt{x}$ and 1. Solve for x.

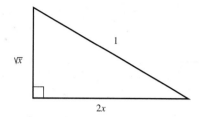

23. The right triangle shown in the figure has hypotenuse of length 1 and legs of length $\sqrt{x}$ and $2x$. Solve for x.

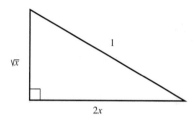

24. The right triangle shown in the figure has hypotenuse of length 1 and legs of length x and $\sqrt{x}$. Solve for x.

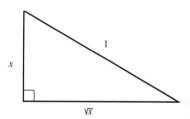

25. Rectangle A is 10 by 20. Rectangle B is a gnomon to rectangle A. What are the dimensions of rectangle B?

26. Rectangle A is 2 by 3. Rectangle B is a gnomon to rectangle A. What are the dimensions of rectangle B?

27. Find the length c of the shaded rectangle so that it is a gnomon to the white rectangle with sides 3 and 9.

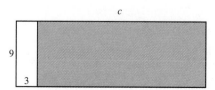

28. Find the value of x so that the shaded figure is a gnomon to the white rectangle.

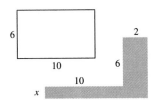

29. Find the value of x so that the shaded "rectangular ring" is a gnomon to the white rectangle.

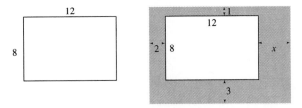

30. Find the value of x so that the shaded "rectangular ring" is a gnomon to the white rectangle.

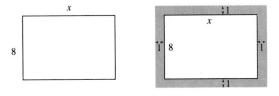

31. Find the values of x and y so that the shaded figure is a gnomon to the white triangle.

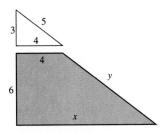

32. Find the values of x and y so that the shaded triangle is a gnomon to the white triangle ABC.

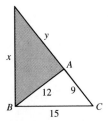

Jogging

33. Find the values of x, y, and z so that the shaded figure has an area that is eight times the area of the white triangle and at the same time is a gnomon to the white triangle.

34. (a) Which of the following O-rings is similar to I? Explain your answer.

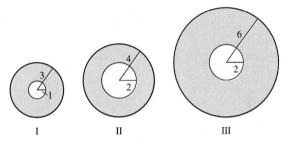

I   II   III

(b) Explain why an O-ring cannot have a gnomon.

35. Find the values of x and y so that the shaded figure has an area of 75 and at the same time is a gnomon to the white rectangle.

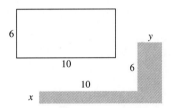

36. Under what conditions is a triangle its own gnomon?

37. Suppose you are given that $\Phi^N = a\Phi + b$. Show that $\Phi^{N+1} = (a + b)\Phi + a$.

38. In the following figure $ABCD$ is a square and the three triangles I, II, and III have equal areas. Show that x/y is the golden ratio.

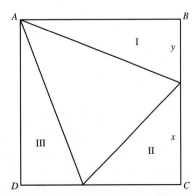

39. Show that the L-shaped object in the following figure is a gnomon for rectangle A as long as the ratios b/h and y/x are equal.

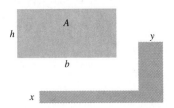

40. Triangle ABC has angles of $72°$, $36°$, and $72°$ as shown in the following figure. Let M be the point on side BC so that AM and AC have equal lengths. The line segment AM divides the triangle into two triangles (I and II) as shown in the figure. Explain why triangle II is a gnomon to triangle I.

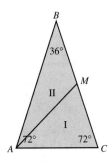

41. (a) Use the explicit description of the Fibonacci sequence

$$F_N = \frac{\left(\dfrac{1 + \sqrt{5}}{2}\right)^N - \left(\dfrac{1 - \sqrt{5}}{2}\right)^N}{\sqrt{5}}$$

and a good calculator to compute F_{10}.

(b) Compare your answer in (a) with the known value $F_{10} = 55$. Can you explain any discrepancy?

42. Let

$$a = \frac{1 + \sqrt{5}}{2} \quad \text{and} \quad b = \frac{1 - \sqrt{5}}{2}.$$

Without using a calculator, expand and simplify

(a) $a + b$
(b) ab
(c) $a^2 + b^2$
(d) $a^3 + b^3$.

43. Consider the following sequence of numbers: 5, 5, 10, 15, 25, 40, 65, ... If A_N is the Nth term of this sequence, write A_N in terms of F_N.

44. Consider the following sequence of numbers: 4, 7, 11, 18, 29, 47, 76, ... If B_N is the Nth term of this sequence, write B_N in terms of F_N.

45. Let *ABCD* be an arbitrary rectangle as shown in the following figure. Let *AE* be perpendicular to the diagonal *BD* and *EF* perpendicular to *AB* as shown. Show that the rectangle *BCEF* is a gnomon to rectangle *ADEF* (*Hint:* Show that the rectangle *ADEF* is similar to the rectangle *ABCD*.)

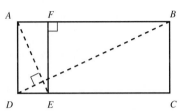

Running

46. Find the values of x, y, and z so that the shaded "triangular ring" is a gnomon to the white triangle.

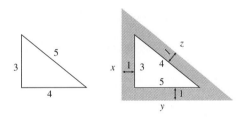

47. Show that $F_1 + F_2 + F_3 + \cdots + F_N = F_{N+2} - 1$. (See Exercise 11.)

48. Show that $F_1 + F_3 + F_5 + \cdots + F_N$ (here we are adding the Fibonacci numbers with odd subscripts up to N) $= F_{N+1}$.

49. Show that every positive integer greater than 2 can be written as the sum of distinct Fibonacci numbers.

50. Consider the following equation relating various terms of the Fibonacci sequence.

$$F_{N+2}^2 - F_{N+1}^2 = F_N \cdot F_{N+3}.$$

Using the algebraic identity $A^2 - B^2 = (A - B)(A + B)$, show that the equation is true for every positive integer N.

51. Show that the sum of any 10 consecutive Fibonacci numbers is a multiple of 11.

52. (a) The triangle *ABC* in the following figure has angles of 36°, 72°, and 72° and sides of length 1, x, and x. (Triangles such as these are called **golden triangles**.) Show that $x = \Phi$.

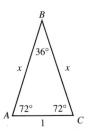

(b) The regular pentagon in the following figure has sides of length 1. Show that the length of any one of its diagonals is Φ.

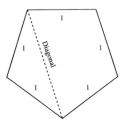

53. During the time of the Greeks the star pentagram was a symbol of the Brotherhood of Pythagoras. A typical diagonal of the large outside regular pentagon is broken up into three segments of lengths x, y, and z as shown in the following figure.

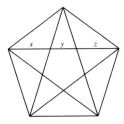

(a) Show that $\dfrac{x}{y} = \Phi$, $\quad \dfrac{x+y}{z} = \Phi$, and $\quad \dfrac{x+y+z}{x+y} = \Phi$.

(b) Show that if $y = 1$, then $x = \Phi$, $x + y = \Phi^2$, and $x + y + z = \Phi^3$.

54. The puzzle of the missing area. Consider a square 8 units on a side and cut into four pieces as shown in the following figure.

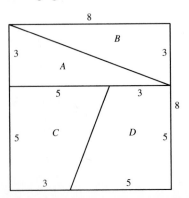

If we rearrange the pieces into a rectangle as shown in the next figure, we see that although the square has area $8 \times 8 = 64$, the rectangle has area $13 \times 5 = 65$.

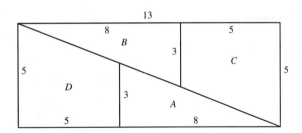

(a) Draw similar figures using other (larger) Fibonacci numbers. How does the area of the square compare with the area of the rectangle?

(b) Explain the discrepancies in the areas.

(c) Consider the following figures.
What are the conditions on a and b so that this puzzle is not a puzzle, that is, the areas are the same?

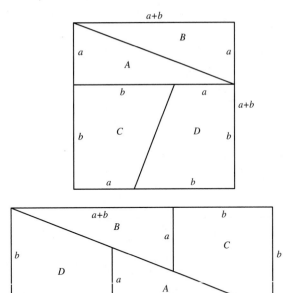

REFERENCES AND FURTHER READINGS

1. Cook, Theodore, *The Curves of Life.* New York: Dover Publications, Inc., 1979.

2. Coxeter, H. S. M., *Introduction to Geometry.* New York: John Wiley & Sons, Inc., 1961, chap. 11.

3. Erickson, R. O., "The Geometry of Phyllotaxis," in *The Growth and Functioning of Leaves,* eds, J. E. Dale and F. L. Milthrope. New York: Cambridge University Press, 1983.

4. Gardner, Martin, "About Phi, an Irrational Number that Has Some Remarkable Geometrical Expressions," *Scientific American,* 201 (August 1959), 128–134.

5. Ghyka, Matila, *The Geometry of Art and Life.* New York: Dover Publications, Inc., 1977.

6. Huntley, H. E., *The Divine Proportion: A Study in Mathematical Beauty.* New York: Dover Publications, Inc., 1970.

7. Jean, R. V., *Mathematical Approach to Pattern Form in Plant Growth.* New York: John Wiley & Sons, Inc., 1984.

8. Kappraff, Jay, *Connections: The Geometric Bridge Between Art and Science.* New York: McGraw-Hill Book Company, 1991.

9. Stevens, P., *Patterns in Nature.* London: Penguin Books, 1976.

10. Thompson, D'Arcy, *On Growth and Form.* New York: Macmillan Publishing Co., Inc., 1942, chaps. 11, 13, and 14.

CHAPTER 10

There Is Strength In Numbers

The Growth of Populations

*. . . and you, be ye
fruitful and multiply.*
. . .

GENESIS 9:7

Can the California condor be saved from extinction? How long will it be before every garbage dump in New York is full? How much money will you have in your savings account in two years? What will the population of the earth be in the year 2100? All these questions fall under the heading of problems of **population growth**.

The breadth of meaning the expression "population growth" has acquired in modern times stems from the generous way in which both the word "population" and the word "growth" are used. The Latin root of the word "population" is *populus* (which means "people"), so that in its original interpretation the word refers to human populations. This original scope has been expanded, however, so that now the word can be applied to any collection of objects (animate or inanimate) that changes over time and about which we want to make a numerical or quantitative statement. Thus, we can speak of a population of birds, of tires, of aluminum cans, of dollars and cents, and needless to say, of people.

Second, we normally think of the word "growth" as being applied to things that get bigger, but in this chapter we will ascribe a slightly more technical meaning to it: "Growth" can mean *decay* (i.e., getting smaller) as well as real growth (i.e., getting bigger). This is convenient, because often we don't know ahead of time what is going to happen when studying a population. Is it going to go up or down? By allowing "growth" to mean up or down, we do not need to concern ourselves with making this distinction.

THE DYNAMICS OF POPULATION GROWTH

The growth of a population is a **dynamical process**, meaning that it represents values that change over time. Mathematicians distinguish two kinds of situations: continuous growth and discrete growth. In **continuous growth** the dynamics of change are in effect all the time—every hour, every minute, every second, there is growth. A typical example of this kind of growth is represented

The many faces of "population growth." (Top right, Eugene Gordon; bottom left, Bernard Pierre Wolff/Photo Reseachers; right, Carl B. Koford/Photo Researchers)

by money left in an account that is drawing interest on a continuous basis (yes, there are banks that offer such accounts). We will not study this type of growth in this chapter.

The second type of growth, **discrete growth**, is the most common and natural way by which populations change. We can think of it as a *stop-and-go* type of situation: For a while nothing happens, then there is a sudden change in the population (we will call this a **transition**), then for a while nothing happens again, then another transition takes place, and so on. Of course the "for a while" (i.e., the period between transitions) can be 100 years, an hour, a second, or a millionth of a second. To us the length of time between transitions will not make a difference. The human population of our planet is an example of what we mean: Nothing happens until someone is born or someone dies, at which point there is a change ($+1$ or -1); then, again there is no change until the next birth or death. Since, however, someone is born every fraction of a second and someone dies slightly less often, it is somewhat tempting to think that the world's human population is for all practical purposes changing in a continuous way. On the other hand, the laws of growth affecting the world's population may be only quantitatively different from the laws of growth affecting the population of Hinsdale County, Colorado (population 408),[1] where a change in the population may not come about for months or even years.

The basic problem of population growth is to figure out what happens to a given population over time. Sometimes we talk about a specific period of time ("The Hispanic population of the United States will grow by 30% before the end of the century."[2]), and sometimes we may talk about the long-term behavior of the population ("The black rhino population is heading for extinction."[3]).In either case, the most basic way to deal with the question of growth of a particular population is to find the rules that govern the transitions. We will call these the **transition rules**. After all, if we have a way to figure out how the population changes each time there is a transition, then (with a little help from mathematics) we can usually figure out how the population changes after many transitions. In this sense, the behavior of a particular population over time can be conveniently thought of as an unending list of numbers called the **population sequence**. Figure 10-1 is a schematic illustration of how a population sequence can be generated.

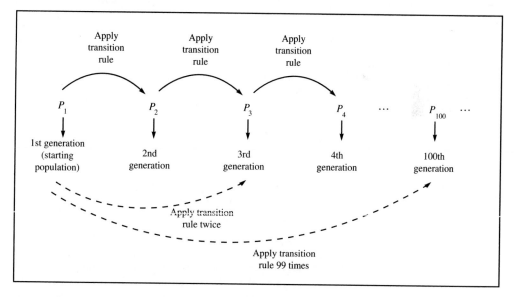

FIGURE 10-1 A model of a population sequence.

[1] Source: *The World Almanac and Book of Facts,* 1990.

[2] U.S. Bureau of the Census, *Statistical Abstract of the United States: 1990,* 110th edition, Washington D.C., 1990, 14.

[3] *New York Times,* May 7, 1991, B5.

Just to get our feet wet, we will start with one of the oldest and best-known examples of a problem in population growth: the Fibonacci rabbit problem.

Example 1. A Fibonacci rabbit is a very unusual breed of bunny. Like clockwork, at the end of each month a mature male-female couple produces as offspring a male-female pair of babies. It takes one month for a newborn baby rabbit to become a mature rabbit and start producing offspring of its own.

In his famous book *Liber Abaci*, Fibonacci raised the following question: If you start with a single pair of male-female newborn Fibonacci rabbits, how many rabbits will there be at the end of one year?

For the sake of convenience, we will count rabbits in male-female pairs, and following the notation we have just adopted, we will let P_1, P_2, P_3, etc. represent the number of pairs in the first, second, third, etc. generations. Figure 10-2 illustrates the pattern of growth for the first six months.

	1st generation	2nd generation	3rd generation	4th generation	5th generation	6th generation	7th generation
Elapsed time		1 month	2 months	3 months	4 months	5 months	6 months
Number of baby pairs	1	0	1	1	2	3	5
Number of mature pairs	0	1	1	2	3	5	8
Total number of pairs	$P_1 = 1$	$P_2 = 1$	$P_3 = 2$	$P_4 = 3$	$P_5 = 5$	$P_6 = 8$	$P_7 = 13$

FIGURE 10-2

We can see from Fig. 10-2 that $P_1 = 1$; $P_2 = 1$; $P_3 = P_2 + P_1$; $P_4 = P_3 + P_2$; etc. Thus, in each generation, the total number of pairs is the corresponding Fibonacci number (see Chapter 9), i.e.,

$$P_N = F_N.$$

The rest is easy. At the end of one year (13th generation) we will have $P_{13} = F_{13} = 233$ pairs of rabbits—a grand total of 466 rabbits! More importantly, we can describe the situation in very general terms. The transition rule for passing from one generation to the next is

$$P_N = P_{N-1} + P_{N-2}$$

number of pairs in the Nth generation	number of mature pairs in the Nth generation equals the total number of pairs in the preceding generation	number of baby pairs in the Nth generation equals the total number of pairs in the $(N-2)$nd generation

Because Fibonacci bunnies are so methodical and precise, and because they keep breeding and breeding and breeding, people tend to confuse them with the Energizer bunnies, but one must keep in mind that there were Fibonacci bunnies long before there were Energizer bunnies. OK, now that we have cleared that issue, let's get real. Clearly, real rabbits are not as accommodating as Fibonacci rabbits—they live and breed by considerably more complicated rules, rules that we could never hope to capture in a simple equation. And what's true about rabbits is generally true about most other types of populations.

In view of this, is there any use for simplistic mathematical models of how populations grow? The answer is yes! We can make excellent predictions about the growth of a population over time even when we don't have a completely realistic set of transition rules. The secret is to capture the variables that are really influential in determining how the population grows, put them into a few transition rules that describe how the variables interact, and forget about the small things. This, of course, is easier said than done. In essence, it is what population biologists and mathematical ecologists do for a living, and it is as much an art as it is a science.

For the rest of this chapter we will scratch the surface of this kind of activity. We will discuss three of the most basic models of population growth: linear growth, exponential growth, and logistic growth.

LINEAR GROWTH

Example 2. A newly opened garbage dump has collected 8000 tons of garbage since its opening. It is projected that the dump will collect 120 tons of garbage per month over the next several years. At this rate how much garbage will the dump have collected 5 years from now?

In this example, the population we are studying is the garbage collected by the particular dump in question. Since we are looking at monthly averages, by definition the transitions occur once a month. (This, of course, is only the way we look at it on paper; in reality, the transitions occur every time a garbage truck dumps a load, but for our purposes that's just a stinking detail.) The essential fact about this population growth problem is that the monthly garbage at the dump grows by a constant amount of 120 tons a month.

Our starting population (P_1) is 8000 tons. Thus, we have the following population sequence:

$$P_1 = 8000;\ P_2 = 8120;\ P_3 = 8240;\ P_4 = 8360;\ \ldots$$

In 5 years we will have had 60 transitions, each of which will represent an increase of 120 tons. In short, the answer to our question is

$$P_{61} = 8000 + 60(120) = 15{,}200 \text{ tons.}$$

Example 2 is typical of a general situation called **linear growth**. In a linear growth situation there is a starting population P_1 and the transition rule is: In each transition, add a constant amount d to the existing population. Any sequence of numbers generated by linear growth is commonly known as an **arithmetic sequence**. The number d is called the **common difference** for the arithmetic sequence because any two consecutive values of the arithmetic sequence will always differ by the amount d. Mathematically, the transition rule for linear growth can be described as follows:

■ Starting Population: P_1

■ $\underbrace{P_N}_{\substack{\text{population} \\ \text{in the } N\text{th} \\ \text{generation}}} = \underbrace{P_{N-1}}_{\substack{\text{population in} \\ \text{the preceding} \\ \text{generation}}} + \underbrace{d}_{\substack{\text{common} \\ \text{difference}}}$

The above description is called a **recursive description** of the population sequence because it calculates values of the population sequence using earlier values of the population sequence. While recursive descriptions tend to be nice and tidy, they have one major drawback: To calculate one value in the population sequence we essentially have to first calculate all the earlier values. As we learned in Chapter 9 when discussing this same problem with regards to the Fibonacci numbers, this can be quite an inconvenience.

Fortunately, in the case of populations that grow according to a linear growth model, there is another very convenient way to describe the population sequence. Figure 10-3 essentially explains what's going on.

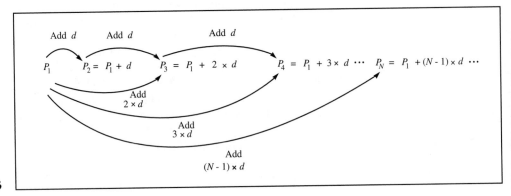

FIGURE 10-3

In short, Fig. 10-3 shows us that

$$P_N = P_1 + (N - 1) \times d.$$

This rule allows us to calculate any value of the population sequence using just the initial population P_1 and the common difference d. A description of a population sequence which allows one to calculate any value of the sequence directly is called an **explicit description** of the sequence.

Example 3. A population grows according to a linear growth model. The starting population is $P_1 = 37$ and the common difference is $d = 6$. (a) What is the population in the 16th generation? (b) What is the population after 25 transitions?

Question (a) asks for P_{16}. Using the explicit description of linear growth we immediately find $P_{16} = 37 + 15 \times 6 = 127$.

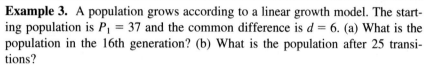

Question (b) asks for P_{26}, and $P_{26} = 37 + 25 \times 6 = 187$.

Plotting Population Growth

A very convenient way to describe population growth is by means of a **plot** or **graph**. The horizontal axis usually represents time (with the tick marks generally corresponding to the transitions), and the vertical axis usually represents the size of the population. Because we have complete freedom in choosing both the horizontal and vertical scales, plots can be misleading. Consider Figs. 10-4(a) and (b). Which population grows faster?

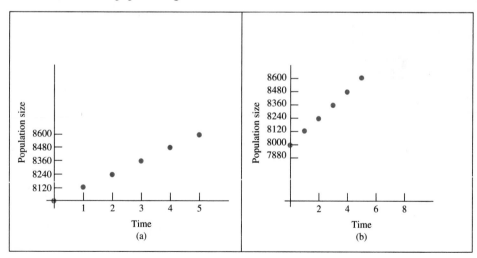

FIGURE 10-4

Actually, both plots represent the growth of the same population: the garbage problem in Example 2. These plots illustrate why linear growth is called linear growth—no matter how we plot it, the values of the population line up in a straight line.

Adding Terms of an Arithmetic Sequence

Example 4. Jane Doe is a company that manufactures tractors. The company has decided to start up a new plant. On the first of each month (for a period of 6 years), a module of equipment that will produce 3 tractors per month will be installed in this plant. What is the total number of tractors that the company will produce over the 6 years the program is in effect?

In this problem, the total production of each module conforms to a linear growth pattern, but the number of months each module works is different. Let's make a list:

- ■ Module installed the 1st month works for 72 months, producing $3 \times 72 = 216$ tractors.

- ■ Module installed the 2nd month works for 71 months, producing $3 \times 71 = 213$ tractors.

- ■ Module installed the 3rd month works for 70 months, producing $3 \times 70 = 210$ tractors.

$$\vdots$$

- ■ Module installed the 72nd month works for 1 month, producing 3 tractors.

The total number of tractors produced at the end of 72 months is

$$216 + 213 + 210 + \cdots + 3.$$

This sum is the sum of consecutive terms of the arithmetic sequence 3, 6, 9, . . . , 213, 216. We could, of course, add these numbers up, with or without a calculator, but that's dull. Let's take a slightly more elegant tack. Let's write our total twice, once forward and once backward:

$$\text{Total} = 216 + 213 + 210 + \cdots + 6 + 3.$$

$$\text{Total} = 3 + 6 + 9 + \cdots + 213 + 216.$$

If we add each term in the first row to each term in the second row, we get

$$2 \times \text{Total} = 219 + 219 + 219 + \cdots + 219 + 219.$$

Since there are 72 such terms, we end up with

$$2 \times \text{Total} = 219 \times 72,$$

and therefore

$$\text{Total} = \frac{219 \times 72}{2} = 7884.$$

The same approach works with *any* arithmetic sequence: We can add up any number of consecutive terms in a very convenient way. The formula

---

**Adding *N* Consecutive Terms
of an Arithmetic Sequence**

$$\text{First term} + \cdots + \text{last term} = \frac{(\text{first term} + \text{last term}) \times N}{2}.$$

Example 5. $\underbrace{5 + 12 + 19 + 26 + 33 + \cdots}_{132 \text{ terms}} = ?$

Here we are adding 132 consecutive terms of an arithmetic sequence. The first term is $P_1 = 5$; the common difference is $d = 7$. We need to find the 132nd term P_{132}. We already know how to do this: $P_{132} = 5 + 131 \times 7 = 922$. We can now apply the formula:

$$5 + 12 + 19 + 26 + 33 + \cdots + 922 = \frac{(5 + 922) \times 132}{2} = 61,182.$$

Example 6. $4 + 13 + 22 + 31 + 40 + \cdots + 922 = ?$

Here we are adding the terms of an arithmetic sequence with $P_1 = 4$ and common difference $d = 9$. To apply the formula we need to find the number of terms N. To find N we set up an equation: $922 = 4 + 9(N - 1)$. From it we get $9(N - 1) = 918$, and therefore $N - 1 = 102$ and $N = 103$. It follows that

$$4 + 13 + 22 + 31 + 40 + \cdots + 922 = \frac{(4 + 922) \times 103}{2} = 47,689.$$

**EXPONENTIAL
GROWTH**

Before we start our discussion of exponential growth in earnest, let's look at a couple of preliminary examples.

Example 7. A firm manufactures an item at a cost of C dollars. The item is marked up 10% and sold to a distributor. The distributor then marks the item up 20% (based on the price he paid) and sells the item to a retailer. The retailer marks the price up 50% and sells the item to the public. By what percent has the item been marked up over its original cost?

- Original cost of item: C
- Price to distributor after 10% markup (D): $D = 110\%$ of $C = (1.1)C$.
- Price to retailer after 20% markup (R): $R = 120\%$ of $D = (1.2)D = (1.2)(1.1)C = (1.32)C$
- Price to the public after 50% markup (P): $P = 150\%$ of $R = (1.5)R = (1.5)(1.32)C = (1.98)C$.

Therefore, the markup over the original cost is 98%.

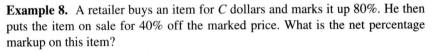

Example 8. A retailer buys an item for C dollars and marks it up 80%. He then puts the item on sale for 40% off the marked price. What is the net percentage markup on this item?

- Original cost of item: C
- Price after 80% markup (P): $P = 180\%$ of $C = (1.8)C$
- Sale price after 40% discount (S): $S = 60\%$ of $P = (0.6)P = (0.6)(1.8)C = (1.08)C$

The net markup is 8%.

We are now ready to tackle exponential growth.

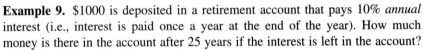

Example 9. $1000 is deposited in a retirement account that pays 10% *annual* interest (i.e., interest is paid once a year at the end of the year). How much money is there in the account after 25 years if the interest is left in the account?

This example is a typical case of a situation involving **exponential growth**: The money draws interest; then the money plus the interest draw interest; and so on. While we will concentrate on examples from the world of finance, exponential growth represents a common pattern of growth in many other areas as well. The essence of exponential growth is *recursive multiplication*: Each transition consists of a multiplication by some fixed amount called the **growth rate**. We will illustrate these ideas by means of Example 9. Table 10-1 will help us get started.

	Account Balance at Beginning of Year	Interest Earned for the Year	Account Balance at End of Year
Year 1	$1000	$100	$1100
Year 2	1100	110	1210
Year 3	1210	121	1331
.	.	.	.
.	.	.	.
.	.	.	.
Year 24	???	???	???
Year 25	???	???	???

▄ **TABLE 10-1**

The critical observation is that the account balance at the end of year 1 is obtained by adding the *principal* ($1000) and the interest earned for the year (10% of $1000). We know that this is equivalent to taking 110% of $1000—in other words $1000 × 1.1. Likewise the account balance at the end of year 2 is

$$(\text{Account balance at beginning of year 2}) \times (1.1) = \$1000 \times (1.1)^2.$$

$$\$1000 \times (1.1)$$

It isn't hard to see that each transition (which occurs at the end of each year) corresponds to taking 110% of the money at the start of that year. This, of course, is the same as a multiplication by 1.1. In our terminology, we will say that the rate of growth in Example 9 is 1.1 (not 10% as is often mistakenly thought). The answer to the question raised in Example 9 is now easy: The account balance at the end of year 25 (and at the beginning of year 26) is

$$\$1000 \times (1.1)^{25} = \$10,834.71.$$

In this example the period between transitions is 1 year; the population (the money in the account) is given by the sequence

$$P_1 = \$1000, P_2 = \$1100, P_3 = \$1210, \ldots, P_{26} = \$10,834.71.$$

In general, at the end of the Nth year—and, of course, also at the beginning of the $(N + 1)$st year—the balance in the account is the $(N + 1)$st value of the population sequence:

$$P_{N+1} = \$1000 \times (1.1)^N.$$

Figure 10-5 plots the growth of the money in the account for the first 8 years.

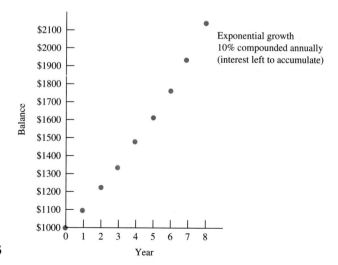

Exponential growth
10% compounded annually
(interest left to accumulate)

FIGURE 10-5

A sequence of the kind described in Example 9, where every term is obtained by multiplying the preceding term by a fixed amount, is called a **geometric sequence**. In any exponential growth problem, the size of the population is always a term in some geometric sequence, and it can therefore be expressed as

$$P_{N+1} = P_N \times \text{growth rate}$$

(which is exactly the same statement as $P_N = P_{N-1} \times$ growth rate). This is the *recursive description*. Alternatively, we can describe the sequence by the *explicit description*

$$P_{N+1} = P_1 \times (\text{growth rate})^N$$

(which is exactly the same statement as $P_N = P_1 \times (\text{growth rate})^{N-1}$). We will use the letter r to denote the growth rate (also known as the **common ratio**) of the geometric sequence.

A common misconception is that exponential growth implies that the population always gets bigger. The truth of the matter is that this need not be the case.

Example 10. A population grows exponentially with an annual growth rate of $r = 0.3$ and starting population $P_1 = 1,000,000$. What is the size of the population at the end of 6 years? (Remember this means we want to find P_7!)

In this case, we have $P_7 = 1,000,000 \times (0.3)^6$. We don't need a calculator to compute this number:

$$P_7 = 10^6 \times \frac{3^6}{10^6} = 3^6 = 729.$$

Fig. 10-6 plots the growth of this population for the first 6 years, and we can clearly see that it is heading toward extinction.

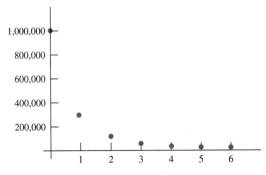

FIGURE 10-6

It is often convenient to distinguish between exponential growth situations in which populations get bigger (as in Example 9) and exponential growth situations in which populations actually decrease (as in Example 10). The latter situation is commonly referred to as **exponential decay**. The difference, of course, is in the growth rate: *If it is positive but less than 1, we have decay; if it is more than 1, we have real growth.*

Putting Your Money Where Your Math Is

Let's return to the world of savings accounts. In Example 9 we discussed a special case of a general situation: A certain sum of money P_1 (called the *principal*) is deposited in an account that draws interest at an annual interest rate r (the interest is paid once a year at the end of the year). How much money is there in the account at the end of N years assuming that the interest is left in the account? We now know that the answer is given by the explicit formula

$$P_{N+1} = P_1 \times (\text{growth rate})^N.$$

How do we compute the growth rate? When the annual interest was 10% (Example 9), we computed the growth rate as 1.1 (110%). If the annual interest had been 12%, the growth rate would have been 1.12 (112%), and if the annual interest had been $6\frac{3}{4}$%, the annual growth rate would have been 1.0675 (106.75%). In general, if we write the annual interest rate as a decimal i (rather than a percent), then the growth rate r will be $1 + i$. Placing this in the preceding formula gives

$$P_{N+1} = P_1(1 + i)^N$$

Example 11. If we deposit $367.51 in a savings account yielding an annual interest of $9\frac{1}{2}$% a year and we leave both the principal and the interest in the account, how much money will there be in the account at the end of 7 years?
Here $P_1 = 367.51$; $i = 0.095$; and $r = 1.095$. The answer is

$$P_8 = 367.51 \times (1.095)^7 = 693.69.$$

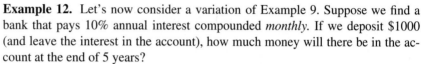

Example 12. Let's now consider a variation of Example 9. Suppose we find a bank that pays 10% annual interest compounded *monthly*. If we deposit $1000 (and leave the interest in the account), how much money will there be in the account at the end of 5 years?
This problem is still a problem in exponential growth. The big difference now is that the period between transitions is a month (instead of a year). At the end of 5 years we will have gone through 60 transitions, so in this example we are after P_{61}. Just as before (remember we are dealing with a geometric sequence), we have

$$P_{61} = \$1000 \times (\text{growth rate})^{60},$$

and just as before, it all boils down to finding the growth rate. Since the period between transitions is a month, we can't use the 10% annual interest to compute the growth rate. Instead, we must divide the annual interest by 12, which gives the **periodic interest**

$$\frac{10\%}{12} = 0.83333 \ldots \%.$$

The growth rate is then $100.83333 \ldots \%$ which, written in decimal form, is $r = 1.0083333 \ldots$, and therefore

$$P_{61} = \$1000 \times (1.008333 \ldots)^{60} \approx \$1645.31.$$

Just out of curiosity, what would happen if we left the money in the account for 25 years? Everything is the same as in our last computation, except that the number of transitions is $25 \times 12 = 300$. It follows that after 25 years the amount of money in the account will be

$$P_{301} = \$1000 \times (1.008333 \ldots)^{300} \approx \text{S}12, 056, 94.$$

Example 13. Now suppose that we find a bank that pays 10% annual interest compounded *daily*. If we deposit the same $1000 for 5 years (just as we did in Example 12), how much will we have at the end of the 5 years?

In this case the period between transitions is 1 day. The total number of transitions in 5 years is $365 \times 5 = 1825$, and the periodic interest is 10%/365, which we write as $0.10/365 \approx 0.00027397$. The growth rate for this exponential growth problem is

$$r = \left(1 + \frac{0.10}{365}\right) \approx 1.00027397,$$

so that the final answer is

$$1000\left(1 + \frac{0.10}{365}\right)^{365 \times 5} \approx 1000(1.00027397)^{1825} \approx 1648.61.$$

Examples 12 and 13 illustrate the general rules for growth under compound interest after N transitions. They are, in short: At the end of N compounding periods, the total amount of money in the account is

$$P_{N+1} = P_1 \times (\text{growth rate})^N,$$

where

growth rate $= 1 +$ periodic interest,

and

$$\text{periodic interest} = \frac{\text{annual interest}}{\text{number of periods in 1 year}}.$$

Example 14 (Shopping for a bank). You have an undisclosed amount of money to invest. Bank *A* offers savings accounts that pay 10% annual interest compounded yearly. Bank *B* offers accounts that pay 9.75% annual interest compounded monthly. Bank *C* offers accounts that pay 9.5% annual interest compounded daily. Which bank offers the best deal?

Note that the problem does not indicate the amount of money we invest or the length of time we plan to leave the money in the account. The answer to the problem depends only on the annual interest and the compounding period. The way to compare these different accounts is to use a common yardstick, for example, How much does $1 grow in 1 year?

With bank *A*, in 1 year $1 becomes $1.10.
With bank *B*, in one year $1 becomes

$$\$\left(1 + \frac{0.0975}{12}\right)^{12} \approx \$1.102.$$

And with bank *C*, in one year $1 becomes

$$\$\left(1 + \frac{0.095}{365}\right)^{365} \approx \$1.0996.$$

We can now see that bank *B* offers the best deal.

These same calculations are described by banks in a slightly different form called the annual yield. The **annual yield** is the percentage increase that the account will produce in 1 year. In Example 14, the annual yield for bank *A* is 10%, for bank *B*, 10.2%, and for bank *C*, 9.96%. These numbers can be read off directly from the preceding calculations.

Adding Terms in a Geometric Sequence

One of the tidbits we learned in Example 4 is that there is a straightforward formula that allows us to add up the beginning terms of any arithmetic sequence. We would like to be able to do something similar with a geometric sequence. The basic fact that we need is given by the following formula:

**Adding *N* Consecutive Terms
of a Geometric Sequence**

$$a + ar + ar^2 + \cdots + ar^{N-1} = \frac{a(r^{N+1} - 1)}{r - 1}.$$

Example 15. $8 + 8 \times 3 + 8 \times 3^2 + 8 \times 3^3 + \cdots + 8 \times 3^{13} = ?$

Here, $a = 8$, $r = 3$, and $N - 1 = 13$. Plugging these values into the formula gives

$$\frac{8 \times (3^{14} - 1)}{3 - 1} = 19,131,872.$$

(The reader is encouraged to try Exercises 32 and 33 at this point.)

Example 16. A mother decides to set up a college trust fund for her newborn child. The plan is to deposit $100 a month for the next 18 years (i.e., 216 months) in a savings account that pays 6% annual interest compounded monthly. How much money will there be in the account at the end of 18 years?

This is a problem of exponential growth with a twist: Each $100 deposit grows at the same rate of 1.005 [growth rate $= 1 + (0.06/12) = 1.005$] but the number of periods is different for each deposit. Let's make a list:

■ 1st deposit of $100 draws interest compounded for 216 months, producing $100(1.005)^{216}$.

■ 2nd deposit of $100 draws interest compounded for 215 months, producing $100(1.005)^{215}$.

■ 3rd deposit of $100 draws interest compounded for 214 months, producing $100(1.005)^{214}$.

$$\vdots$$

■ 216th deposit of $100 draws interest for 1 month, producing $100(1.005)^{1}$.

The total amount in the account at the end of 18 years will be

$$100(1.005)^{216} + 100(1.005)^{215} + \cdots + 100(1.005).$$

This is the sum of the terms of a geometric sequence. Using the formula for adding consecutive terms of a geometric sequence we get (see Exercise 46)

$$\$ \frac{100(1.005)[(1.005)^{216} - 1]}{0.005} \approx \$38,929.$$

Eighteen years from now this amount of money might cover tuition for about one semester!

LOGISTIC GROWTH

When dealing with animal populations, the two models we have studied so far are mostly inadequate. As we now know, *linear growth* represents the case in which there is a fixed amount of growth during each period between transitions. This model might work with inanimate objects (garbage, production goods, sales figures, etc.) but fails completely when there is an element of breeding that must be taken into account. *Exponential growth,* on the other hand, represents the case in which there is a constant rate of growth, which is appropriate when there is unrestrained breeding (as with money left to compound in a bank account). In population biology, however, it is generally the case that the rate of growth of an animal population is not always the same. It depends on the relative sizes of other interacting populations (predators, prey, etc.) and, even more importantly, on the relative size of the population itself. When the relative size of the population is small (we will define more precisely what we mean by this soon) and there is plenty of room to grow, then the rate of growth is high. As the population gets larger, there is less room to grow and the growth rate starts to taper off. Sometimes the population gets too large for its own good, leading toward decay and possibly to extinction.

A well-known experiment with rats in a cage illustrates some of these ideas. Put a few rats in a cage with plenty of food. If the cage is big enough, the rats will start breeding in an unrestrained fashion, and for a while the growth of the rat population will follow an exponential growth model. As the cage gets more crowded, the rate of growth will slow down dramatically. The force that regulates this slowdown is competition for the resources that are essential for growth: food, sex, and space. Eventually, the competition gets so keen that the rats start killing each other off—it is their own quick fix for dealing with the overcrowding problem. Often when the population gets back down to an acceptable level, the cannibalism stops. Sometimes nature's growth-regulating mechanism may get out of kilter—the killing frenzy may not quite stop in time, and the rats will wipe each other out in a total rodent holocaust.

The above scenario applies (with variations) to almost every situation in which there is a limited environment for a population to live in. Population biologists call such an environment the **habitat**. The habitat might be a cage (as in the example of the rats), a lake (as for a population of fish), a garden (as for a population of snails), and of course the planet itself, which is everyone's habitat.

Of the many mathematical models that attempt to deal with the basic principle of a variable growth rate in a fixed habitat, the simplest is the **logistic growth model**. To put it in a very informal way, the key idea in the logistic growth model is that the rate of growth of the population is directly proportional to the amount of elbow room available in the population's habitat. Thus, lots of elbow room means a high growth rate; little elbow room means a low growth rate (possibly less than 1, which, as we know, means that the population is actually going down); and finally, if it ever gets to be the case that the habitat is completely saturated, the population will die out.

There are two equivalent ways we can describe the situation mathematically. If C is some constant that describes the total saturation point of the habitat (population biologists call C the **carrying capacity** of the habitat), then for a population of size P_N we can say that the amount of elbow room is the difference between the carrying capacity and the population size, namely $(C - P_N)$. Then, if the growth rate is proportional to the amount of elbow room (as described above) we have

growth rate for period $N = R(C - P_N)$

(where R is a constant of proportionality that depends only on the particular population we are studying). Using the fact that (population at period N) $\times$ (growth rate for period N) = population for period $(N + 1)$, we get the following recursive transition rule for the logistic growth model:

$$P_{N+1} = R(C - P_N)P_N.$$

There are two constants in the above transition rule: R which depends on the animal population we are studying, and C which depends on the habitat.

A computationally more convenient way to describe exactly the same thing is to put everything in relative terms: The maximum of the population is 1 (i.e., 100% of the habitat is taken up by the population); the minimum is 0 (i.e., the population is extinct); and every possible population size P_N is represented by some fraction between 0 and 1 which we will denote by p_N (to distinguish it from P_N). The relative amount of elbow room then is $(1 - p_N)$, and the transition rules for the logistic model can be rewritten in the form of the following equation, called the **logistic equation**:[4]

$$p_{N+1} = r(1 - p_N)p_N.$$

In this equation the value p_N represents the fraction of the habitat's carrying capacity taken up by the actual population P_N ($p_N = P_N/C$), and the constant r depends on both the original growth rate R and the habitat's carrying capacity C. We will call r the **growth parameter**.

Because it looks at population growth using a single common yardstick (the fraction of its habitat's carrying capacity taken up by the population), the second description is particularly convenient when making growth comparisons between populations and is the one preferred by ecologists and population biologists. We will stick to it ourselves. In the examples that follow we will look at

[4] This equation is sometimes known as the *Verhulst equation* after the Belgian Pierre François Verhulst who proposed it in the late nineteenth century.

the growth pattern of an imaginary population under the logistic growth model. In each case, all we need to get started is the original population p_1 (given as a fraction of the habitat's carrying capacity) and the value of the growth parameter r (p_1 should always be between 0 and 1 and, for mathematical reasons we will restrict r to be between 0 and 4). The logistic equation and a good calculator will do the rest. (However, the reader should be forewarned that the calculations shown in the examples that follow were done with a computer and carried to 16 decimal places before being rounded off to 3 or 4 decimal places—they may not match exactly with identical calculations done with a hand calculator.)

Example 17. Suppose that we have a pond we wish to stock with a particular type of fish—let's say for the sake of argument, a particular kind of trout. Let's assume that the growth parameter for this breed of trout is $r = 2.5$. Let's say that the starting population is given by $p_1 = 0.2$ (remember, this means that we stocked the pond to 20% of its carrying capacity). Now let's see what the logistic growth model predicts. After one breeding season[5] we have

(The population of the pond has doubled. Things are looking good!)

$$p_2 = 2.5 \times (1 - 0.2) \times (0.2) = 0.4.$$

After the next breeding season we have

(The growth rate has slowed down a bit.)

$$p_3 = 2.5 \times (1 - 0.4) \times (0.4) = 0.6.$$

By the third generation, we have

$$p_4 = 2.5 \times (1 - 0.6) \times (0.6) = 0.6.$$

Let's try one more:

$$p_5 = 2.5 \times (1 - 0.6) \times (0.6) = 0.6.$$

It is quite clear that the trout population has now stabilized at 60% of the pond's carrying capacity, and unless some external change is made, it will remain at the same level for all future generations. This may be somewhat disappointing for the grower, but it is a perfect situation for the trout. We have here an example of a population that is in stable balance with its habitat.

[5] In animal populations, the transitions usually correspond to breeding seasons.

Example 18. Suppose that we have the same pond and the same trout as in Example 17 (in other words, we still have $r = 2.5$), but we wonder what would happen if we stocked the pond differently—let's say we started with $p_1 = 0.3$. We now have

$$p_2 = 2.5 \times (1 - 0.3) \times (0.3) = 0.525$$

(Things are looking good!) $$p_3 = 2.5 \times (1 - 0.525) \times (0.525) \approx 0.6234$$

(A bit of a disappointment!) $$p_4 = 2.5 \times (1 - 0.6234) \times (0.6234) \approx 0.5869$$

(Up again!) $$p_5 = 2.5 \times (1 - 0.5869) \times (0.5869) \approx 0.6061$$

(And down.) $$p_6 = 2.5 \times (1 - 0.6061) \times (0.6061) \approx 0.5968$$

$$p_7 = 2.5 \times (1 - 0.5968) \times (0.5968) \approx 0.6016.$$

We leave it to the reader to verify that the population of the pond is once again moving closer and closer to the value 0.6 but in an oscillating (up, down, up, down, . . .) manner.

Example 19. What happens in Example 18 if $p_1 = 0.7$? After the first generation the population behaves identically with that in Example 18. This follows from the fact that in both cases we get the same value for p_2:

$$p_2 = 2.5 \times 0.3 \times 0.7 = 2.5 \times 0.7 \times 0.3 = 0.525.$$

A useful general rule about logistic growth can be spotted here: If we replace p_1 with its complement $(1 - p_1)$, then after the first generation the populations will behave identically.

Example 20. Let's consider a different population of fish—let's say catfish—and let's say that for them the growth parameter is $r = 3.1$. We start again as in Example 17 with $p_1 = 0.2$. For the sake of brevity we will write the values of the population in sequence form and leave the calculations to the reader.

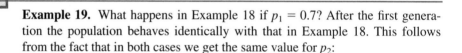

$$p_1 = 0.2, \qquad p_2 = 0.496, \qquad p_3 \approx 0.775, \qquad p_4 \approx 0.541,$$

$$p_5 \approx 0.770, \qquad p_6 \approx 0.549, \qquad p_7 \approx 0.767, \qquad p_8 \approx 0.553,$$

$$p_9 \approx 0.766, \qquad p_{10} \approx 0.555, \qquad p_{11} \approx 0.766, \qquad p_{12} \approx 0.556,$$

$$p_{13} \approx 0.765, \qquad p_{14} \approx 0.557, \qquad p_{15} \approx 0.765, \qquad p_{16} \approx 0.557, \; . . .$$

An interesting pattern emerges here: The population of the pond seems to have settled into a two-period cycle alternating between a high population period at 0.765 (famine) and a low population period at 0.557 (feast). There are many animal populations whose growth is cyclical—a lean period followed by a boom period followed by a lean period, etc.

Example 21. Next let's try a population of beetles. For this particular type of beetle we are told that $r = 3.5$. Let's assume that $p_1 = 0.56$ and use the logistic equation to analyze the growth pattern of this beetle population. We leave it to the reader to verify these numbers and fill in the missing details (a calculator is all that is needed).

$$p_1 = 0.560, \quad p_2 = 0.862, \quad p_3 \approx 0.415, \quad p_4 \approx 0.850,$$

$$p_5 \approx 0.446, \quad p_6 \approx 0.865, \quad \ldots \quad p_{21} \approx 0.497,$$

$$p_{22} \approx 0.875, \quad p_{23} \approx 0.383, \quad p_{24} \approx 0.827, \quad p_{25} \approx 0.501,$$

$$p_{26} \approx 0.875, \quad \ldots \ldots$$

It took a while, but we can now see a pattern: Since $p_{26} = p_{22}$, the population will repeat itself in a four-period cycle ($p_{27} = p_{23}$, $p_{28} = p_{24}$, $p_{29} = p_{25}$, $p_{30} = p_{26} = p_{22}$, etc.). There are many populations that follow more complicated cyclical patterns of growth: Locusts are the most notorious example.

Example 22. Our last and most remarkable example is given by the case $r = 4$. Let's start with $p_1 = 0.2$. The first 20 values of the population sequence are given by

$$p_1 = 0.2000, \quad p_2 = 0.640, \quad p_3 \approx 0.9216, \quad p_4 \approx 0.2890,$$

$$p_5 \approx 0.8219, \quad p_6 \approx 0.5854, \quad p_7 \approx 0.9708, \quad p_8 \approx 0.1133,$$

$$p_9 \approx 0.4020, \quad p_{10} \approx 0.9616, \quad p_{11} \approx 0.1478, \quad p_{12} \approx 0.5039,$$

$$p_{13} \approx 0.9999, \quad p_{14} \approx 0.0002, \quad p_{15} \approx 0.0010, \quad p_{16} \approx 0.0039,$$

$$p_{17} \approx 0.0157, \quad p_{18} \approx 0.0617, \quad p_{19} \approx 0.2317, \quad p_{20} \approx 0.7121.$$

Figure 10-7 shows the behavior of the population for the first 20 generations. The reader is encouraged to chart this population for a few additional generations. The surprise here is the absence of any predictable pattern. Even though the population sequence is governed by a very precise rule (the logistic equation), to an outside observer the pattern of growth appears to be quite erratic and seemingly random.

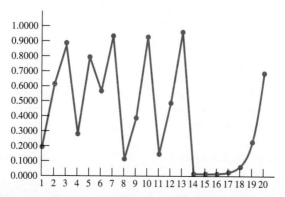

FIGURE 10-7

The behavior of populations under the logistic growth model exhibits many interesting surprises. In addition to doing Exercises 34 through 38 at the end of this chapter, the reader is encouraged to experiment on his or her own in a manner similar to the work we did in the preceding examples. (Choose a p_1 between 0 and 1, choose an r between 0 and 4, and fire up both your calculator and your imagination!) An excellent nontechnical account of the surprising patterns produced by the logistic growth model can be found in reference 2. More technical accounts of the logistic equation can be found in references 7 and 8.

CONCLUSION

In this chapter we studied three simple models that describe the way that populations grow.

In the *linear model* of population growth, the population sequence is described by an arithmetic sequence, and at each transition period the population grows by a constant amount. Linear growth is most common with populations consisting of inanimate objects.

In the *exponential model* of population growth, the population is described by a geometric sequence. Here in each transition period the population is multiplied by a constant amount called the growth rate. Exponential growth is typical of situations in which there is unrestrained breeding. Money drawing interest in a bank account is one such example.

The *logistic model* of population growth represents situations in which the rate of growth of the population varies from one season to the next depending on the amount of space available in the population's habitat. Many animal populations are governed by the logistic model or simple variations of it.

It goes without saying that most serious studies of population growth involve models with much more complicated mathematical descriptions, but to us that is neither here nor there. Ultimately, the details are not as important as the overall picture: a realization that mathematics can be useful even in its most simplistic forms to describe and predict the rise and fall of populations in many fields—from the artificial world of industry and finance to the natural world of population biology and animal ecology.

KEY CONCEPTS

annual yield	expotential decay
arithmetic sequence	exponential growth
carrying capacity	geometric sequence
common difference	growth parameter
continuous growth	growth rate
discrete growth	habitat
dynamical process	linear growth
explicit description	logistic equation

logistic growth

logistic growth model

periodic interest

plot (graph)

population growth

population sequence

recursive description

transition

transition rule

EXERCISES

Walking

1. A population of laboratory rats grows according to the following transition rule: $P_N = P_{N-1} + P_{N-2}$. The starting population is $P_1 = 6$ and the population in the second generation is $P_2 = 10$.
   (a) Find the population in the fourth generation.
   (b) Find the population after five transitions.
   (c) Can the size of the rat population ever be an odd number? Explain.

2. A population of guinea pigs grows according to the following transition rule: $P_N = 2P_{N-1} + P_{N-2}$. The starting population is $P_1 = 3$ and the population in the second generation is $P_2 = 5$.
   (a) Find the population in the third generation.
   (b) Find the population after four transitions.
   (c) Can the size of the guinea pig population ever be an even number? Explain.

3. Mr. GQ is a snappy dresser and has an incredible collection of neckties. Each month at the end of the month, he buys himself five new neckties. Let P_N represent the number of neckties in his collection during the Nth month. Assuming he started out with just three neckties, and that he never throws anything away,
   (a) give a recursive description for P_N
   (b) give an explicit description for P_N
   (c) find P_{300}.

4. A nuclear power plant produces 12 lbs. of radioactive waste every month. The radioactive waste must be stored in a special storage tank. At present, the tank holds 25 lbs. of radioactive waste. Let P_N represent the amount of radioactive waste (in lbs.) in the storage tank in the Nth month.
   (a) Give a recursive description for P_N.
   (b) Give an explicit description for P_N.
   (c) If the maximum capacity of the storage tank is 500 lbs., how long will it take before the tank has reached its maximum capacity?

5. Consider a population that grows according to the linear growth model $P_N = P_{N-1} + 125$, with starting population $P_1 = 80$.
   (a) Find P_2, P_3, and P_4.
   (b) Find P_{100}.
   (c) Give an explicit description of the population sequence.

6. Consider a population that grows according to the linear growth model $P_N = P_{N-1} + 23$, with starting population $P_1 = 57$.
   (a) Find P_2, P_3, and P_4.
   (b) Find P_{201}.
   (c) Give an explicit description of the population sequence.

7. Consider a population that grows according to a linear growth model. The starting population is $P_1 = 8$, and the population in the tenth generation is $P_{10} = 35$.
   (a) Find the common difference d.
   (b) Find P_{51}.
   (c) Give an explicit description of the population sequence.

8. Consider a population that grows according to a linear growth model. The population in the sixth generation is $P_6 = 37$, and the population in the seventh generation is $P_7 = 42$.
   (a) Find the common difference d.
   (b) Find the starting population P_1.
   (c) Give an explicit description of the population sequence.

9. Find $\underbrace{2 + 7 + 12 + \cdots}_{100 \text{ terms}}$.

10. Find $\underbrace{21 + 28 + 35 + \cdots}_{57 \text{ terms}}$.

11. Find $12 + 15 + 18 + \cdots + 309$.

12. Find $1 + 10 + 19 + \cdots + 2701$.

13. Consider a population that grows according to a linear growth model. The starting population is $P_1 = 23$ and the common difference is $d = 7$.
    (a) Find $P_1 + P_2 + \cdots + P_{1000}$.
    (b) Find $P_{101} + P_{102} + \cdots + P_{1000}$.

14. Consider the arithmetic sequence with the first four terms 7, 11, 15, 19.
    (a) Find P_{100}.
    (b) Find P_N.
    (c) Find $P_1 + P_2 + \cdots + P_{1000}$.
    (d) Find $P_{101} + P_{102} + \cdots + P_{1000}$.

15. The city of Lightsville currently has 137 street lights. As part of an urban renewal program the city council has decided to install and have operational 2 additional street lights at the end of each week for the next 52 weeks. Each street light costs $1 to operate for 1 week.
    (a) How many street lights will the city have at the end of 38 weeks?
    (b) How many street lights will the city have at the end of N weeks? ($N \le 52$)
    (c) What is the cost of operating the original 137 lights for 52 weeks?
    (d) What is the additional cost for operating the newly installed lights for the 52-week period during which they are being installed?

16. A manufacturer currently has on hand 387 widgets. During the next 2 years, the manufacturer will be increasing his inventory by 37 widgets per week. Each widget costs 10 cents a week to store.
    (a) How many widgets will the manufacturer have on hand 21 weeks from today?
    (b) How many widgets will the manufacturer have on hand N weeks from today? ($N \le 104$)
    (c) What is the cost of storing the original 387 widgets for 2 years?
    (d) What is the additional cost for storing the increased inventory of widgets for the next 2 years?

17. You have a coupon worth 15% off any item in the store (including sale items). The particular item you want is on sale at 30% off the marked price of $100. The store policy allows you to use your coupon before the 30% discount or after the 30% discount (i.e., you can take 15% off the marked price first and then take 30% off the resulting price, or you can take 30% off the marked price first and then take 15% off the resulting price).
  (a) What is the dollar amount of the discount in each case?
  (b) What is the total percentage discount in each case?
  (c) Suppose the article cost P dollars (instead of $100). What is the percentage discount in each case?

18. You have $1000 to invest in one of two competing banks (bank A or bank B) both of which are paying 10% annual interest on deposits left for 1 year. Bank A is offering a 5% bonus credited to your account at the time of the initial deposit, provided the funds are left in the account for a year. Bank B is offering a 5% bonus paid on your account balance at the end of the year after the interest has been credited to your account.
  (a) How much money would you have at the end of the year if you invested in bank A? In bank B?
  (b) What is the total percentage gain (interest plus bonus) at the end of the year for each of the two banks?
  (c) Suppose you invested P dollars (instead of $1000). What is the total percentage gain (interest plus bonus) at the end of the year for each bank?

19. At Tasmania State University, during a three-year period, the tuition increased by 10%, 15%, and 10% respectively each year. What was the total percentage increase overall during the three-year period? (*Hint*: The answer is *not* 35%!)

20. A membership store gives a 10% discount on all purchases to its members. If the store marks each item up 50% (based on its cost), what is the markup actually realized by the store when an item is sold to a member?

21. The amount of $3250 is deposited in a savings account that draws 9% annual interest with interest credited to the account at the end of each year. Assuming no withdrawals are made, how much money will be in the account after 4 years?

22. The amount of $1237.50 is deposited in a savings account that draws $8\frac{1}{4}$% annual interest with interest credited to the account at the end of each year. Assuming no withdrawals are made, how much money will be in the account after 3 years?

23. **(a)** The amount of $5000 is deposited in a savings account that pays 12% annual interest compounded monthly. Assuming no withdrawals are made, how much money will be in the account after 5 years?
  (b) What is the annual yield on this account?

24. **(a)** The amount of $874.83 is deposited in a savings account that pays $7\frac{3}{4}$% annual interest compounded daily. Assuming no withdrawals are made, how much money will be in the account after 2 years?
  (b) What is the annual yield on this account?

25. You have some money to invest. The Great Bulldog Bank offers accounts that pay 6% annual interest. The First Northern Bank offers accounts that pay 5.75% annual interest compounded monthly. The Bank of Wonderland offers 5.5% annual interest compounded daily. What is the annual yield for each bank?

26. Complete the following table:

Annual interest rate	Compounded	Annual yield
12%	Yearly	12%
12%	Semiannually	?
12%	Quarterly	?
12%	Monthly	?
12%	Daily	?

27. You decide to open a Christmas Club account at a bank that pays 6% annual interest compounded monthly. You deposit $100 on the first of January and on the first of each succeeding month through November. How much will you have in your account on the first of December?

28. You decide to save money to buy a car by opening a special account at a bank that pay 8% annual interest compounded monthly. You deposit $300 on the first of each month for 36 months. How much will you have in your account at the end of the thirty-sixth month?

29. You are interested in buying a car 5 years from now, and you estimate the future cost will be $10,000. You decide to deposit money today in an account that pays interest, so that 5 years later you will have the $10,000 necessary to purchase your "dream" car. How much money do you need to deposit if the account you deposit your money in
   (a) has an interest rate of 10% compounded annually?
   (b) has an interest rate of 10% compounded quarterly?
   (c) has an interest rate of 10% compounded monthly?

30. A population grows according to an exponential growth model. The starting population is $P_1 = 8$ and the growth rate is $r = 1.5$.
   (a) Find P_2.
   (b) Find P_{10}.
   (c) Give an explicit description for the population sequence.

31. A population grows according to an exponential growth model. The starting population is $P_1 = 11$ and the growth rate is $r = 1.25$.
   (a) Find P_2.
   (b) Find P_{10}.
   (c) Give an explicit description for the population sequence.

32. Consider the geometric sequence with first four terms 1, 3, 9, 27.
   (a) Find P_{100}.
   (b) Find P_N.
   (c) Find $P_1 + P_2 + \cdots + P_{100}$.
   (d) Find $P_{50} + P_{51} + \cdots + P_{100}$.

33. Consider the geometric sequence with first term $P_1 = 3$ and common ratio $r = 2$.
   (a) Find P_{100}.
   (b) Find P_N.
   (c) Find $P_1 + P_2 + \cdots + P_{100}$.
   (d) Find $P_{50} + P_{51} + \cdots + P_{100}$.

Exercises 34 through 38 refer to the logistic growth model $p_{N+1} = r(1 - p_N)p_N$. For most of these exercises, a calculator with a memory register is suggested.

34. A population grows according to the logistic growth model, with growth rate $r = 1.5$ and $p_1 = 0.8$.
   (a) Find p_2.
   (b) Find p_3.
   (c) Find the population after four transitions.

35. A population grows according to the logistic growth model, with growth rate $r = 2.8$ and $p_1 = 0.15$.
   (a) Find p_2.
   (b) Find p_3.
   (c) Find the population after four transitions.

36. Suppose we know that under the logistic growth model the growth parameter for a particular colony of birds is $r = 2.5$. Suppose also that the starting population is $p_1 = 0.8$.
   (a) What is p_2?
   (b) What is the population after the second transition?
   (c) What does the logistic growth model predict in the long term for this population?
   (d) What other starting population would result in the same population sequence after the first transition?

37. A population grows according to the logistic growth model, with growth rate $r = 3.0$ and $p_1 = 0.15$. Find the values of p_2 through p_{10}.

38. A population grows according to the logistic growth model, with growth rate $r = 3.8$ and $p_1 = 0.23$. Find the values of p_2 through p_{10}.

Jogging

39. How much should a retailer mark up her goods so that when she has a 25% off sale, the resulting prices will still reflect a 50% markup (on her cost)?

40. What annual interest rate compounded semiannually gives an annual yield of 21%?

41. Before Annie set off for college, Daddy Warbucks offered her a choice between the following two incentive programs:

   ■ *Option 1.* A $100 reward for every A she gets in a college course.

   ■ *Option 2.* One cent for her first A, 2 cents for the second A, 4 cents for the third A, 8 cents for the fourth A, etc.

   Annie chose option 1. After getting a total of 30 A's in her college career, Annie is happy with her reward of $100 × 30 = $3000. Unfortunately, Annie did not get an A in math. Help her figure out how much she would have made had she chosen option 2.

42. Consider a population that grows according to the logistic growth model with growth parameter $r(r > 1)$. Show that for $p_1 = (r - 1)/r$ the population is constant.

43. Suppose the habitat of a population of snails has a carrying capacity of $C = 20,000$ and the current population size is 5000. Suppose also that the growth parameter for this particular type of snail is $r = 3.0$. What does the logistic growth model predict for this population after four transition periods?

Exercises 44 and 45 refer to the following situation. If B dollars are borrowed at a periodic interest rate I and N equal periodic payments are to be made, then the periodic payment p is given by the formula

$$p = \frac{BI(1 + I)^N}{(1 + I)^N - 1}.$$

44. (a) You buy a house for $120,000 with $20,000 down and finance the balance over 30 years at 9% annual interest (with equal monthly payments). What is your monthly payment?

(b) What is the monthly payment if the loan described in (a) is financed over 40 years instead of 30 years?

45. You decide that you can afford a $1000-per-month house payment. The current going interest rate on 30-year home loans is 11%. How much money can you borrow at this rate so that your payment will not exceed $1000 per month?

46. Find $100(1.005)^{216} + 100(1.005)^{215} + \cdots + 100(1.005)$. Use the formula for adding the terms of a geometric sequence and a calculator. (*Hint:* Read the sum from right to left. What is a? What is r?)

47. $1 + 5 + 3 + 8 + 5 + 11 + 7 + 14 + \cdots + 99 + 152 = ?$

48. $1 + 1 + 2 + \frac{1}{2} + 4 + \frac{1}{4} + 8 + \frac{1}{8} + \cdots + 4096 + \frac{1}{4096} = ?$

Running

49. (a) Show that

$$1 + r + r^2 + \cdots + r^{100} = (r^{101} - 1)/(r - 1).$$

[*Hint:* Do the multiplication

$$(1 + r + r^2 + \cdots + r^{100})(r - 1)$$

and see what you get!]

(b) Show that

$$1 + r + r^2 + \cdots + r^N = (r^{N+1} - 1)/(r - 1).$$

(c) Show that

$$a + ar + ar^2 + \cdots + ar^N = a(r^{N+1} - 1)/(r - 1).$$

50. Show that the sum of the first N terms of an arithmetic sequence with first term c and common difference d is

$$\frac{N}{2} \quad [2c + (N \cdot 1)d].$$

51. You are purchasing a home for $120,000 and are shopping for a loan. You have a total of $31,000 to put down, including the closing costs of $1000 and any loan fee that might be charged. Bank *A* offers a 10% annual interest loan amortized over 30 years with 360 equal monthly payments. There is no loan fee. Bank *B* offers a 9.5% annual interest loan amortized over 30 years with 360 equal monthly payments. There is a 3% loan fee (i.e., a one-time up-front charge of 3% of the loan). Which loan is better?

52. A friend of yours sold his car to a college student and took a personal note (cosigned by the student's rich uncle) for $1200 with no interest, payable at $100 per month for 12 months. Your friend immediately approaches you and offers to sell you this note. How much should you pay for the note if you want an annual yield of 12% on your investment?

53. Suppose $r > 3$. Using the logistic growth model, find a population p_1 such that $p_1 = p_3 = p_5 = \ldots$, but $p_1 \neq p_2$. [*Hint:* Exercise 42 shows that $p_1 = (r - 1)/r$ would give $p_1 = p_2 = p_3 = \ldots$.]

REFERENCES AND FURTHER READINGS

1. Clark, C. W., *Bioeconomics Modeling and Fishery Management.* New York: John Wiley & Sons, 1985.

2. Gleick, James, *Chaos: Making a New Science.* New York: Viking Penguin, Inc., 1987, chap. 3.

3. Hoppensteadt, Frank, *Mathematical Methods of Population Biology.* Cambridge: Cambridge University Press, 1982.

4. Hoppensteadt, Frank, *Mathematical Theories of Population: Demographics, Genetics and Epidemics.* Philadelphia: Society for Industrial and Applied Mathematics, 1975.

5. Hoppensteadt, Frank, and Charles Peskin, *Mathematics in Medicine and the Life Sciences.* New York: Springer-Verlag, 1992.

6. Kingsland, Sharon E., *Modeling Nature: Episodes in the History of Population Ecology.* Chicago: University of Chicago Press, 1985.

7. May, Robert M., "Biological Populations with Nonoverlapping Generations: Stable Points, Stable Cycles and Chaos," *Science,* 186 (1974), 645–647.

8. May, Robert M., "Simple Mathematical Models with Very Complicated Dynamics," *Nature,* 261 (1976), 459–467.

9. May, Robert M., and George F. Oster, "Bifurcations and Dynamic Complexity in Simple Ecological Models," *American Naturalist,* 110 (1976), 573–599.

10. Smith, J. Maynard, *Mathematical Ideas in Biology.* Cambridge: Cambridge University Press, 1968.

CHAPTER 11

Mirror, Mirror, off The Wall ...

Symmetry of Motion

Symmetry, as wide or as narrow as you may define its meaning, is one idea by which man through the ages has tried to comprehend and create order, beauty and perfection.
HERMANN WEYL*

It is said that Eskimos have over fifty different words for ice. Ice is a pervasive and fundamental element in the Eskimo's world. In this sense, we would expect mathematics to have at least as many words to describe the notion of symmetry. Unfortunately, just the opposite is the case. In mathematics, "symmetry" is a rather general all-encompassing term, and even in common usage it's a word that has different meanings for different people. (Before reading on, the reader might want to stop and write down his or her own definition of symmetry.)

Regardless of how we define it, in its everyday use symmetry is usually an attribute we give to a physical object, and in this sense symmetry is a distinctly geometric concept. We should mention in passing that this need not be the only possible setting for the concept of symmetry. In a more general way symmetry is a notion that applies equally well to music, literature, and many areas of mathematics besides geometry.

Be that as it may, in this chapter we will only discuss traditional symmetry in the setting of physical objects. We will refer to this kind of symmetry by the more specific term of symmetry of motion. In Chapter 12 we will discuss a non-traditional but still geometric type of symmetry which we will call symmetry of scale.

SYMMETRY OF MOTION

One of the foremost popularizers of symmetry in our times was the Dutch artist M. C. Escher. The popularity of his work proved that the notion of symmetry, when creatively packaged, can be of interest to everyone.

* Hermann Weyl, *Symmetry* (Princeton, NJ: Princeton University Press, 1952).

357

Day and Night by M. C. Escher. (© 1938 M. C. Escher/Cordon Art, Baarn, Holland) "At the left, the white silhouettes merge and form a dark sky; to the right, the black ones melt together and form a nocturnal background. The two landscapes are each other's mirror image and flow together through the fields, from which, once again, the birds take shape."—M. C. Escher

Hand with Reflecting Sphere by M. C. Escher. (© 1935 M. C. Escher/Cordon Art, Baarn, Holland) ". . . (a) spherical mirror, resting on (my) *left* hand. But as print is the reverse of the original stone drawing, it is my *right* hand that you see depicted. (Being left-handed, I needed my left hand to make the drawing.)"—M. C. Escher

Let's start with a somewhat informal definition: We will say that an object has **symmetry** if it looks exactly the same when seen from two or more different vantage points. Imagine a very tiny, nearsighted observer standing at one of the vertices of triangle I (Fig. 11-1) and looking inward. The observer would see exactly the same thing whether standing at vertex *A, B,* or *C.* In fact, if the vertices were not labeled and without any other frame of reference, the observer would be unable to distinguish one position from the other. In triangle II, the observer would see the same thing when standing at *B* or *C* but not when standing at *A.* In triangle III, the observer would see a different thing at each vertex. Informally, we might say that triangle I has more symmetry than triangle II, which in turn has more symmetry than triangle III.

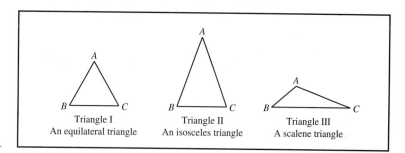

FIGURE 11-1

All of this is fine, but still rather vague. Our goal in this chapter is to develop a systematic set of tools for analyzing *how much* and *what kind* of symmetry a given object has.

A different, but equivalent, way to think about symmetry of motion is that rather than moving the observer, we can move the object itself. To say, for example, that triangle II in Fig. 11-1 looks exactly the same to an observer standing at vertex *B* as it does to the same observer standing at vertex *C* is the same as saying that there must be some way to move triangle II so that vertex *B* is where vertex *C* used to be and yet the triangle as a whole is in exactly the same place as it was before. Thus, the term **symmetry of motion.**

Having agreed that we will turn our attention to symmetry as it applies to physical objects, let's make one further restriction and agree to focus our attention to two-dimensional (i.e., planar) objects and shapes. Needless to say, three-dimensional objects and shapes are equally deserving of our attention, but it just happens that the analysis is a little more complicated. Moreover, we can gain most of the important insights by looking at the two-dimensional case.

RIGID MOTIONS

The first step in a detailed discussion of symmetry involves the concept of a **rigid motion.** When we take an object and move it from some starting position to some ending position without altering its shape or size, that's a rigid motion. When in the process of moving the object we stretch it, tear it, or generally alter its shape or size, that's *not* a rigid motion. While in general a rigid motion might be a complicated sequence of moves, we will discover that the net result isn't.

Take a quarter sitting on top of a dresser. In the morning we might pick it up, put it in a pocket, drive around town with it, take it out of the pocket and flip it into the air, put it back in a different pocket, go home, take it out of the pocket, and lay it on the same dresser again. While the actual trip taken by the quarter was a complicated one, the end result is all that we care about: The quarter started somewhere on top of the dresser and ended somewhere else on the dresser. We could have accomplished the whole thing by a much simpler action—perhaps a short slide, perhaps a little rotation, possibly a single flip over (if the starting and final positions had opposite faces up).

The example of the quarter illustrates a little known but fundamental fact. The *net effect* of any rigid motion, no matter how complicated the actual motion is, can be boiled down to just one of a few possibilities. If we consider the case of two-dimensional objects that start and end in the same plane, the list of possibilities is only four: *reflection, rotation, translation,* and *glide reflection.* For convenience, we will call these four rigid motions the **basic rigid motions of the plane.**[1]

A rigid motion (let's call it M) of the plane moves each point in the plane from its starting position P to an ending position P', also in the plane. We will call the point P' the **image** of the point P under the rigid motion M and describe this informally by saying that M **sends P to P'**. (Throughout the chapter we will stick to the convention that the image point has the same label as the original point but with a prime symbol added.) It is possible for a point P to end up back where it started under M ($P' = P$) in which case we call P a **fixed point** of the rigid motion.

It would appear that to "completely know" a rigid motion, one would need to know where the rigid motion sends every point of the plane, but fortunately, we will see that all it takes is to know where the rigid motion sends just a very few points (three at the most). Because the motion is rigid, the behavior of those few points forces all the rest of the points to follow their lead. (We can think of this as a sort of "Pied Piper effect.")

We will now discuss each of the basic rigid motions of the plane in a little more detail.

REFLECTIONS

A **reflection** is often called a *mirror reflection.* It is the kind of motion that moves an object into a new position that is a mirror image of the starting position. The line corresponding to the position of the mirror is called the **axis** of the reflection, and any reflection can be completely described by specifying its axis. Figures 11-2 through 11-4 show examples of reflections. For convenience, the

---

[1] The reader may be interested to know that in the case of three-dimensional objects moving in space, the number of basic rigid motions is only six (*reflection, rotation, translation,* and *glide reflection,* plus two other motions called *rotary reflection* and *screw displacement*).

original figure is shown in black, and its reflected or mirror image in red. Note that there is no prohibition against the axis of reflection cutting through the object itself, as in Fig. 11-4.

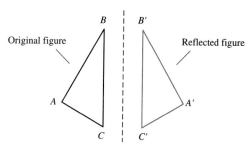

FIGURE 11-2

Axis of reflection

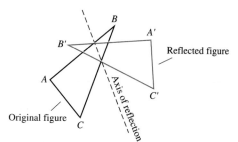

FIGURE 11-3   Axis of reflection

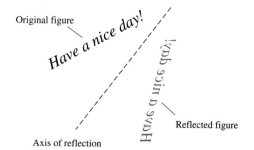

FIGURE 11-4

We will find it convenient to have a more mathematical description of reflection, so here it is: Given the axis of the reflection, we can find the image of a point P by drawing a line through P perpendicular to the axis and finding the point P' that is on this line and at the same distance as P from the axis (Fig. 11-5). Of course, any point on the axis itself is a fixed point of the reflection.

The process can also be reversed, so that if we know any point P and its image P' under the reflection, we can find the axis. It is the perpendicular bisector of the segment joining the two points (Fig. 11-5). A useful consequence of this is that a point P and its image P' under the reflection completely specify the reflection.

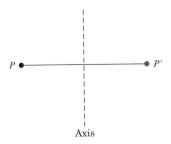

FIGURE 11-5

P •————————• P'

Axis

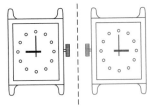

FIGURE 11-6

An important characteristic of a reflection is that it reverses all the traditional frames of reference one uses for orientation. As illustrated in Fig. 11-6, in a reflection, left is interchanged with right, and clockwise, with counterclockwise. We will say that reflection is an **improper** rigid motion to indicate the fact that it reverses the left-right and clockwise-counterclockwise orientations.

Another important fact about reflections is the following: If we apply the same reflection twice, every point ends up exactly where it started. In other words, the net effect of applying the same reflection twice is the same as not having moved the object at all. This leads us to an interesting semantic question: Should not moving an object at all be considered in itself a rigid motion? On the one hand, it seems rather absurd to say yes. If we are talking about motion, then there should be some kind of movement, however small. On the other hand, we are equally compelled to argue that the result of combining two (or more) consecutive rigid motions should itself be a rigid motion. If this is the case, then combining two consecutive reflections with the same axis (which produces the same result as no motion at all) should be a rigid motion. We will opt for the latter alternative because it is the mathematically correct way to look at things. We will formally agree therefore that not moving an object at all is itself a rigid motion of the object, and we will call it the **identity motion**.

ROTATIONS

The second type of rigid motion we will consider is a **rotation**. A rotation is described by giving the **center** of the rotation and the **angle** or amount of the rotation. (It is important to note that in two dimensions there is a center rather than an axis of rotation.) Figures 11-7 through 11-9 show examples of rotations. Again we show the original position in black and the final position in red.

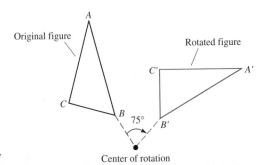

Original figure

Rotated figure

75°

FIGURE 11-7   Center of rotation

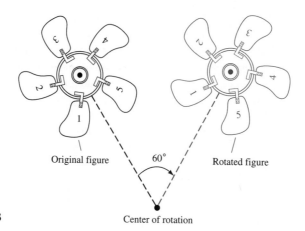

Original figure   60°   Rotated figure

FIGURE 11-8   Center of rotation

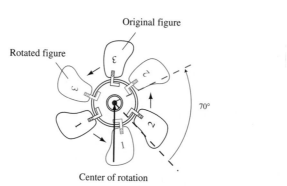

Original figure

Rotated figure

70°

FIGURE 11-9   Center of rotation

A few comments about the examples in the figures are in order. In all three examples we have specified the angle of rotation in degrees. This is strictly a matter of personal choice—some people prefer degrees, others radians. We will be consistent and stick with degrees, while at the same time reminding the reader that one can always change degrees to radians, or radians to degrees, by using the following relations:

$$\text{Radians} = \frac{\text{degrees}}{180} \times \pi \quad ; \quad \text{Degrees} = \frac{\text{radians}}{\pi} \times 180.$$

Our second observation starts with the well-known fact that a rotation by 360° is the same as no motion at all—in other words, it is the identity motion. This has several useful consequences. First, any rotation by an angle that is more than 360° is equivalent to another rotation with the same center by an angle that is between 0° and 360°. All we have to do is divide the angle by 360 and take the remainder. For example, as a rigid motion, a clockwise rotation by 759° is the same as a clockwise rotation by an angle of 39°, because 759 divided by 360 gives a quotient of 2 and a remainder of 39. Second, any rotation that is specified in a clockwise orientation can just as well be specified in a counterclockwise orientation. In Fig. 11-8 for example, we gave the angle of rotation as 60° clockwise, but we could have just as well said 300° counterclockwise. While it is possible to eliminate the need to specify the orientation (clockwise or counterclockwise) of the angle of rotation by choosing one or the other and sticking to it, we prefer the freedom of choosing the orientation that is most convenient.

One rotation for which we do not need to specify the orientation is a rotation by 180°. Here clockwise or counterclockwise produces the same result. The fact that two consecutive 180° rotations produce the same result as two consecutive reflections on the same axis gives rise to the frequent misconception that a 180° rotation is the same as a reflection. A moment's reflection (no pun intended) and/or Fig. 11-10 can make it clear that this is not the case.

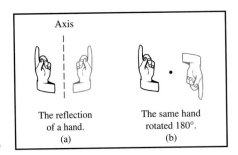

FIGURE 11-10

A second way to argue that a 180° rotation cannot be the same as a reflection is to note that a rotation is a rigid motion that leaves all original orientations (left, right, clockwise, counterclockwise) unchanged. Any rigid motion that does this is called a **proper** rigid motion. We have already observed that a reflection is an *improper* rigid motion.

Our final comment about rotations is that, unlike reflections, they cannot be completely described by giving a single point P and its image P'. There are infinitely many rotations that send P to P'. Any point on the perpendicular bisector of the segment PP' can be the center of a possible rotation sending P to P' (Fig. 11-11[a]). A second pair Q, Q' will allow us to nail down the rotation: The center is the point where the perpendicular bisectors of PP' and QQ' meet (Fig. 11-11[b]). In the special case where PP' and QQ' are parallel, the center of rotation is the intersection of PQ and $P'Q'$ (Fig. 11-11[c]).

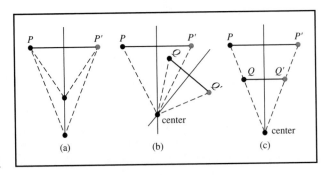

FIGURE 11-11

TRANSLATIONS

A **translation** is essentially a slide of an object in the plane. A translation is completely specified by the direction and amount of the slide. These two pieces of information are combined in the form of a **vector**. A vector can be represented by an arrow giving its direction and length. As long as the arrow points in the proper direction and has the right length, its actual placement is immaterial. Figures 11-12 and 11-13 show examples of translations. As usual, the original position is in black and the final position is in red. In both figures, any one of the arrows labeled ①, ②, or ③ (and for that matter infinitely many others) represent the same vector of translation.

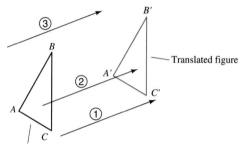

FIGURE 11-12   Original figure

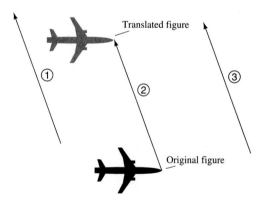

FIGURE 11-13

Translations, unlike reflections, are *proper* rigid motions of the plane: They do not change the left-right or clockwise-counterclockwise orientations. There is one characteristic that translations do share with reflections: A translation can be completely described by giving a point P and its image P'. The arrow joining P to P' gives us the vector of the translation. Once we have the vector, we know where the translation sends any other point.

GLIDE REFLECTIONS

A **glide reflection**, as the name suggests, is a rigid motion consisting of a translation (the glide part) followed by a reflection. The axis of the reflection *must* be parallel to the direction of the translation. The wording "translation followed by a reflection" is somewhat misleading: We can just as well do the reflection first and the translation second, and the end result will be the same. Figure 11-14 shows the stages in a glide reflection with the starting position shown in black, the ending position in red, and the intermediate position in gray.

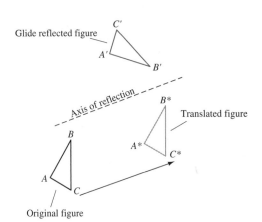

FIGURE 11-14

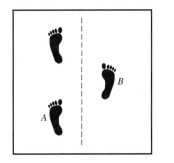

FIGURE 11-15 Footprints: A glide reflection moves footprint *A* to footprint *B*.

A glide reflection is an *improper* rigid motion—it changes left-right and clockwise-counterclockwise orientations. We can thank the reflection part of the glide reflection for that.

A glide reflection cannot be determined by just one point *P* and its image *P'*. As with a rotation, another point *Q* and its image *Q'* are needed. Given the two pairs *P*, *P'* and *Q*, *Q'*, the axis of the reflection can be found by joining the midpoints of the segments *PP'* and *QQ'* (Fig. 11-16[b]). This follows from the fact that in any glide reflection the midpoint between a point and its image belongs to the axis of the reflection (Fig. 11-16[a]). Once the axis of reflection is known, the vector of the translation can be determined by locating the intermediate point *P** which is the image of *P'* under the reflection (Fig. 11-16[c]).

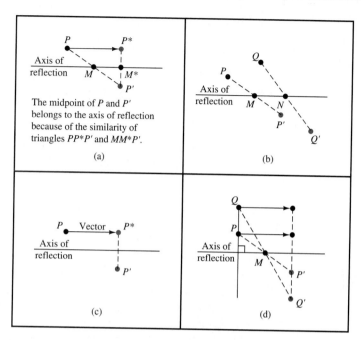

FIGURE 11-16

In the unlikely event that the midpoints of PP' and QQ' are the same point M, then the line passing through P and Q must be perpendicular to the axis of reflection (Fig. 11-16[d]), so the axis of reflection is obtained by taking a line perpendicular to the line PQ and passing through the common midpoint M.

SYMMETRY OF MOTION REVISITED

We have already discussed the fact that in rigid motions, unlike sightseeing trips, *how* we get to the final destination is irrelevant. If rigid motion A moves every point in the plane to exactly the same point as rigid motion B, then we say that A and B are **equivalent** rigid motions. We have already mentioned that the result of applying two consecutive reflections with the same axis is a rigid motion equivalent to the identity rigid motion. Likewise, we have also noted that the result of applying two consecutive 180° rotations with the same center is also equivalent to the identity rigid motion. Here are a few additional facts about combining rigid motions:

■ The result of applying two consecutive rotations with the same center is equivalent to another rotation with the same center (Exercise 26).

■ The result of applying the same glide reflection twice is equivalent to a translation (Exercise 27).

■ The result of applying two consecutive reflections with parallel axes is equivalent to a translation (Exercise 28).

■ The result of applying two consecutive reflections with intersecting axes is a rotation (Exercise 29). A particularly interesting special case of this is when the axes of reflection are perpendicular to each other, in which case we get a 180° rotation.

■ The result of applying two consecutive translations is another translation (Exercise 30).

We can play this game for quite a while, and it gets increasingly complicated, especially when we start mixing different types of rigid motions. What, for example, is the result of applying a rotation, followed by a translation, followed by a glide reflection? The remarkable thing is that *whatever it is, it has to be equivalent to one of the four basic rigid motions: it's either a reflection, a rotation, a translation, or a glide reflection.*

We are finally in a position to give a formal definition of **symmetry of motion**: A symmetry of motion of an object is any rigid motion that moves the object back onto itself. It is important to note that this does not necessarily force the rigid motion to be the identity motion. Individual parts of the object may be moved to different starting and ending positions, even while the whole object is moved back into itself. Of course, the identity motion is itself a symmetry.

Based on the fact that there are only four basic types of rigid motion in the plane, we can classify the symmetries of planar objects accordingly.

Reflection Symmetry

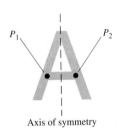

P_1 P_2

Axis of symmetry

FIGURE 11-17

Any object that falls back onto itself after a reflection is said to have **reflection symmetry**, more commonly referred to as **bilateral symmetry**. The axis of the reflection is called the **axis of symmetry**. In our illustrations we will use a dashed line to indicate an axis of symmetry. Figure 11-17 shows an example of reflection symmetry. Note that while the object looks exactly the same after the reflection, the pair of points P_1 and P_2 interchange positions.

The equilateral triangle shown in Figure 11-18(a) has three different reflection symmetries—the axes are the perpendicular bisectors of the three sides. On the other hand, the scalene triangle shown in Fig. 11-18(b) has no reflection symmetries. A circle has infinitely many reflection symmetries; any line passing through the center is an axis of symmetry (Fig. 11-18[c]).

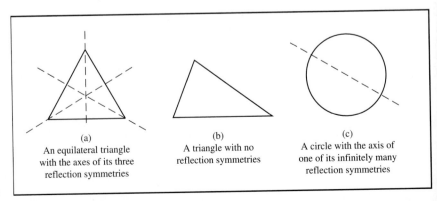

(a)
An equilateral triangle
with the axes of its three
reflection symmetries

(b)
A triangle with no
reflection symmetries

(c)
A circle with the axis of
one of its infinitely many
reflection symmetries

FIGURE 11-18

Patterns

Let's take a brief detour from our study of symmetry to discuss patterns. A **pattern** is an abstraction representing an infinite figure with infinitely repeating themes. In two dimensions, the most common occurrences of patterns are on wallpaper, printed fabrics, friezes (the decorations marking the ceiling lines of older buildings), ribbons, etc.

For planar patterns we will distinguish two different situations. On the one hand, we have the case exemplified by ribbons, borders, and friezes in which the repetition of the theme occurs in only one direction. We will call these **border patterns** (Fig. 11-19).

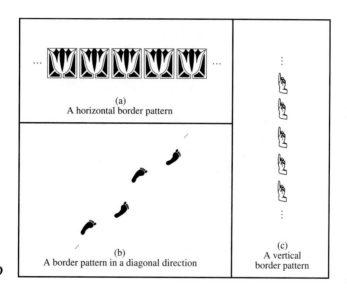

(a)
A horizontal border pattern

(b)
A border pattern in a diagonal direction

(c)
A vertical
border pattern

FIGURE 11-19

Note that the pattern can repeat itself in any direction. For us, it is typographically convenient to make the direction of the pattern horizontal (it wastes the least amount of page space), and we will do so from now on. Thus, when we say "horizontal direction" we mean the direction of the pattern, and it follows that when we say "vertical direction" we mean the direction perpendicular to the direction of the pattern.

When it comes to reflection symmetries, a border pattern can have

- No reflection symmetries at all (Fig. 11-20[a])

- Horizontal (i.e., in the direction of the pattern) reflection symmetry only (Fig. 11-20[b])

- Vertical (i.e., perpendicular to the direction of the pattern) reflection symmetry only (Fig. 11-20[c]). (Note that here there are infinitely many different choices for the axis of reflection.)

- Horizontal and vertical reflection symmetries (Fig. 11-20[d]).

In a border pattern, a theme repeats itself in just one direction, be it in a straight line (as in a paper pattern) or wrapping itself into a circle (as in a piece of pottery).

··· S S S S S S S ···	··· B B B B B B B ···
(a)	(b)
··· A A A A A A A ···	··· X X X X X X X ···
(c)	(d)

FIGURE 11-20

The second type of pattern we will consider is one that takes up the entire plane. Such is the situation with wallpaper and printed fabrics, and we call these patterns **wallpaper patterns.** In wallpaper patterns the repetition of the theme occurs in more than one direction.

In a wallpaper pattern a theme repeats itself in two nonparallel directions, filling up the entire plane.

The possible reflection symmetries of wallpaper patterns are a little more complicated than we would expect. The possibilities are:

■ No reflections (Fig. 11-21[a])

■ Reflections in one direction only (Fig. 11-21[b])

■ Reflections in two directions only (Fig. 11-21[c])

■ Reflections in three directions only (Fig. 11-21[d])

■ Reflections in four directions only (Fig. 11-21[e])

■ Reflections in six directions only (Fig. 11-21[f]).

There are no other possibilities. Note that particularly conspicuous in its absence is the case of reflections in exactly five different directions.

Now let's return to our main theme, the analysis of symmetries of motion.

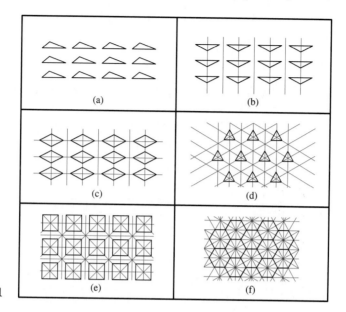

FIGURE 11-21

Rotation Symmetry

FIGURE 11-22 A five-bladed fan has 72°, 144°, 216°, 288°, and 360° rotation symmetry about the center. It is enough to specify the 72°.

Any object that falls back onto itself after a rotation by a certain angle is said to have **rotation symmetry**. We already know that a 360° rotation is the identity motion and that any rotation by an angle more than 360° can be reduced to one by an angle less than 360°, so we can assume that in rotation symmetry the angle of rotation is less than 360°. When an object has rotation symmetry by an angle, then it automatically has rotation symmetry by any multiple of that angle. For this reason, we usually just specify the *smallest* positive angle for which the object has rotation symmetry (Fig. 11-22).

A circle is the one two-dimensional shape that has infinitely many rotation symmetries about its center, and any angle, no matter how small, is an appropriate angle of rotation. While the fan shown in Fig. 11-22 has only rotation symmetry (no reflections), in nature it is often the case that rotation symmetry is accompanied by reflection symmetry, as in Fig. 11-23.

FIGURE 11-23

What kinds of rotation symmetries are possible in patterns? The answers are a bit surprising:

■ In a *border pattern,* the only possible angle of rotation is 180°.

■ In a *wallpaper pattern,* the only possible angles of rotation are 180°, 120°, 90° and 60°. These correspond to what are known respectively as twofold (Fig. 11-24[a]), threefold (Fig. 11-24[b]), fourfold (Fig. 11-24[c]), and six-fold (Fig. 11-24[d]) rotation symmetry.

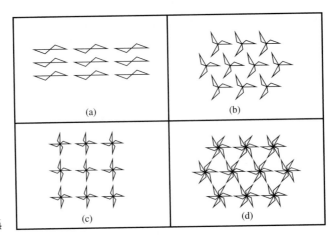

FIGURE 11-24

It is worth noting that in wallpaper patterns there is no fivefold rotation symmetry (angle of rotation = 72°). An interesting consequence of this is that while we can tile our bathroom floor with tiles that are triangles, squares, or regular hexagons, we can't tile a floor (without leavings any gaps, that is) with a tile that is a regular pentagon.

Translation Symmetry

Any object that falls back onto itself after a translation is applied to it is said to have **translation symmetry**. The first thing to note is that a finite figure cannot possibly have translation symmetry—*only patterns* can have this type of symmetry. In fact, we can define patterns as exactly those figures that have translation symmetry. Border patterns are those that have translation symmetry in only one direction (the direction of the pattern) (Fig. 11-25[a]); wallpaper patterns have translation symmetry in more than one direction (Fig. 11-25[b]).

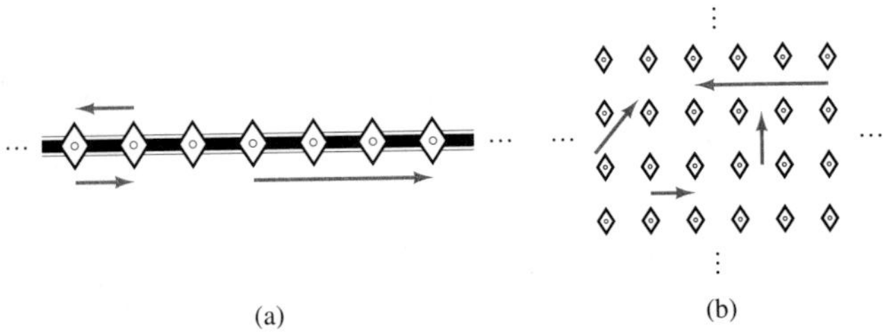

FIGURE 11-25 (a) Three (of the infinitely many) translation symmetries are indicated by the arrows. The translations are always in the horizontal direction. (b) Four (of the infinitely many) translation symmetries are indicated by the arrows.

Glide Reflection Symmetry

A glide reflection can also move a figure onto itself, in which case it is a symmetry. Since there is a translation involved, the figure cannot be finite. If a figure has glide reflection symmetry then it must be a pattern. If a pattern has a glide reflection symmetry, then applying the same glide reflection twice results in a translation symmetry. The direction of the glide and the direction of the translation are the same, and the amount of translation is double the amount of the glide (see Exercise 27). Figure 11-26 shows a border pattern with glide reflection symmetry. The point A moves to the position A', and the point B moves to the position B' under the glide reflection. Applying the glide reflection again moves A' to A'' and B' to B''. The combined sequence moving A to A'' and B to B'' is a translation symmetry.

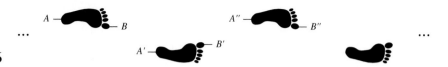

FIGURE 11-26

Consider now the pattern in Fig. 11-27. It has both translation symmetry and horizontal reflection symmetry, and these symmetries can be combined into a glide reflection that moves the pattern back onto itself. In this case, however, the glide reflection is a symmetry by default, in contrast to the situation in Fig. 11-26 where the glide reflection is a symmetry in its own right. We will find it convenient to draw a distinction between these two situations. From now on when we say that a pattern has a glide reflection symmetry, we will mean only the situation exemplified by Fig. 11-26.

FIGURE 11-27

With wallpaper patterns, the possibilities for glide reflection symmetries are:

■ No glide reflections (Fig. 11-28[a])

■ Glide reflections in one direction only (Fig. 11-28[b])

■ Glide reflections in two directions only (Fig. 11-28[c])

■ Glide reflections in three directions only (Fig. 11-28[d])

■ Glide reflections in four directions only (Fig. 11-28[e])

■ Glide reflections in six directions only (Fig. 11-28[f]).

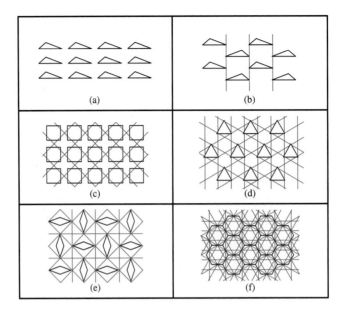

FIGURE 11-28

CONCLUSION: CLASSIFYING PATTERNS

Patterns play an important role in many areas of science. Archaeologists can recognize cultures and cultural influences from the types of patterns showing up in pottery, wall paintings, textiles, etc. Chemists and crystallographers are interested in the patterns formed by crystals, because they say a great deal about the chemical structure and properties of a particular material.

One of the fundamental tools used in any analysis of patterns is the classification of patterns according to their symmetries. Take, for example, wallpaper patterns. There are infinitely many different possible wallpaper designs. (In fact, it often seems that there are that many in a single wallpaper store!) Two very different looking wallpaper designs, however, can be obtained by just replacing one particular theme (say a doll) by another (say a flower). In this sense, we might say that Figs. 11-29(a) and (b) represent two different designs based on the same pattern type. In fact, we can see that mathematically what really makes the pattern special is not the doll or the flower but the fact that it has a very well defined set of symmetries: translations in more than one direction, a reflection in only one direction, and a glide reflection in only one direction.

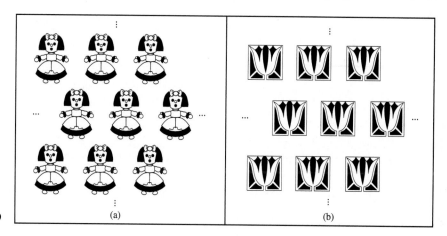

FIGURE 11-29   (a)   (b)

This particular set of symmetries describes a **pattern type**, and any pattern having exactly these symmetries (no more and no less) is said to be of the same type. Thus, classifying patterns means spelling out the various combinations of symmetries that can come together to form a pattern type. When confronted with this problem, a frequent response is, Can't we put together any combination of symmetries we want to form a pattern type? (sort of like picking symmetries out of a buffet line). While the answer (no) is only mildly surprising, the actual details almost always come as a real shock:

■ **Fact 1.**   Any *border* pattern is of one of only *seven* different possible pattern types.

■ **Fact 2.**   Any *wallpaper* pattern is of one of only *seventeen* different possible pattern types.

A very complete and accurate proof of these facts (especially fact 2) requires some very abstract, sophisticated mathematical ideas. In fact, while it is believed that these facts were informally understood as far back as the time of the Egyptians (who used all seventeen of the wallpaper patterns in their ornamental designs), a rigorous mathematical proof of fact 2 was given only in 1924 by the Hungarian mathematician George Polya.

In Appendix 1 we illustrate and briefly discuss the seven pattern types for border patterns. The seventeen wallpaper pattern types are somewhat more briefly illustrated and described in the table shown in Appendix 2. It is certainly not intended for the student to memorize the information in these appendices but rather that they be used as a handy reference guide by those whose interest has been kindled and who wish to pursue the subject further.

We conclude this chapter with a brief quote from the great mathematician Hermann Weyl:

Symmetry is a vast subject, significant in art and nature. Mathematics lies at its root, and it would be hard to find a better one on which to demonstrate the working of the mathematical intellect.

KEY CONCEPTS

axis (of reflection, of symmetry)
basic rigid motions of the plane
bilateral symmetry
border pattern
equivalent rigid motion
fixed point
glide reflection
glide reflection symmetry
identity motion
image
improper rigid motion
pattern

pattern type
proper rigid motion
reflection
reflection symmetry
rigid motion
rotation
rotation symmetry
symmetry of motion
translation
translation symmetry
vector of translation
wallpaper pattern

EXERCISES

Walking

1. Given a reflection that sends the point P to the point P' as shown in the figure, find

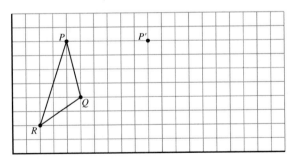

(a) the axis of reflection
(b) Q' (the image of Q) under the reflection
(c) the image of triangle PQR under the reflection.

2. Given a reflection with the axis of reflection as shown in the figure, find

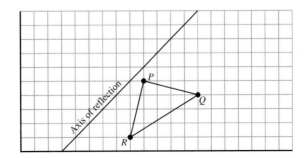

   (a) P' (the image of P) under the reflection
   (b) the image of triangle PQR under the reflection.

3. In each of the following give an answer between 0° and 360°.
   (a) A clockwise rotation by an angle of 500° is equivalent to a clockwise rotation by an angle of .
   (b) A clockwise rotation by an angle of 3681° is equivalent to a counterclockwise roation by an angle of .

4. In each of the following give an answer between 0° and 360°.
   (a) A clockwise rotation by an angle of 500° is equivalent to a counterclockwise rotation by an angle of .
   (b) A clockwise rotation by an angle of 3681° is equivalent to a clockwise rotation by an angle of .

5. Given a rotation that sends the point B to the point B' and the point C to the point C' as shown in the figure, find

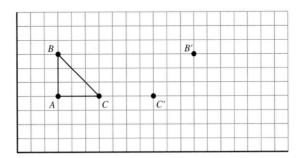

   (a) the center of the rotation
   (b) the image of triangle ABC under the rotation.

6. Given a rotation that sends the point B to the point B' and the point C to the point C' as shown in the figure, find

6. Given a rotation that sends the point B to the point B' and the point C to the point C' as shown in the figure, find

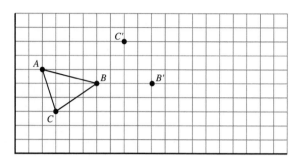

(a) the center of the rotation
(b) the image of triangle ABC under the rotation.

7. Given a translation that sends the point E to the point E' as shown in the figure, find

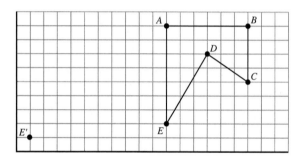

(a) the image A' of A under the translation
(b) the image of figure $ABCDE$ under the translation.

8. Given a translation that sends the point Q to the point Q' as shown in the figure, find

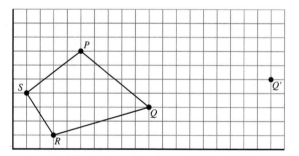

(a) the image P' of P under the translation
(b) the image of figure $PQRS$ under the translation.

9. Given a glide reflection that sends the point B to the point B' and the point D to the point D' as shown in the figure, find

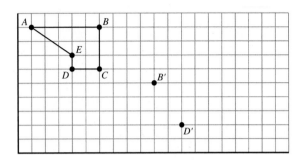

    (a) the axis of the glide reflection
    (b) A' (the image of A) under the glide reflection
    (c) the image of figure $ABCDE$ under the glide reflection.

10. Given a glide reflection that sends the point A to the point A' and the point C to the point C' as shown in the figure, find

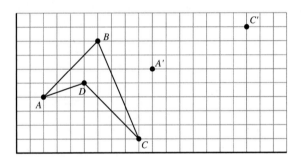

    (a) the axis of the glide reflection
    (b) B' (the image of B) under the glide reflection
    (c) the image of figure $ABCD$ under the glide reflection.

11. What "word" do you get if you reflect each of the given "words" about a vertical axis?
    (a) T A d
    (b) T I M
    (c) b A d
    (d) Y o d

12. What "word" do you get if you rotate each of the given "words" 180° about the center of the "O"?
    (a) S O S
    (b) W O W
    (c) M O N
    (d) N O S

13. For each of the following symbols describe all the symmetries. (You don't have to mention the identity symmetry.)
   (a) A
   (b) D
   (c) +
   (d) Z
   (e) Q.

14. For each of the following shapes describe all the symmetries (other than the identity symmetry).
   (a) equilateral triangle
   (b) square
   (c) rectangle
   (d) parallelogram (not a rectangle)
   (e) circle.

15. Give an example of a capital letter of the alphabet that has
   (a) horizontal reflection only
   (b) vertical reflection only
   (c) horizontal and vertical reflection
   (d) 180° rotation but no reflection
   (e) no symmetries other than the identity symmetry.

16. List all the digits (0, 1, 2, 3, 4, 5, 6, 7, **8**, 9) that have
   (a) horizontal reflection
   (b) vertical reflection
   (c) horizontal and vertical reflections
   (d) 180° rotation.

17. (a) Give an example of a figure that has a 120° rotation symmetry and three different reflection symmetries.
   (b) Give an example of a figure that has a 120° rotation symmetry and *no* reflection symmetries.

18. (a) Give an example of a figure that has 45° rotation symmetry and reflection symmetries.
   (b) Give an example of a figure that has 45° rotation symmetry and *no* reflection symmetries.

19. Describe all the symmetries of each of the following figures.

   (a)

   (b)

   (c)

   (d)

20. Describe all the symmetries of each of the following figures.

    (a)

    (b)

    (c)

    (d)

21. Describe all the symmetries of each of the following border patterns.

    (a) ⋯ A A A A A A A A ⋯

    (b) ⋯ D D D D D D D ⋯

    (c) ⋯ S S S S S S S ⋯

    (d) ⋯ L L L L L L L L ⋯

22. Describe all the symmetries of each of the following border patterns.

    (a) ⋯ ⧖ ⧖ ⧖ ⧖ ⧖ ⧖ ⧖ ⋯

    (b) ⋯ ✚ ✚ ✚ ✚ ✚ ✚ ⋯

    (c) ⋯ ❀ ❀ ❀ ❀ ❀ ❀ ❀ ⋯

    (d) ⋯ ✳ ✳ ✳ ✳ ✳ ✳ ⋯

23. Describe all the symmetries of each of the following border patterns.

    (a) ⋯ ⋯

    (b) ⋯ ♣ ♣ ♣ ♣ ♣ ♣ ⋯

    (c) ⋯ ▶ ▶ ▶ ▶ ▶ ⋯

    (d) ⋯ " " " " ⋯

24. Explain why any proper rigid motion other than the identity that has a fixed point must be equivalent to a rotation.

25. Explain why any rigid motion other than the identity that has two or more fixed points must be equivalent to a reflection.

Jogging

26. **(a)** Rotation 1 has center C and a clockwise angle of $30°$, and rotation 2 has the same center C and a clockwise angle of $50°$. Show that the result of applying rotation 1 followed by rotation 2 is equivalent with center C and a clockwise angle of $80°$.

   (b) Show that if rotation 1 has center C and clockwise angle α and rotation 2 has center C and clockwise angle β, then the result of applying rotation 1 followed by rotation 2 is equivalent to a rotation with center C and clockwise angle $\alpha + \beta$.

   (c) Show that for the two rotations described in (b) the result of applying rotation 1 followed by rotation 2 is equivalent to applying rotation 2 followed by rotation 1.

27. **(a)** Given a glide reflection with axis and vector as shown, find the image Q'' of the point Q when the glide reflection is applied twice.

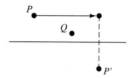

   (b) Show that the result of applying the same glide reflection twice is equivalent to a translation. Describe the direction and amount of the translation in terms of the direction and amount of the original glide.

28. Reflection 1 has axis l_1; reflection 2 has axis l_2; l_1 and l_2 are parallel, and the distance between them is d.

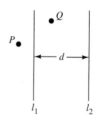

   (a) Find the images of P and Q when we apply reflection 1 followed by reflection 2.

   (b) Show that the result of applying reflection 1 followed by reflection 2 is a translation. Describe the direction and amount of the translation.

   (c) Show that the result of applying reflection 1 followed by reflection 2 is *not* the same as the result of applying reflection 2 first followed by reflection 1. Describe the difference.

29. Reflection 1 has axis l_1; reflection 2 has axis l_2; l_1 and l_2 intersect at C. The angle between l_1 and l_2 as shown in the figure is α.

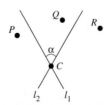

30. Translation 1 moves point P to point P'; translation 2 moves point Q to point Q' as shown in the figure.

    (a) Find the images of P and Q when we apply translation 1 followed by translation 2.

    (b) Find the images of P and Q when we apply translation 2 followed by translation 1.

    (c) Show that the result of applying translation 1 followed by translation 2 is a translation. Give a geometric description of the vector of the translation.

31. **(a)** Find all the reflection symmetries of a regular pentagon.

    (b) Find all the reflection symmetries of a regular hexagon.

    (c) Describe all the reflection symmetries of a regular polygon with N sides.

32. Describe all the symmetries of each of the following figures.

    (a) ☆ (5-point star)

    (b) ✶ (6-point star)

    (c) N-point star (*Hint:* Consider separately the cases N even and N odd).

33. Describe all the symmetries of each of the following border patterns.

    (a) ⋯ MOWMOWMO ⋯

    (b) ⋯ p d p d p d ⋯

    (c) ⋯ p d b q p d b q ⋯

34. For each of the following sets of symmetries, give an example of a border pattern that has exactly those symmetries (no more and no less). Do not use any of the patterns in Exercise 33. You can use letters of the alphabet, numbers, or symbols to create the patterns.

    (a) Translations only

    (b) Translations and vertical reflections

    (c) Translations and horizontal reflections

    (d) Translations and 180° rotations

    (e) Translations and glide reflections

    (f) Translations, vertical reflections, glide reflections, and 180° rotations

    (g) Translations, vertical reflections, horizontal reflections, and 180° rotations.

35. A rigid motion moves the triangle PQR into the triangle $P'Q'R'$ as shown in the figure.

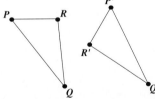

35. A rigid motion moves the triangle PQR into the triangle $P'Q'R'$ as shown in the figure.

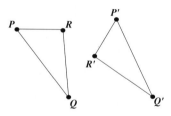

(a) Explain why the rigid motion cannot possibly be a rotation or a translation. (*Hint:* Is the rigid motion proper or improper?)
(b) Explain why the rigid motion cannot possibly be a reflection.
(c) What kind of rigid motion is it then?

36. What single rigid motion is the result of applying two consecutive reflections with parallel axes followed by a translation?

37. What single rigid motion is the result of applying two consecutive reflections with intersecting axes followed by a rotation about the point of intersection of the axes?

38. What single rigid motion is the result of applying two consecutive reflections with parallel axes followed by two more consecutive reflections with parallel axes? (Note: The second pair of parallel axes need not be parallel to the first pair of parallel axes.)

39. What single rigid motion is the result of applying the same glide reflection three times?

40. A *palindrome* is a word that is the same when read forward or backward. MOM is a palindrome, and so is ANNA. (For simplicity we will assume all letters are capitals.)
(a) Explain why if a word has vertical reflection symmetry, then it must be a palindrome.
(b) Give an example of a palindrome (other than ANNA) that doesn't have vertical reflection symmetry.
(c) If a palindrome has vertical reflection symmetry, what can you say about the symmetries of the individual letters in the word?
(d) Find a palindrome with 180° rotational symmetry.

Running

41. Suppose that a rigid motion moves points P to P', Q to Q', and R to R' (as in the figure). We do not know what kind of rigid motion it is.

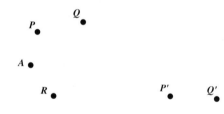

(a) For an arbitrary point A in the plane, find the image A'. (*Hint:* $AP = A'P'$, $AQ = A'Q'$, and $AR = A'R'$.)

(b) Describe a general procedure for finding the image of a point A under some rigid motion when we know three points P, Q, and R (not on a straight line) and their images P', Q', and R'.

42. Rotation 1 is a rotation of 90° clockwise about center A. Rotation 2 is a rotation of 60° clockwise about center B.

(a) Show that the result of applying rotation 1 followed by rotation 2 is another rotation, and find the center and angle of this rotation.

(b) Show that the result of applying rotation 2 followed by rotation 1 is another rotation, and find the center and angle of this rotation.

(c) Generalize the results of (a) and (b) to the case where the angles of rotation are any angles between 0° and 180°.

43. Find all the symmetries of the following wallpaper pattern.

44. Find all the symmetries of the following wallpaper pattern.

45. Find all the symmetries of the following wallpaper pattern.

46. Find all the symmetries of the following wallpaper pattern.

47. Find all the symmetries of the following wallpaper pattern.

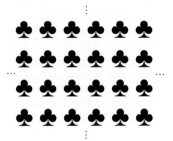

48. Find all the symmetries of the following wallpaper pattern.

49. Find all the symmetries of the following wallpaper pattern.

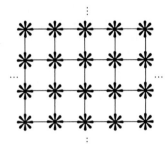

50. Find all the symmetries of the following wallpaper pattern.

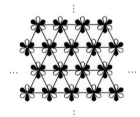

APPENDIX 1: The Seven Border Pattern Types

In this appendix we illustrate and describe the seven different pattern types that are possible with border patterns. There are several different ways to organize this classification. We will adopt one that is consistent with the standard notation used in crystallography.

Let's start with the fact that every border pattern has translations in the direction of the pattern. As we have already observed, when dealing with patterns, translations come with the territory. We will now divide the possibilities into two categories.

- **Category *m*.**   The pattern has vertical reflections (remember that this really means reflections in a direction perpendicular to the direction of the pattern).
- **Category 1.**   The pattern has no vertical reflections.

In category *m* there are three possibilities.

- ***m*1.**   Translations, vertical reflections, and nothing else
- ***mg*.**   Translations, vertical reflections, glide reflections, and 180° rotations
- ***mm*.**   Translations, vertical reflections, horizontal reflections, and 180° rotations.

In category 1 we have the other four possible pattern types.

- **11.** Translations and nothing else
- **1*m*.**   Translations plus horizontal reflections.
- **12.**   Translations plus 180° rotations
- **1*g*.**   Translations plus glide reflections.

The table on page 388 summarizes the seven pattern types with their symmetries.

Type	Translation	Reflection Along Direction of Pattern	Reflection Along Perpendicular to Pattern	180° Rotation	Glide Reflection	Example
11	✓					··· △ △ △ △ ···
1m	✓	✓				··· ⟁ ⟁ ⟁ ⟁ ···
m1	✓		✓			··· △ △ △ △ ···
12	✓			✓		··· △ ▽ △ ▽ ···
1g	✓				✓	··· △ ▽ △ ▽ ···
mg	✓		✓	✓	✓	··· △ △ ▽ ▽ ···
mm	✓	✓	✓	✓		··· ⟁ ⟁ ⟁ ⟁ ···

Why are these the only possibilities? Without going into a complete proof, a clue as to what happens can be found by looking at the impossible combination of symmetries consisting of just translations, vertical reflections, and horizontal reflections. Such a combination is impossible because the vertical reflection combined with the horizontal reflection automatically forces a 180° rotation (and therefore forces the pattern to be of type *mm*). In a similar fashion we can show that any other combination not on the list is impossible. (Some of them are a little bit harder to show than the one we picked.)

APPENDIX 2: The Seventeen Wallpaper Pattern Types

In this appendix we will just show the seventeen wallpaper pattern types with the standard notation used in crystallography.

Type	Translation	Rotations				Reflections (Number of Directions)					Glide Reflections (Number of Directions)					Example
		60°	90°	120°	180°	1	2	3	4	6	1	2	3	4	6	
p2gg	✓				✓	✓					✓					
p211	✓				✓							✓				
p2mg	✓				✓											
p4mm	✓		✓		✓				✓			✓				
p4gm	✓		✓		✓	✓								✓		
p4	✓		✓		✓											

Type	Translation	Rotations				Reflections (Number of Directions)					Glide Reflections (Number of Directions)					Example
		60°	90°	120°	180°	1	2	3	4	6	1	2	3	4	6	
c1m1	✓					✓					✓					
p1m1	✓					✓										
p1g1	✓										✓					
p1	✓															
p2mm	✓				✓		✓									
c2mm	✓				✓		✓					✓				

Type	Translation	Rotations				Reflections (Number of Directions)					Glide Reflections (Number of Directions)					Example
		60°	90°	120°	180°	1	2	3	4	6	1	2	3	4	6	
p3m1	✓			✓				✓					✓			
p31m	✓			✓				✓					✓			
p3	✓			✓												
p6mm	✓	✓		✓	✓					✓					✓	
p6	✓	✓		✓	✓											

**REFERENCES
AND FURTHER
READINGS**

1. Bunch, Bryan, *Reality's Mirror: Exploring the Mathematics of Symmetry.* New York: John Wiley & Sons, Inc., 1989.

2. Coxeter, H. S. M., *Introduction to Geometry* (2d ed.). New York: John Wiley & Sons, Inc., 1967.

3. Crowe, Donald W., "Symmetry, Rigid Motions and Patterns," *UMAP Journal,* 8 (1987), 206–236.

4. Field, M., and M. Golubitsky, *Symmetry in Chaos.* New York: Oxford University Press, 1992.

5. Gardner, Martin, *The New Ambidextrous Universe: Symmetry and Asymmetry from Mirror Reflections to Superstrings* (3rd ed.). New York: W. H. Freeman & Co., 1990.

6. Grunbaum, Branko, and G. C. Shephard, *Tilings and Patterns: An Introduction.* New York: W. H. Freeman & Co., 1989.

7. Hofstadter, Douglas R., *Gödel, Escher, Bach: An Eternal Golden Braid.* New York: Vintage Books, 1980.

8. Martin, George E., *Transformation Geometry: An Introduction to Symmetry.* New York: Springer Publishing Co., Inc., 1982.

9. Rose, Bruce, and Robert D. Stafford, "An Elementary Course in Mathematical Symmetry," *American Mathematical Monthly,* 88 (1981), 59–64.

10. Schattsneider, Doris, *Visions of Symmetry: Notebooks, Periodic Drawings, and Related Work of M. C. Escher.* New York: W. H. Freeman & Co., 1990.

11. Shubnikov, A. V., and V. A. Kopstik, *Symmetry in Science and Art.* New York: Plenum Publishing Corp., 1974.

12. Stewart, I., and M. Golubitsky, *Fearful Symmetry.* Cambridge, MA: Blackwell Publishers, 1992.

13. Weyl, Hermann, *Symmetry.* Princeton, NJ: Princeton University Press, 1952.

14. Wigner, Eugene, *Symmetries and Reflections.* Bloomington, IN: Indiana University Press, 1967.

15. Yale, Paul B., *Geometry and Symmetry.* San Francisco: Holden-Day, 1968.

CHAPTER 12

Fractally Speaking

Symmetry of Scale and Fractals

In Chapter 11 we discussed the concept of *symmetry of motion*—the characteristic of shapes, objects, and patterns that look exactly the same after they have been moved in various ways. Symmetry of motion represents the conventional way that people think about symmetry, and in common usage symmetry of motion is referred to as just plain symmetry, as if it were the only kind. In fact, it isn't.

In this chapter we will introduce a different, nontraditional, somewhat more exotic type of symmetry called *symmetry of scale,* and we will use symmetry of scale to discuss some remarkable geometric objects called *fractals.* The fantastic geometric shapes we will come across in our study of symmetry of scale are unlike anything we could imagine based on our experience with Euclidean (high school) geometry.

We will motivate our definition of symmetry of scale with some preliminary examples.

THE KOCH SNOWFLAKE

The Koch snowflake is a remarkable geometric shape discovered by the Swedish mathematician Helge von Koch in 1904. The construction of the Koch snowflake proceeds in steps as follows:

- ■ **Start.** Start with a solid black equilateral triangle of arbitrary size. For simplicity we will assume that the sides of the triangle are of length 1 and call the area of the triangle *A*.

- ■ **Step 1.** Divide each of the sides of the triangle into three equal segments. Attach to the middle segment of each side a small, black equilateral triangle (Fig. 12-1[b]). The resulting star-shaped object has 12 sides, each of length $\frac{1}{3}$ (Fig. 12-1[c]).

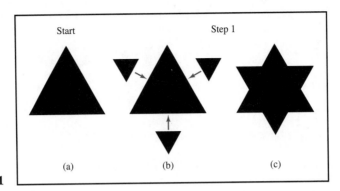

FIGURE 12-1

■ **Step 2.** For each of the 12 sides of the star in Fig. 12-1(c), repeat the process described in step 1: Divide each side into three equal segments and attach a black equilateral triangle to the middle segment. The resulting shape has 48 sides, each of length $\frac{1}{9}$ (Fig. 12-2[a]).

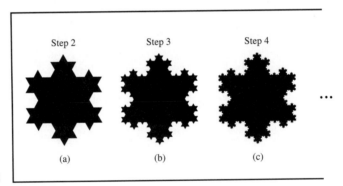

FIGURE 12-2

■ **Steps 3, 4, etc.** Repeat the process (divide sides into thirds, attach appropriate-sized equilateral triangles to the middle thirds) ad infinitum (Fig. 12-2[b], [c], etc.).

Each step in the construction described above takes us closer and closer to the ultimate destiny of this trip: the **Koch snowflake** itself (Fig. 12-3).

The perceptive reader will immediately notice that Fig. 12-3 is not a picture of the real Koch snowflake but that of a useful impostor. The Koch snowflake requires an infinite number of steps in its construction, so a complete and perfect picture of it is impossible. However, the fact that we cannot show a perfect picture of the snowflake should not deter us from drawing good approximations of it (as we did in Fig. 12-3) or from using such drawings to study its mathematical properties. (This is very similar to the situation in high school geometry where we learned a lot about squares, triangles, and circles even when our drawings of them were far from perfect.)

Step   ∞

FIGURE 12-3

Recursive Processes

The construction of the Koch snowflake is an example of a *recursive process,* a process in which the same basic rules are applied over and over again with the end product at each step becoming the starting point for the next step (see Fig. 12-4). The concept of a recursive process is not new to us: We saw it in Chapter 1 (recursive ranking methods), Chapter 9 (the Fibonacci numbers), and Chapter 10 (transition rules for population growth). In this chapter the objects of the recursive process are geometric shapes rather than numbers, but other than that the basic principles are quite similar.

The use of recursion allows us to describe the construction of the Koch snowflake in a very efficient form of shorthand we will call a **recursive replacement rule**.

Recursive Replacement Rule for the Koch Snowflake

■ Start with a black triangle.

■ Wherever you see an edge _____ , replace it with __▲__ .

Recursive Rule for the Fibonacci Numbers

■ Start with $F_1 = 1$ and $F_2 = 1$.

■ For $N > 2$: $F_N = F_{N-1} + F_{N-2}$

FIGURE 12-4 Two examples of recursive processes

When we look at the Koch snowflake from the perspective of Euclidean geometry, we find some very bizarre characteristics. Let's start by asking two typical questions from traditional geometry: What is the perimeter of the Koch snowflake? What is its area?

Perimeter of the Koch Snowflake

The boundary of the Koch snowflake will be of particular interest to us. It is commonly known as the **Koch curve** or the **snowflake curve** (see Fig. 12-5).

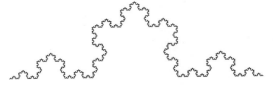

FIGURE 12-5 A section of the Koch curve.

How long is the Koch curve? Figure 12-6 shows what happens when we take the length of the boundary at each of the first few steps of the construction.

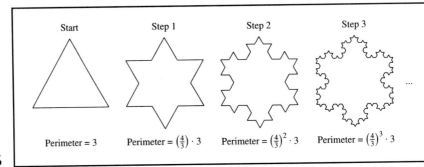

Start — Perimeter = 3

Step 1 — Perimeter = $\left(\frac{4}{3}\right) \cdot 3$

Step 2 — Perimeter = $\left(\frac{4}{3}\right)^2 \cdot 3$

Step 3 — Perimeter = $\left(\frac{4}{3}\right)^3 \cdot 3$

FIGURE 12-6

We can see from Fig. 12-6 that the perimeters grow by a factor of $\frac{4}{3}$. (In the language of Chapter 10, the perimeters grow exponentially with a growth rate of $\frac{4}{3}$.) It follows that the perimeter of the Koch snowflake must be infinite.

> *The boundary of the Koch snowflake has infinite length.*

Area of the Koch Snowflake

> *The area of the Koch snowflake is 1.6 times the area of the starting equilateral triangle.*

The preceding statement is a startling fact! It implies that the Koch snowflake has a finite area enclosed by an infinite boundary—a fact that common geometric intuition finds somewhat difficult to accept. Clearly, this is not an everyday, run-of-the-mill geometric object!

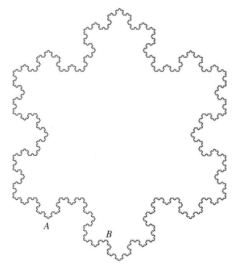

FIGURE 12-7 A tiny ant trying to get from point A to point B traveling along the nooks and crannies would be hard put to make it there soon—the distance to be traveled is infinite!

Perimeter: infinite

To verify that the area of the Koch snowflake is 1.6 times the area of the starting equilateral triangle will take some doing, but the necessary tools are already at our disposal. In what follows, we will give an outline of the argument, leaving the technical details as exercises for the reader. In fact, the reader who wishes to do so may skip the forthcoming explanation without prejudice.

The strategy we will adopt to calculate the area of the Koch snowflake is as follows:

- Calculate a formula that gives the area of the polygon we get at the Nth step (i.e., a generic step) in the construction.

- Determine what happens to the formula obtained above as N gets bigger and bigger.

Figure 12-8 shows how the area changes when we go from one step of the construction to the next. From Fig. 12-8 we observe that the area of the polygon at the Nth step of the construction is

$$A + \left(\tfrac{1}{3}\right)A + \left(\tfrac{4}{9}\right)\left(\tfrac{1}{3}\right)A + \left(\tfrac{4}{9}\right)^2\left(\tfrac{1}{3}\right)A + \cdots + \left(\tfrac{4}{9}\right)^{N-1}\left(\tfrac{1}{3}\right)A$$

(where A is the area of the starting equilateral triangle). Except for the first term, we are looking at the sum of terms of a geometric sequence. Using the formula for adding the consecutive terms of a geometric sequence (Chapter 10), we can simplify the preceding expression to

$$A + \left(\tfrac{3}{5}\right)A\left[1 - \left(\tfrac{4}{9}\right)^N\right].$$

We leave the technical details to the reader (Exercise 34).

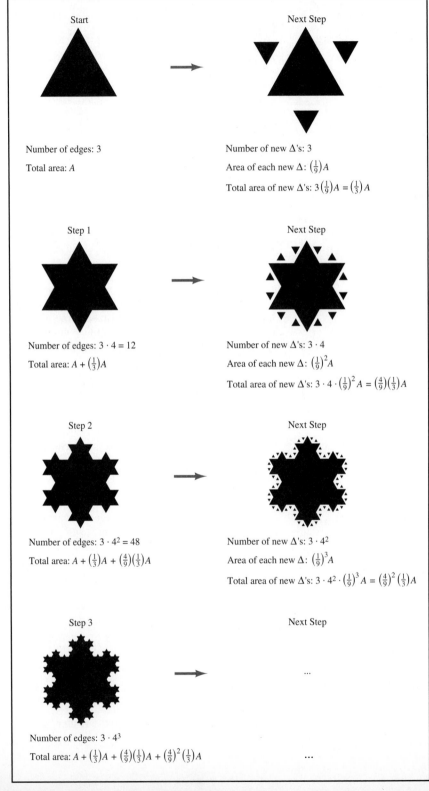

Start

Number of edges: 3

Total area: A

Next Step

Number of new Δ's: 3

Area of each new Δ: $\left(\frac{1}{9}\right)A$

Total area of new Δ's: $3\left(\frac{1}{9}\right)A = \left(\frac{1}{3}\right)A$

Step 1

Number of edges: $3 \cdot 4 = 12$

Total area: $A + \left(\frac{1}{3}\right)A$

Next Step

Number of new Δ's: $3 \cdot 4$

Area of each new Δ: $\left(\frac{1}{9}\right)^2 A$

Total area of new Δ's: $3 \cdot 4 \cdot \left(\frac{1}{9}\right)^2 A = \left(\frac{4}{9}\right)\left(\frac{1}{3}\right)A$

Step 2

Number of edges: $3 \cdot 4^2 = 48$

Total area: $A + \left(\frac{1}{3}\right)A + \left(\frac{4}{9}\right)\left(\frac{1}{3}\right)A$

Next Step

Number of new Δ's: $3 \cdot 4^2$

Area of each new Δ: $\left(\frac{1}{9}\right)^3 A$

Total area of new Δ's: $3 \cdot 4^2 \cdot \left(\frac{1}{9}\right)^3 A = \left(\frac{4}{9}\right)^2 \left(\frac{1}{3}\right)A$

Step 3

Number of edges: $3 \cdot 4^3$

Total area: $A + \left(\frac{1}{3}\right)A + \left(\frac{4}{9}\right)\left(\frac{1}{3}\right)A + \left(\frac{4}{9}\right)^2 \left(\frac{1}{3}\right)A$

Next Step

...

...

FIGURE 12-8

We are now ready to wrap this up. We only need to figure out what happens to $(\frac{4}{9})^N$ when N gets bigger and bigger. In fact, what happens to any positive number less than 1 when we raise it to higher and higher powers? If you know the answer, then you are finished. If you don't, take a calculator, enter a number between 0 and 1, and multiply it by itself repeatedly. (You will readily convince yourself that the result gets closer and closer to 0.) The bottom line is that as N gets bigger and bigger, the expression inside the square brackets gets closer and closer to 1, and therefore the area gets closer and closer to $(1.6)A$.

Symmetry of Scale   Let's say, for the sake of argument, that we are not totally sold on the idea of the Koch snowflake and the Koch curve. We demand a closer look! What does the fine detail look like? In Fig. 12-9 we show a section of the Koch curve magnified nine times and then a portion of the magnified curve magnified nine times again. We can continue this magnification process indefinitely, but it won't help. No matter how much we crank up our imaginary microscope, we will continue seeing exactly the same thing!

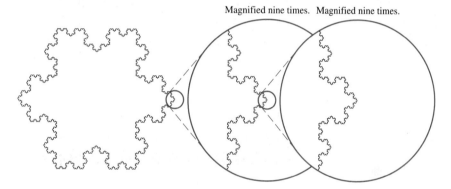

Magnified nine times. Magnified nine times.

FIGURE 12-9

This remarkable characteristic of the Koch curve is called **symmetry of scale** (or sometimes **self-similarity**). As the name suggests, it is a symmetry that carries itself across different scales—a symmetry between the large and the small, the small and the tiny. Thus, we say that the Koch curve has symmetry of scale to indicate that any piece of it can be found again at another level of magnification and that its defining pattern (＿∧＿) occurs at infinitely many scales.

Before we move on, a final point is in order. Suppose we wanted a realistic geometric description of a very jagged piece of coastline (say, something like the Scandinavian fjords). We would be hard put to find among the traditional shapes of Euclidean geometry anything that comes as close to giving the "feel" of that coastline as a piece of the Koch curve. The Koch snowflake and its boundary, the Koch curve, are not mathematical freaks but rather very good mathematical descriptions of the way certain things (such as coastlines) look in nature.

THE SIERPINSKI GASKET

This construction was first suggested (in a slightly modified form) by the Polish mathematician Waclaw Sierpinski[1] around 1915. It involves another recursive construction like that of the Koch snowflake. It also starts with a black triangle ABC, but unlike the Koch snowflake, the starting black triangle does not have to be equilateral—any black triangle will do (Fig. 12-10[a]). In this construction, instead of *adding* smaller copies of the original triangle, we will *remove* smaller copies of the original triangle as follows:

- ■ **Start.** Start with an arbitrary black triangle ABC.

- ■ **Step 1.** Join the midpoints M_1, M_2, and M_3 of sides AB, AC, and BC, respectively. This gives us four triangles (AM_1M_2, BM_1M_3, CM_2M_3, and the *middle* triangle $M_1M_2M_3$ as shown in Fig. 12-10[b]). Each of the four triangles is similar to the original black triangle. We now remove the interior of the middle triangle $M_1M_2M_3$, leaving a triangular white hole. If we call the interior of the middle triangle $M_1M_2M_3$ the *heart* of triangle ABC, we can use a cruel but convenient metaphor—we have "cut out the heart" of triangle ABC!

- ■ **Step 2.** Cut out the heart of each of the 3 black triangles in step 1. This leaves us with 9 black triangles (all similar to the original triangle ABC) and 4 triangular white holes (Fig. 12-10[c]).

- ■ **Steps 3, 4, etc.** Repeat the process (cut out the heart of every black triangle) ad infinitum.

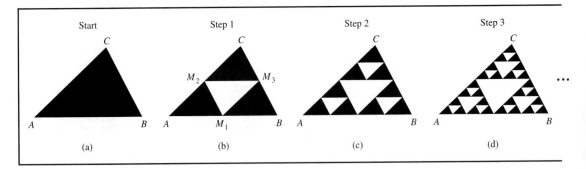

Start	Step 1	Step 2	Step 3

(a) (b) (c) (d)

FIGURE 12-10

After infinitely many steps of this recursive construction one gets a bizarre kind of geometric Swiss cheese called the **Sierpinski gasket** (Fig. 12-11).

---

[1] Waclaw Sierpinski (1882–1969) was one of the great mathematicians of the first half of this century. He is also one of the lucky few to have one of the craters of the moon named after him.

Once again we observe that Fig. 12-11 is just an approximation of the Sierpinski gasket. The tiny black triangles that we see are the result of poor eyesight and the inadequacies of printing. The Sierpinski gasket has no solid black triangles! If we were to magnify any of these seemingly black triangles we would see a replica of what we see in Fig. 12-11 (Fig. 12-12). We now have a name for this phenomenon: The Sierpinski gasket has symmetry of scale.

Just like the Koch snowflake, the Sierpinski gasket can be described in a very convenient way by a recursive replacement rule.

FIGURE 12-11 The Sierpinski gasket.

Recursive Replacement Rule for the Sierpinski Gasket

■ Start with an arbitrary black triangle ▲ .

■ Wherever you see a ▲ , replace it with a ▲ .

We leave the following two facts as exercises to be verified by the reader:

■ The Sierpinski gasket has zero area (Exercise 3).

■ The Sierpinski gasket has an infinitely long boundary (Exercise 4).

(By now, nothing surprises us anymore!)

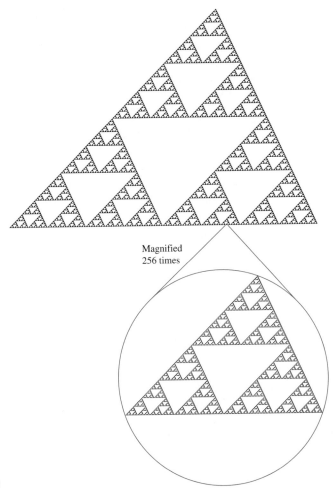

Magnified
256 times

FIGURE 12-12

THE CHAOS GAME   This example involves an arbitrary triangle *ABC* and an honest die. To each of the vertices of the triangle we assign two of the six possible outcomes of rolling the die—say, *A* is assigned the numbers 1 and 2, *B* is assigned 3 and 4, and *C* is assigned 5 and 6. (The object is to give each of the three vertices an equal chance of being chosen. Instead of rolling a die, we could just as easily draw the name *A*, *B*, or *C* out of a hat.) We are now ready to play the game.

■ **Start.**   Roll the die. Mark the vertex corresponding to the roll. Say we roll a 5—we then mark vertex *C* (Fig. 12-13[a]). This is our starting position.

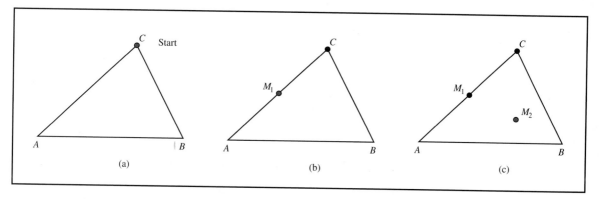

FIGURE 12-13 The chaos game after three rolls of the die.

- **Step 1.** Roll the die again. Say we roll a 2 (i.e., we have picked vertex *A*). We now mark the point M_1 halfway between the starting position *C* and the chosen vertex *A* (Fig. 12-13[b]).

- **Step 2.** Roll the die again. Mark the point M_2 halfway between the previous position M_1 and the chosen vertex. If the roll is 3, for example, mark M_2 halfway between M_1 and *B* (Fig. 12-13[c]).

- **Steps 3, 4, etc.** Continue playing the chaos game ad infinitum, each time marking the point halfway between the preceding position and the chosen vertex.

What kind of a picture does the chaos game produce? Figure 12-14(a) shows the results after 50 rolls of the die, just a bunch of scattered dots. Figure 12-14(b) shows the results after 500 rolls, and we begin to detect the vague outlines of something familiar. Figure 12-15 shows the results after 5000 rolls—the pattern is unmistakably a Sierpinski gasket! We won't show the picture of what we get after a billion rolls, but Fig. 12-11 is a very good approximation of it.

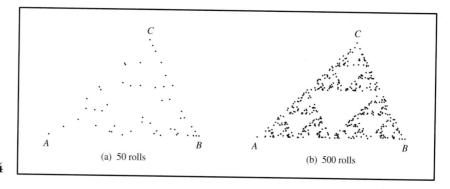

FIGURE 12-14

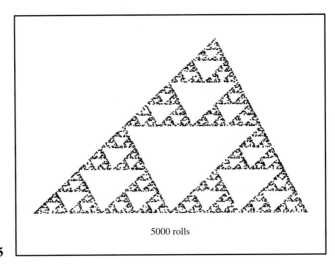

5000 rolls

Figure 12-15

This is a rather surprising turn of events: The chaos game is ruled by the roll of the die and thus by the laws of chance. While our natural expectation is that no predictable pattern should appear, in fact an extremely predictable pattern emerges. There are no ifs or buts about it, the longer you play the chaos game the closer you get to the Sierpinski gasket!

Thus, the chaos game gives us a different way to construct the Sierpinski gasket. While the process is still recursive, an element of chance has been added.

SYMMETRY OF SCALE IN ART AND LITERATURE

The notion of symmetry of scale as exemplified by the Koch snowflake and the Sierpinski gasket is not unique to formal mathematical or geometric constructions. Artists, poets, writers, and even philosophers have used it in their own ways and for their own purposes. In this section we will briefly mention just a few such examples.

Figure 12-16 shows a typical artistic expression of symmetry of scale. It is the cover of an issue of "TV Week" in which there is a man sitting in an armchair holding a remote control in his left hand and an issue of "TV Week" in his right hand. It so happens that the issue he is holding shows on its cover the same scene: a man sitting in an armchair holding in his right hand an issue of "TV Week," and so on ad infinitum. Just as in our previous examples, there is a recursive construction behind the picture which, in fact, can be described very precisely in geometric terms: Start with the picture of the man with the TV, the remote, the armchair, etc. (everything except the issue of the "TV Week" in his right hand), shrink it down to 35% of the original size, and then apply a rotation and a translation to the new picture so that it "slips" right into the man's right

The Fresno Bee

TV Week

Complete listings for Continental Cable subscribers

Channel your energy

Click through our biggest fall preview issue

Week of Sept. 17 - Sept. 23

FIGURE 12-16 (Reproduced by permission of the *Fresno Bee.*)

hand in the old picture. If we repeat this process again and again, we get the effect of Fig. 12-16.

As with any other artificial rendition of symmetry of scale, the symmetry of scale in Fig. 12-16 is only illusory: The recursive process is supposed to go on to infinity, but in this case it stops after four steps. (The artist, after all, has to get on with his life.) The message, however, is unmistakable and clear—emotionally all of us accept that the recursion goes on forever even if we know that in reality the process must stop at some point.

More ingenious artistic examples of symmetry of scale can be found in the work of the Dutch artist M. C. Escher and in the literary works of such diverse writers as Lewis Carroll, Aldous Huxley, and e. e. cummings. Even philosophers have dealt with the concept of symmetry of scale. The German mathematician and philosopher G. W. Leibniz believed that within each drop of water lived an entire universe containing all the elements of our universe (including of course, more drops of water).

We conclude this section with a poem by Jonathan Swift. The theme is unmistakable symmetry of scale.

> *So Nat'ralists observe, A Flea*
> *Hath Smaller Fleas that on him prey*
> *and these have smaller Fleas to bite 'em*
> *And so proceed, ad infinitum*

Smaller and Smaller by M. C. Escher. (© 1956 M. C. Escher/Cordon Art, Baarn, Holland) "In this woodcut I have consistently and almost maniacally continued the reduction down to the limit of practical execution. I was dependent on four factors: the quality of my wood material, the sharpness of my tool, the steadiness of my hand, and especially my keen-sightedness. . . ."—M. C. Escher

THE MANDELBROT SET

We now return to a more mathematical example. In fact, the mathematics in this example goes a bit beyond the level of this book (it requires an understanding of *complex numbers*), so we will describe the overall idea in rather general, oversimplified terms. The actual purpose of this example is not to deal with mathematical details but rather to illustrate an important variation of symmetry of scale we haven't seen before. Besides, it gives us an excuse to introduce some incredibly exotic pictures.

The Mandelbrot set is named after the mathematician Benoit Mandelbrot, an IBM Fellow and professor at Yale University. Mandelbrot was the first person to extensively study and fully appreciate the importance of this beautiful and complex mathematical object.

The construction we are going to describe results in an actual geometric shape called the Mandelbrot set. Figure 12-17(a) is a picture of the Mandelbrot set in black and white. To most people, the Mandelbrot set looks like some sort of bug; a cockroach from another planet might be an apt description. The structure of the Mandelbrot set can be described as consisting of a body, a head, and an antenna coming out of the middle of the head. Both the head and the body are surrounded by warts of various sizes. In Fig. 12-17(a) we notice that the larger warts have a structure similar (but not identical) to the original object (body, head, and antennas). When we blow up one of the warts (Fig. 12-17[b]), we can clearly see the repetition of the original overall structure, including the wart's many warts and subwarts. But we can also see new and interesting patterns, such as the swirling clusters appearing on the upper right-hand side and in the lower center of Fig. 12-17(b), as well as the structure resembling a snowflake on the upper left. We also see a few scattered, small, black blots which appear to resemble the original Mandelbrot set itself. Figure 12-17(c) is a close-up of the area around one of the small warts in Fig. 12-17(b). It shows something similar to a jewel-studded brooch shaped like a seahorse's tail in the upper center and a new kind of swirling pattern in the lower center. Smaller versions of both of these patterns can be spotted throughout the picture. In the midst of these new structures we can again detect what appears to be small replicas of the original starting picture. Figure 12-17(d) is a close-up of one of the seahorse tails in Fig. 12-17(c). It is made up of familiar-looking clusters of various sizes as well as new structures we haven't seen before. The process goes on and on and on. At every new level of magnification, familiar and previously unseen formations mingle in a breathtaking dance of infinite repetition and infinite variety—structures repeating themselves at infinitely many different scales but never in exactly the same way.

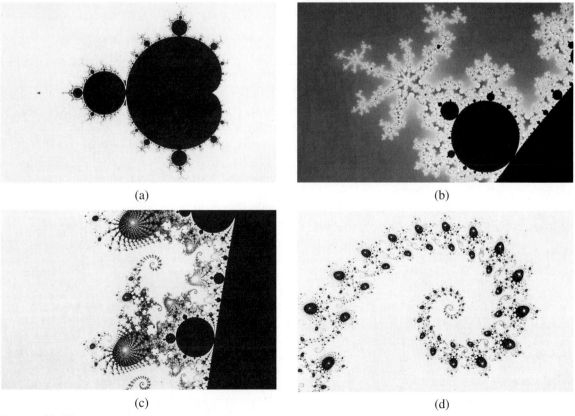

(a)

(b)

(c)

(d)

FIGURE 12-17

What sort of symmetry of scale do we find in the Mandelbrot set? In our original examples (the Koch snowflake, the Sierpinski gasket, the cover of "TV Week"), symmetry of scale implied that exact copies of a structure occur across infinitely many scales. This is sometimes described as **exact** symmetry of scale. In the Mandelbrot set, however, there is only an **approximate** symmetry of scale where images repeat themselves across infinitely many scales but are never exactly alike.

Constructing the Mandelbrot Set

When faced with the beauty and complexity of the Mandelbrot set (see color plates), it is easy to forget its origin: The Mandelbrot set is a completely mathematical object which, just like the Koch snowflake and the Sierpinski gasket, is defined by a relatively simple recursive replacement rule. The recursive process itself works like this: Start by picking a number we will refer to as the **seed** of the recursive process and then apply infinitely many times a recursive rule that at each step squares the number in the preceding step and then adds the seed to the result. The following box summarizes the recursive process, which we will refer to as the **Mandelbrot replacement process**.

Mandelbrot Replacement Process

■ **Start:**   Seed (s).

■ **Recursive step:**   Replace x by $x^2 + s$.

The choice of the name "seed" for our starting value gives us a convenient metaphor: Each seed, when planted, produces a different sequence of numbers (the tree). The recursive step is like a rule telling the tree how to grow from one season to the next. Each tree has its own rule for growth, which is encoded in the seed, but the rules for growth for different trees all fall under one general pattern.

Well, let's take a trip to the botanical garden and look at some examples of the Mandelbrot's replacement process.

Example 1. (Seed: $s = 1$)

	Start	Step 1	Step 2	Step 3	Step 4	. . .
Seed	$s = 1$	$s = 1$	$s = 1$	$s = 1$	$s = 1$	
Input		$x = 1$	$x = 2$	$x = 5$	$x = 26$	
Recursive step		$x^2 + s = 2$	$x^2 + s = 5$	$x^2 + s = 26$	$x^2 + s = 677$	
Output		$x = 2$	$x = 5$	$x = 26$	$x = 677$	

We can see that, with this particular seed, as the process continues, we get bigger and bigger numbers. We say that in this situation the Mandelbrot replacement process *goes off to infinity.* (Alternatively, we can call this the *Jack-and-the-beanstalk syndrome.*)

Example 2. (Seed: $s = -1$)

	Start	Step 1	Step 2	Step 3	Step 4 $\ldots$
Seed	$s = -1$	$s = -1$	$s = -1$	$s = -1$	$s = -1$
Input		$x = -1$	$x = 0$	$x = -1$	$x = 0$
Recursive step		$x^2 + s = 0$	$x^2 + s = -1$	$x^2 + s = 0$	$x^2 + s = -1$
Output		$x = 0$	$x = -1$	$x = 0$	$x = -1$

We can readily see that with the seed $s = -1$ the numbers hop back and forth between 0 and -1. We say in this case that the Mandelbrot replacement process is *periodic* (i.e., it goes around and around in a cycle).

Example 3. (Seed: $s = -0.75$)

	Start	Step 1	Step 2	Step 3	Step 4 $\ldots$
Seed	$s = -0.75$	$s = -0.75$	$s = -0.75$	$s = -0.75$	$s = -0.75$
Input		$x = -0.75$	$x = -0.1875$	$x = -0.714844$	$x = -0.238998$
Recursive step		$x^2 + s = -0.1875$	$x^2 + s = -0.714844$	$x^2 + s = -0.238998$	$x^2 + s = -0.69288$
Output		$x = -0.1875$	$x = -0.714844$	$x = -0.238998$	$x = -0.69288$

As an exercise (Exercise 35), the reader may want to carry this example out, say, for another 100 or so steps and try to figure out what happens. Here is a hint as to how one can do this (as well as any other example) very efficiently with a calculator that has a memory and a square function key:

- **Start:** Store -0.75 ($=s$) in memory

- **Recursive step:** Press $\boxed{x^2}$ (squares the current value)
  Press $\boxed{+}$
  Press $\boxed{MR}$ (memory recall)
  Press $\boxed{=}$ (adds the contents of the memory)

Complex Numbers

In Examples 1 through 3 we were careful to choose some fairly simple types of numbers (integers and decimals) for our seeds. In reality, the Mandelbrot replacement process is most interesting when applied to a more complicated category of numbers called complex numbers. You may recall having seen these numbers before (probably in intermediate algebra). These are the numbers that allow us to take square roots of negative numbers, solve quadratic equations of any kind, etc. The basic building block for complex numbers is the number $\sqrt{-1} = i$. Using i we can build all other complex numbers such as $(3 + 2i)$, $(\frac{5}{3} - \frac{4}{3}i)$, and the generic complex number $(a + bi)$.

Let's look at one more example of the Mandelbrot replacement process using a seed that is a complex number.

Example 4. (Seed: $s = i$)

	Start	Step 1	Step 2	Step 3	Step 4 . . .
Seed	$s = i$	$s = i$	$s = i$	$s = i$	$s = i$
Input		$x = i$	$x = -1 + i$	$x = -i$	$x = -1 + i$
Recursive step		$x^2 + s = i^2 + i$	$x^2 + s = (-1 + i)^2 + i$	$x^2 + s = (-i)^2 + i$	$x^2 + s = -i$
		$= -1 + i$	$= 1 - 2i + i^2 + i$	$= -1 + i$	(see step 2)
			$= -i$		
Output		$x = -1 + i$	$x = -i$	$x = -1 + i$	$x = -i$

This situation is almost identical to the one in Example 2 (the process goes around and around in a cycle), except that the cycle does not go all the way back to the seed. ▬

For our purposes, the most relevant fact about complex numbers is that each complex number can be drawn as a point in the plane. Figure 12-18 is self-explanatory. Thus, we can talk about complex numbers and points in the plane as being one and the same.

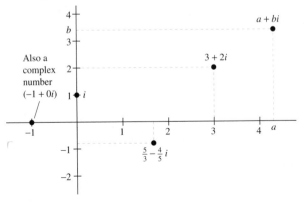

FIGURE 12-18 Because of this geometric interpretation of complex numbers, it is common practice to identify complex numbers with points in the plane.

**Back to the
Mandelbrot Set**

We are ready (finally) to explain how the Mandelbrot set comes about. Examples 1 through 4 illustrate the fact that the Mandelbrot replacement process behaves in different ways depending on the choice of the seed. In particular, we want to make a distinction between seeds for which the process goes off to infinity (What we informally called in Example 1 the Jack-and-the-beanstalk syndrome) and seeds where the process *does not* go off to infinity (as in Examples 2 through 4). If we color the former seeds white and the latter black, we get the Mandelbrot set!

Let's take a deep breath now and consider what we have done. In essence, it boils down to this: We have a recursive process (the Mandelbrot replacement process) that accepts any point in the plane (i.e., any complex number) as a seed. For some seeds, the process goes off to infinity, and we color such seeds white. All other seeds we color black. The collection of all black seeds turns out to be the Mandelbrot set. Not quite as simple as one-two-three but simple enough!

The simplicity of the Mandelbrot replacement process stands in marked contrast to the complexity of the results. The Mandelbrot set has been called the "most complex mathematical object known by man." It has also become one of the most popular and fascinating mathematical discoveries of this century.

FRACTALS

The word **fractal** (from the Latin *fractus*, "broken up, fragmented") was coined by the mathematician Benoit Mandelbrot in the mid-1970's to describe under one term objects as diverse as the Koch curve, the Sierpinski gasket, and the Mandelbrot set, as well as many shapes in nature such as those of clouds, coastlines, lightning, mountains, the vascular system in the human body, and the lungs.

All of these objects are fractals. As such, they share several characteristics. One of them is that they all have some form of symmetry of scale. (Some of the other mathematical characteristics of fractals, such as fractional dimensions, are beyond the scope of this book. (Any reader who wishes to investigate other mathematical properties of fractals in depth is encouraged to look at some of the many references listed at the end of this chapter.)

A word of caution is in order: Symmetry of scale as we defined it does not by itself ensure that the object is a fractal. The picture in Fig. 12-16, for example, has symmetry of scale but is not a fractal. All fractals, however, have some form of symmetry of scale, be it exact symmetry of scale (as in the Koch curve and the Sierpinski gasket) or approximate symmetry of scale (as in the Mandelbrot set).

The discovery and study of new fractal shapes has become arguably one of the hottest mathematical topics of the last 20 years. As a field of study, fractal geometry is a mathematician's dream come true: It combines complex and interesting mathematics, beautiful graphics (see color plates 9-13), and extreme relevance to the real world.

**CONCLUSION:
FRACTAL
GEOMETRY,
A NEW FRONTIER**

In his classic book, *The Fractal Geometry of Nature*, Benoit Mandelbrot[2] wrote:

> *Why is [standard] geometry often described as "cold" and "dry"? One reason lies in its inability to describe the shape of a cloud, a mountain, a coastline, or a tree. Clouds are not spheres, mountains are not cones, coastlines are not circles, and bark is not smooth, nor does lightning travel in a straight line.*

> *. . . many patterns of Nature are so irregular and fragmented, that compared with [standard geometry] Nature exhibits not only a higher degree but an altogether different level of complexity. The number of distinct scales of length of natural patterns is for all practical purposes infinite.*

There is a striking visual difference between the shapes of traditional geometry and the geometric shapes we discussed in this chapter. It is difficult to mistake one for the other. The shapes of traditional geometry (squares, circles, cones, etc,) and the objects we build based on them (bridges, machines, buildings, etc.) have a distinct artificial look, and no amount of artistic license can disguise this. To paraphrase Mandelbrot, one cannot make a realistic looking cloud or mountain using spheres and cones. For a long time, it was assumed that making realistic geometric models of the natural world was pretty much impossible. Consider, however, color plates 14a, 15a, and 15b. These artificial landscapes were created using tools of a new geometry called **fractal geometry**. In this new geometry, mathematically defined fractals replace the traditional squares, circles, cones, etc., just as the computer replaces the traditional ruler, compass, protractor, etc.

Mandelbrot, who is the father of fractal geometry, was the first to realize that fantastic geometric constructions that imitate nature are possible using fractals. This realization has revolutionized many fields in which modeling nature is important. One we can all easily relate to is the field of computer animation. Viewers of the movie *Star Trek II: The Wrath of Khan* are always impressed by the Genesis planetary sequence. In this sequence an ecologically benign bomb is dropped on a barren planet and a lush tropical world emerges in front of our very eyes. This entire special effect was created in a computer using fractal geometry. Since then, fractally created landscapes have appeared in science fiction movies (*The Last Starfighter*, etc.) and in many dazzling pieces of computer art.

---

[2] Benoit Mandelbrot, *The Fractal Geometry of Nature* (New York: W. H. Freeman & Co., 1983).

"Clouds are not spheres, mountains are not cones, coastlines are not circles, and bark is not smooth, nor does lightning travel in a straight line." — B. Mandelbrot (Top left and right, Myron Wood/Photo Researchers Inc. Bottom left, Max and Kit Hunn/Photo Researchers. Bottom right, National Center for Atmospheric Research/National Science Foundation.)

Because symmetry of scale is a pervasive and fundamental characteristic of many objects in nature, fractals have become an essential tool in their study—be it to predict, analyze, or imitate either the natural objects themselves or their functions. Important applications of fractal geometry have been discovered in recent years in such diverse fields as materials science, population biology, and human physiology. For details, the reader is referred to references 5, 8, 9, 11, 17, 24, and 27 at the end of the chapter.

Geometry as we have known it in the past was developed by the Greeks about 2000 years ago and passed on to us essentially unchanged. It was (and still is) a great triumph of the human mind, and it has allowed us to develop much of our technology, engineering, architecture, etc. As a tool and language for modeling and representing nature, Greek geometry has by and large been a failure. The discovery of fractal geometry seems to have given science the right mathematical language to remedy this failure and as such it promises to be one of the great achievements of twentieth-century mathematics.

KEY CONCEPTS

chaos game
fractal
fractal geometry
Koch curve (snowflake curve)
Koch snowflake

Mandelbrot replacement process
Mandelbrot set
recursive replacement rule
Sierpinski gasket
symmetry of scale

EXERCISES

Walking

The Sierpinski Gasket. *Exercises 1 through 4 refer to the Sierpinski gasket as defined in the chapter. The following figure shows the first few steps of the construction.*

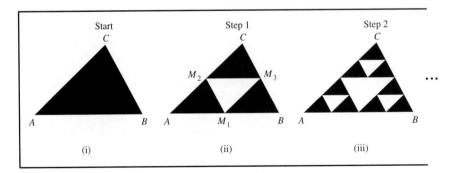

1. Suppose the area of the starting black triangle ABC is X.
   (a) Find the area of the black triangle AM_1M_2 in (ii). Explain.
   (b) Find the area of the black triangle M_1BM_3 in (ii). Explain.

    (c) Find the area of the white triangle $M_1M_2M_3$ in (ii). Explain.

    (d) Find the area of the black region in (ii).

2. Suppose the perimeter of the starting black triangle ABC is P.

    (a) Find the perimeter of the black triangle AM_1M_2 in (ii).

    (b) Find the perimeter of the black triangle M_1BM_3 in (ii).

    (c) Find the perimeter of the white triangle $M_1M_2M_3$ in (ii).

    (d) Find the length of the boundary of the black region in (ii).

3. Suppose the area of the starting black triangle ABC is X.

    (a) Find the area of the black region obtained at step 2 in the construction of the Sierpinski gasket.

    (b) Find the area of the black region obtained at step 3 in the construction of the Sierpinski gasket.

    (c) Find the area of the black region obtained at step N (a generic step) in the construction of the Sierpinski gasket.

    (d) Explain why the area of the Sierpinski gasket is zero. (Use your answer in [c].)

4. Suppose the perimeter of the starting black triangle ABC is P.

    (a) Find the length of the boundary of the black figure obtained at step 2 in the construction of the Sierpinski gasket.

    (b) Find the length of the boundary of the black figure obtained at step 3 in the construction of the Sierpinski gasket.

    (c) Find the length of the boundary of the black figure obtained at step N in the construction of the Sierpinski gasket.

    (d) Explain why the Sierpinski gasket has an infinitely long boundary.

The Mitsubishi Gasket. *Exercises 5 through 8 refer to the Mitsubishi gasket, a fractal defined by the following recursive construction:*

    ■ ***Start .*** *Start with a black equilateral triangle ABC as shown in (i).*

    ■ ***Step 1.*** *Subdivide the black triangle into nine equal subtriangles and remove the three "interior" subtriangles, as shown in (ii).*

    ■ ***Step 2.*** *For each of the remaining black triangles repeat the process (subdivide each black triangle into nine equal triangles and remove the three "interior" triangles). The result is shown in (iii).*

    ■ ***Steps 3, 4, etc.*** *Repeat the process described in the previous steps ad infinitum.*

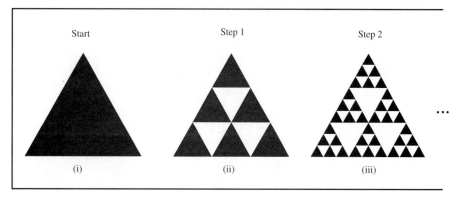

5. How many solid black triangles are there at the start of step 3 in the construction? How about at step 10? How about at step N?

6. Suppose the area of the starting black triangle ABC is X.
   (a) Find the area of one of the white triangles in (ii).
   (b) Find the area of the black region in (ii).
   (c) Find the area of the black region in (iii).

7. Suppose the perimeter of the starting black triangle ABC is P.
   (a) Find the perimeter of one of the white triangles in (ii).
   (b) Find the length of the boundary of the black region in (ii).
   (c) Find the length of the boundary of the black region in (iii).

8. Suppose the area of the starting black triangle ABC is X.
   (a) Find the area of the black region at step N in the construction of the Mitsubishi gasket.
   (b) Explain why the area of the Mitsubishi gasket is zero. (Use your answer in [a]).

The Sierpinski Carpet.   *Exercises 9 through 12 refer to the Sierpinski carpet, a fractal defined by the following recursive construction:*

   ■ *Start.   Start with a solid black square.*

   ■ *Step 1.   Subdivide the square into nine equal subsquares and remove the central subsquare.*

   ■ *Step 2, 3, etc.   Continue removing the central square of every black subsquare ad infinitum.*

9. Using graph paper, carefully draw the figures that result at steps 1, 2, and 3 of the Sierpinski carpet.

10. Describe a recursive replacement rule for the Sierpinski carpet. (It has the following form: Start with a _____ . Wherever you see a _____, replace it with a _____.)

11. Suppose the area of the starting square is X.
    (a) How many black squares are there in the figure obtained at step 3 in the construction?
    (b) What is the area of the figure obtained at step 3 in the construction?
    (c) How many black squares are there in the figure obtained at step N in the construction?
    (d) What is the area of the figure obtained at step N in the construction?
    (e) Explain why the area of the Sierpinski carpet is zero.

12. Suppose that the starting square has sides of length 1.
    (a) Find the length of the boundary of the figure at steps 1 and 2 in the construction.
    (b) Find the length of the boundary of the figure at step N in the construction.
    (c) Explain why the length of the boundary of the Sierpinski carpet is infinite.

The Square Snowflake.   *Exercises 13 through 16 refer to the "square snowflake," an object defined by the following recursive construction:*

   ■ *Start.   Start with a solid black square.*

   ■ *Step 1.   Divide each side of the square into three equal segments. Attach to the middle segment of each side a small, black square with sides one-third the sides of the starting square.*

■ *Steps 2, 3, etc.* *Repeat the process (divide sides into thirds and attach appropriate-sized squares to the middle third) ad infinitum.*

13. Describe a recursive replacement rule for the "square snowflake." (It has the following form: Start with a _____. Wherever you see a _____, replace it with a _____.)

14. Using graph paper, carefully draw the figures that result at steps 1, 2, and 3 of the "square snowflake."

15. Suppose that the starting square has sides of length 1.
    (a) Find the length of the boundary of the figure at steps 1 and 2 in the construction.
    (b) Find the length of the boundary of the figure at step N in the construction.

16. Suppose the area of the starting square is X.
    (a) What is the area of the figure obtained at step 3 in the construction?
    (b) What is the area of the figure obtained at step N in the construction?

The Checkered Flag. *Exercises 17 through 20 refer to the "checkered flag," an object defined by the following recursive replacement rule:*

■ *Start with a white rectangle* ▭ .

■ *Whenever you see a white rectangle* ▭ *, replace it with a* ▞▚ .

17. Using graph paper, carefully draw the figures that result at steps 1, 2, and 3 of a " checkered flag."

18. Describe in words the process for construction of a "checkered flag."

19. Suppose the area of the starting rectangle is X. Find the area of the black region at step N in the construction of the "checkered flag."

20. Suppose the perimeter of the starting rectangle is P. Find the length of the boundary of the black region at step N in the construction of the "checkered flag."

The Quadratic Koch Island. *Exercises 21 through 24 refer to the "quadratic Koch Island," a fractal defined by the following recursive replacement rule:*

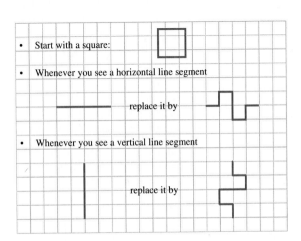

21. Using graph paper, draw the figures that result at steps 1 and 2 in the construction of the quadratic Koch island.

22. The figure obtained at step N in the construction of the quadratic Koch island is a polygon. How many sides does it have?

23. (a) Find the perimeter of each of the polygons obtained at steps 1 and 2 in the construction of the quadratic Koch island.
    (b) Find the perimeter of the polygon obtained at step N in the construction of the quadratic Koch island.

24. Suppose that the starting square encloses an area X.
    (a) Find the areas enclosed by the figures at steps 1 and 2 in the construction of the quadratic Koch island.
    (b) Find the area enclosed by the figure at step N in the construction of the quadratic Koch island.

Exercises 25 through 27 refer to the chaos game as described in the chapter. Assume that we start with an arbitrary triangle ABC and that we will roll an honest die. Vertex A is assigned numbers 1 and 2; vertex B is assigned numbers 3 and 4; and vertex C is assigned numbers 5 and 6.

25. Suppose the die is rolled 16 times and the outcomes are 3, 4, 2, 3, 6, 1, 6, 5, 5, 3, 1, 4, 2, 2, 2, 3. Draw the points P_1 through P_{16} corresponding to these outcomes.

26. Suppose the die is rolled 12 times and the outcomes are 5, 5, 1, 2, 4, 1, 6, 3, 3, 6, 2, 5. Draw the points P_1 through P_{12} corresponding to these outcomes.

27. Suppose that the die is rolled 6 times and that by sheer coincidence the outcomes are 1, 2, 3, 4, 5, 6. Show the trajectory of the trip, as one travels from point to point.

Exercises 28 through 30 refer to the Mandelbrot replacement process described in the chapter. (You may need a calculator to do the computations.)

28. (a) Apply the Mandelbrot replacement process to the seed $s = 2$ and calculate the numbers obtained in the first five steps of this process.
    (b) What happens to the numbers when we repeat the process indefinitely?

29. (a) Apply the Mandelbrot replacement process to the seed $s = -2$ and calculate the numbers obtained in the first five steps of this process.
    (b) What happens to the numbers when we repeat the process indefinitely?

30. (a) Apply the Mandelbrot replacement process to the seed $s = -0.25$ and calculate the numbers obtained in the first twenty steps of this process.
    (b) What happens to the numbers when we repeat the process indefinitely?

Jogging

31. This exercise refers to the construction of the Sierpinski gasket. Explain why there are $(3^N - 1)/2$ white triangles in the figure obtained at step N in the construction.

32. This exercise refers to the construction of the Sierpinski carpet discussed in Exercises 9 through 12. How many white squares does the figure obtained at step N in the construction have?

33. Explain why the construction of the Sierpinski gasket does not end up in an all–white triangle.

34. Use the formula for adding consecutive terms of a geometric sequence (See Chapter 10) to show that
    (a) $1 + \left(\frac{4}{9}\right) + \left(\frac{4}{9}\right)^2 + \cdots + \left(\frac{4}{9}\right)^{N-1} = \frac{9}{5}\left[1 - \frac{4}{9}^N\right]$
    (b) $\left(\frac{1}{3}\right)A + \left(\frac{4}{9}\right)\left(\frac{1}{3}\right)A + \left(\frac{4}{9}\right)^2\left(\frac{1}{3}\right)A + \cdots + \left(\frac{4}{9}\right)^{N-1}\left(\frac{1}{3}\right)A = \frac{3}{5}A\left[1 - \left(\frac{4}{9}\right)^N\right]$.

Exercises 35 through 40 refer to the Mandelbrot replacement process described in the chapter. (You will need a calculator.)

35. Apply the Mandelbrot replacement process to the seed $s = -0.75$. What is the long-term behavior of the numbers in this process?

36. Apply the Mandelbrot replacement process to the seed $s = 0.2$. What is the long-term behavior of the numbers in this process?

37. Apply the Mandelbrot replacement process to the seed $s = 0.25$. What is the long-term behavior of the numbers in this process?

38. Apply the Mandelbrot replacement process to the seed $s = -1.25$. What is the long-term behavior of the numbers in this process?

39. Apply the Mandelbrot replacement process to the seed $s = \sqrt{2}$. What is the long-term behavior of the numbers in this process?

40. Apply the Mandelbrot replacement process to the seed $s = -\sqrt{2}$. What is the long-term behavior of the numbers in this process?

Running

41. **The Koch antisnowflake.** This fractal shape is obtained by essentially reversing the process for constructing the Koch snowflake. At each stage, instead of adding a black equilateral triangle to the outside on the middle third of each side, we remove a black equilateral triangle from the inside on the middle third of each side. The following figure shows the first three steps in the construction of the Koch anti-snowflake.

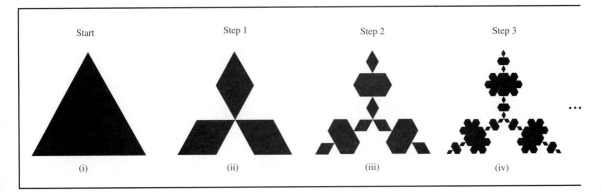

Start   Step 1   Step 2   Step 3

(i)   (ii)   (iii)   (iv)

If the area of the starting equilateral triangle is X, what is the area of the Koch anti-snowflake?

42. Find the area of the "square snowflake" defined in Exercises 13 through 16.

43. Suppose that we play the chaos game using triangle ABC and M_1, M_2, M_3 are the midpoints of the three sides of the triangle. Explain why it is impossible at any time during the game to land inside triangle $M_1M_2M_3$.

44. Apply the Mandelbrot replacement process to the following seeds:
  (a) $s = 1 + i$. What is the long-term behavior of the numbers in this process?
  (b) $s = 1 - i$. What is the long-term behavior of the numbers in this process?

45. Apply the Mandelbrot replacement process to the following seeds:
  (a) $s = -0.25 + 0.25i$. What is the long-term behavior of the numbers in this process?
  (b) $s = -0.25 - 0.25i$. What is the long-term behavior of the numbers in this process?

46. Show that the Mandelbrot set has a reflection symmetry. (*Hint*: See Exercises 44 and 45.)

REFERENCES AND FURTHER READINGS

1. Briggs, John, *Fractals: The Patterns of Chaos*. New York: Simon and Schuster, 1992.

2. Dewdney, A. K., " Computer Recreations: A computer microscope zooms in for a look at the most complex object in mathematics," *Scientific American*, 253 (August 1985), 16–24.

3. Dewdney, A. K., " Computer Recreations: A tour of the Mandelbrot set aboard the Mandelbus," *Scientific American*, 260 (February 1989), 108–111.

4. Dewdney, A. K., " Computer Recreations: Beauty and profundity. The Mandelbrot set and a flock of its cousins called Julia," *Scientific American*, 257 (November 1987), 140–145.

5. Dewdney, A. K., " Computer Recreations: Of fractal mountains, graftal plants and other computer graphics at Pixar," *Scientific American*, 255 (December 1986), 14–20.

6. Field, M., and M. Golubitsky, *Symmetry in Chaos*. New York: Oxford University Press, 1992.

7. Fryde, M. M., "Waclaw Sierpinski—Mathematician," *Scripta Mathematica*, 27 (1964), 105–111.

8. Gardner, M., *Penrose Tiles to Trapdoor Ciphers*. New York: W. H. Freeman & Co., 1988, chap. 3.

9. Gleick, James, *Chaos: Making a New Science*. New York: Viking Penguin, Inc., 1987, chap. 4.

10. Gleick, James, "The Man Who Reshaped Geometry," *New York Times Magazine*, 135 (December 8, 1985), 64.

11. Goldberger, A. L., D. R. Rigney, and B. J. West, "Chaos and Fractals in Human Physiology," *Scientific American*, 262 (February 1990), 44–49.

12. Jürgens, H., H. O. Peitgen , and D. Saupe, "The Language of Fractals," *Scientific American,* 263 (August 1990), 60–67.

13. Krantz, Steven, "Fractal Geometry," *Mathematical Intelligencer*, 11 (Fall 1989), 12–16.

14. La Brecque, Mort, "Fractal Symmetry," *Mosaic*, 16 (January-February 1985), 14–23.

15. Lauwerier, Hans, *Fractals*, Princeton, NJ: Princeton University Press, 1991.

16. Lord, E. A., and C. B. Wilson, *The Mathematical Description of Shape and Form.* New York: John Wiley & Sons, Inc., 1984, chap. 8.

17. Mandelbrot, Benoit, *The Fractal Geometry of Nature.* New York: W. H. Freeman & Co., 1983.

18. McDermott, Jeanne, "Dancing to Fractal Time," *Technology Review,* 92 (January 1989), 6–9.

19. McWorter, W. A., Jr., and J. M. Tazelaar, "Creating Fractals," *Byte,* 12 (August 1987). 123–130.

20. Peitgen, H. O., H. Jürgens, and D. Saupe, *Chaos and Fractals: New Frontiers of Science.* New York: Springer-Verlag, Inc., 1992.

21. Peitgen, H. O., H. Jürgens, and D. Saupe, *Fractals for the Classroom.* New York: Springer-Verlag, Inc., 1992.

22. Peitgen, H. O., and P. H. Richter, *The Beauty of Fractals.* New York: Springer-Verlag, Inc., 1986.

23. Peterson, Ivars, *The Mathematical Tourist.* New York: W. H. Freeman & Co., 1988, chap. 5.

24. Prusinkiewicz, P., and A. Lindenmayer, *The Algorithmic Beauty of Plants.* New York: Springer-Verlag, Inc., 1990.

25. Schechter, Bruce, " A New Geometry of Nature," *Discover*, 3 (June 1982), 66–68.

26. Schroeder, Manfred, *Fractals, Chaos, Power Laws: Minutes from and Infinite Paradise.* New York: W. H. Freeman & Co., 1991.

27. Sorensen, Peter, "Fractals," *Byte*, 9 (September 1984), 157–172.

28. Steen, Lynn Arthur, " Fractals: A World of Nonintegral Dimensions," *Science News*, 112 (August 20, 1977), 122–123.

PART

IV

Statistics

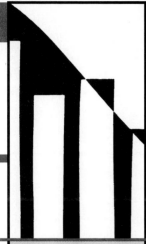

CHAPTER 13

Censuses, Surveys, and Clinical Studies

Collecting Data

"Data! data! data!" he cried impatiently. / "I can't make bricks without clay."
SHERLOCK HOLMES*

- More than 63% of All Americans Support New Tax Cuts
- SAT Scores Up This Year! National Averages Improve to 424 in Verbal and 478 in Math
- Consumer Prices Rose Three-tenths of One Percent in the Last Quarter
- Nielsen Ratings Show " Wheel of Fortune" Inches Ahead of Other Game Shows with an 18.8 Rating for February
- New Study Shows that Eating More than One Pound of Chocolate a Day Increases the Risk of Diabetes by 40% (but what a way to go!)

All of the above headlines are fictitious, but typical of what one frequently sees in the newspaper. They all have one thing in common: They give information that is *numerical* or *quantitative*. Numerical or quantitative information is **data**[1] and, to put it in a nutshell, statistics is the science of dealing with data.

Behind every statistical headline there is a story, and as in any story there is a beginning, a middle, an end, and a moral. This chapter is about the beginning, which in statistics typically means the process of *collecting the data*. Data are the raw material with which statistical information is built and if the data are flawed, then the conclusions one can draw from such data may be of little or no value.

There is a deceptive simplicity to the idea of collecting data, but in fact just the opposite is true. Collecting good data in an efficient and timely manner

---

*Arthur Conan Doyle, "The Adventure of the Copper Beeches," *Adventures of Sherlock Holmes.* New York: Doubleday, Doran & Co. Inc., (1930), p. 297.

[1] The proper usage is **data** (plural) and **datum** (singular). However, the use of data as both a singular and a plural noun is accepted in common usage.

423

is often the most difficult part of the statistical story. At every turn in the process there are serious issues to deal with and potential pitfalls to avoid.

In this chapter we will discuss several systematic methods of collecting data and the issues associated with them. We will use actual case studies to illustrate the main ideas (both good and bad) in the chapter.

SIZING UP THE POPULATION

Every statistical statement refers, directly or indirectly, to some group of individuals or objects. In statistical terminology, this collection of individuals or objects is called the **population**. The population in the first headline is "all Americans." Likewise, the population for the second headline is the group of students taking the Scholastic Aptitude Test (SAT) that particular year. The population for the third headline is made up of numbers rather than people (the prices that people pay for goods and services). For the last two headlines it's hard to tell directly from the wording exactly what the populations are. Often one has to read the fine print before an exact determination of the population can be made.

Defining the Population

Having an exact and precise definition of who or what is the population is essential. This may seem self-evident, but it is often not done, either because of carelessness or because it is in fact impossible. What exactly is meant by the phrase "all Americans" as used in the first headline? Does it include children? How about resident aliens? How about Americans living abroad? How about people in American Samoa? Clearly, the phrase "all Americans" is very vague and subject to many possible interpretations, each of which can affect the meaning and significance of the statistical information one wants to convey.

There are occasions in which a precise definition of the population is in fact impossible. Most public opinion polls preceding an election refer to the population of "voters." But who exactly are these voters? People who are registered to vote and who may even be planning to vote may end up not voting. Is a potential voter a voter?

How Big Is the Population?

Along with the issue of defining the population comes the issue of its size. The size of the population is often used to draw statistical conclusions. But how reliable is the information on size? When populations are small and accessible one can actually get an exact "head count" by simply counting heads the way one counts pennies in a jar. When an instructor reports to a class the average score on an exam, the population consists of all students who took the exam. The instructor can easily determine the exact size of this population—all he has to do is to check his roster.

For large populations, an accurate count is always expensive, usually difficult, and sometimes impossible. How many elephants are there in the wild? In the absence of this information, how much faith can we put in statistical statements about wild elephants? Here is another one: How many people live in the United States? Presumably this information is available from the national census which is mandated by the Constitution and conducted every ten years. According to the 1990 U.S. census, there are 248,709,873 people living in the United States. As we will find out later in the chapter, there is very little reason to believe that this number is accurate. In spite of this, all kinds of other important statistical information are based on this figure.

CENSUS VERSUS SURVEY

Once a population is defined, each member of the population becomes a potential source of data and there are basically two ways to go about collecting the data: (1) collect data from each and every member of the population, or (2) collect data only from a selected subgroup of the population. Method 1 is called a **census**; method 2 is called a **survey**.

A word of caution: The word "census" is associated by most people with the U.S. census which is conducted every ten years, but any method for collecting data that uses the entire population is also a census. For example, it is common practice for an instructor to report to the class the average score of an exam. This piece of statistical information is based on a census of the entire population (the students in the class). To avoid confusion, we will speak of "the U.S. census" when we refer to the national census and use "a census" for any other example.

A census is most commonly used when the population is small and accessible. Of course "small" is a relative term, but generally speaking, if the size of the population is more than a few thousand, we wouldn't want to use a census unless there is a compelling reason. At the same time, a census can be a bad idea even when the population is small. Suppose we want to collect data on the weights and sizes of a population of fish in a small pond. Conducting a census would require catching, weighing, and measuring each and every fish in the pond. But fish are elusive and clever creatures, and the time and effort required to do this may not be justified by the situation. Moreover, short of draining the pond (a bad idea), how would we ever know when the census is complete?

The alternative to collecting data by means of a census is to collect data using a survey. In a survey the data are collected using only a selected subgroup of the population called a **sample**. The basic idea is that questions about the entire population can be answered based on the information collected from the sample. We can think of the members of the sample (whatever they might be) as "speaking" for the population, not unlike the idea of representative government where elected officials act as spokespersons for the people.

The critical issue in surveys (just as in politics) is to choose a sample that is truly representative of the entire population. This is easier said than done unless the population is made up of identical individuals. For a population in which the members are not all alike, collecting data by means of survey as opposed to a census involves a tradeoff: A survey is always subject to some error, but at the same time it requires a lot less work. The exact nature of this tradeoff depends on the situation, but if the sample is properly chosen, the margin of error can be guaranteed by statistical theory to be small and the lack of accuracy can be more than offset by the savings in cost, time, and effort. In many real–world situations, surveys are the only way to go, and their practical value has been proven time and time again.

Some of the best known types of surveys are *public opinion polls* such as the Gallup poll and the Harris poll so often mentioned in the press. Such polls are used to report statistical information on national or regional issues ranging from voters' preferences before an election to national opinions on issues such as the environment, abortion, and the economy. Considering the fact that most modern polls are based on samples consisting of less than two thousand individuals to draw conclusions about populations of many millions, the accuracy of such polls is remarkable. (A typical Gallup poll, for example, is based on samples consisting of less than 1500 individuals.) How can such small samples give accurate information about such large populations? The secret lies in getting a good blend. George Gallup, one of the fathers of modern public opinion polls, puts it this way[2]:

> *Whether you poll the United States or New York State or Baton Rouge... you need only the same number of interviews or samples. It's no mystery really—if a cook has two pots of soup on the stove, one far larger than the other, and thoroughly stirs them both, he doesn't have to take more spoonfuls from one than the other to sample the taste accurately.*

Some Basic Terminology

Before we move on to our case studies we will formalize some of the basic concepts and terms. We will start with a convention: From now on we will use N to denote the size of the population, and n to denote the size of the sample. Since the sample is by definition part of the population, it must always be the case that $n < N$. The ratio n/N is called the **sampling rate**. When expressed as a percentage, a sampling rate of $x\%$ tells us that the sample is $x\%$ of the population.

Example 1. Say we have a population of size $N = 500,000$. If we pick a sample of size $n = 1000$, we get a sampling rate of $1000/500,000 = 1/500$ or 0.2%.

---

[2]Quoted in "The Man Who Knows How We Think," *Modern Maturity*, 17, no. 2 (April–May, 1974), 11.

Essentially, this tells us that each member of the sample "represents" 500 members of the population, and alternatively, that the sample is 0.2% of the population.

Example 2. Once again, suppose the size of the population is $N = 500,000$ and suppose that we need a sampling rate that is about 3%. In order to accomplish this we need a sample size of about 15,000 (3% of 500,000).

As we know now, using a sample is one way to collect statistical information about an entire population. Statisticians use the terms *parameter* and *statistic* to distinguish between exact statistical information about a population and statistical information based on a sample. Any piece of data about the population obtained from a sample is called a **statistic**. A statistic is always an estimate for some fixed but unknown quantity called a **parameter**. Let's put it this way: A parameter is always the statistical information we are after—the pot of gold at the end of the rainbow, so to speak. Calculating a parameter is often difficult, and sometimes impossible. The only hope for getting an exact value for it is to use a census. If we use a sample, then we can get only an estimate for the parameter, and this estimate is called a statistic.

We will use the term **sampling error** to describe the difference between a parameter and a statistic used to estimate that parameter. Sampling error can be attributed to two factors: *chance error* and *sample bias*. **Chance error** is the result of a basic fact about surveys: A statistic cannot give exact information about the population because it is by definition based on partial information (the sample). In surveys, chance error comes with the territory. While chance error is unavoidable, with careful selection of the sample and the right choice of sample size, it can be kept to a minimum.

A parallel but different cause of errors in surveys is due to sample bias. **Sample bias** is the result of having a poorly chosen sample. Even with the best intentions, getting a sample that is representative of the entire population can be very difficult, and many subtle factors can affect the "representativeness" of the sample. Sample bias is the result. As opposed to chance error, sample bias can be eliminated by using proper methods of sample selection. In this chapter, we will focus our attention on the issue of sample bias. We will touch on the concept of chance error and its applications to sampling in Chapter 16.

The essential facts about using samples to collect data about a population can be summarized as follows:

■ A parameter is an exact measurement of some attribute of a population; a statistic is an estimate of the parameter obtained from a sample.

■ A parameter is a fixed quantity, whereas a statistic depends on the choice of the sample. If one chooses a different sample, one is almost certain to get a different statistic (and this is true even when the samples are chosen using exactly the same procedures). This variability among samples is called, not surprisingly, **sampling variability**.

■ Sampling error is the difference between the parameter and the statistic obtained from the sample. Sampling error is made up of two elements: chance error and sample bias.

■ Chance error is an inevitable consequence of sampling variability.

■ Sample bias is a systematic source of error caused by using improper methods for selecting the sample.

■ While chance error cannot be eliminated, it can be brought under control by choosing a suitable sample. As the size of the sample increases, chance error decreases, but surprisingly, the decrease is not in proportion to the size of the sample. After a certain point, a big increase in the size of the sample may have a very minor effect on reducing the chance error.

■ As opposed to the situation with chance error, increasing the size of the sample does not guarantee a reduction in sample bias. In fact, with a poorly chosen sample, a large sample may actually increase sample bias.

This last proposition will be best illustrated in our first case study, one of the worst bungles in the history of public opinion polling.

CASE STUDY 1. BIASED SAMPLES: THE 1936 *LITERARY DIGEST* POLL

The presidential election of 1936 pitted Alfred Landon, the Republican governor of Kansas, against the incumbent President, Franklin D. Roosevelt. The year 1936 marked the tail end of the Great Depression, and economic issues such as unemployment and government spending were the dominant themes of the campaign. The *Literary Digest* was one of the most respected magazines of the time and had a history of accurately predicting the winners of presidential elections that dated back to 1916. For the 1936 election, the *Literary Digest* poll predicted that Landon would get 57% of the vote against Roosevelt's 43% (these were the *statistics*). The actual results of the election were 62% for Roosevelt against 38% for Landon (these were the *parameters* the poll was after). The sampling error in the *Literary Digest* poll was a whopping 19%, the largest ever in a major public opinion poll. Practically all of the sampling error was the result of sample bias.

The irony of the situation is that the *Literary Digest* poll was also one of the largest and most expensive polls ever conducted, with a sample size of approximately 2.4 million people. At the same time that the *Literary Digest* poll was making its fateful mistake, George Gallup was able to predict a victory for Roosevelt using a much smaller sample of approximately 50,000 people.

These facts illustrate the proposition that bad sampling methods cannot be cured by increasing the size of the sample, which in fact just compounds the mistakes. *The critical issue in sampling is not sample size but how best to reduce sample bias.* There are many different ways that bias can creep into the sample selection procedure. Two of the most common ones occurred in the case of the *Literary Digest* poll.

The *Literary Digest* method for choosing its sample was as follows: Based on every telephone directory in the United States, lists of magazine subscribers, rosters of clubs and associations, and other sources, a mailing list of about *10 million* names was created. Every name on this list was mailed a mock ballot and asked to return the marked ballot to the magazine.

One cannot help but be impressed by the sheer ambition of such a project. Neither is it surprising that the magazine's optimism and confidence were in direct proportion to the magnitude of the effort. In its August 22, 1936, issue (p.3), the *Literary Digest* crowed:

> *Once again, [we are] asking more than ten million voters—one out of four, representing every county in the United States—to settle November's election in October.*
>
> *Next week, the first answers from these ten million will begin the incoming tide of marked ballots, to be triple-checked, verified, five-times cross-classified and totaled. When the last figure has been totted and checked, if past experience is a criterion, the country will know to within a fraction of 1 percent the actual popular vote of forty million [voters].*

We are now ready to analyze the sources of sample bias in the *Literary Digest* poll. There were two basic causes of the *Literary Digest's* downfall[3]: *selection bias* and *nonresponse bias*.

The first major problem with the poll was in the selection process for the names on the mailing list which, as we mentioned, were taken from telephone directories, club membership lists, lists of magazine subscribers, etc. Such a list is guaranteed to be slanted toward middle- and upper-class voters and, by default, to exclude lower-income voters. One must remember that in 1936 telephones were much more of a luxury than they are today. Furthermore, at a time when there were still 9 million people unemployed, it is obvious that the names of a very significant element of the population would not show up on lists of club memberships and magazine subscribers. At least with regard to economic status, the *Literary Digest* mailing list was far from being a representative cross section of the population. This is always a critical problem because voters are generally known to vote their pocketbooks, and it was magnified in the 1936 election when economic issues were preeminent in the minds of the voters.

When the method for choosing the sample has a built-in tendency to exclude certain segments of the population (whether intentional or not), we say that they survey suffers from **selection bias**. While it is obvious that to make

[3]This is no figure of speech—the magazine stopped publication soon after the poll.

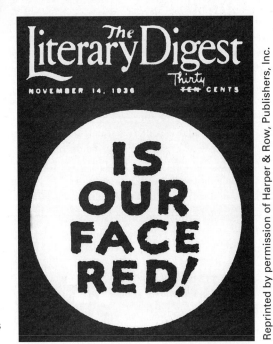

Owning up to a bad case of sample bias: The cover of the *Literary Digest* issue after the 1936 election says it all.

good surveys selection bias must be avoided at all costs, it is not always easy to detect it ahead of time. Even the most scrupulous attempts to get a sample that is a representative cross section of the population can fall short. This fact will become quite apparent in our next case study.

The second problem with the *Literary Digest* poll was that out of the 10 million people whose names were on the original mailing list, only about 2.4 million responded to the survey. Thus, the size of the sample ($n = 2.4$ million) was about one-fourth of what was originally intended. It is a well-known fact that people who respond to surveys are very different from people who don't, not only in the obvious (their attitude toward the usefulness of surveys) but also in more subtle and significant ways: They tend to be better educated and in higher economic brackets (and are in fact more likely to vote Republican).

The ratio between the number of people responding to a survey and the number of people asked to participate in the survey is called the **response rate**. For the *Literary Digest* poll, the response rate was $2,400,000/10,000,000 = 0.24$, which is extremely low. When the response rate is low,[4] a survey is said to suffer from **nonresponse bias**. Nonresponse bias is a special type of selection bias, since it means that reluctant and nonresponsive people have been automatically excluded from the sample.

Dealing with nonresponse bias presents its own set of difficulties. In a free

---

[4] While it is impossible to specify an ironclad rule for what constitutes an acceptable response rate in all situations, it is generally safe to say that any response rate below 0.5 is too low.

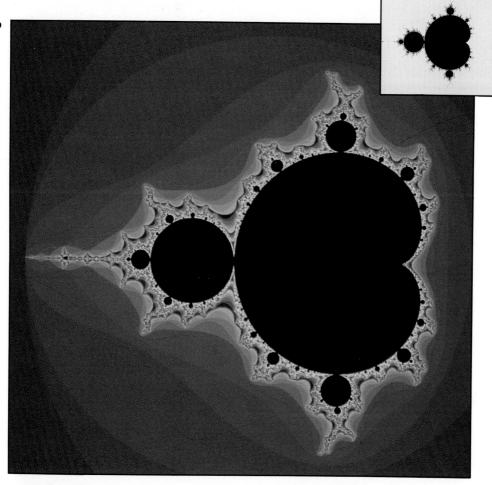

A black and white rendering of the Mandelbrot set (upper insert) comes to life when color is added. What does the picture depict? A cosmic bug? An eclipse in some distant galaxy? A scene from *Star Wars?* Hardly. The Mandelbrot set is a mathematical *fractal*—the result of applying a simple recursive process to the numbers in the complex plane (for details see chapter 12). The plates in the next four pages show the exquisite details and strange beauty of what has been called "the most complex mathematical object known by man." (Computer images by Rollo Silver.)

10a

10b

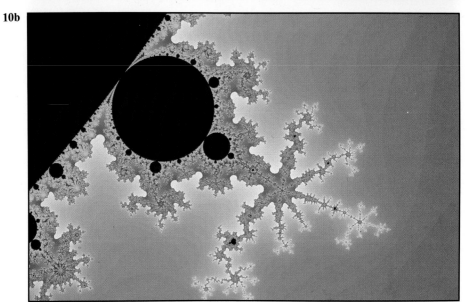

Hundreds of *Mandelbuds* surround the boundary of the Mandelbrot set. Each of the buds, in turn, is surrounded by hundreds of smaller buds and the process repeats itself ad infinitum. They are all similar to each other and to the Mandelbrot set itself. Striking differences show up, however, when we look at the "reefs" around each bud. "Urchins" and "seahorses" (top) are replaced by "starfish" and "snails" (bottom). This blending of infinite repetition and infinite variety is one of the hallmarks of the Mandelbrot set.

11a

11b

A closer look at a "seahorse tail" (top) and a detail of it (bottom). Similar but smaller seahorse tails can be seen throughout the bottom plate sometimes in singles and sometimes in pairs, and if we look deeper, in sets of four (see next page), eight, sixteen,...

Old themes and new themes come together. Top: *Seahorse Quadrille.* A foursome of seahorse tails stand guard around a small replica of the original Mandelbrot set. The magnification is approximately 250,000 times the original set (plate 9b). If plate 9b was done at the same scale as this plate, it would be approximately 20 miles wide. Bottom: *Rainbow vortex.* An entirely new scene, deep inside the labyrinth that surrounds the Mandelbrot set.

13b

13a

Left: *The Monkey's Tail.* A new spiral, different in shape but similar in makeup to the seahorse tail (plate 11a), found fishing in deeper waters on the eastern coast of the Mandelbrot set. Right: *The Bird of Paradise.* A closeup of one of the small pieces that make up the monkey's tail. This picture is at a magnification of over 700,000 times the original Mandelbrot set.

14a

14b

A case of life imitating art. Top: *Carolina* by Ken Musgrave and Benoit Mandelbrot). This artificial landscape was created with a computer using fractal constructions not unlike the ones discussed in chapter 12. Bottom: *Sunset over Mt. Annapurna* (by Scott Woolums). A real landscape, photo- graphed in an expedition to the Himalayas. The remarkable resemblance between the two scenes is purely accidental (the photog- rapher had never seen the computer land- scape) but it reinforces the fact that fractals are the natural tools for describing nature.

15a

15b

A case of art imitating life. Fractal geometry can be used to create remarkable forgeries of natural scenes. Top: This beautiful scene depicting the rugged Sandia mountains in New Mexico is a fractal forgery created entirely in a computer. Bottom: A blending of man made and natural elements. Hot air balloons (created using elements from traditional euclidean geometry such as spheres, cones and ellipsoids) approach a fractally created island. (Both images were created by John Mareda, Gary Mastin and Peter Wattenberg of Sandia National Laboratories.) The mathematical techniques used to create these images are used not only for fanciful pictures but they are important in many applications, from the training of airplane pilots to the modeling of high energy reactions in particle physics.

16a

16b

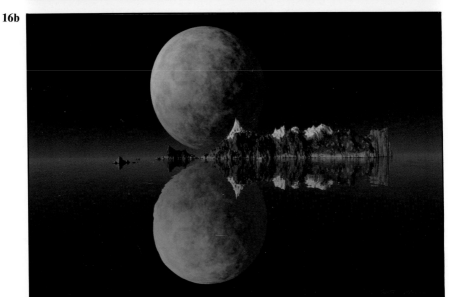

The ability to create natural landscapes using fractal geometry (see preceding pages) can be carried one step further to create "otherwordly" landscapes—fantastic but compelling images of imaginary worlds. Top: *Lethe*—the river of forgetfulness and oblivion in Greek and Roman mythology.

Drinking its waters caused one to forget all former life. Across the river, a vision of Hades. Bottom: *Blessed State*. A giant moon rising over untroubled waters produces a surreal effect. (Both images created by Ken Musgrave and Benoit Mandelbrot of Yale University.)

country we cannot force people to participate in a survey, and paying them is hardly ever a solution since it can introduce other forms of bias. There are known ways, however, of minimizing nonresponse bias. The *Literary Digest* survey was conducted by mail. This approach is the most likely to magnify non-response bias because people often consider a mailed questionnaire just another form of junk mail. Of course, considering the size of the mailing list, the *Literary Digest* really had no other choice. Here again is another illustration of how a big sample size can be more of a liability than an asset.

Nowadays, almost all legitimate public opinion polls are conducted either by telephone or by personal interviews. Telephone polling is subject to slightly more nonresponse bias than personal interviews, but it is considerably cheaper. Even today, however, a significant segment of the population has no telephone in its homes (in fact, a significant segment of the population has no homes), so that selection bias can still be a problem in telephone surveys.[5]

The *Literary Digest* story has two morals: (1) A badly chosen big sample is much worse than a well-chosen small sample, and (2) watch out for selection bias and nonresponse bias.

Our next case study illustrates how difficult it can be to get rid of selection bias even with the very best intentions.

CASE STUDY 2. QUOTA SAMPLING: THE 1948 PRESIDENTIAL ELECTION

Soon after the 1936 fiasco, the *Literary Digest* went out of the business of polling. At the same time, the practice of using public opinion polls to measure the pulse of the American electorate was thriving. By 1948 there were several major polls competing for the big prize, that of accurately predicting the outcome of presidential elections. The best known was the Gallup poll, and the two main competitors were the Roper poll and the Crossley poll.

By this time, all major polls were using what was believed to be a much more scientific method for choosing their samples called **quota sampling**. Quota sampling had been introduced by George Gallup as early as 1935 and had been successfully used by him to predict the winner of the 1936, 1940, and 1944 elections. Quota sampling is nothing more than a systematic effort to force the sample to fit a certain national profile by using quotas: The sample should have so many women, so many men, so many blacks, so many whites, so many under 40, so many over 40, etc. The numbers in each category are taken to represent the same proportions in the sample as are in the electorate at large.

---

[5] The most extreme form of nonresponse bias occurs when the sample consists only of those individuals who step forward and actually "volunteer" to be in the sample. A blatant example of this is the Area Code 900 telephone polls. Here an individual not only has to step forward and volunteer to be in the sample, *he or she actually has to pay* (50 cents or more) to do so. It goes without saying that people who are willing to pay to express their opinions are hardly representative of the general public and that information collected from such polls should be considered useless. Unfortunately, the use and misuse of Area Code 900 polls is proliferating.

"Ain't the way I heard it," Truman gloats while holding an early edition of the *Chicago Daily Tribune* in which the headline erroneously claimed a Dewey victory based on the predictions of all the polls. (UPI/Bettmann)

If we assume that every important characteristic of the population is taken into account when setting up the quotas, it is reasonable to expect that quota sampling will produce a good cross section of the population and therefore lead to accurate predictions. For the 1948 election between Thomas Dewey and Harry Truman, Gallup conducted a poll with a sample size of approximately 3250. Each individual in the sample was interviewed in person by a professional interviewer to minimize nonresponse bias, and each interviewer was given a very detailed set of quotas to meet. For example, an interviewer could have been given the following quotas: seven white males under 40 living in a rural area, five black males under 40 living in a rural area, six black females under 40 living in a rural area, six white females over 40 living in a rural area, etc. Other than meeting these quotas, the ultimate choice of who was interviewed was left to each interviewer.

Based on the results of this poll, Gallup predicted a victory for Dewey, the Republican candidate. The predicted breakdown of the vote was 50% for Dewey, 44% for Truman, and 6% for third-party candidates Strom Thurmond and Henry Wallace. The actual results of the election turned out to be almost exactly reversed: 50% for Truman, 45% for Dewey, and 5% for the third-party candidates.

Truman's victory was a great surprise to the nation as a whole. So convinced was the *Chicago Daily Tribune* of Dewey's victory that it went to press on its early edition for November 4, 1948, with the headline "Dewey defeats Truman"—a blunder that led to Truman's famous retort, "Ain't the way I heard it." The picture of Truman holding aloft a copy of the *Tribune* (see photo) has become part of our national folklore. To pollsters and statisticians, the results of this election were a clear indication that as a method for selecting a representative sample, quota sampling can have some serious flaws.

The basic idea of quota sampling is on the surface a good one: Force the sample to be a representative cross section of the population by having each important characteristic of the population proportionally represented in the sample. Since income is an important factor in determining how people vote, the sample should have all income groups represented in the same proportion as the population at large. Ditto for sex, race, age, etc. Right away we can see one of the potential problems: Where do we stop? No matter how careful one might be, there is always the possibility that some criterion that would affect the way people vote might be missed and the sample could be deficient in this one regard.

An even more serious flaw in the method of quota sampling is the fact that ultimately the choice of who is in the sample is left to the human element. Recall that other than meeting the quotas the interviewers were free to choose whom they interviewed. Looking back over the history of quota sampling, one can see a clear tendency to overestimate the Republican vote. In 1936, using quota sampling, Gallup predicted the Republican candidate would get 44% of the vote, but the actual number was 38%. In 1940 the prediction was 48%, and the actual vote was 45%; in 1944 the prediction was 48%, and the actual vote was 46%. But in spite of the errors, Gallup was able to predict the winner correctly in 1936, 1940, and 1944. This was merely due to luck—the spread between the candidates was large enough to cover the error. In 1948 Gallup (and all the other pollsters) simply ran out of luck. It was time to ditch quota sampling.

The failure of quota sampling as a method for getting representative samples has a moral: Even with the most carefully laid plans, human intervention in choosing the sample is always subject to bias.

RANDOM SAMPLING

If human intervention in choosing the sample is always subject to bias, what are the alternatives? The answer is to let the laws of chance decide who is in the sample. There is a certain amount of irony in the fact that the best way to choose a representative cross section of the population is to, in a manner of speaking, draw the names out of a hat. In fact, people find this very hard to believe. Isn't it possible, the argument goes, to get by sheer chance a sample that is very biased (say for example, all Republican?) The answer is yes in theory, but when the sample is large enough, the odds of it happening are so low that in practice we can pretty much rule it out. Most present-day methods of product quality control in industry, corporate audits in business, and public opinion polling are based on **random sampling** methods, that is, methods for choosing the sample in which chance intervenes in one form or another. The validity of using random sampling methods to collect reliable data is supported by both practical experience and mathematical theory. We will discuss some of the details of this theory in Chapter 16.

The most basic form of random sampling is called **simple random sampling**. Simple random sampling is based on the principle that any group of members of the population has the same chance of being in the sample as any other group as long as the two groups are of the same size. To put it another way, any member of the population has the same chance of being in the sample as any other member; any two members of the population have the same chance of being in the sample as any other two members; any three members have the same chance of being in the sample as any other three members; etc.

In principle, simple random sampling can be carried out like this: We put the name of each individual in the population in a hat, mix the names well, and then draw as many names as we need for our sample. Of course "a hat" is just a metaphor. If our population is 100 million voters and we want to choose a simple random sample of 50,000, putting all the names in a real hat and then drawing 50,000 names one by one may not be such a good idea. The modern way to do any serious simple random sampling is by computer: Make a list of members of the population, enter it in the computer, and then let the computer randomly select the names.

While simple random sampling works well in many cases, for national surveys and polls the method has some serious practical problems. To begin with, it requires us to have a list of all the members of the population. As we noted at the beginning of the chapter, the population itself may not be clearly defined and, even when it is, a complete list of its members may not be available. Can we tag all wild elephants in Africa so that we can make a list for a computer? Better yet, is there a list of all people living in the United States? The answer to both of these questions is no.

Another serious problem in implementing simple random sampling on a national scale is expediency and cost. Interviewing several thousand people chosen by simple random sampling means chasing people all over the country. This

Simple random sampling.
(The New York State
Lottery)

requires an inordinate amount of time and money. For most public opinion polls, especially those that are done on a regular basis, the time and money needed to do this are simply not available.

Our next case study illustrates the sampling method currently used for polling public opinion.

CASE STUDY 3. STRATIFIED SAMPLING: MODERN PUBLIC OPINION POLLS

Present-day methods for conducting public opinion polls need to take into account two different sets of considerations: (1) minimizing sample bias, and (2) choosing a sample that is accessible in a cost-efficient, timely manner. A random sampling method that deals in a satisfactory way with both these issues is **stratified sampling**. The basic idea of stratified sampling is to break the population into categories called **strata** and then randomly choose a sample from these strata. The chosen strata are then further divided into categories called substrata, and a random sample is taken from these substrata. The selected substrata are further subdivided, a random sample is taken from them, and so on. The process goes on for a predetermined number of layers.

In public opinion polls, the strata and substrata are usually defined by criteria that involve a combination of geographic and demographic elements. For example, at the first level, the nation is divided into "size of community" strata (big cities, medium cities, small cities, villages, rural areas, etc.). Each of these strata is then subdivided by geographical region (New England, Middle Atlantic, East Central, etc.). Within each geographical region and within each size of community stratum, some communities are selected by simple random sampling. The selected communities (called *sampling locations*) are the only places where interviews will be conducted. To further randomize things, each of the selected sampling locations is subdivided into geographical units called *wards* and within each sampling location some of its wards are once again selected by simple random sampling. The selected wards are then divided into smaller units called *precincts*, and within each ward some of its precincts are selected by simple random sampling. At the last stage, *households* are selected for interviewing by simple random sampling within each precinct. The interviewers are then given specific instructions as to which households in their assigned area they must conduct interviews in, and the order which they must follow.

The efficiency of stratified sampling compared to simple random sampling in terms of cost and time is clear. The members of the sample are clustered in well-defined and easily manageable areas, significantly reducing the cost of conducting interviews as well as the response time needed to get the data together. At the same time, every single member of the population has an equal chance of being in the sample, which guarantees a small sample bias. Note, however, that this is not simple random sampling because not all groups of individuals have an

The first step in a public opinion poll: collecting the data.
(© 1991 The Gallup Organization, Inc., Princeton, New Jersey)

equal chance of being selected. (Two people living in the same precinct, for example, have a better chance of both being in the sample than two people living in different precincts.)

For a large, heterogeneous nation like the United States, stratified sampling has proven to be an effective and highly reliable method for collecting national data. Most major public opinion polls today use stratified sampling methods that result in sampling errors of less than 3%. Stratified sampling methods are also used with excellent results by the Bureau of Labor Statistics to collect extremely important statistical information for business and government such as data for the monthly Consumer Price Index (CPI) and the monthly unemployment figures given by the Current Population Survey (CPS). An excellent account of how stratified sampling is used for the CPI and CPS can be found in references 7 and 3 respectively.

THE GALLUP POLL

Design of the Sample

The design of the sample used by the Gallup Poll for its standard surveys of public opinion is that of a replicated area probability stratified sample down to the block level in the case of urban areas and to segments of townships in the case of rural areas.

After stratifying the nation geographically and by size of community in order to insure conformity of the sample with the 1990 Census distribution of the population, over 360 different sampling locations or areas are selected on a mathematically random basis from within cities, towns, and counties which have in turn been selected on a mathematically random basis. The interviewers have no choice whatsoever concerning the part of the city, town, or county in which they conduct their interviews.

Approximately five interviews are conducted in each randomly selected sampling point. Interviewers are given maps of the area to which they are assigned and are required to follow a specified travel pattern on contacting households. At each occupied dwelling unit, interviewers are instructed to select respondents by following a prescribed systematic method. This procedure is followed until the assigned number of interviews with male and female adults have been completed . . . .

Since this sampling procedure is designed to produce a sample which approximates the adult civilian population (18 and older) living in private households (that is, excluding those in prisons and hospitals, hotels, religious and educational institutions, and on military reservations) the survey results can be applied to this population for the purpose of projecting percentages into numbers of people. The manner in which the sample is drawn also produces a sample which approximates the population of private households in the United States. Therefore, survey results also can be projected in terms of numbers of households.

Sampling Error

In interpreting survey results, it should be borne in mind that all sample surveys are subject to sampling error, that is, the extent to which the results may differ from those that would be obtained if the whole population surveyed had been interviewed.

Choosing the sample for the Gallup poll. Source: *The Gallup Report*. Princeton, NJ: American Institute of Public Opinion, 1991.

**CASE STUDY 4.
COUNTING THE
UNCOUNTABLE:
THE U.S. CENSUS**

The most ambitious (and most expensive) data collecting project in the history of mankind was the 1990 U.S. census; before that it was the 1980 U.S. census, and before that it was the 1970 U.S. census.

Article 1, Section 2, of the U.S. Constitution mandates that a national census be conducted every 10 years. The original intent of the census was to "count heads" for a twofold purpose: taxes and political representation. Like everything else in the Constitution, Article 1, Section 2, was a compromise of many competing interests: The count was to exclude "Indians not taxed" and to count slaves as "three-fifths of a free Person." Today, the scope of the U.S. census has been expanded by the Fourteenth Amendment and the courts to count all persons physically present and permanently residing in the United States: citizens, legal residents, and illegal aliens.[6]

Besides counting heads, the modern U.S. census collects additional information about the population: sex, age, race, and ethnicity, marital status, housing, income, and employment. Surprisingly, some of this information is collected by means of a survey. In the 1990 U.S. census, two types of census forms were designed. Eighty-three percent of households received a short form consisting of twenty basic questions on sex, age, marital status, type of housing, etc.; the remaining 17% randomly selected households received longer forms asking for more detailed information on employment, family income, etc.

The importance and pervasiveness of U.S. census data in all walks of American life cannot be overestimated. The data collected by the U.S. census is used as the basis for

- The apportionment of seats to states in the House of Representatives

- The redrawing of legislative districts within each state

- The distribution of billions of federal tax dollars to states, counties, cities, and municipalities

- The collection of other official government statistics such as the Consumer Price Index and the Current Population Survey

- The strategic planning of production and services by business and industry.

Given the fact that there is no question about the value and importance of the U.S. census and given the tremendous resources put behind the effort (over 500,000 workers, 500 field offices, and a budget in excess of 2.5 billion dollars

---

[6] A new law implemented with the 1990 U.S. census requires that military personnel and other federal workers stationed overseas also be included in the census count.

were used for the 1990 U.S. census), one would expect it to be a resounding success. In fact, the exact opposite is true. The reliability of the data collected by the U.S. census has been challenged by several states, most large cities, professional statisticians, and even the public. The U.S. census has become, in the opinion of many, a national boondoggle.

We can understand a census being expensive; that is by definition one of its drawbacks. But inaccurate and unreliable? How can one possibly go wrong when the source of the data is the entire population? Unfortunately, the notion that all individuals living in the United States can be counted like pennies in a jar is totally out of tune with the times. In 1790, when the first U.S. census was carried out, the population was smaller and relatively homogeneous, people tended to stay in one place, and by and large felt comfortable in their dealings with the government. Under these conditions it might have been possible for census takers to accurately count heads.

Today's conditions are completely different. Getting an exact head count for a population as large and as diverse as that of the United States at present, where people are constantly on the move, where a significant number of people are trying to hide from the law, where there are hundreds of thousands of homeless, and where there is a general distrust of the government,[7] is a hopeless endeavor no matter how much money and effort one puts into it.

To make matters worse, the inability of the U.S. census to accurately count the population and measure the nation's demographics is definitely biased against the urban poor, who were undercounted by as much as 20% in some large cities. After the 1990 U.S. census, the cities of New York, Los Angeles, Chicago, Houston, the states of New York and California, the Leagues of Cities, the U.S. Conference of Mayors, the N.A.A.C.P., and the League of United Latin American Citizens were all parties to various lawsuits against the Census Bureau and its parent agency, the Commerce Department, claiming that the undercount of some ethnic minorities was depriving some cities and states of billions of dollars in federal funds. Most of these lawsuits have yet to be resolved, but regardless of the outcome the credibility of the data collected by the U.S. census has been irreparably damaged.

Based on the 1990 U.S. census, the official figure for the population of the United States is 248,709,873, but the spurious accuracy of such a figure is extremely misleading. Statisticians believe that this count could be off by as much as 5 million people and that the notion that the U.S. census can produce an exact head count of the population is unrealistic. Many statisticians now suggest that the constitutional requirement to count the nation's population should be implemented by means of a statistically designed survey based on a method similar to the one used by population biologists called the "capture-recapture method" (see Exercise 31 and reference 5).

---

[7]In some large cities the response rate for the 1990 U.S. census was under 50%.

CLINICAL STUDIES

All our examples so far have dealt with situations in which the issues centered around the question, What is the *source* of the data? Should it come from every member of the population, or from some selected sample? If the latter, should the sampling rate be small or large? Should the sample be chosen by human design or by chance?

Once we got to the source of the data all our examples pretty much assumed that the data itself was available to the observer in a direct and objective manner. If the election were held today, would you vote for candidate X or candidate Y? How many teenagers live in this household? Which TV programs did you watch last night? etc.

A very different and important type of data collection involves questions for which there is no clear, immediate answer. Is smoking hazardous to your health? Will taking aspirin reduce your chances of having a heart attack? Does regular class attendance in mathematics courses help improve SAT scores? All these questions have two things in common: (1) They involve a cause and an effect, and (2) the answers require observation over an extended period of time.

The standard approach for answering questions of this sort is to set up some sort of *study* or *experiment*. When one wants to know if a certain cause X produces a certain effect Y, one sets up an experiment in which cause X is produced and its effects are observed. If the effect Y is observed, then it is possible that X was indeed the cause of Y. The problem, however, is the nagging possibility that some other cause Z different from X sneaked in and produced the effect Y and that X had nothing to do with it.

Let's illustrate with an example. This one is fictitious but not all that far-fetched. Suppose we want to find out if too much chocolate in one's diet can increase one's chance of becoming diabetic. Here the cause X is eating too much chocolate, and the effect Y is diabetes. We set up an experiment in which 100 rats are fed a pound of chocolate a day for a period of six months. At the end of a six-month period, 15 of the 100 rats have diabetes. Since in the general rat population only 3% are diabetic, we are tempted to conclude that the diabetes in the rats is indeed caused by the excessive chocolate diet. The problem is that there is no certainty that the chocolate diet was the cause. Could there be another unknown reason for the observed effect? Could it be the lack of exercise of the confined rats? Could it be that it's not chocolate itself but rather the high calorie intake that caused the diabetes? Could it be a genetic predisposition of this particular group of rats?

For most cause-and-effect situations, especially those complicated by the involvement of human beings, a single effect can have many possible and actual causes. What causes heart attacks? Unfortunately, there is no single cause—diet, lifestyle, stress, and heredity are all known to be contributory causes. And the extent to which each of these causes contributes individually and the extent to which they interact with each other are extremely difficult questions that can be answered only by means of carefully designed statistical experiments.

For the remainder of this chapter we will illustrate an important type of experiment called a **clinical study**. Generally, clinical studies are concerned with determining whether a single variable or treatment (usually a vaccine, drug, therapy, etc.) can cause a certain effect (a disease, a symptom, a cure, etc.). The importance of such clinical studies is self-evident: Every new vaccine, drug, or treatment must "prove" itself by means of clinical study before it is officially approved for public use. Likewise, almost everything that is bad for us (cigarettes, caffeine, cholesterol, etc.) gets its "official" certification of badness by means of a clinical study. (A recent newspaper headline asking "Are Clinical Studies Bad for Your Health?" was only partly meant to be a joke.)

Properly designing a clinical study can be both difficult and controversial, and as a result we are often bombarded with conflicting information produced by different studies examining the same cause-and-effect question. The basic principles guiding a clinical study, however, are pretty much established by statistical practice and are almost always followed. We will discuss them next.

The first and most important issue in any clinical study is to isolate the cause (treatment, drug, vaccine, therapy, etc.) that is under investigation from all other possible contributing causes (called **confounding variables**) that could produce the same effect. The way that this is accomplished is by performing the experiment on two different groups: a **treatment group** and a **control group**. The treatment group receives the treatment, and the control group does not. The control group is there for *comparison* purposes only: If a cause-and-effect relationship exists, then the treatment group should show the effects of the treatment and the control group should not. The comparison is most effective when the treatment and control groups are identical to each other in all other respects (except that one group is receiving the treatment and the other one isn't). If this is accomplished and the groups show differences, then the differences can be safely attributed to the treatment. In addition, both the treatment and control groups should, as much as possible, be representative of the entire population to which the experiment applies.

Any experiment in which a cause-and-effect relationship is established by comparing the results in a treatment group with the results in a control group is called a **controlled** (or **comparative**) **experiment**. Our example of a controlled experiment is a famous clinical study carried out in 1954 to determine the effectiveness of a new vaccine against polio.

CASE STUDY 5. CONTROLLED EXPERIMENTS: THE 1954 SALK POLIO VACCINE FIELD TRIALS

Polio (infantile paralysis) has been practically eradicated in the western world. In the first half of the twentieth century, however, it was a major public health problem. Over one-half million cases of polio were reported between 1930 and 1950, and the actual number may have been considerably higher.

Because polio attacks mostly children and because its effects can be so serious (paralysis or death), eradication of the disease became a top public health priority in the United States. By the late 1940s it was known that polio is a virus and as such can best be treated by a vaccine which is itself made up of a virus. The vaccine virus can be a closely related virus that does not have the same harmful effects, or it can be the actual virus that produces the disease but which has been killed by a special treatment. The former is known as a *live-virus vaccine*, and the latter as a *killed-virus vaccine*. In response to either vaccine the body is known to produce *antibodies* which remain in the system and give the individual immunity against an attack by the real virus.

Both the live-virus and the killed-virus approaches have their advantages and disadvantages. The live-virus approach produces a stronger reaction and better immunity, but at the same time it is also more likely to cause a harmful reaction and in some cases even to produce the very disease it is supposed to prevent. The killed-virus approach is safer in terms of the likelihood of producing a harmful reaction, but it is also less effective in providing the desired level of immunity.

These facts are important because they help us understand the extraordinary amount of caution that went into the design of the experiment that tested the effectiveness of the polio vaccine. By 1953, several potential vaccines had been developed, one of the more promising of which was a killed-virus vaccine developed by Jonas Salk at the University of Pittsburgh. The killed-virus approach was chosen because there was a great potential risk in testing a live-virus vaccine in a large-scale experiment and a large-scale experiment was needed to collect enough information on polio (which in the 1950s had a rate of incidence among children of about 1 in 2000).

The testing of any new vaccine or drug creates many ethical dilemmas which have to be taken into account in the design of the experiment. With a killed-virus vaccine the risk of harmful consequences produced by the vaccine itself is small, so one possible approach could have been to distribute the vaccine widely among the population (ideally giving it to every child, but this was not possible because supplies were limited) and then follow up on whether there was a decline in the national incidence of polio in subsequent years. This is called the *vital statistics* approach and is the simplest way to test a vaccine. This is essentially the way the smallpox vaccine was determined to be effective. The problem with such an approach for polio is that polio is an epidemic type of disease, which means that there is a great variation in the incidence of the disease from one year to the next. In 1951, there were close to 60,000 reported cases of polio in the United States, but in 1952 the number of reported cases had dropped to almost half (about 35,000). Since no vaccine or treatment was used, the cause of the drop can only be attributed to the natural variability that is typical of epidemic diseases. If a totally ineffective vaccine had been tested in 1951, the observed effect—a significant drop in the incidence of polio in 1952—could have been incorrectly interpreted as a proof that the vaccine worked rather than attributed to the real cause. The serious consequences of such a mistake are obvious.

The final decision on how best to test the effectiveness of the Salk vaccine was left to an advisory committee of doctors, public officials, and statisticians convened by the National Foundation for Infantile Paralysis and the Public Health Service.[8] In order to isolate the cause under investigation (the Salk vaccine) from other possible causes of the desired effect (a reduction in the incidence of polio), it was decided that the experiment would be a controlled experiment involving a treatment group (those receiving the actual vaccine) and a control group (those receiving a shot of harmless salt solution). An experiment of this kind is called a **controlled placebo experiment**. The word "placebo" refers to the fact that the members of the control group receive a fake version of the treatment (a fake vaccine, a fake pill, etc.) which is called a **placebo**. The reasons for using placebos go back to our desire that the treatment and control groups be as equal as possible in all respects, except of course that one group is receiving the vaccine and the other one isn't. It is a well-known fact that just thinking that one is getting a treatment can actually affect the way the body responds, and this effect (called the **placebo effect**) is one of the primary reasons for using controlled placebo experiments to conduct many clinical studies.

It is obvious from our preceding discussion that in a controlled placebo experiment it is always desirable that neither the members of the treatment group nor the members of the control group know to which of the two groups they belong. When this is the case the experiment is called a **blind experiment**. It is also desirable that the scientists conducting the experiment not know which individuals are given the actual treatment and which are given the placebo. The purpose of this is to make the observation, analysis, and interpretation of the results of the experiment as impartial as possible. A controlled experiment in which neither the subjects nor the scientists conducting the experiment know which individuals are in the treatment group and which are in the control group is called a **double-blind experiment.**

Making the Salk vaccine experiment double-blind was particularly important because polio is not an easy disease to diagnose—it comes in many different forms and degrees. Sometimes it can be a borderline call, and if the doctor collecting the information had had prior knowledge of whether the subject had received the real vaccine or not, the diagnosis could have been subjectively tipped one way or the other.

With all this background we can now describe the actual details of the experiment. Approximately 750,000 children were randomly selected to participate in the study. Of these, about 340,000 declined to participate and another 8500 dropped out in the middle of the experiment. The remaining children were divided into two groups—a treatment group and a control group—with approximately 200,000 children in each group. The choice of which children were selected for the treatment group and which for the control group was made by *random selection*. (An experiment in which the treatment group and control

---

[8]Because of disagreements as to the best way to carry out the experiment, there were actually two different but parallel experiments carried out. In this case study we will discuss only one of these, the one conducted by the Public Health Service. A complete description of both experiments and the results is given in reference 2.

	Number of Children	Number of Reported Cases of Polio	Number of Paralytic Cases of Polio	Number of Fatal Cases of Polio
Treatment group	200,745	82	33	0
Control group	201,229	162	115	4
Declined to participate in the study	338,778	182[a]	121[a]	0[a]
Dropped out in the middle	8,484	2[a]	1[a]	0[a]
Total	749,236	428	270	4

[a] These figures are not a reliable indicator of the actual number of cases—they are only self-reported cases.

■■■ **TABLE 13-1**
(Adapted from Thomas Francis, Jr., et al., "An Evaluation of the 1954 Poliomyelitis Vaccine Trials—Summary Report," *American Journal of Public Health*, 45 (1955) 25, Table 2b.)

group are chosen by random selection is called a **randomized controlled experiment**.) Some of the figures and results of the experiment are shown in Table 13-1.

While Table 13-1 shows only a small part of the data collected by the Salk vaccine experiment, it can be readily seen that the difference between the treatment and control groups was significant and could rightfully be interpreted as a clear indication that the vaccine was indeed effective.

Based on the data collected by the 1954 field trials, a massive inoculation campaign was put into effect. Today, all children are routinely inoculated against polio,[9] and polio has essentially been eradicated in the United States. A statistically designed experiment played a key role in this breakthrough.

CONCLUSION

In this chapter we discussed different methods for collecting data. In principle, the most accurate way to collect data is by means of a *census*, a method that relies on collecting data from each member of the population. In most cases, because of considerations of cost and time, a census is a completely unrealistic strategy for collecting data. When data are collected from only a subset of the population (called a *sample*), the data collection method is called a *survey*. The most important rule in designing good surveys is to eliminate or minimize *sample bias*. Today, almost all strategies for collecting data are based on surveys in which the laws of chance are used to determine how the sample is selected, and these methods for collecting data are called *random sampling* methods. Random sampling is the best way known to minimize or eliminate sample bias. Two of

[9] The Salk vaccine, which was used for many years, has been replaced some years ago by the Sabin vaccine, a more effective oral vaccine based on the live-virus approach.

the most common random sampling methods are *simple random sampling* and *stratified sampling*. In some special situations other, more complicated types of random sampling can be used.

Sometimes identifying the sample is not enough. In cases in which cause-and-effect questions are involved, the data may come to the surface only after an extensive study has been carried out. In these cases isolating the cause variable under consideration from other possible causes (called *confounding variables*) is an essential prerequisite for getting reliable data. The standard strategy for doing this is a *controlled experiment* in which the sample is broken up into a *treatment group* and a *control group*. Controlled experiments are used (and sometimes abused) in almost all clinical and scientific studies carried out today. We can thank this area of statistics for many breakthroughs in social science, medicine, and public health, as well as for the many depressing health and diet reports about all the things that are supposed to be bad for us but which we can't seem to live without.

KEY CONCEPTS

blind experiment
census
chance error
clinical study
confounding variable
control group
controlled experiment
controlled placebo experiment
data
double-blind experiment
nonresponse bias
parameter
placebo
placebo effect
population
quota sampling

randomized controlled experiment
random sampling
response rate
sample
sample bias
sample variability
sampling error
sampling rate
selection bias
simple random sampling
statistic
strata
stratified sampling
survey
treatment group

EXERCISES

Walking

Exercises 1 through 4 refer to the following survey. In 1988 "Dear Abby" asked her readers to let her know whether they had cheated on their spouses or not. The readers' responses are summarized in the accompanying table.

	Women	Men
Faithful	127,318	44,807
Unfaithful	22,468	15,743
Total	149,786	60,550

Based on the results of this survey, Dear Abby concluded that the amount of cheating among married couples is much less than people believe (in her words, " . . . the results were astonishing. There are far more faithfully wed couples than I had surmised.")

1. (a) Describe as specifically as you can the population for this survey.
   (b) What was the size of the sample?
   (c) How was the sample chosen?
   (d) Eighty-five percent of the women who responded to this survey claimed to be faithful. Is the number 85% a parameter? A statistic? Neither? Explain your answer.

2. (a) Explain why this survey was subject to selection bias.
   (b) Explain why this survey was subject to nonresponse bias.

3. (a) Based on the Dear Abby data, estimate the percentage of married men who are faithful to their spouses.
   (b) Based on the Dear Abby data, estimate the percentage of married people who are faithful to their spouses.
   (c) How accurate do you think these estimates are? Explain.

4. If money were no object, could you devise a survey that might give more reliable results than the Dear Abby survey? Describe briefly what you would do.

Exercises 5 through 8 refer to the following hypothetical situation. The Cleansburg Planning Department is trying to determine what percent of the people in the city want to spend public funds to revitalize the downtown mall. In order to do so, they decide to conduct the following survey: Five professional interviewers (A, B, C, D, and E) are hired, and each is asked to pick a street corner of their choice within the city limits. Everyday between 4:00 and 6:00 P.M. the interviewers are to ask each passerby if he or she wishes to respond to a survey sponsored by Cleansburg City Hall and make a record of their response. If the response is yes, the person is asked to respond to the next question: Are you in favor of spending public funds to revitalize the downtown mall? Yes or no? The interviewers are asked to return to the same street corner as many days as are necessary until each one has conducted a total of 100 interviews. The data collected are seen in Table 13-2.

Interviewer	Yes[a]	No[b]	Nonrespondents[c]
A	35	65	321
B	21	79	208
C	58	42	103
D	78	22	87
E[d]	12	63	594

[a] In favor of spending public funds to revitalize the downtown mall.

[b] Opposed to spending public funds to revitalize the downtown mall.

[c] Declined to be interviewed.

[d] Got frustrated and quit.

■ TABLE 13-2

5. **(a)** Describe as specifically as you can the population for this survey.
   (b) What is the size of the sample?

6. **(a)** Calculate the response rate in this survey.
   (b) Explain why this survey was subject to nonresponse bias.

7. **(a)** Can you explain the big difference in the data from interviewer to interviewer?
   (b) One of the interviewers conducted the interviews at a street corner downtown. Which one? Explain.
   (c) Do you think the survey was subject to selection bias? Explain.
   (d) Was the sampling method used in this survey the same as quota sampling? Explain.

8. **(a)** Do you think this was a good survey? If you were a consultant to the Cleansburg Planning Department, could you suggest some improvements? Be specific.

Exercises 9 through 12 refer to the following survey. The dean of students at Tasmania State University wants to determine the percent of undergraduates living at home during the current semester. There are 15,000 undergraduates at TSU, so it is decided that the cost of checking with each and every one would be prohibitive. The following method is proposed to choose a representative sample of undergraduates to interview: Start with the registrar's alphabetical listing containing the names of all undergraduates. Pick randomly a number between 1 and 100 and count that far down the list, taking the name and every 100th name after it. (For example, if the random number chosen is 73, then pick the 73rd, 173rd, 273rd, etc., names on the list.) Assume the survey has a response rate of 0.95.

9. **(a)** Describe the population for this survey.
   (b) Give the exact value of N.

10. **(a)** Find the size n of the sample.
    (b) Find the sampling rate.

11. **(a)** Was this survey subject to selection bias? Explain.
    (b) Explain why the method used for choosing the sample is not simple random sampling.

12. Do you think the results of this survey will be reliable? Explain.

Exercises 13 through 16 refer to the following hypothetical study. The manufacturer of a new vitamin (vitamin X) decides to sponsor a study to determine the effectiveness of vitamin X in curing the common cold. Five hundred college students in the San Diego area who are suffering from colds are paid to participate as subjects in this study. They are all given two tablets of vitamin X a day. Based on information provided by the subjects themselves, 457 out of the 500 subjects are cured of their colds within 3 days. The average number of days a cold lasts is 4.87 days. As a result of this study, the manufacturer launches an advertising campaign claiming that "vitamin X is more than 90% effective in curing the common cold."

13. **(a)** Describe as specifically as you can the population for this study.
    (b) How was the sample selected?
    (c) What was the size n of the sample?
    (d) Was this health study a controlled experiment?

14. **(a)** Do you think the placebo effect could have played a role in this study?
    (b) List three possible causes other than the effectiveness of vitamin X itself that could have confounded the results of this study.

15. List four different problems with this study that indicate poor design.

16. Make some suggestions for improving the study.

Exercises 17 through 20 refer to the following. A study by a team of Harvard University scientists [Science News, 138, no. 20 (November 17, 1990), 308] found that regular doses of beta carotene (a nutrient common in carrots, papayas, and apricots) may help prevent the buildup of plaque-producing arteriosclerosis (clogging of the arteries), which is the primary cause of heart attacks. The subjects in the study were 333 volunteer male doctors, all of whom had shown some early signs of coronary artery disease. The subjects were randomly divided into two groups. One group was given a 50-milligram beta carotene pill every other day for six years, and the other group was given a similar-looking placebo pill. The study found that the men taking the beta carotene pills suffered 50% fewer heart attacks and strokes than the men taking the placebo pills.

17. Describe as specifically as you can the population for this study.

18. (a) Describe the sample.
   (b) What was the size *n* of the sample?
   (c) Was the sample chosen by random sampling? Explain.

19. (a) Explain why this study can be described as a controlled placebo experiment.
   (b) Describe the treatment group in this study.
   (c) Explain why this study can be described as a randomized controlled experiment.

20. (a) Mention two possible confounding variables in this study.
   (b) Carefully state what a legitimate conclusion from this study might be.

Exercises 21 through 24 refer to the following hypothetical situation. A college professor has a theory that a dose of about 10 milligrams of caffeine a day can actually improve students' performance in their college courses. To test his theory, he chooses the 13 students in his Psychology 101 class that got an "F" in the first midterm and asks them to come to his office three times a week for "individual tutoring." When the students come to his office, he engages them in friendly conversation, while at the same time pouring them several cups of strong coffee. After a month of doing this, he observes that of the 13 students, 8 show significant improvement in their second midterm scores; 3 show some improvement, and 2 show no improvement at all. Based on this, he concludes that his theory about caffeine is correct.

21. Which of the following terms best describes the professor's study: (i) randomized controlled experiment, (ii) double-blind experiment, (iii) controlled placebo experiment, or (iv) clinical study? Explain your choice and why you ruled out the other choices.

22. (a) Describe the population and the sample of this study.
   (b) What was the value of *n*?
   (c) Which of the following percentages best describes the sampling rate for this study: (i) 10%, (ii) 1%, (iii) 0.1%, (iv) 0.01%, or (v) less than 0.01%? Explain.

23. List at least three possible causes other than caffeine that could have confounded the results of this study.

24. Make some suggestions to the poor professor as to how he might improve the study.

Jogging

25. Informal surveys. In everyday life, we are constantly involved in activities that can be described as *informal surveys*, often without even realizing it. Here are some examples:

(i) Al gets up in the morning and wants to know what kind of day it is going to be, so he peeks out the window. He doesn't see any dark clouds, so he figures it's not going to rain.

(ii) Betty takes a sip from a cup of coffee and burns her lips. She concludes the coffee is too hot and decides to add a tad of cold water to it.

(iii) Carla goes to the doctor to have a checkup. The nurse draws 5 ml. of blood from Carla's right arm and sends it to the lab. The lab report comes out negative for all diseases tested.

For each of the above examples,
(a) Describe the population.
(b) Discuss whether the sample is random or not.
(c) Discuss the validity of the conclusions drawn in each case. (There is no right or wrong answer to this question, but you should be able to make a reasonable case for your position.)

26. Read the examples of informal surveys given in Exercise 25. Give three more examples of your own. Make them as different as possible from the ones given in Exercise 25. (Changing coffee to tea or soup in (ii) is not acceptable.)

27. Leading question bias. In many surveys, the way the questions in the survey are phrased can itself be a source of bias. When a question is worded in such a way as to predispose the respondent to provide a particular response, the results of the survey are tainted by a special type of bias called leading question bias. The following is an extreme hypothetical situation intended to drive the point home.

The American Self-Righteous Institute is a conservative think tank. In an effort to find out how the American taxpayer feels about a tax increase, the institute conducts a "scientific" poll. The main question in the poll is phrased as follows:

Are you in favor of paying higher taxes to bail the federal government out of its disastrous economic policies and its mismanagement of the federal budget? Yes____ No____ .

Ninety-five percent of the respondents answered no. The results of the survey are announced by the sponsors with the statement:

Public opinion polls show that 95% of American taxpayers oppose a tax increase.

(a) Explain why the results of this survey might by invalid.
(b) Rephrase the question in a neutral way. Pay particular attention to "highly charged" words.
(c) Make up your own (more subtle) example of leading question bias. Analyze the critical words that are the cause of bias.

28. Consider the following hypothetical survey designed to find out what percentage of people cheat on their income taxes. Fifteen hundred taxpayers are randomly selected from the Internal Revenue Service (IRS) rolls. These individuals are then interviewed in person by representatives of the IRS and read the following statement:

This survey is for information purposes only. Your answer will be held in strict confidence. Have you ever cheated on your income taxes? Yes____ No____.

Twelve percent of the respondents answered yes.

(a) Explain why the above figure might be unreliable.

(b) Can you think of ways in which a survey of this type might be designed so that more reliable information could be obtained? In particular, discuss who should be sponsoring the survey and how the interviews should be carried out.

29. Listing bias. Today, most consumer marketing surveys are conducted by telephone. In selecting a sample of households that are representative of all the households in a given geographical area the two basic techniques used are (i) randomly selecting telephone numbers to call from the local telephone directory or directories, and (ii) using a computer to randomly generate seven-digit numbers to try that are compatible with the local phone numbers.

(a) Briefly discuss the advantages and disadvantages of each of the two techniques. In your opinion, which of the two techniques will produce the more reliable data? Explain.

(b) Suppose that you are trying to market burglar alarms in New York City. Which of the two techniques for selecting the sample would you use? Explain your reasons.

30. The following two surveys were conducted in January 1991 in order to assess how the American public viewed media coverage of the Persian Gulf war.

Survey 1 was an Area Code 900 telephone poll survey conducted by "ABC News." Viewers were asked to call a certain 900 number if they felt the media was doing a good job of covering the war, and a different 900 number if they felt the media was not doing a good job in covering the war. Each call cost 50 cents. Of the 60,000 respondents, 83% felt the media was not doing a good job.

Survey 2 was a telephone poll of 1500 randomly selected households across the United States conducted by the Times-Mirror survey organization. In this poll 80% of the respondents indicated that they approved of the press coverage of the war.

(a) Briefly discuss survey 1, indicating any possible types of bias.

(b) Briefly discuss survey 2, indicating any possible types of bias.

(c) Can you explain the discrepancy between the results of the two surveys?

(d) In your opinion, which of the two surveys gives the more reliable data?

31. The capture-recapture method. For many wildlife populations (including in some cases humans), it is difficult, if not impossible, to get an exact count of the size of the population. To estimate the size of wildlife populations, population biologists often use a technique called the *capture-recapture method*. The basic idea of this method is to first capture a sample of size n_1, tag the animals without harming them, and let them go. After an appropriate amount of time a new sample of size n_2 is captured, and the number of tagged animals is counted (we'll call this number k). Under the right conditions, the size of the entire population can be estimated from the values of n_1, n_2, and k. The purpose of this exercise is to illustrate how this method works.

(a) You have a small pond stocked with fish and want to estimate how many fish there are in the pond. Let's suppose that you capture $n_1 = 500$ fish, tag them, and throw them back in the pond. After a couple of days you go back to the pond and capture $n_2 = 120$ fish, of which $k = 30$ are tagged. Using these values of n_1, n_2, and k, estimate the fish population N in the pond.

(b) Give a formula that allows us to estimate N based on the numbers n_1, n_2, and k.

(c) For the capture-recapture method to give a reasonable estimate of N, list all the assumptions you think should be made about the two samples.

(d) Give reasons why in many situations the assumptions in (c) may not hold true.

(e) The following real example is based on data given in D. G. Chapman and A. M. Johnson, "Estimation of Fur Seal Pup Populations by Randomized Sampling," *Transactions of the American Fisheries Society,* 97 (July 1968), 264–270. To estimate the population of fur seal pups in a rookery, 4965 fur seal pups were captured and tagged in early August. In late August, 900 fur seal pups were captured. Of these, 218 had been tagged. Based on these figures, estimate the population of fur seal pups in the rookery to the nearest hundred.

32. (Open-ended question.) Consider the following hypothetical situation. A potentially effective new drug for treating AIDS patients must be tested by means of a clinical study. Based on experiments conducted with laboratory animals, the drug appears to be extremely effective in treating the more serious effects of AIDS, but it also appears to have caused many side effects, including serious kidney disorders in about 20% of the laboratory animals tested.

(a) Discuss the ethical and moral issues you think should be considered in designing a clinical study to test this drug.

(b) Taking into account the ethical and moral issues discussed in (a), describe how you would design a clinical study for this new drug. (In particular, how would you choose the participants in the study, the treatment and the control groups, etc.)

REFERENCES AND FURTHER READINGS

1. Anderson, Margo, "According to their Respective Numbers . . . for the Twenty-First Time," *Chance,* 3 (Winter 1990), 12–18.

2. Francis, Thomas, Jr., et al., "An Evaluation of the 1954 Poliomyelitis Vaccine Trials—Summary Report," *American Journal of Public Health*, 45 (1955), 1–63.

3. Freedman, D., R. Pisani, R. Purves, and A. Adhikari, *Statistics (2nd ed.).* New York: W. W. Norton, Inc., 1991, chaps. 19 and 20.

4. Gallup, George, *The Sophisticated Poll Watchers Guide.* Princeton, NJ: Princeton Public Opinion Press, 1972.

5. Glieck, James, "The Census: Why We Can't Count," *New York Times Magazine* (July 15, 1990), 22–26, 54.

6. Hansen, Morris, and Barbara Bailar, "How to Count Better: Using Statistics to Improve the Census," in *Statistics: A Guide to the Unknown (3rd ed.),* ed. Judith M. Tanur, et al. Belmont CA: Wadsworth, Inc., 1989, 208–217.

7. McCarthy, Philip J., "The Consumer Price Index," in *Statistics: A Guide to the Unknown (3rd ed.)*, ed. Judith M. Tanur, et al. Belmont, CA: Wadsworth, Inc., 1989, 198–207.

8. Meier, Paul, "The Biggest Public Health Experiment Ever: The 1954 Field Trial of the Salk Poliomyelitis Vaccine," in *Statistics: A Guide to the Unknown (3rd ed.)*, ed. Judith M. Tanur, et al. Belmont, CA: Wadsworth, Inc., 1989, 3–14.

9. Mosteller, F., et al., *The Pre-election Polls of 1948.* New York: Social Science Research Council, 1949.

10. Parten, Mildred, *Surveys, Polls and Samples.* New York: Harper and Row, 1950.

11. Paul, John, *A History of Poliomyelitis.* New Haven, CN: Yale University Press, 1971.

12. Scheaffer, R. L., W. Mendenhall, and L. Ott, *Elementary Survey Sampling.* Boston: PWS-Kent, 1990.

13. Stephan, F. F., and P. J. McCarthy, *Sampling Opinions: An Analysis of Survey Procedure.* New York: John Wiley & Sons, Inc., 1958.

14. Warwick, D. P., and C. A. Lininger, *The Sample Survey: Theory and Practice.* New York: McGraw-Hill Book Co., 1975.

15. Yates, Frank, *Sampling Methods for Censuses and Surveys.* New York: Macmillan Publishing Co., Inc., 1981.

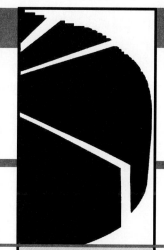

CHAPTER 14

Graphing and Summarizing Data

Descriptive Statistics

It is a proof of high culture to say the greatest matters in the simplest way.
RALPH WALDO EMERSON

In Chapter 13 we discussed the ins and outs of collecting statistical data. And once the data have been collected, what should we do with all of these numbers?

One of the primary purposes of collecting data is to use the data to communicate some sort of a message, a message whose purpose is usually to inform and occasionally to persuade. It is a bit ironic, therefore, that after all the care and attention to detail invested in collecting data, the typical situation one is confronted with is that of having *too much* data. Having too much data presents a peculiar dilemma. There is a point past which the human mind cannot absorb or comprehend the information presented because the amount of data is much too large. There are two ways to deal with this dilemma: One is to present the data in the form of graphs (we can usually understand pictures better than numbers); the other is to use numerical summaries that serve as "snapshots" of the data. Sometimes we even combine the two strategies, using pictures and numerical summaries together.

In this chapter we will discuss all of these approaches to describing data. This area of statistics is called **descriptive statistics**. The main purpose of descriptive statistics is to convey the information embodied by the data in an efficient and accurate way. No more and no less.

GRAPHICAL REPRESENTATION OF DATA

The old saying "A picture is worth a thousand words" is particularly appropriate in descriptive statistics, where a picture can be worth thousands of individual pieces of data. (Statisticians call these individual pieces of data **data values** or **data points**.)

We will start our discussion of *graphical descriptions* of data with a hypothetical situation not unlike one that can routinely be found on every college campus across the land.

Example 1. (Blackbeard's Stat 101 test scores). Professor Groucho Blackbeard has a well-deserved reputation at Tasmania State University for tough (some say brutal) exams. The results of the most recent midterm exam in his Stat 101 course are shown in Table 14-1.

ID	Score	ID	Score	ID	Score	ID	Score	ID	Score
1257	12	2651	10	4355	8	6336	11	8007	13
1297	16	2658	11	4396	7	6510	13	8041	9
1348	11	2794	9	4445	11	6622	11	8129	11
1379	24	2795	13	4787	11	6754	8	8366	13
1450	9	2833	10	4855	14	6798	9	8493	8
1506	10	2905	10	4944	6	6873	9	8522	8
1731	14	3269	13	5298	11	6931	12	8664	10
1753	8	3284	15	5434	13	7041	13	8767	7
1818	12	3310	11	5604	10	7196	13	9128	10
2030	12	3596	9	5644	9	7292	12	9380	9
2058	11	3906	14	5689	11	7362	10	9424	10
2462	10	4042	10	5736	10	7503	10	9541	8
2489	11	4124	12	5852	9	7616	14	9928	15
2542	10	4204	12	5877	9	7629	14	9953	11
2619	1	4224	10	5906	12	7961	12	9973	10

■ TABLE 14-1   **Stat 101 midterm exam scores (25 points possible).** $N = 75$.

In this example we have a population of size $N = 75$ (the students taking the exam). Each of the 75 students is identified by a student ID number (rights of privacy prohibit the use of names), and the data values are the midterm scores (whole numbers between 0 and 25) shown to the right of each student's ID number. The collection of all the data values in a specific statistical problem is called the **data set**. Thus, the data set for Example 1 consists of the set of all test scores. (Note that in this example, the student IDs are not data.) We will use the data set in this example several times in this and later chapters. For ease of reference we will call it the *Stat 101 data set*.

Like students everywhere, students in Professor Blackbeard's Stat 101 class have two questions foremost on their minds regarding the exam: (1) How did I do? and (2) How did the class as a whole do? The answer to question 1 can be found directly in Table 14-1, but the answer to question 2 requires a little extra effort. How can all the information given by Table 14-1 be packaged into a single intelligible whole? Let us count the ways.

Bar Graphs and Variations Thereof

Our first approach is to put the scores into a **frequency table** as shown in Table 14-2.

Exam score	1	6	7	8	9	10	11	12	13	14	15	16	24
Frequency	1	1	2	6	10	16	13	9	8	5	2	1	1

■ TABLE 14-2   **Frequency Table for the Stat 101 Data Set.**

The number below each score represents the **frequency** of
is, the number of students getting that score. In this example the
dents with a score of 10, 2 students with a score of 15, etc. Note th
are no students getting a particular score (i.e., the frequency is 0), we can omit
the score from the table.

While Table 14-2 is a considerable improvement over Table 14-1 visually,
it leaves something to be desired. Figure 14-1 shows the same information in a
much more visually striking way called a **bar graph**.

In a bar graph the frequencies are displayed by the *heights* of the columns.
This has the decided advantage that one can see in one fell swoop the overall
picture. Consider, for example, the problem of detecting **outliers**. An outlier is a
data value that stands out from the crowd, that is to say, a value that is notice-
ably larger or smaller than the rest of the data. One of the advantages of a graph
such as the one in Fig. 14-1 is that outliers can be easily picked up by the naked
eye. In this case there are two students, one with an abnormally low score (1)
and one with an abnormally high score (24), and they are both outliers. Notice,
however, that the bar graph does not identify who the outliers are (to find this
out Professor Blackbeard would have to go back to Table 14-1).

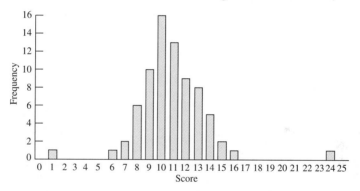

FIGURE 14-1 Bar graph for the Stat 101 data set.

Sometimes it is more convenient to show frequencies as percentages of the
total population rather than as exact counts. This is particularly true when we are
dealing with very large data sets. Figure 14-2 shows a bar graph for the Stat 101
data set in which the column height for each score represents the percentage of
the class getting that score. The notation on the vertical axis clearly indicates

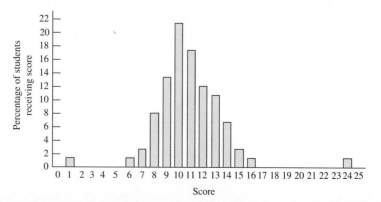

FIGURE 14-2 Bar graph for the Stat 101 data set using percentages of the population. $N = 75$.

that we are dealing with percentages rather than total counts. The change from total counts to percentages does not change the shape of the graph—it is basically a change in scale.

While the term "bar graph" is most commonly used for graphs like the ones in Figs. 14-1 and 14-2, devices other than bars can be used to add a little extra flair or to subtly influence the content of the information given by the raw data. Professor Blackbeard, for example, may have chosen to display the test data using a graph like the one shown in Fig. 14-3, which gives the observer all the information given by the more staid version (Fig. 14-1) and at the same time sends a subtle little message about what he thought of the students' performance.

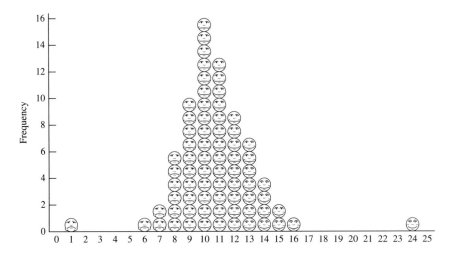

FIGURE 14-3 Frequency chart for the Stat 101 data set.

The general point here is that a bar graph is often used not only to inform but also to impress and persuade, and in such cases a clever design for the frequency columns can be more effective than just a bar. It is for this reason that we prefer to use the term **frequency chart** for graphs such as the ones in Fig. 14-3 and Fig.14-4 in the next example.

Example 2. Figure 14-4 is a frequency chart showing the "impressive" growth in the yearly sales of the XYZ Corporation over the period from 1986 to 1991.

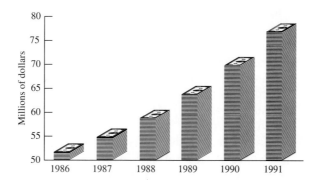

FIGURE 14-4 Frequency chart. XYZ Corporation annual sales (in millions of dollars).

Figure 14-5 is a more sobering bar graph displaying the same information in a more realistic way. The noticeable difference between the two graphs can be attributed to several factors: (1) the choice of the starting value on the vertical axis, (2) the choice of scale on the vertical axis, and (3) the use of graphical devices (the stacks of bills).

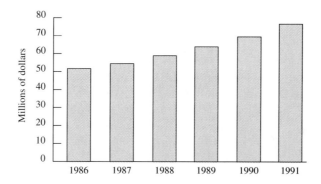

FIGURE 14-5 Bar graph. XYZ Corporation annual sales (in millions of dollars).

The point of Example 2 is that the graphic display of data is as much an art as it is a science and the impact that a graph can have on its intended audience is not only a function of the data values themselves but also of the way they are presented. This is what makes descriptive statistics so challenging and at the same time so dangerous. The line separating objectivity and propaganda is particularly fine in this area of statistics.

VARIABLES: QUANTITATIVE AND QUALITATIVE; CONTINUOUS AND DISCRETE

Before we continue with our discussion of graphs, we need to briefly analyze the concept of a **variable**. In statistical usage, a variable is any characteristic that varies with the members of a population. The students in Professor Blackbeard's Stat 101 course (the population) do not all perform equally on the exam. Thus, the *test score* is a variable which in this particular example is a whole number between 0 and 25. In some instances, such as when the instructor gives partial credit or when there is subjective grading, a test score may be a variable that takes on a fractional value such as $18\frac{1}{2}$ or even $18\frac{1}{4}$. Even in these cases, however, the possible increments for the values of the variable are given by some minimum amount: a quarter-point, a half-point, whatever. In contrast to this situation, consider a different variable, the *length of time* it takes a student to complete the exam. In this case the variable can take on values that differ by arbitrary small increments—a second, a tenth of a second, a hundredth of a second, etc.

When a variable represents a measurable quantity, it is called a **quantitative or numerical variable**. When the difference between the values of a quantitative variable can be arbitrarily small, we call the variable a **continuous variable**; whereas when possible values of the quantitative variable change by minimum increments, the variable is called a **discrete variable**. Examples of discrete variables are IQ, pulse, shoe size, family size, number of automobiles owned, and points scored in a basketball game. Examples of continuous variables are height, weight, foot size (as opposed to shoe size), and the time it takes to run a mile.

Sometimes the distinction between continuous and discrete variables is blurred in the real world. Height, weight, and age are all continuous variables in theory, but in practice they are frequently rounded off to the nearest inch, ounce, and year (or month in the case of babies) respectively, at which point they become discrete variables. On the other hand, money, which is in theory a discrete variable (the difference between two values of this variable cannot be less than a penny) is almost always thought of as continuous because in most real-life situations, a penny can be thought of as an infinitesimally small amount of money.

Variables can also describe characteristics that cannot be measured numerically: nationality, sex, hair color, brand of automobile owned, etc. Variables of this type are called **qualitative**.[1]

In some ways, qualitative variables must be treated differently from quantitative variables: They cannot, for example, be added, multiplied, or averaged. In other ways, qualitative variables can be treated very much like discrete quantitative variables, particularly when it comes to graphical descriptions such as bar graphs and frequency charts.

Example 3. The bar graph in Fig. 14-6 shows undergraduate enrollments at Tasmania State University by school. In this example the variable being described is the school in which the student is enrolled.

---

[1] Another frequently used name is **categorical variable**.

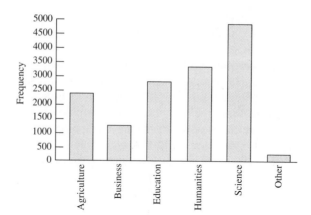

FIGURE 14-6
Undergraduate enrollments at Tasmania State University by school. $N = 15,000$.

The bar graph in Fig. 14-7 gives exactly the same information in percentages of the total undergraduate student population. Conceptually, there is hardly any difference between these two graphs and the graphs in Figs. 14-1 and 14-2 other than the fact that the **categories** (sometimes also called **classes**) that describe the "values" of the data are not numbers.

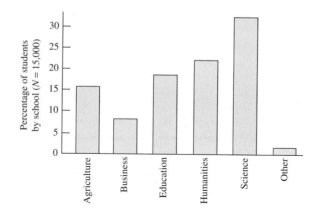

FIGURE 14-7 Percentage of undergraduate population at Tasmania State University by school.

The category labeled "Other" is one that frequently appears on bar graphs for qualitative data and deserves special attention. "Other" is used as a convenient place to put members of the population that do not fit neatly into one of the main categories (in Example 3 it is students with undeclared majors) or a place to lump together several categories that contain fractions of the population that are too small to be listed on their own.

When the number of categories is small, as it is in Example 3, another commonly used way to describe relative frequencies of a population by categories is the **pie chart**. In a pie chart, the "pie" represents the entire population (100%), and the "slices" represent the categories or classes, with the size (area) of each slice being proportional to the relative frequency of the corresponding category. Some relative frequencies such as 50%, 25%, etc, are very easy to describe; but how do we accurately draw the slice corresponding to a more complicated frequency—say 32.47%? Here a little high school geometry comes in handy. Since the area of each slice is proportional to the angle at the center, the problem boils down to finding the appropriate angle for the slice. To do this, we use the fact that 100% equals 360°, which means that 1% is given by 360°/100 = 3.6°. It follows that a frequency such as 32.47% is given by (32.47)(3.6°) = 117° (rounded to the nearest degree, which is generally good enough for most practical purposes).

Figure 14-8 shows a pie chart with exactly the same information as the bar graph in Fig. 14-7. Note that the value of N (the size of the entire population) is listed along with the pie chart itself. This allows the reader to know the total from which the percentages in the pie chart are taken and, if necessary, to convert these percentages to actual counts. When the size of the population N is available, it is good statistical practice to provide this information along with the pie chart.

Bar graphs and pie charts are an excellent way to graphically display qualitative data, but, as always, we should be wary of jumping to hasty conclusions based on what we see on a graph. Our next example illustrates this point.

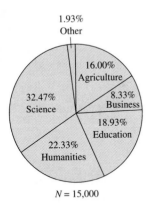

FIGURE 14-8 Percentage of undergraduate population at Tasmania State University by school.

Example 4. (Who's watching the boob tube tonight?) According to Nielsen Media Research data, the percentage of TV audience watching TV during prime time (8:00 P.M. to 11:00 P.M.), broken up by age group is: adults (18 years and over), 83%; teenagers (12–17 years), 7%; children (2–11 years), 10%.[2] The pie chart in Fig. 14-9 shows this breakdown of audience composition by age group.

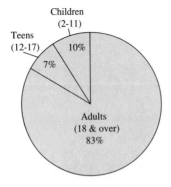

FIGURE 14-9 Audience composition for prime-time TV viewership by age group.

---

[2] These figures are rough approximations based on information taken from the *World Almanac* and averaged over several years. The exact figures vary from year to year.

When looking at this pie chart, one is tempted to conclude during prime time, children and teenagers do not watch much TV. reports we read about how much young people watch TV be wrong?

The problem with a pie chart such as the one in Fig. 14-9 is that while accurate, it is also very misleading: Children (2–11 years) make up only 15% of the population at large, teens (12–17) make up only 8% of the population, and adults make up the rest. Given that there are more than 5 times as many adults as there are children, is it any wonder that there are more prime-time TV-viewing adults than there are prime-time TV-viewing children? Likewise, in absolute terms there are more TV-viewing children than teenagers, but that is only because there are almost twice as many children as there are teenagers. In relative terms, a higher percentage of teenagers (taken out of the total teenage population) watches prime-time TV than do children. (This is not all that surprising, given that most children's bedtime is around 8:00 P.M.)

The moral of this example is that using absolute percentages as we did in Fig. 14-9 can be quite misleading. When comparing characteristics of a population that is broken up into categories it is essential to take into account the relative sizes of the various categories.

Class Intervals

While the distinction between qualitative and quantitative data is important in many aspects of statistics, when it comes to deciding how best to display graphically the frequencies of a population, a critical issue is the number of categories into which the data can fall. When the number of categories is too big (say, for example, in the hundreds), then a bar graph or frequency chart can become too muddled and thereby be rendered ineffective. With qualitative data this is generally not a problem, but with quantitative data this problem does come up quite frequently: Both continuous and discrete variables can take on infinitely many values, and even when they don't, the number of values can be too large for any reasonable graph.

Example 5. Suppose that, as part of a special research project, we want to look at the cumulative SAT test scores for the population of students discussed in Example 1 (those in Professor Blackbeard's Stat 101 course). Just as in Example 1, our data represents a discrete quantitative variable (in this case cumulative SAT scores). While in theory the situation is no different from that in Example 1, in practice, because of the extremely large number of possible SAT scores (cumulative SAT scores are given in 10-point increments and range between 400 and 1600), we must deal with such data differently. The standard way to display frequency graphs in this situation is to break up the range of scores into intervals called **class intervals**. The decision as to how the class intervals are defined and how many there should be is a matter of personal choice. In this example, a sensible thing to do might be to break up the SAT scores into twelve class intervals. In this case our bar graph would look something like Fig. 14-10.

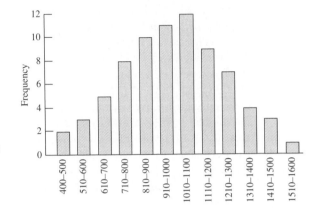

FIGURE 14-10 Cumulative SAT scores for $N = 75$ students in Professor Blackbeard's Stat 101 course.

Note that in Example 5 we made it a point to create class intervals of the same size,[3] and this should be done as much as possible. Sometimes, however, it might make more sense to define class intervals of different lengths, as illustrated by our next example.

Class Interval	Grade
18–25	A
14–17	B
11–13	C
9–10	D
0–8	F

■ **TABLE 14-3**

Example 6. The moment of truth for Professor Blackbeard's Stat 101 students has arrived! It is now time to turn the numerical scores on the exam into letter grades. In our terminology, this entails converting a quantitative variable (test score) into a qualitative one (letter grade) by defining class intervals associated with each grade category. In this case there is a good reason not to use class intervals of equal length. Following his own mysterious way of doing things, Professor Blackbeard defines the class intervals for this particular exam according to the breakdown shown in Table 14-3.

If we combine the Stat 101 exam scores in Table 14-2 with the class intervals for grades as defined in Table 14-3, we will get a new frequency table (Table 14-4) and a corresponding bar graph for the grade distribution in the exam (Fig. 14-11).

Grade	F	D	C	B	A
Frequency	10	26	30	8	1
Percentage	13.33%	34.67%	40%	10.67%	1.33%

■ **TABLE 14-4**

[3] A tiny exception was made for the class interval 400–500, which has one more possible test score than the others.

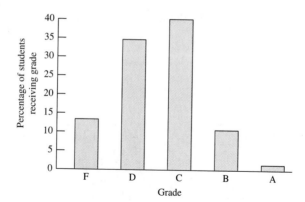

FIGURE 14-11 Grade distribution for the Stat 101 data set.

Histograms

When a quantitative variable is continuous, then the possible values of the variable can vary by infinitesimally small increments. As a consequence of this, there are no gaps between the class intervals, and our old way of doing things (using black columns or other types of cute stacks separated by gaps for clear viewing) will no longer work. In this case we will use a variation of a bar graph called a **histogram**. We illustrate the concept of a histogram in the next example.

Example 7. Suppose we want to use a graph to display the distribution of starting salaries for last year's graduating class at Tasmania State University.

The starting salaries of the $N = 3258$ graduates ranged from a low of $20,350 to a high of $54,800. Based on this range and the amount of detail we want to show, we must decide on the length of the class intervals. While there are no hard-and-fast rules, any breakdown with up to 20 class intervals will produce a reasonably clean graph. Too many more than that, and the graph starts getting cluttered. Let's decide on class intervals of $5000. Table 14-5 is a frequency table for the data based on these class intervals. The third column in the table shows the data as a percentage of the population.

Salary	Number of Students	Percentage
20,000 – 25,000	228	7%
25,000⁺– 30,000	456	14%
30,000⁺– 35,000	1043	32%
35,000⁺– 40,000	912	28%
40,000⁺– 45,000	391	12%
45,000⁺– 50,000	163	5%
50,000⁺– 55,000	65	2%
Total	3258	100%

■ **TABLE 14-5**

The histogram for these data is shown in Fig. 14-12.

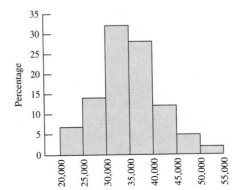

FIGURE 14-12   Histogram
showing starting salaries
for first-year graduates of
Tasmania State University.
$N = 3258$.

As Fig. 14-12 shows, a histogram is very similar to a bar graph. There are, however, several important distinctions that must be made. To begin with, because a histogram is used with continuous variables, there can be no gaps between the class intervals, and it follows therefore that the columns of a histogram must touch each other. Among other things, this forces us to make an arbitrary decision as to what happens to a value that falls exactly on the boundary between two class intervals. Should it always belong to the class interval to the left or to the one to the right? This is called the *endpoint convention*. The + marks in Table 14-5 indicate how we chose to deal with the endpoint convention in Fig. 14-12: A starting salary of $30,000, for example, should be assigned to the second rather than to the third class interval.

As with regular bar graphs, in creating histograms we should try, as much as possible, to define class intervals of equal length. When the class intervals are of unequal lengths, the rules for creating a histogram are considerably more complicated, since it is no longer appropriate to use the heights of the columns to indicate the frequencies of the class intervals. We will not discuss the details of this situation here but refer the interested reader to Exercises 41 and 42 at the end of the chapter.

NUMERICAL SUMMARIES OF DATA

As we have seen, pictures can be an excellent tool for summarizing large amounts of data. Unfortunately, circumstances do not always lend themselves equally well to the use of pictures. If Professor Blackbeard were walking across campus and a student were to stop him and in casual tones ask, "How were the scores on the Stat 101 exam?" one could hardly expect Professor Blackbeard to flash a bar graph. The fact of the matter is that *numerical summaries* of data can often be used to substitute for or improve on the visual effect of graphical presentations.

In this section we will discuss several ways in which large sets of data can be summarized by a few well-chosen numbers. Before we go into specifics we will find it useful to draw an important distinction. Numerical summaries of data fall into two categories: numbers that tell us something about where the values of the data fall, and numbers that tell us something about how spread-out the values of the data are. The former are called; **measures of location**, and the latter are called **measures of spread**. The most important measures of location are the **mean** or **average**, the **median**, and the **quartiles**. The most important measures of spread are the **range**, the **interquartile range**, and the **standard deviation**. We will discuss each of these in order.

The Average

The best known of all numerical summaries of data is the *average*, sometimes also called the *mean*. (As much as possible, we will stick to the word "average"—it is a good down-to-earth word.) The **average** of a set of N numbers is obtained by adding the numbers and dividing by N. When the set of numbers is small, one can often calculate the average in one's head; for larger data sets, pencil and paper or a calculator is helpful. In either case, the idea is very straightforward.

Example 8. In 10 games a basketball player scores 8, 5, 11, 7, 15, 0, 7, 4, 11, and 14 points, respectively. The total is 82 points in 10 games. The average is 8.2 points per game. Note that it is actually impossible for a player to score 8.2 points. As it often happens, the average taken by itself can be an impossible data value.

Example 9. We will now calculate the average score for the Stat 101 data set. For convenience we show once again the frequency table (Table 14-6).

Exam Score	1	6	7	8	9	10	11	12	13	14	15	16	24
Frequency	1	1	2	6	10	16	13	9	8	5	2	1	1

■ **TABLE 14-6 Frequency Table for the Stat 101 Data Set.**

The 75 data values can be totaled by taking each score and multiplying it by its corresponding frequency and adding. In this case we get

$$\text{Total} = (1 \times 1) + (6 \times 1) + (7 \times 2) + (8 \times 6) + (9 \times 10) +$$
$$(10 \times 16) + (11 \times 13) + (12 \times 9) + (13 \times 8) + (14 \times 5) +$$
$$(15 \times 2) + (16 \times 1) + (24 \times 1) = 814.$$

The average score on the midterm exam (rounded off to two decimal places) is

$$814 \div 75 \approx 10.85 \text{ points.}$$

Intuitively, we think of this average as representing a "typical" student's score. If all test scores had been about the same, then, given the same total, each score would have been "about" 10.85 points.

Data Value	s_1	s_2	$\cdots$	s_k
Frequency	f_1	f_2	$\cdots$	f_k

■ **TABLE 14-7**

Table 14-7 shows a generic frequency table. To find the average of the data we do the following:

■ **Step 1.**   Calculate the total of the data.

   Total $= (s_1 \times f_1) + (s_2 \times f_2) + \cdots + (s_k \times f_k).$

■ **Step 2.**   Calculate N.

   $N = f_1 + f_2 + \cdots + f_k$

■ **Step 3.**   Calculate the average.

   Average $=$ Total $\div$ N.

As the reader can see, this is the kind of algorithm for which computers are particularly well suited, and even many calculators have a built-in key for calculating averages.

So far, all our examples have involved numbers that are positive, but negative data values are also possible, and when both negative and positive data values are averaged, the results can be a little misleading.

Example 10. The monthly savings (monthly income minus monthly spending) of a family over a one-year period is shown in Table 14-8. A negative amount indicates that, rather than saving money, the family spent more that month than their monthly income.

Month	Jan.	Feb.	Mar.	Apr.	May	Jun.	Jul.	Aug.	Sept.	Oct.	Nov.	Dec.
Savings (in $)	-732	-158	-71	-238	1839	-103	-148	-162	-85	-147	-183	500
	Christmas bills				$2000 lottery winnings							Christmas bonus from work

■ **TABLE 14-8**

The average monthly savings of this family over the year is

$$\frac{-732 - 158 - 71 - 238 + 1839 - 103 - 148 - 162 - 85 - 147 - 183 + 500}{12} = 26$$

The $26 monthly savings paints a deceptively rosy picture of this family's finances. The truth of the matter is that this is a family living beyond its means which was bailed out by a one-shot lottery prize.

Our next example illustrates a useful concept—the concept of *deviation from the mean.*

Example 11. We're back to the Stat 101 data set. What we would now like to know is how far the test scores deviated from the average test score of 10.85 points (in other words, the difference between each test score and 10.85). Table 14-9 summarizes these deviations, commonly known as **deviations from the mean**.

Exam Score	1	6	7	8	9	10	11	12	13	14	15	16	24
Deviation: Test Score -10.85	-9.85	-4.85	-3.85	-2.85	-1.85	-0.85	0.15	1.15	2.15	3.15	4.15	5.15	13.15
Frequency	1	1	2	6	10	16	13	9	8	5	2	1	1

■ **TABLE 14-9**

If we average the above deviations, we get

Average deviation from 10.85 (the average test score)

$$= \frac{-9.85 - 4.85 - (3.85 \times 2) - (2.85 \times 6) + \cdots + 13.15}{75} = \frac{0.25}{75} \approx 0.0033$$

The fact that the average deviation from the average test score in Example 11 is so small could lead someone who hasn't seen the details to conclude that the test scores were very close to each other and to the average score, which is not the case at all. What really happened here is that the negative deviations and the positive deviations from the average cancelled each other out. In fact, the only reason this average deviation did not come out to be exactly 0 (see Exercise 44) is because we rounded off the average test score (10.85) to two decimal places. Once again, the problem lies in the fact that averaging numbers that are positive and negative can present a misleading picture of what the relative sizes of these numbers are. We call this the **cancellation effect**.

There are two ways out of this dilemma, and both are based on the idea that we need to get rid of the negative signs. One way to do this is to take absolute values (i.e., drop the minus signs); the other is to square the numbers (squared real numbers cannot be negative). For practical reasons the second approach is preferable and is the standard strategy used by statisticians.

The preceding comments can be summarized as follows:

1. When dealing with data values that are both positive and negative, averages can be misleading because of the possible cancellation between the positive and negative values.

2. To avoid the cancellation effect a common strategy is to square the data values first and then take the average of the squares.

Example 12. Let's apply this strategy to the deviations from the mean on the Stat 101 data set (Table 14-9).

Frequency	1	1	2	6	10	16	13	9	8	5	2	1	1
Deviation from the Mean	−9.85	−4.85	−3.85	−2.85	−1.85	−0.85	0.15	1.15	2.15	3.15	4.15	5.15	13.15
Squared Deviation	97.02	23.52	14.82	8.12	3.42	0.72	0.02	1.32	4.62	9.92	17.22	26.52	172.92

▪ **TABLE 14-10**

When we average the data in Table 14-10 we get

Average of squared deviations

$$= \frac{97.02 + 23.52 + 14.82 \times 2 + 8.12 \times 6 + \cdots + 172.92}{75} = \frac{577.20}{75} = 7.70.$$

While the squaring strategy makes all quantities positive and therefore eliminates the problem created by the cancellation effect, it creates a problem of its own: Squaring changes the order of magnitude of numbers. Whatever the original units were, we are now dealing with square units. In our last example the units were points on the exam, but the last average we calculated was square points (whatever they are). This problem can be eliminated by taking the square root (of the average of the squared data). Thus, in Example 12, if we take $\sqrt{7.70}$, we have a value that is of the same order of magnitude as the original values of the data.

The Root Mean Square

The process we have just described may appear at first sight to be both tortuous and confusing. In reality it gives an important numerical summary for a set of data called the **root mean square** (commonly denoted by rms). The rms is a useful substitute for the average only in cases where the cancellation effect must be avoided. In most typical situations the rms should not be used.

The procedure for calculating the rms consists of three steps.

■ **Step 1.** Square each of the data values. (This gets rid of the negatives.)

■ **Step 2.** Take the average (mean) of the values obtained in step 1. (This produces an average, but it is distorted because of the squaring.)

■ **Step 3.** Take the square root of the value obtained in step 2. (This brings the average back to the right order of magnitude.)

The most important use of the rms is when it its applied not to the data values themselves but to the deviations between the data values and their average. This is exactly what we did in Example 12. Starting with the Stat 101 data set, we calculated the deviations from the mean and for this new set of numbers we calculated the rms. This rms (which in the example turned out to be $\sqrt{7.70} \approx 2.77$) is called the **standard deviation**, and we will come back to it toward the end of the chapter.

The Median

The median is another important, commonly used numerical summary of a set of data. To find the median of a set of numbers we must first sort the numbers by size. That is to say, we must rewrite the numbers in increasing order from left to right (or right to left—it makes no difference). We must then find the number in the "middle" of the sorted list. That number is the **median**.

Example 13. To find the median of the numbers 4.8, −2, 3.1, −6.5, 1.6, 0.5, and 4.25, we first sort the numbers by size and get

−6.5, −2, 0.5, 1.6, 3.1, 4.25, 4.8.

Of these seven numbers, the number in the middle is the one in the fourth position in the list, so the median is 1.6.

Example 14. How do we find the median of the numbers 2.2, −5.1, −2.7, 4.1, 6.2, −3, 1.2, 3.8? When we sort these numbers by size we get

−5.1, −3, −2.7, 1.2, 2.2, 3.8, 4.1, 6.2.

Which is the "middle" number in this list of eight numbers? Actually one could say that there are two numbers—the one in the fourth position (1.2) and the one in the fifth position (2.2)—with an equal claim to being in the "middle." In the spirit of compromise we settle the dispute by taking the number halfway between these two numbers and calling it the median. Thus, in this example the median is $\dfrac{1.2 + 2.2}{2} = 1.7$.

We can generalize what we discovered in Examples 13 and 14 in the following set of rules.

FINDING THE MEDIAN OF N NUMBERS

■ **Step 1.** Sort the numbers by size.

■ **Step 2.** Find the number(s) located in the "middle" position(s) in the sorted list. There are two cases:
(a) N odd. The "middle" position is position $(N + 1)/2$. The number in position $(N + 1)/2$ is the median.
(b) N even. There are two "middle" positions. They are $N/2$ and $(N/2) + 1$. The number halfway between the numbers in position $N/2$ and $(N/2) + 1$ is the median.

Example 15. We will now find the median score for Professor Blackbeard's Stat 101 data set. For the reader's convenience we display the frequency table (Table 14-11).

Exam Score	1	6	7	8	9	10	11	12	13	14	15	16	24
Frequency	1	1	2	6	10	16	13	9	8	5	2	1	1

■ Table 14-11 **Frequency Table for the Stat 101 Data Set.**

Having the frequency table available eliminates the need for sorting the scores—the frequency table has in fact done this for us. The total number of scores is $N = 75$, which means that the list has a single middle position. It is position $(75 + 1)/2 = 38$. We now know that the median is in the 38th position in the sorted list of the data. (It is important to understand that the median is not 38, but rather the 38th score in the sorted list.) To find the 38th score, we start tallying the frequencies as we move from left to right along the frequency row of Table 14-11: $1 + 1 = 2, 1 + 1 + 2 = 4, 1 + 1 + 2 + 6 = 10, 1 + 1 + 2 + 6 + 10 = 20, 1 + 1 + 2 + 6 + 10 + 16 = 36$. At this point we know that the 36th test score on the list is a 10 (the last of the 10's) and the next 13 scores are all 11's. We can conclude that the 38th test score (the median test score) is 11. In essence this says that half of the class had a score of 11 or less and half of the class had a score of 11 or more.

Finding a median is not nearly as complicated as it seems, and with a little practice the reader will find this to be the case. (Start by trying Exercises 2, 5, and 11.) As in finding the average, when large amounts of data are involved, letting a computer do the work is the most practical way to go.

A fairly common mistake is to confuse the median and the mean. The two words are quite similar, and they both define related concepts. This is one more good reason why (as much as possible) the term "average" should be used instead of "mean." Even among those who can keep the two concepts straight, a frequent misconception is to assume that the median and the average must be close to each other in value. While this is indeed the case in many types of real-life data, it is not true in general. Take, for example, the numbers, 1, 1, 1, and 97. The median of these numbers is 1, while the average is 25, a much larger number. On the other hand, if we take the numbers 1, 1, 100, 101, and 102, then the median (100) is much larger than the average (61). We can see that it is a mistake to assume, as many people do, that the median and the average are "about the same."

The median has the effect of separating the sorted data into two halves: a **lower half** and an **upper half**. Fig. 14-13 illustrates exactly what we mean.

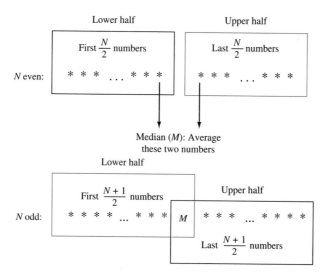

FIGURE 14-13

Example 16. The heights in inches of 10 children are 38, 41, 36, 49, 28, 44, 35, 43, 39, and 41. To find the lower and upper halves of these data values, we first sort them by size. This results in the list 28, 35, 36, 38, 39, 41, 41, 43, 44, 49. The lower half is then 28, 35, 36, 38, 39, while the upper half is 41, 41, 43, 44, 49. The median height of this group is 40 inches.

Example 17. Suppose that one more child of height 32 inches joins the group in Example 16. The new sorted data values are 28, 32, 35, 36, 38, 39, 41, 41, 43, 44, 49. The lower half now is 28, 32, 35, 36, 38, 39, while the upper half is 39, 41, 41, 43, 44, 49. Note that in this example the median (39) is in both halves.

The Quartiles

Now that we understand the formal meaning of the "lower half" and the "upper half" of a set of numbers, we can define the quartiles in a very straightforward manner: The **first quartile** (Q_1) is the median of the lower half; the **third quartile** (Q_3) is the median of the upper half.

Example 18. Consider the same set of 10 children's heights discussed in Example 16. The lower half of these numbers is 28, 35, 36, 38, 39. The median of these five numbers is 36. It follows that the first quartile is 36 $(Q_1 = 36)$. The upper half of the data is 41, 41, 43, 44, 49, and the median of these five numbers is 43. It follows that the third quartile is 43 $(Q_3 = 43)$.

Example 19. For the 11 heights discussed in Example 17, we found that the lower half of the data consists of the numbers 28, 32, 35, 36, 38, 39. The first quartile is the median of these six numbers, which is 35.5. The upper half of the data consists of the numbers 39, 41, 41, 43, 44, 49. The third quartile is the median of these six numbers, which is 42.

Just as the median serves to break up the data into halves, the purpose of quartiles is to break up the data into quarters. Saying, for example, that the first

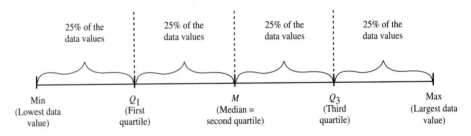

| 25% of the data values | | 25% of the data values | | 25% of the data values | | 25% of the data values |

| Min (Lowest data value) | Q_1 (First quartile) | M (Median = second quartile) | Q_3 (Third quartile) | Max (Largest data value) |

Figure 14-14

quartile of a set of numbers is 35.5 means that 25% of the numbers are smaller than or equal to 35.5. Likewise, saying that the third quartile of a set of numbers is 42 means that 75% of the numbers are smaller than or equal to 42. The median itself is the second quartile, which means that 50% of the numbers are smaller than or equal to the median. It is not hard to see why the first quartile, the median, and the third quartile are sometimes referred to as the 25th percentile, the 50th percentile, and the 75th percentile, respectively.

We will now describe the general procedure for finding the quartiles of a set of numbers.

FINDING THE QUARTILES OF N NUMBERS

■ **Step 1.**   Sort the numbers by size.
■ **Step 2.**   Find the median M.
■ **Step 3.**   Find the lower half and upper half of the data set (see Fig. 14-13).
■ **Step 4.**   Find the median of the lower half. This is the first quartile Q_1.
■ **Step 5.**   Find the median of the upper half. This is the third quartile Q_3.

The Five-Number Summary

A reasonably good summary for a large set of data can be provided numerically by giving the lowest value of the data (Min), the first quartile (Q_1), the median (M), the third quartile (Q_3), and the largest value of the data (Max). These five numbers constitute the **five-number summary** of the data.

Example 20. Let's go back to the Stat 101 data set and see if we can come up with the five-number summary for it. We already know that the lowest score is Min = 1, the median score is $M = 11$, and the highest score is Max = 24. We need to find the first and third quartiles. Since the total number of scores is 75 (odd), we know that the first half of the data consists of the first 38 numbers. Since 38 is even, the median of the first 38 numbers is halfway between the 19th and 20th scores. To find the 19th score we go to the frequency table and start counting frequencies beginning on the left. We leave it to the reader to verify that the 19th and 20th scores are both 9. It follows that $Q_1 = 9$. To find the third quartile we need to find the median of the upper half of the scores. These are the 38th through the 75th scores. The median of these 38 numbers is the number halfway between the two middle numbers. To find it, we can count in the frequency table from left to right (starting at the 38th number) or, what is much easier, count from the top down until we find the 19th and 20th numbers from the end. Using this last approach, we find that the 19th and 20th scores from the top are both 12, so we have $Q_3 = 12$. The five-number summary for the Stat 101 data set is

$$\text{Min} = 1, Q_1 = 9, M = 11, Q_3 = 12, \text{Max} = 24.$$

Note that without Q_1 and Q_3 we would have a very distorted picture of the Stat 101 data set, since both Min = 1 and Max = 24 are outliers. The scores were not evenly spread out in the range between 1 and 24, in fact just the opposite was true. With the quartiles we can get a much better idea of what happened: The middle half of test scores fell in a very narrow range of between 9 and 12 points; only one quarter of the test scores were 9 or less and only one quarter of the test scores were 12 or more.

MEASURES OF SPREAD

An important element in summarizing data numerically is to give an idea of how spread-out the data values are. Consider the following two data sets: Set 1 = {45, 46, 47, 48, 49, 51, 52, 53, 54, 55}, and Set 2 = {1, 11, 21, 31, 41, 59, 69, 79, 89, 99}. We leave it to the reader to verify that for both data sets the average is 50 and the median is 50. At the same time, it is obvious that these two data sets are very different. The numbers in Set 2 are much more spread-out than those in Set 1. What can we do to measure the amount of spread in a data set? The most obvious approach is to take the difference between the highest and lowest values of the data (Max − Min). This difference is called the **range**. For Set 1 = {45, 46, . . . , 55} the range is 10, and for Set 2 = {1, 11, . . . , 89, 99} the range is 98.

As a measure of spread, the range is useful only if there are no outliers, since outliers can significantly affect the range. For example, on the Stat 101 data set the range of the exam scores is 24 − 1 = 23 points, but without the two outliers the range would be 16 − 6 = 10.

To eliminate the possible distortion caused by outliers, a common practice is to use the **interquartile range** (IQR). The interquartile range is the difference between the third quartile and the first quartile (IQR = $Q_3 - Q_1$), and its significance is that it tells us how spread-out the middle 50% of the data values are. For many types of real-world data the interquartile range is a useful measure of spread. When the five-number summary is used, both the range and the interquartile range come, essentially, free in the bargain.

Example 21. The five-number summary for a set of 200 test scores is Min = 4, $Q_1 = 23$, $M = 27$, $Q_3 = 32$, Max = 61. Here the range is Max − Min = 57, and the interquartile range is IQR = $Q_3 - Q_1 = 9$. What does the five-number summary tell us about this set of test scores? We know that the lowest 50 scores (the bottom 25%) are between 4 and 23; the next 50 scores are tightly bunched up between 23 and 27; likewise, the next 50 are between 27 and 32; scores are spread out between 32 and 61. (Could these be score one of Professor Blackbeard's exams?)

Box Plots

A **box plot** is a graphical device that displays in a single picture all the information contained in the five-number summary of a data set. Figure 14-15 shows a generic box plot for a data set. The box plot consists of a rectangular box that goes from the first quartile Q_1 to the third quartile Q_3. A line crosses the box indicating the position of the median M. On both sides of the box are whiskers extending to the smallest value Min and largest value Max of the data.

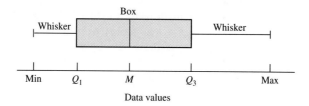

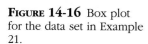

FIGURE 14-15 A typical box plot for a data set.

Figure 14-16 shows a box plot for the data given in Example 21. (Min = 4, $Q_1 = 23$, $M = 27$, $Q_3 = 32$, Max = 61.)

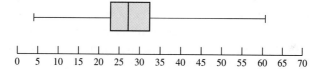

FIGURE 14-16 Box plot for the data set in Example 21.

Box plots are particularly useful when comparing data for two or more populations. This fact is illustrated in the next example.

Example 22. Figure 14-17 shows box plots for the starting salaries of two different populations: first-year agriculture and engineering graduates of Tasmania State University. Superimposing the two box plots on the same scale allows us to make comparisons in a particularly efficient way. It is clear, for instance, that engineering graduates are doing better overall than agriculture graduates, even though at the very top levels agriculture graduates are better paid. Another interesting point is that the median salary of agriculture graduates is less than the first quartile of the salaries of engineering graduates. The very short whisker on the left side of the agriculture box plot tells us that the bottom 25% of agriculture salaries are concentrated in a very narrow salary range. We can also see that agriculture salaries are much more spread-out than engineering salaries, even though most of the spread occurs at the higher end of the salary scale.

It is clear that we can get a lot of mileage out of a simple picture like the one in Fig. 14-17. (See Exercises 30 and 31.)

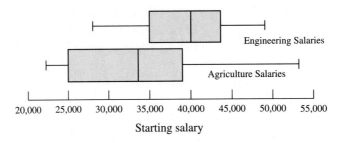

FIGURE 14-17
Comparison of starting
salaries for Tasmania State
University first-year graduates in agriculture and engineering using box plots.

The Standard Deviation

Perhaps the most commonly used measure of spread for a set of data is the **stan-dard deviation**. The reader may recall that in Examples 11 and 12 we carried out a series of calculations for the Stat 101 data set that involved first computing the deviations of the test scores from the average score and then computing the rms of these deviations. We called this value the standard deviation of the data set.

For any data set, we can calculate the standard deviation using the follow-ing set of rules.

FINDING THE STANDARD DEVIATION OF N NUMBERS

■ **Step 1.** Find the average of the N numbers. Call it A.

■ **Step 2.** For each number x in the data set, compute $x - A$. We call this x's *deviation from the mean (average)*.

■ **Step 3.** Using the deviations from the mean as our new data set, we compute their rms (remember, this means square the data, average the squares,[4] and then take the square root). This value is called the stan-dard deviation (SD).

Example 23. Let's find the standard deviation of the numbers 45, 46, 47, 48, 49, 51, 52, 53, 54, 55.

■ **Step 1.** $A = 50$.

■ **Step 2.** Taking the deviations from the mean gives the data set -5, -4, -3, -2, -1, 1, 2, 3, 4, 5.

---

[4] In many statistics books and statistical computer programs, this step in the algorithm is changed so that when calculating the average of the squared deviations instead of dividing by N, the division is by $N - 1$. Everything else is exactly the same. There are good reasons why this definition makes good sense, especially when the data values are obtained from a sample, but to explain them would go beyond the purpose and scope of this chapter. In any case, except for small values of N the an-swers provided by either method are very close.

■ **Step 3.**

(a) Square: 25, 16, 9, 4, 1, 1, 4, 9, 16 , 25.

(b) Mean: [total from step 3(a)]/N = 110/10 = 11.

(c) Root: $\sqrt{11}$.

The answer is SD = $\sqrt{11} \approx 3.317$

It is important for the reader to understand the difference between the rms of a data set and the standard deviation for the same data set. They are not the same thing! The rms is a substitute for the average that is used only when the positive values of the data and the negative values of the data tend to cancel each other out. The standard deviation is a special use of the rms. Here we are not finding the rms of the data set itself but rather the rms of the deviations from the mean. For the data set consisting of the deviations from the mean the cancellation effect is at its peak: The negative deviations and the positive deviations always cancel out completely (see Exercise 44).

Although the standard deviation is a measure of spread that can be used with any set of numbers, it is particularly important in many situations involving real-life data. We will discuss this point in greater detail in Chapter 16.

CONCLUSION

The basic theme of this chapter is "putting things in a nutshell." More specifically, when we have relatively large data sets, there is a need to summarize the information provided by the data in a meaningful way. In this chapter we discussed ways to summarize data sets where each data value consists of a single number (or a single category in the case of categorical variables).

Graphical summaries of data can be produced by bar graphs, frequency charts, pie charts, histograms, etc. (There are many other types of graphical descriptions that we did not discuss in the chapter.) Which kind of graph is the most appropriate for which situation depends on many factors, and there is no substitute for experience in making these kinds of choices. One of the challenging things about descriptive statistics is that there is plenty of room for creativity.

Numerical summaries of data fall into two categories: (1) *measures of location* such as the average, the root mean square, the median, and the quartiles, and (2) *measures of spread* such as the range, the interquartile range, and the standard deviation.

The famous science fiction writer H. G. Wells once said,

Statistical thinking will one day be as necessary for efficient citizenship as the ability to read and write.

The concepts introduced in this chapter are to statistics what the alphabet is to reading and writing.

KEY CONCEPTS

average (mean)
bar graph
box plot
cancellation effect
category (class)
class interval
continuous variable
data set
data values (data points)
descriptive statistics
deviations from the mean
discrete variable
five-number summary
frequency
frequency chart
frequency table

histogram
interquartile range
lower half
measures of location
measures of spread
median
outlier
pie chart
qualitative (categorical) variable
quantitative (numerical) variable
quartiles
range
root mean square (rms)
standard deviation
upper half

EXERCISES

Walking

Exercises 1 through 4 refer to the final exam in Chem 103 which consisted of 10 questions worth 10 points each with no partial credit given. The scores on the exam are given in Table 14-12.

Student ID	Score	Student ID	Score	Student ID	Score	Student ID	Score
1362	50	2877	80	4315	70	6921	50
1486	70	2964	60	4719	70	8317	70
1721	80	3217	70	4951	60	8854	100
1932	60	3588	80	5321	60	8964	80
2489	70	3780	80	5872	100	9158	60
2766	10	3921	60	6433	50	9347	60

▬ TABLE 14-12   **Chem 103 Final Exam Scores**

1. **(a)** Make a frequency table for the Chem 103 final exam scores.
   (b) Make a bar graph showing the actual frequencies of the scores on the exam.
   (c) Make a bar graph showing the relative frequencies of the scores on the exam (i.e., the percentage of students receiving each score).

2. For the Chem 103 final exam scores, find
   (a) the range
   (b) the median
   (c) the first and third quartiles
   (d) the interquartile range.

3. For the Chem 103 final exam scores,
   (a) find the five-number summary
   (b) draw a box plot.

4. For the Chem 103 final exam scores, find
   (a) the average
   (b) the standard deviation.

Exercises 5 through 8 refer to a midterm exam in History 3B. The percentage scores on the exam are given in Table 14-13.

Student ID	Score	Student ID	Score	Student ID	Score	Student ID	Score	Student ID	Score
1075	74%	1998	75%	3491	57%	4713	83%	6234	77%
1367	83%	2103	59%	3711	70%	4822	55%	6573	55%
1587	70%	2169	92%	3827	52%	5102	78%	7109	51%
1877	55%	2381	56%	4355	74%	5381	13%	7986	70%
1946	76%	2741	50%	4531	77%	5717	74%	8436	57%

■ TABLE 14-13   **History 3B Midterm Exam Scores**

5. For the data set in Table 14-13, find
   (a) the range
   (b) the median
   (c) the first and third quartiles
   (d) the interquartile range.

6. For the data set in Table 14-13,
   (a) find the five-number summary
   (b) draw a box plot.

7. For the data set in Table 14-13, find
   (a) the average
   (b) the standard deviation.

8. Suppose the class intervals for letter grades on the History 3B midterm exam are

   90%–100% A

   80%–89% B

   70%–79% C

   60%–69% D

   Below 60% F

(a) Make a frequency table for the letter grades on the exam.

(b) Draw a pie chart for the percentage of students receiving each letter grade.

Exercises 9 and 10 refer to Table 14-14 which gives the distance from home to school (measured to the closest $\frac{1}{2}$ mile) for each kindergarten student at Cleansburg Elementary School.

Student ID	Distance to School (miles)	Student ID	Distance to School (miles)	Student ID	Distance to School (miles)	Student ID	Distance to School (miles)
1362	1.5	2877	1.0	4355	1.0	6573	0.5
1486	2.0	2964	0.5	4454	1.5	8436	3.0
1587	1.0	3491	0.0	4531	1.5	8592	0.0
1877	0.0	3588	0.5	5482	2.5	8854	0.0
1932	1.5	3711	1.5	5533	1.0	8964	2.0
1946	0.0	3780	2.0	5717	8.5		
2103	2.5	3921	5.0	6307	1.5		

■ **TABLE 14-14   Distance From Home to School for Cleansburg Elementary School Kindergarten students**

9. **(a)** Make a frequency table for the data set in Table 14-14.

   (b) Draw a bar graph showing the relative frequencies for the data set in Table 14-14.

10. Suppose that class intervals for the distances from home to school for the kindergarteners at Cleansburg Elementary School are defined by

    Very close: Less than 1 mile.

    Close: 1 mile up to and including 1.5 miles.

    Nearby: 2 miles up to and including 2.5 miles.

    Not too far: 3 miles up to and including 4.5 miles.

    Far: 5 miles or more.

    (a) Make a frequency table for the class intervals.

    (b) Draw a pie chart for the percentage of students in each category.

11. The following bar graph shows the frequencies of various scores received by students in Math A on a 10-point pop quiz.

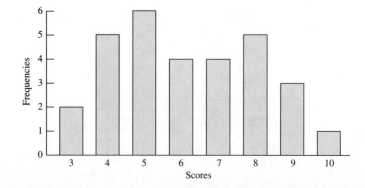

(a) How many students took the quiz?

(b) Make a frequency table for the scores on the pop quiz.

(c) What is the average score?

(d) What is the median score?

Exercises 12 and 13 refer to the pie chart shown in the margin. The pie chart shows the percentage of the undergraduate student body at Tasmania State University for each ethnic group.

12. (a) Give a frequency table showing the actual frequencies for each category.

(b) Draw the bar graph corresponding to the frequency table in (a).

13. Calculate the size of the angle (to the nearest degree) for each of the slices shown in the pie chart.

14. A pie chart is made up of four slices representing four different categories in a population (*A, B, C,* and *D*).

Slice *A* has an angle of 52°.

Slice *B* has an angle of 108°.

Slice *C* has an angle of 125°.

Slice *D* has an angle of 75°.

Find the percentage of the population in each category.

15. The pie chart in the margin shows the percentage of babies born at each of the four hospitals in the city of Cleansburg in the last year.

(a) How many babies were born at Downtown Hospital?

(b) How many babies were born outside one of the four hospitals (at home, on the way to the hospital, etc.)?

(c) Draw a bar graph showing the frequency for each category.

16. Calculate the size of the angle (in degrees) for each of the slices shown in the pie chart used in Exercise 15.

Exercises 17 through 20 refer to the data in Table 14-15, which shows the weights (in ounces) of the 625 babies born in the city of Cleansburg in the last year.

WEIGHT IN OUNCES

More than	Less than or Equal to	Frequencies
48	60	15
60	72	24
72	84	41
84	96	67
96	108	119
108	120	184
120	132	142
132	144	26
144	156	5
156	168	2

■ **TABLE 14-15**

7%
Other

11%
Asian

19%
Hispanic

39%
Caucasian

24%
African
American

N = 15,000

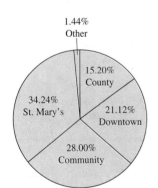

1.44%
Other

15.20%
County

34.24%
St. Mary's

21.12%
Downtown

28.00%
Community

N = 625

17. **(a)** Give the length of each class interval (in ounces).
    (b) Suppose a baby weighs exactly 5 pounds 4 ounces. What class interval does she belong to? Describe the endpoint convention.

18. Write a new table for these data values using class intervals of length equal to 24 ounces.

19. Draw the histogram corresponding to these data values using the class intervals as shown in the original table.

20. Draw the histogram corresponding to the same data when class intervals of 24 ounces are used.

21. Find the average and the standard deviation for each of the following three sets of numbers: {5, 5, 5, 5}, {0, 5, 5, 10}, {−5, 0, 0, 25}. Note the similarities and differences in your answers and explain.

22. Find the average and the standard deviation for each of the following three sets of numbers: {10, 10, 10, 10}, {1, 6, 13, 20}, {1, 1, 18, 20}. Note the similarities and differences in your answers and explain.

23. For the data set {0, 1, 2, 3, 4, 5, 6, 7, 8, 9}, find
    (a) the average
    (b) the median
    (c) the standard deviation.

24. For the data set {1, 2, 3, 4, 5, 6, 7, 8, 9, 10}, find
    (a) the average
    (b) the median
    (c) the standard deviation.

25. For the data set {0, 1, 2, 3, 4, 5, 6, 7, 8, 9}, find
    (a) the first quartile
    (b) the third quartile
    (c) the interquartile range.

26. For the data set {1, 2, 3, 4, 5, 6, 7, 8, 9, 10}, find
    (a) the first quartile
    (b) the third quartile
    (c) the interquartile range.

27. For the data set {1, 2, 3,..., 99, 100}, find
    (a) the average
    (b) the median.

28. For the data set {1, 2, 3,..., 99, 100}, find
    (a) the first quartile
    (b) the third quartile
    (c) the interquartile range.

29. **(a)** Find the five-number summary, the average, and the standard deviation for the data given in the following frequency table.

Value	9	10	11	12	13	14	15	16	17	18	19	20	21	22	23	24	25
Frequency	3	5	7	4	12	10	13	11	15	13	11	4	3	0	0	0	1

(b) Draw a box plot for this data.

Exercises 30 and 31 refer to the following figure comparing the starting salaries for Tasmania State University first-year graduates in Agriculture and Engineering using box plots. (These are the two box plots discussed in Example 22.)

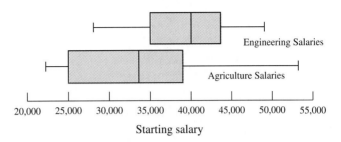

30. **(a)** Approximately how much is the median salary for agriculture majors?
    (b) Approximately how much is the median salary for engineering majors?
    (c) Explain how we can tell that the median salary for engineering majors is more than the third quartile of the salaries for agriculture majors.

31. **(a)** If the number of engineering majors was 612, how many of them had a starting salary of $35,000 or more?
    (b) If the number of agriculture majors was 960, approximately how many of them made less than $25,000?

32. Mike's average on the first five exams in Econ 1A is 88. What must he earn on the next exam in order to raise his overall average to 90?

33. Sarah's overall average in Physics 101 was 93%. Her average was based on four exams each worth 100 points and a final exam worth 200 points. What is the lowest possible score she could have made on the first exam?

Jogging

34. The average annual starting salary for the 125 speech majors that graduated this year from Tasmania State University is $50,752. This is an impressive figure, but before we all rush out to change majors, consider the fact that this average is seriously distorted by the presence of an outlier—basketball star "Hoops" Tallman who just signed a professional contract for $2.5 million a year. If we were to disregard this one outlier, what would the average annual starting salary be for the remaining speech majors?

35. Josh and Ramon each have an 80% average on the five exams given in Psychology 4. Ramon, however, did better than Josh on all of the exams except one. Give an example that illustrates this situation.

36. Kelly and Karen each have an average of 75 on the six exams given in Botany 1. Kelly's scores have a small standard deviation and Karen's scores have a large standard deviation. Give an example that illustrates this situation.

37. **(a)** Give an example of 10 numbers such that their average is less than their median,
    (b) Give an example of 10 numbers such that their median is less than their average.

(c) Give an example of 10 numbers such that their average is less than the first quartile.

(d) Give an example of 10 numbers such that their average is more than the third quartile.

38. Suppose that the average of 10 numbers is 7.5 and that the smallest of them is Min = 3.

(a) What is the smallest possible value of Max?

(b) What is the largest possible value of Max?

39. What happens to the five-number summary of the Stat 101 data set (Example 20) if

(a) 2 points are added to each score?

(b) 10% is added to each score?

40. Under what conditions would the standard deviation of a set of numbers be 0?

Exercises 41 and 42 refer to histograms with unequal class intervals. When drawing histograms with class intervals of unequal lengths, the columns must be drawn so that the frequencies or percentages are proportional to the area of the column.

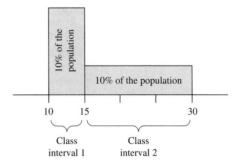

The figure illustrates what we mean. If the column over class interval 1 represents 10% of the population, then the column over class interval 2, also representing 10% of the population, must be one-third as high because the class interval is three times as large.

41. If the height of the column over the class interval 20–30 is 1 unit and the column represents 25% of the population, then

(a) How high should the column over the interval 30–35 be if 50% of the population falls into this class interval?

(b) How high should the column over the interval 35–45 be if 10% of the population falls into this class interval?

(c) How high should the column over the interval 45–60 be if 15% of the population falls into this class interval?

42. Two-hundred senior citizens are tested for fitness and rated on their times on a 1-mile walk. These ratings and associated frequencies are given in the following table.

Time	Rating	Frequency
6^+– 10 minutes	Fast	10
10^+–16 minutes	Fit	90
16^+– 24 minutes	Average	80
24^+– 40 minutes	Slow	20

Draw a histogram for these data based on the categories given by the ratings in the table.

43. Tasmania State University has been accused by the news media of discriminating against women in the admission policies in the schools of architecture and engineering. The Tasmania Gazette states, "...68% of all male applicants to the schools of architecture or engineering are admitted while only 51% of the female applicants to these same schools are admitted." The actual data by school is given in the following table,

	School of Architecture		School of Engineering	
	Applied	Admitted	Applied	Admitted
Male	200	20	1000	800
Female	500	100	400	360

(a) What percent of the male applicants to the School of Architecture were admitted? What percent of the female applicants to this same school were admitted?

(b) What percent of the male applicants to the School of Engineering were admitted? What percent of the female applicants to this same school were admitted?

(c) How did the Tasmania Gazette come up with its figures?

(d) Explain how it is possible for (a), (b), and the Tasmania Gazette statement to all be true.

Running

44. Given the numbers $x_1, x_2, x_3, \ldots, x_N$ with average m, show that $(x_1 - m) + (x_2 - m) + (x_3 - m) + \cdots + (x_N - m) = 0$.

45. (a) Find two numbers whose average is m and standard deviation is s.

(b) Find three equally spaced numbers whose average is m and standard deviation is s.

(c) Generalize the preceding by finding N equally spaced numbers whose average is m and standard deviation is s. (*Hint:* Consider N even and N odd separately.)

46. Show that the median and the average of the numbers $1, 2, 3, \ldots, N$ are always the same.

47. Suppose that the average of the numbers $x_1, x_2, x_3, \ldots, x_N$ is m and that the standard deviation of these same numbers is s. Suppose also that the average of the numbers $x_1^2, x_2^2, x_3^2, \ldots, x_N^2$ is M. Show that $s^2 = M - m^2$. (In other words, for any data set, if we take the average of the squared data values and subtract the square of the average of the data values we get the square of the standard deviation.)

48. Given that the numbers $x_1, x_2, x_3, \ldots, x_N$ have standard deviation s, explain why the numbers $x_1 + c, x_2 + c, x_3 + c, \ldots, x_N + c$ also have standard deviation s.

49. (a) Using the formula $1^2 + 2^2 + 3^2 + \cdots + N^2 = \dfrac{N(N+1)(2N+1)}{6}$, find the standard deviation of the data set $\{1, 2, 3, \ldots, 98, 99\}$. (*Hint:* Use Exercise 47.)

(b) Find the standard deviation of the data set $\{315, 316, \ldots, 412, 413\}$. (*Hint:* Use Exercise 48.)

REFERENCES AND FURTHER READINGS

1. Berry, D. A., and B. W. Lindgren, *Statistics: Theory and Methods.* Pacific Grove, CA: Brooks/Cole Publishing Co., 1990, chap. 7.

2. Cleveland, W. S., *The Elements of Graphing Data.* Pacific Grove, CA: Brooks/Cole Publishing Co., 1985.

3. Cleveland, W. S., and M. E. McGill (eds.), *Dynamic Graphics for Statistics.* Pacific Grove, CA: Brooks/Cole Publishing Co., 1988.

4. Freedman, D., R. Pisani, R. Purves, and A. Adhikari, *Statistics* (2nd. ed.). New York: W. W. Norton, Inc., 1991, chaps. 3 and 4.

5. Groeneveld, Richard A., *Introductory Statistical Methods.* Boston: PWS-Kent Publishing Co., 1988, chap. 2.

6. Mosteller, F., W. Kruskal, et al., *Statistics by Example: Exploring Data,* Reading, MA: Addison-Wesley Publishing Co., Inc., 1973.

7. Sincich, Terry, *Statistics by Example.* San Francisco, CA: Dellen Publishing Co., 1990, chaps. 2 and 3.

8. Tanner, Martin, *Investigations for a Course in Statistics.* New York: Macmillan Publishing Co., Inc., 1990.

9. Tufte, Ed, *Envisioning Information.* Cheshire, CN: Graphics Press, 1990.

10. Tufte, Ed, *The Visual Display of Quantitative Information.* Cheshire, CN: Graphics Press, 1983.

11. Tukey, J. W., *Exploratory Data Analysis.* Reading, MA: Addison-Wesley Publishing Co., Inc., 1977.

12. Wainer, H., "How to Display Data Badly," *The American Statistician,* 38 (1984), 137–147.

13. Wildbur, Peter, *Information Graphics,* New York: Van Nostrand Reinhold Co., 1986.

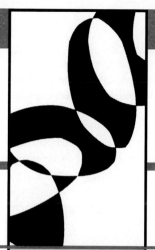

CHAPTER 15

Chances Are ...

Probability Is ...

> *The most important questions of life are, for the most part, really only problems of probability*
> *PIERRE SIMON DE LAPLACE**

- ■ "There is a 90% chance of rain tomorrow." (Weather report.)

- ■ "Each year the average driver has one chance in three of being in an automobile accident." (Insurance company report.)

- ■ "The odds that the Cincinnati Reds will win the World Series are 5 to 2." (Nevada oddsmaker.)

- ■ "The probability of tossing a coin heads three times in a row is 0.125." (Gambler's manual.)

"Probabilities," "chances," "odds"—these words are as much a part of our everyday vocabulary as "mother," "baseball," and "apple pie." While we all probably (there it is again!) have an intuitive idea of what each of the four statements above means, giving a precise definition of terms such as "probability," "chances," and "odds" is surprisingly difficult.

In this chapter we will discuss probabilities, chances, and odds (they are just different ways of looking at the same thing) using both an informal and a more formal mathematical approach. Before we start our discussion in full, let's clear the air a little regarding the terminology.

The word "probability" is commonly used in two different ways. On the one hand probability is a *discipline*—in fact an important branch of mathematics—and it is in this context that we have used it in the subtitle of this chapter. On the other hand, we will talk about the probability of something happening, and in this context "the probability" is a number such as the one in the statement "The probability of tossing a coin heads three times in a row is 0.125."

---

* Pierre Simon, Marquis de Laplace (1749–1827) was a renowned French mathematician and astronomer. Among his many works, Laplace wrote *Théorie Analytique des Probabilités (The Analytical Theory of Probabilities)*, one of the earliest treatises in the mathematical theory of probability.

When used in the latter context, probability can also be expressed by using the word "chance" as in "the chances of tossing a coin heads three times in a row are 12.5%." [1] The term "odds" is used to express the same concept as probability in a slightly different form. Later in the chapter we will explain how to convert odds to probabilities and probabilities to odds.

Our discussion in this chapter is broken up into two parts. In the first part we lay down the basic concepts needed for a meaningful discussion of probability, and it isn't until the second part that we actually define and calculate probabilities using a formal approach.

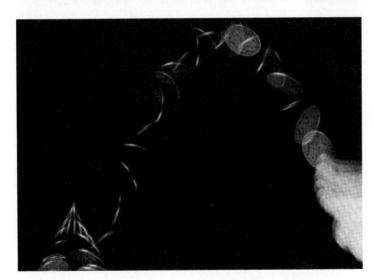

Top left, courtesy National Center for Atmospheric Research/National Science Foundation: top right, Doreen E. Pugh, Photo Researchers; bottom left, © Estate of Harold Edgerton, 1965; bottom right, AP/Wide World Photos.

[1] It is customary to express probabilities in decimals (or fractions) and chances in percentages, and we will follow that custom in this chapter.

WHAT IS PROBABILITY? RANDOM EXPERIMENTS AND SAMPLE SPACES

Probability is the mathematical study of *unpredictable events*. As usual, a few formal introductions are in order.

We will use the term **random experiment** to describe an activity or process whose outcome cannot be predicted ahead of time. Typical examples of random experiments are tossing a coin, rolling a pair of dice, shooting a free throw, predicting the outcome of a baseball game, etc. Notice that in this setting, the word "experiment" is used in a very liberal way: Unlike chemistry or physics experiments, random experiments do not require elaborate setups or fancy equipment.

Associated with every random experiment is the *set* of all its possible outcomes, called the **sample space**. We illustrate this concept (and some unexpected subtleties) by means of several examples. We will use the letter S to denote the sample space, and N to denote the number of outcomes in S.

Example 1. Random experiment: Toss a coin.

Sample space: $S = \{H, T\}$ (From now on we will abbreviate heads as H and tails as T.) The curly brackets are there merely to indicate that the sample space is a set.

Size of sample space: $N = 2$.

Example 2. Random experiment: Toss a coin twice.

Sample space: $S = \{HH, HT, TH, TT\}$. (Here HT means the first toss came up H and the second toss came up T.)

Size of sample space: $N = 4$.

Example 3. Random experiment: Shoot a pair of free throws.

Sample space: $S = \{ss, sf, fs, ff\}$. (Here s means success and f means failure.)

Size of sample space: $N = 4$.

Notice the similarity between Examples 2 and 3. In fact, if we were to identify H with success and T with failure, the sample spaces would be exactly the same. Examples 2 and 3 illustrate the fact that seemingly different random experiments (tossing a pair of coins, shooting a pair of free throws) can turn out to have essentially the same sample space (the symbols may be different, but the essence is the same).

Example 4. Random experiment: Shoot as many free throws as needed until you make one, and then stop.

Sample space: $S = \{s, fs, ffs, fffs, \ldots\}$. The " $\ldots$ " here is significant. It is an indication that we are dealing with an infinite sample space. There is, after all, no limit to the number of consecutive free throws that someone could miss.

While infinite sample spaces are important in many applications of probability theory, they are beyond the level of our presentation in this chapter. The last example was given simply to illustrate the fact that infinite sample spaces are also possible. We will now return to finite sample spaces.

Example 5.   Random experiment: Roll a single die.[2]
Sample space: $S = \{\;\boxdot,\;\boxdot,\;\boxdot,\;\boxdot,\;\boxdot,\;\boxdot\;\}$.
Size of sample space: $N = 6$.

Example 6.   Random experiment: Roll a pair of dice.
Sample space: $S = \{$ $\}$.

Size of sample space: $N = 36$.
Notice that in this example we are treating the dice as distinguishable objects (as if one were white and the other red), so that and are considered different outcomes.

Example 7.   In many games of chance involving the roll of a pair of dice (such as Monopoly, craps, etc.), the outcome of the game depends on the total sum rolled. The roll corresponds to the total 8, and so on. In this situation we are not interested in the individual rolls of the dice but rather in their combination. What is the sample space in this case? Since the possible outcomes are the possible totals $2, 3, \ldots, 12$, we have
Random experiment: Roll a pair of dice and check for the total sum rolled.
Sample space: $S = \{2, 3, 4, 5, 6, 7, 8, 9, 10, 11, 12\}$.
Size of sample space: $N = 11$.

---

[2] Singular, die; plural, dice.

There is no contradiction between Examples 6 and 7. While the random experiment in both cases can be said to be the same (roll a pair of dice), the *observation* we make is not. In Example 6, we just roll, in Example 7 we roll and add, and so the sample spaces reflect this difference. This is our second lesson: *The same basic random experiment can result in different sample spaces* depending on exactly what aspect of the experiment one is observing. It is important to notice that the only unpredictable part in rolling the dice and adding the total is in the rolling of the dice (let's agree that adding is not a random experiment), and yet the combination of the two (roll, then add) results in a new sample space.

Our next example reiterates the previous point.

Example 8.   Two friends, Thelma and Mabel, go to the racetrack. There are five horses entered in race 1—let's call them A, B, C, D, and E. Thelma makes a straight bet on a winner (if her horse wins, she wins the bet). For Thelma the sample space is $S_T = \{A, B, C, D, E\}$. Mabel buys a trifecta ticket, so she must correctly get the first-, second-, and third-place finishers and in exactly that order to win her bet. For Mabel, the sample space is $S_M = \{ABC$ (ABC means horse A comes in first, horse B comes in second, and horse C comes in third), ACB, BAC, BCA, CAB, CBA, ABD, ADB, . . . $\}$. Obviously, S_M is a pretty huge sample space, and for the time being we won't worry too much about writing it all out. The main point once again is that the same random experiment (the horse race) can produce more than one sample space depending on the nature of the observation that interests us.

SIZING THE SAMPLE SPACE: THE MULTIPLICATION PRINCIPLE AND VARIATIONS THEREOF

Example 8 illustrates an important situation: a sample space (S_M) that is too big to write down in its entirety. This situation is not unusual. Many finite sample spaces are incredibly large, and it is inconceivable that in such cases we would ever try to write down a complete list of all the outcomes. Fortunately, to us the critical question is the *size* of the sample space, and we can answer this question even when we don't write the sample space down. We discuss how to do this next.

Example 9.   Random experiment: Toss a coin three times.
Sample space: $S = \{HHH, HHT, HTH, HTT, THH, THT, TTH, TTT\}$.
Size of sample space: $N = 8$.

Example 10. Random experiment: Toss a coin eight times.

In this example the sample space S is too big to write down. One may ask what a typical outcome might be like. Taking our cue from Example 9, we can say that a typical outcome can be described as a string of eight consecutive letters where the letters can be only H's or T's. For example, the string *THHTHTHH* represents the following outcome: First toss came up T, second toss came up H, third toss came up H, etc. It is obvious that there are many other possible strings and that writing them all down and then counting them one by one like we count sheep would be an exhausting and very unmathematical thing to do. There is, however, another way:

■ Number of possibilities for the first toss = 2 (H or T).

■ Number of possibilities for the second toss = 2 (ditto).

.

.

.

■ Number of possibilities for the eighth toss = 2.

Total number of outcomes = $2 \times 2 \times 2 \times 2 \times 2 \times 2 \times 2 \times 2 = 256$.

We have now found in a relatively painless way that the size of our sample space is $N = 256$.

The basic principle that we used in Example 10 is called (for obvious reasons) the **multiplication principle**.[3] Informally stated, it says that when something takes place in several stages, to find the total number of ways it can occur we find the number of ways each individual stage can occur and then multiply them.

We can justify *why* the multiplication principle works with a simple example.

Example 11. You want a single scoop of ice cream. There are two types of cones available (sugar and regular) and three flavors to choose from (vanilla, chocolate chip, and chocolate). Figure 15-1 shows all the possible combinations.

---

[3] The multiplication principle is also discussed (in somewhat lesser detail) in Chapter 2.

Cones   Flavors   Combinations

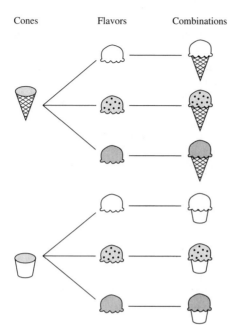

FIGURE 15-1 The multiplication principle at work.

The multiplication principle is the fundamental tool used in solving **counting problems**, that is, problems that ask in how many different ways can one thing or another happen. These are exactly the kinds of questions one needs to answer to carry out the basic probability calculations that we will want to do later in this chapter. Before we get to that, we will take a brief detour to explore a few of the subtleties involved in working with counting problems. This is an important and rich subject, full of interesting twists and turns, but due to necessity, our detour will be brief.

Our detour starts with a straightforward application of the multiplication principle. Each successive example shows some variation of the basic theme.

Example 12. Dolores goes on a business trip. She packs three different pairs of shoes, four different skirts, six different blouses, and two different jackets. If all the items are color-coordinated, how many different outfits are possible?

To answer this question we must first define what we mean by an outfit. Let's assume that an "outfit" consists of a pair of shoes, a skirt, a blouse, and a jacket. We can now use the multiplication principle to calculate the total number of possible outfits. It is $3 \times 4 \times 6 \times 2 = 144$. Color coordination obviously pays: Dolores can go for over four months and never have to wear the same outfit twice!

A few subtleties here and there can make the use of the multiplication principle a little more of a challenge.

Example 13. Once again, Dolores goes on a business trip. This time she packs three pairs of shoes, four skirts, three pairs of slacks, six blouses, three turtlenecks, and two jackets. As usual, everything goes with everything else, and we want to know how many different outfits are possible. This time let's define an "outfit" as consisting of a pair of shoes, some "lower wear" (either a skirt or a pair of slacks), some "upper wear" (either a blouse or a turtleneck, or both), and she may or may not choose to wear a jacket (it's springtime). Let's count the possibilities for each stage that makes up the outfit separately:

■ Number of possibilities for the shoes = 3.

■ Number of possibilities for lower wear = 4 + 3 = 7.
   Skirt   Slacks

■ Number of possibilities for upper wear
   = 6 + 3 + 18 = 27.
   Blouse   Turtleneck   Blouse/Turtleneck
   alone   alone   combination (per
                         multiplication principle)

■ Number of possibilities for jacket = 2 + 1 = 3.
   Jacket   No jacket

According to the multiplication principle, the total number of different outfits is $3 \times 7 \times 27 \times 3 = 1701$.

Example 14. In Example 8 we discussed a sample space S_M corresponding to all possible ways in which five horses (A, B, C, D, and E) could finish 1-2-3 in a horse race (remember Mabel's trifecta bet?). In Example 8 we started writing out the sample space:

$$S_M = \{ABC, ACB, BAC, BCA, CAB, CBA, ABD, ADB, \ldots\}$$

but we gave up because we realized it was too big. Using the multiplication principle, we can now determine exactly how big S_M is. We have

■ Number of possibilities for the 1st place horse = 5.

■ Number of possibilities for the 2nd place horse = 4 (any other horse except the one in first place).

■ Number of possibilities for the 3rd place horse = 3 (any horse except the ones in first and second place).

The total number of possible outcomes = $5 \times 4 \times 3 = 60$.

More Sophisticated Counting Problems

In counting problems, it is often the case that the multiplication principle by itself is not enough, and some subtle adaptations may be necessary. Take for example the question of "double deckers" at Baskin-Robbins.™

Example 15. Baskin-Robbins offers 31 different flavors of ice cream. A "double decker" is the name some kids use for a double (two scoops) of two *different* flavors. How many different double deckers can be ordered at Baskin-Robbins?
   Taking our cue from Example 14, we say

- Number of possibilities for first flavor = 31.

- Number of possibilities for second flavor = 30 (remember, we want two different flavors!).

   It would appear at first glance that the total number of possible double deckers is $31 \times 30 = 930$, but when we look at the double deckers with a little care, we find that we have double counted each of them. Double counting double deckers? Why? When we put down strawberry as the first flavor and chocolate as the second flavor we get a double decker, but when we put down chocolate first and strawberry second, we get the same double decker. When we use the multiplication principle, each one of these comes up separately, but they are really one and the same double decker. Fortunately, once we realize this, we can make the appropriate correction: Since 930 counts each double decker twice, the exact number of double deckers is half of that (465). To summarize:
   Total number of double deckers at Baskin-Robbins = $(31 \times 30)/2$.
   In the above expression, the numerator is obtained from the multiplication principle, whereas the denominator comes from the fact that the order in which we give the flavors is irrelevant.

Example 16. Let's carry the ideas of Example 15 one step further. Let's say that a "triple decker" consists of three scoops of ice cream, each of the three scoops being of a different flavor. How many different triple deckers can be ordered at Baskin-Robbins?
   Starting with the multiplication principle, we have 31 choices for the "first" flavor, 30 choices for the "second" flavor, and 29 choices for the "third" flavor, for a grand total of $31 \times 30 \times 29 = 26{,}970$ combinations. But once again, this is not the correct answer—in fact, the correct answer is the above number divided by 6. In other words, the exact number of possibilities for ordering a triple decker at Baskin-Robbins is $26{,}970/6 = 4495$.
   The critical question is, Why did we divide by 6? If we look at a particular combination of three flavors, say chocolate (*C*), strawberry (*S*), and vanilla (*V*), there are six ways in which these flavors can be mentioned when we give our order (*CSV*, *CVS*, *SCV*, *SVC*, *VCS*, and *VSC*). Each one of these is counted separately when we use the multiplication principle, but in reality they are all the

same triple decker. Needless to say, the same thing happens with any other combination of three flavors: one triple decker, six different ways to count it. This explains why dividing our original count of 26,970 by 6 gives us the correct answer.

It is helpful to look at the final answer (4495 triple deckers) in terms of its pedigree: $4495 = (31 \times 30 \times 29)/(3 \times 2 \times 1)$. We already know where the numerator comes from (the multiplication principle). The denominator $3 \times 2 \times 1$ comes from the fact that there are that many ways to shuffle around three things (in this case the three ice cream flavors in a particular triple decker).

One more point about the answer we got in Example 16: the denominator $3 \times 2 \times 1$ is the factorial of 3. The *factorial* of a positive integer N is denoted by $N!$, and is the number $N \times (N - 1) \times (N -2) \times \cdots \times 2 \times 1$. It represents, among other things, the number of ways in which N objects can be reordered or, if you will, reshuffled. We discussed the concept of the factorial of a number in Chapters 2 and 6, so we won't dwell on it at this point.

Example 17 (The California Lottery). To play the California Lottery, a person has to pick 6 out of 51 numbers (after paying $2 for the privilege). If the pigeon, er, person picked exactly the same 6 numbers as the ones drawn by the lottery, he or she can win mountains of money (usually a few million but it can be as much as forty or fifty million). To us, the relevant question is, How many different lottery tickets are possible?

Without the benefit of Examples 15 and 16, this would be a fairly difficult question, but at this point we are in pretty good shape to answer it. The first useful observation is that this is the same type of problem as the ones in Examples 15 and 16. After all, choosing 6 out of 51 numbers is just like ordering a "six decker" at an ice cream parlor that offers 51 flavors.

We start with the multiplication principle:

- Number of ways to choose the first number = 51.
- Number of ways to choose the second number = 50.
- Number of ways to choose the third number = 49.
- Number of ways to choose the fourth number = 48.
- Number of ways to choose the fifth number = 47.
- Number of ways to choose the sixth number = 46.

The total number of ways to choose the six numbers in sequence is given by $51 \times 50 \times 49 \times 48 \times 47 \times 46$, a very big number indeed. Since the lottery ticket does not depend on the particular order in which the numbers are chosen we need to divide this number by $6! = 6 \times 5 \times 4 \times 3 \times 2 \times 1$ (the number of different orders in which the same six numbers can be chosen). In conclusion, the total number of possible lottery tickets in the California lottery is

$$\frac{51 \times 50 \times 49 \times 48 \times 47 \times 46}{6 \times 5 \times 4 \times 3 \times 2 \times 1} = 18{,}009{,}460$$

We now leave the wonderful world of counting problems and return to the main theme of the chapter—the mathematical theory of probabilities.

RANDOM VARIABLES

In Chapter 14 we discussed the concept of a variable associated with a given population. Variables can also be associated with a random experiment in a natural way since, after all, in a random experiment there must be, by definition, variability in the outcomes. When the value of a *quantitative* variable is determined by a random experiment, we call such a variable a **random variable**. To put it in a slightly different way, if the population we are studying is a sample space (why not?), then any quantitative variable associated with that population is called a random variable.

Let's illustrate the idea of a random variable with a few simple examples. We will follow traditional usage and describe random variables by capital letters such as X, Y, etc.

Example 18. Random experiment: Roll a pair of dice.
Sample space: $S = \{$ $\}$.

Random variable: X = sum of the faces showing.
Possible values of X: 2, 3, . . . 12.

Example 19. Random experiment: Toss a coin eight times.
Sample space: too big to write down. $N = 256$ (see Example 10).
Random variable: X = number of heads tossed.
Possible values of X: 0, 1, 2, . . . , 8.

Example 20. Random experiment: Toss a coin eight times.
Random variable: W = total winnings if you win $1 for every head tossed and lose $1 for every tail tossed.
Possible values of W: -8, -6, -4, -2, 0, 2, 4, 6, 8. (See Exercise 48.)

While the use of random variables appears at first glance to be a somewhat formal way to look at things that are in general quite natural, the idea of a random variable has a tremendous advantage: It allows all the tools of descriptive statistics (graphs, frequency distributions, numerical summaries, etc.) to be used in the study of random experiments.

PROBABILITIES (THE NUMBERS)

So far, we have talked about random experiments, sample spaces, the multiplication principle, and random variables, but have said little about probabilities. It is now time to say more.

Suppose that we toss a coin in the air. What is the probability that it will land heads? This is not a deep mathematical question, and almost everybody has an opinion on the matter. Typical answers are 50%, 1 out of 2, about one-half, etc. When asked the reason for such answers, the responses given usually fall into two categories.

One type of argument is that since the coin toss can result in two possible outcomes and heads is one of the two possibilities, the probability of heads must be 1 out of 2. We will call this type of argument the *objective approach* to defining probabilities.

The other type of argument essentially goes like this: If we toss the coin over and over again, many, many times, in the long run about half of the tosses will turn out to be heads and about half will turn out to be tails, so the probability of heads is about one-half. We call this type of argument the *frequency approach* to defining probabilities.

While both of these are legitimate arguments (assuming the coin is an honest coin), there are difficulties generalizing either of them to cover all situations that involve probabilities.

Suppose that a person who has never played basketball is asked to shoot a free throw. What is the probability that he or she will make the free throw? It is clear that the objective approach cannot be used in this case. Once again there are only two outcomes (success and failure), but quite obviously the probability of success cannot be forced to be 1 out of 2. We cannot split the probability pie into equal shares because success and failure are not equally likely outcomes! In this situation, any probability assigned to a successful free throw would be nothing more than the subjective opinion of an observer expressed in numerical form.

The frequency approach to defining probabilities (the one that uses expressions like "if we toss the coin many times" or "in the long run") has a different kind of drawback: There are many random experiments that are not repeatable and yet can be discussed in terms of probabilities. "What are the chances that inflation will be less than 2% this year?" Here is a perfectly legitimate probability question even though the random experiment is definitely not repeatable. Even an apparently precise answer to this question ("The chances are 15%") is nothing more than a subjective opinion formulated in numerical form.

The debate over what is the correct way to define probabilities has been around for many years. There are *objectivists*, those who believe that probability is an intrinsic characteristic of the random experiment, and there are *subjectivists*, those who believe that probabilities are basically opinions expressed as numbers and are therefore a characteristic of the observer and not of the random experiment itself.

The mathematical way out of this dilemma is what we will discuss next. We will present only the basic ideas behind the formal mathematical treatment of probabilities because the subject is deep and covers a great deal of ground. The basic ideas originated in the early 1930s and are attributed to the Russian mathematician A. N. Kolmogorov.

Probability Assignments

Let's go back to the free throw example discussed earlier.

Example 21. An individual shoots a free throw. We know nothing about his or her abilities (for all we know, the person could be Michael Jordan or it could be Joe Schmoe). What is the probability that he or she will make the free throw?

We know (we feel it in our bones) that the answer need not be 0.5. In fact, we know that the probability could be just about any number. Any number? Well, let's back off a little bit. It could not be a negative number (that's obvious), and it could not be a number bigger than 1 (100% chance), so we can cut our original boast down to "The probability could be any number between 0 and 1." How about 0 and 1? A probability of 0 would mean no chance of success at all; a probability of 1 would mean guaranteed success. While both of these situations are unlikely, they are nevertheless technically possible and we will allow them.

We can summarize what we have so far as follows: The probability of a successful free throw is a number between 0 and 1 inclusive. We don't know the exact value because we don't know anything about the person shooting the free throw, but that's no problem—we just make it a variable (say p).

One final comment about this example. The probability of an unsuccessful free throw is also a number between 0 and 1 inclusive. In fact, the two probabilities (successful and unsuccessful) are related by the fact that they must add up to 1 (the free throw is either successful or unsuccessful—there are no other alternatives!). This means that the probability of missing the free throw must be $1 - p$.

Table 15-1 is a summary of all the preceding comments.

Outcomes	Probabilities
Success (s)	Probability of success = p
Failure (f)	Probability of failure = $1 - p$

■ **TABLE 15-1　Random Experiment: Shooting a Free Throw**

Table 15-2 is the same thing in mathematical shorthand (the new notation is self-explanatory).

Outcomes	Probabilities
Sample space $\begin{cases} s \\ f \end{cases}$	$\Pr(s) = p$ $\Pr(f) = 1 - p$

■ **TABLE 15-2**

It is worth mentioning that our new description is incredibly flexible. It works just as well as when the free throw shooter is Michael Jordan (make $p = 0.85$) or when it is Joe Schmoe (make $p = 0.15$) or when it is any other Tom, Dick, or Harry in between. Each one of the choices results in a different **probability assignment** for the sample space.

Michael Jordan and Joe Schmoe shoot a freethrow. Same sample space, different probabilities. (Jordan from Barry Gossage/ National Basketball Association)

Example 22. Five players, Boris, Martina, Andre, Gabriela, and Monica, enter a tennis tournament. We are interested in who is going to be the winner of the tournament. The sample space is $S = \{$Boris, Martina, Andre, Gabriela, Monica$\}$.

According to one expert, the probability assignment for this sample space is $\Pr(\text{Boris}) = .25$, $\Pr(\text{Martina}) = .22$, $\Pr(\text{Andre}) = .14$, $\Pr(\text{Gabriela}) = .18$. The value of $\Pr(\text{Monica})$ is not given, but we can determine that $\Pr(\text{Monica}) = .21$ because the total sum of the probabilities must be 1.

A second tennis expert completely disagrees with this assignment, claiming that Martina, Gabriela, and Monica all have an equal chance of winning, Andre has a 10% chance of winning, and Boris has a 27% chance of winning. These opinions translate into a second probability assignment for the sample space. What is it? Since Boris and Andre together account for a 37% chance of winning, Martina, Gabriela, and Monica must account for the remaining 63%.

Since they all have an equal chance of winning, the chance for each of them is 21%. The precise formulation of the second expert's opinion is the following probability assignment for the sample space: Pr(Andre) = .1, Pr(Boris) = .27, Pr(Martina) = .21, Pr(Gabriela) = .21, and Pr(Monica) = .21.   ▱

Examples 21 and 22 illustrate the meaning of a probability assignment for a sample space, which we now formalize. A **probability assignment** for a sample space S is a set of numbers that is assigned to the outcomes in the sample space and which satisfies the following two conditions:

    1. Each number in the set is between 0 and 1 (inclusive).

    2. The numbers add up to 1.

Any set of numbers that satisfies conditions 1 and 2 is a legal probability assignment.

EVENTS

So far we have talked about the probabilities of the individual outcomes in a sample space—these are given by a probability assignment for the sample space. We will now take things one step further and talk about combinations of outcomes and their probabilities.

Example 23. Suppose we want to determine the probability that a female player will win the tennis tournament in Example 22. Intuitively, it is clear that all we have to do is to look up the probability assigned to each female player in the sample space and add up these numbers.
    A slightly more mathematical description is obtained by first defining the appropriate set

        Female = {Martina, Gabriela, Monica}

and then computing its probability

        Pr(Female) = Pr(Martina) + Pr(Gabriela) + Pr(Monica).

    Of course the exact answer depends on the probability assignment. According to the first expert, Pr(Female) = .22 + .18 + .21 = .61, but according to the second expert, Pr(Female) = .21 + .21 + .21 = .63.   ▱

Most probability questions are concerned with combinations of outcomes called events. We will define an **event** as any set of outcomes taken from the sample space.

Example 24. In Example 9 we considered the random experiment of tossing a coin three times and saw that the sample space was $S = \{HHH, HHT, HTH, HTT, THH, THT, TTH, TTT\}$. There are many possible events for this sample space. Table 15-3 shows just a few of them.

	Event	Set of Outcomes	Size of Event
1.	Toss two or more heads	$\{HHT, HTH, THH, HHH\}$	4
2.	Toss more than two heads	$\{HHH\}$	1
3.	Toss two heads or less	$\{TTT, TTH, THT, HTT, THH, HTH, HHT\}$	7
4.	Toss no tails	$\{HHH\}$	1
5.	Toss exactly one tail	$\{HHT, HTH, THH\}$	3
6.	Toss exactly one head	$\{HTT, THT, TTH\}$	3
7.	First toss is heads	$\{HHH, HHT, HTH, HTT\}$	4
8.	Toss same number of heads as tails	$\{\}$	0
9.	Toss at most three heads	S	8
10.	First toss is heads and at least two tails are tossed	$\{HTT\}$	1

■ TABLE 15-3   **Some of the Many Possible Events in a Sample Space**   ▭

As Example 24 illustrates, there are many ways in which outcomes in a sample space can be combined to make an event, and the same event can be described (in English) in more than one way (e.g., events 2 and 4 in Table 15-3). The actual number of outcomes in an event can be as low as 0 and as high as N (the size of the sample space). In the case in which the number of outcomes is 0 (as in event 8 in Table 15-3), the event is called the **impossible event**; in the case in which the number of outcomes is N (as in event 9 in Table 15-3), the event is the whole sample space S and it is called the **certain event**.

Once a probability assignment is made for the sample space, every event is automatically assigned a probability. It is obtained by adding the probabilities of the individual outcomes that make up that event. No matter what the specific probability assignment is, the probability of the impossible event is always 0 $[\Pr(\{\}) = 0]$ and the probability of the certain event is always 1 $[\Pr(S) = 1]$.

We are now ready to describe what we will call a **formal probability model**. This is where the various concepts we have discussed so far finally come together.

FORMAL PROBABILITY MODELS

A probability model starts with a sample space, which represents the set of all possible outcomes of an observed random experiment. Next comes a probability assignment for the sample space. It bestows on each individual outcome a number between 0 and 1 inclusive, which is the probability of that outcome. In principle the mathematical model is not concerned with where these particular probabilities come from. They could come from subjective opinions of an observer, from the results of long-term frequency calculations, or from applying some complicated mathematical formula. In the eyes of the formal probability model, they are all perfectly legal as long as the rules of the game (each of the numbers is between 0 and 1 and they add up to 1) are obeyed. Once a probability assignment for the sample space is made, not only individual outcomes but also combinations of outcomes called events have probabilities assigned to them. The probability of any event is obtained by adding the probabilities of the individual outcomes that make up that event. In particular, the impossible event {} has probability 0 and the certain event S has probability 1. All the above facts are summarized as follows:

FORMAL PROBABILITY MODEL

- **Sample space.** Set of all possible outcomes of a random experiment $S = \{o_1, o_2, \ldots, o_N\}$.
- **Probability assignment for S.** To each individual outcome o_i we assign a number $\Pr(o_i)$. The rules are
  1. $0 \leq \Pr(o_i) \leq 1$, and
  2. $\Pr(o_1) + \Pr(o_2) + \cdots + \Pr(o_N) = 1$.
- **Events.** Any subset of S is an event. In particular, {} (impossible event) and S itself (certain event) are events.
- **Probability of events.** The probability of an event is obtained by adding the probabilities of the individual outcomes that make up that event. In particular, $\Pr(\{\}) = 0$ and $\Pr(S) = 1$.

When Every Outcome Is Equally Likely

One of the most common uses of probability involves situations associated with gambling. While we do not condone gambling, it does provide a rich source of examples and gives rise to many interesting mathematical questions. Many gambling situations involve the use of a physical device, a coin or coins, a die or dice, a deck of cards, a roulette wheel, etc. In these situations it is a given that the device is honest, that is, the coin or the dice are fair, the cards haven't been marked, and the roulette wheel is perfectly balanced. Mathematically speaking, this assumption translates into the fact that each individual outcome in the sample space is equally likely to occur; that is, all the outcomes have equal probabilities. Knowing that the probability of each outcome is the same and that the probabilities add up to 1 (one of the rules in any probability model) tells us (1) that the probability assignment for each outcome must be $1/N$ (where N is the size of the sample space), and (2) the probability of an event is obtained by dividing the number of outcomes in the event by N. We can describe this formal probability model mathematically as follows:

**PROBABILITY MODEL WHEN ALL OUTCOMES ARE
EQUALLY LIKELY**

■ Size of sample space $= N$.

■ Pr(any outcome) $= 1/N$.

■ If E is an event, $\mathrm{Pr}(E) = \dfrac{\text{number of outcomes in } E}{N}$.

In this model, probabilities can be computed exactly by counting the number of outcomes in both the event and the sample space, and this takes us back to the multiplication principle.

Example 25. A card is drawn from an honest deck of 52 cards. What is the probability of drawing an Ace?
   Here $N = 52$. The event

$$\text{Ace} = \left\{ \boxed{A\,\heartsuit}\ \boxed{A\,\diamondsuit}\ \boxed{A\,\clubsuit}\ \boxed{A\,\spadesuit} \right\}$$

is made up of four outcomes, each of which has probability $\frac{1}{52}$. It follows that

$$\mathrm{Pr}(\text{Ace}) = \frac{4}{52} = \frac{1}{13} \approx .077.$$

Example 26. Two cards are drawn in order from an honest deck of 52 cards. What is the probability of drawing a pair of Aces?
   We first need to calculate the size of the sample space. Given that the cards are taken in order (first one, and then the other one), we can argue as follows:

■ The number of ways in which the first card can be drawn $= 52$.

■ The number of ways in which the second card can be drawn $= 51$.

■ $N =$ total number of ways of drawing the two cards $= 52 \times 51$.

   Next, we calculate the size of the event E: drawing two Aces.

■ The number of possibilities for the first Ace $= 4$.

■ The number of possibilities for the second Ace $= 3$.

■ The size of the event $E = 4 \times 3$.

$$\mathrm{Pr}(E) = \frac{4 \times 3}{52 \times 51} = \frac{1}{13 \times 17} = \frac{1}{221} \approx .0045.$$

Example 27. Suppose that we roll a pair of honest dice. What is the probability of rolling a total of 11? What is the probability of rolling a total of 7? What is the probability of rolling a 7 or an 11?

There are 36 ways to roll a pair of dice (Example 6), and since the dice are honest, each of these outcomes has probability $\frac{1}{36}$. There are two ways of rolling a total of 11 ($\{\boxed{\cdot\cdot}\boxed{\vdots}, \boxed{\vdots}\boxed{\cdot\cdot}\}$).

It follows that

$$\Pr(\text{R}11) = \frac{2}{36} = \frac{1}{18} \approx .056.$$

(We will use R11 to denote the event "rolling a total of 11," R7 to denote the event "rolling a total of 7," etc.)

The event R7 has six possible outcomes:

$$\text{R7} = \{\boxed{\cdot}\boxed{\vdots}, \boxed{\cdot}\boxed{\vdots}, \boxed{\cdot\cdot}\boxed{\vdots}, \boxed{\cdot\cdot}\boxed{\vdots}, \boxed{\cdot\cdot}\boxed{\vdots}, \boxed{\vdots}\boxed{\cdot\cdot}\}$$

so

$$\Pr(\text{R7}) = \frac{6}{36} = \frac{1}{6} \approx .167.$$

The event R7 or R11 has eight possible outcomes (the six in R7 and the two in R11), so

$$\Pr(\text{R7 or R11}) = \frac{8}{36} = \frac{2}{9} \approx .22.$$

Example 28. Suppose that we roll a pair of honest dice. What is the probability that we will roll at least one $\boxed{\cdot}$?

We already know (Example 27) that each individual outcome has probability $\frac{1}{36}$. We will show three different ways to solve this problem.

■ **Solution 1 (The Brute Force Approach).** If we just write down the event E, which is "we will roll at least one $\boxed{\cdot}$," we have

$$E = \left\{ \begin{array}{l} \boxed{\cdot}\boxed{\cdot}, \boxed{\cdot}\boxed{\vdots}, \boxed{\cdot}\boxed{\vdots}, \boxed{\cdot}\boxed{\vdots}, \boxed{\cdot}\boxed{\vdots}, \boxed{\cdot}\boxed{\vdots}, \\ \boxed{\cdot}\boxed{\cdot}, \boxed{\vdots}\boxed{\cdot}, \boxed{\vdots}\boxed{\cdot}, \boxed{\vdots}\boxed{\cdot}, \boxed{\vdots}\boxed{\cdot} \end{array} \right\}$$

It follows that $\Pr(E) = \frac{11}{36}$.

■ **Solution 2 (The Roundabout Approach).** Let's say for the sake of argument that we will win if at least one of the two dice comes up a $\boxed{\cdot}$ and we will lose otherwise. This means that we will lose if both dice come up with a number other than ▐. Let's calculate the probability that we will lose first (this is called the roundabout way of doing things). Using the multiplication principle, we can calculate the number of individual outcomes in the event "we lose":

- ■ Number of ways first die can come up (not a ▐) = 5.

- ■ Number of ways the second die can come up (not a ▐) = 5.

- ■ Total number of ways both dice can come up (neither a ▐) = $5 \times 5 = 25$.

Probability that we will lose: $\Pr(\text{lose}) = \frac{25}{36}$.

Probability that we will win: $\Pr(\text{win}) = 1 - \frac{25}{36} = \frac{11}{36}$.

- ■ **Solution 3 (Independent Events).** In this solution we consider each die separately. In fact, we will find it slightly more convenient to think in this case of rolling a single honest die twice (mathematically it is exactly the same thing as rolling a pair of honest dice once).

  Let's start with the first roll. The probability that we won't roll a ▐ is $\frac{5}{6}$ (there are six possible outcomes, five of which are not a ▐). For the same reason, the probability that the second roll will not be a ▐ is also $\frac{5}{6}$.

  Now comes a critical observation: The probability that neither of the first two rolls will be a ▐ is $\frac{5}{6} \times \frac{5}{6} = \frac{25}{36}$. The reason we are able to multiply the probabilities of the two events ("first roll is not a ▐" and "second roll is not a ▐") was that these two events are **independent**: The outcome of the first roll does not in any way affect the outcome of the second roll.

  We finish the problem exactly as in solution 2: $\Pr(\text{lose}) = \frac{25}{36}$ and therefore $\Pr(\text{win}) = 1 - \frac{25}{36} = \frac{11}{36}$.

Of the three solutions to Example 28, solution 3 appears on the surface to be the most complicated, but in fact it is the one that shows us the most useful approach. It is based on what we will call the **probability multiplication principle for independent events**.

> **Independent Events.** Two events are said to be independent if the outcome of one event does not affect the outcome of the other.

> **The Probability Multiplication Principle for Independent Events.** When a complex event E can be broken down into a combination of two simpler events that are *independent* (call them F and G), then we can calculate the probability of E by multiplying the probabilities of F and G.

The probability multiplication principle is an important and useful rule, but it works only when the parts are independent. The next two examples illustrate the usefulness of the probability multiplication principle for independent events.

Example 29. Suppose that we roll an honest die four times. What is the probability that at least once we will roll a ⚀ ?

Let's try to use the same approach we used in Example 28, solution 3. (If we try a brute force approach similar to the one we used in Example 28, solution 1, we will soon realize why we ought to be thankful for our new-found wisdom.) Let's say once again that we will win if we roll a ⚀ at least once, and we will lose if none of the four rolls come up ⚀. We know that

- Pr(first roll not a ⚀) = $\frac{5}{6}$.

- Pr(second roll not a ⚀) = $\frac{5}{6}$.

- Pr(third roll not a ⚀) = $\frac{5}{6}$.

- Pr(fourth roll not a ⚀) = $\frac{5}{6}$.

Because each roll is independent of the preceding ones, we can use the probability multiplication principle.

$$\text{Pr(lose)} = \text{Pr(not rolling any ⚀'s in four rolls)} = \left(\frac{5}{6}\right)^4 \approx .482.$$

It follows that

$$\text{Pr(win)} = \text{Pr(rolling at least one ⚀ in four rolls)} \approx .518.$$

The practical consequence of these calculations is that if we played the above game regularly and always chose to bet that we can roll a ⚀ at least once in four rolls, we would win about 51.8% of the time. Assuming even odds (bet $1 to win $1), this is a good game to play.

Example 30. Suppose that we play a game similar to the one in the preceding example, but instead of rolling a single die we roll a pair of honest dice 24 times. If we roll "snake-eyes" (⚀ ⚀), we will win; otherwise we will lose. Should we play this game? What is the probability of winning? Once again, it would be extremely hard to solve this problem using any method other than the probability multiplication principle.

First note that we will lose if we don't roll snake-eyes on any of the 24 rolls. The probability of not rolling snake-eyes on the first roll is $\frac{35}{36}$ (see Exercise 16). It is exactly the same for the second, third, . . . , twenty-fourth rolls. By the probability multiplication principle (each roll is independent of the others),

$$\text{Pr(lose)} = \text{Pr(not rolling snake-eyes on any of the 24 rolls)}$$

$$= \left(\tfrac{35}{36}\right)^{24} \approx .509.$$

It follows that

$$\text{Pr(win)} = 1 - \text{Pr(lose)} \approx 1 - .509 = .491.$$

If we play this game regularly, we will win 49.1% of the time. Assuming even odds, we may want to think twice about playing this game on a regular basis.[4]

ODDS AND ENDS

Dealing with probabilities as numbers that are always between 0 and 1 is the mathematician's way of having a consistent terminology. To the everyday user consistency is not that much of a concern, and we know that people talk about *chances* (probabilities expressed as percentages) and *odds*, which are most frequently used to describe probabilities associated with gambling situations. In this section we will briefly discuss how to interpret and calculate odds. To simplify our discussion we will only consider the situation in which all outcomes are equally likely.

> **Odds in favor of an event.** The odds in favor of event E are given by the ratio of the number of ways event E can occur to the number of ways in which event E cannot occur.

Example 31. If we roll a pair of honest dice, what are the odds in favor of rolling a total of 7?

We saw in Example 27 that of the 36 different outcomes that are possible

---

[4] There is an interesting historical footnote to Examples 29 and 30. In a sense the beginnings of the mathematical study of probability can be traced back to these problems. Both of the games described in Examples 29 and 30 were popular among the seventeenth-century French nobility. One of the high rollers in that crowd was a certain Antoine Gombauld, Chevalier de Méré, who happened to be acquainted with the philosopher-mathematician Blaise Pascal. De Méré asked Pascal to do a mathematical analysis of these games for him. In solving these problems a theoretical framework for the study of games of chance had to be developed. Pascal, together with two other mathematicians of the time (Pierre Fermat and Abraham de Moivre), began looking at gambling questions as mathematical problems, and it was thus that the mathematical theory of probability was born.

when rolling a pair of dice, 6 are favorable (result in a total of 7) and the other 30 are unfavorable (result in a total that is not 7). It follows that the odds in favor of rolling a 7 are 6 to 30 or, equivalently, 1 to 5.   ▭

Example 32. If we roll a pair of honest dice, what are the odds *against* rolling a total of 7?

This question is essentially the opposite of the one asked in Example 31. Of the 36 possible outcomes, 30 are favorable (result in a total that is not 7) and 6 are unfavorable (result in a total of 7). Thus, the odds against rolling a 7 are 30 to 6 or 5 to 1, the same numbers as in Example 31 but reversed.   ▭

> **Odds against an event.** If the odds in favor of event E are m to n, then the odds against event E are n to m.

Sometimes we want to calculate the odds in favor of an event but all we have to go on is the probability of that event.

Example 33. When Michael Jordan shoots a free throw, the probability that he will make it is .85. In other words, on the average, out of every 100 free throws he attempts, he will make 85 and miss 15. It follows that the odds in favor of his making a free throw are 85 to 15, which reducing to simplest form[5] gives 17 to 3.   ▭

The general rule for converting probabilities into odds is

> If $\Pr(E) = \dfrac{a}{b}$ then the *odds in favor* of E are a to $b - a$ and the *odds against* E are $b - a$ to a.

The general rule for converting odds to probabilities is

> If the *odds in favor* of an event E are m to n, then $\Pr(E) = \dfrac{m}{m + n}$.

---

[5] It is not uncommon to see the numbers reduced even further and find these odds given as $5\frac{2}{3}$ to 1.

Example 34. Consider the tennis tournament first discussed in Example 22. Recall that according to the first expert the probability assignment for winning the tournament was Pr(Boris) = .25, Pr(Martina) = .22, Pr(Andre) = .14, Pr(Gabriela) = .18, and Pr(Monica) = .21. Let's convert all these probabilities into odds.

From Pr(Boris) = .25 = $\frac{1}{4}$ we find that the odds in favor of Boris winning are 1 (numerator) to 3 (denominator minus numerator). The odds against Boris winning are 3 to 1.

Likewise, Pr(Martina) = .22 = $\frac{22}{100}$ = $\frac{11}{50}$, so the odds in favor of Martina winning are 11 to 39. Note that we converted the decimal .22 to a reduced fraction before we computed the odds.

Pr(Andre) = .14 = $\frac{14}{100}$ = $\frac{7}{50}$. The odds in favor of Andre winning are 7 to 43.

Pr(Gabriela) = .18 = $\frac{18}{100}$ = $\frac{9}{50}$. The odds in favor of Gabriela winning are 9 to 41.

Pr(Monica) = .21 = $\frac{21}{100}$. The odds in favor of Monica winning are 21 to 79.

A word of caution: There is a difference between odds as discussed in this section, and the *payoff odds* posted by casinos or bookmakers in sports gambling situations. For example, say we read in the newspaper that the Las Vegas sports books have established that "the odds that the Cincinnati Reds will win the World Series are 5 to 2." What this means is that if you want to bet in favor of the Reds, for every $2 that you bet, you can win $5 if the Reds win. This ratio of 5 to 2 may be taken as some indication of the actual odds in favor of the Reds winning, but there are several other factors affecting payoff odds, and the connection between payoff odds and actual odds is tenuous at best.

CONCLUSION

While the average citizen thinks of probabilities, chances, and odds as vague, informal concepts that are useful primarily when discussing the weather or playing the lottery, scientists, mathematicians, and statisticians think of probability as a formal framework within which the laws that govern chance events can be understood. The basic elements of this framework are a *sample space* (which represents a precise mathematical description of all the possible outcomes of a *random experiment*) and a *probability assignment* (which associates a numerical value with each of these outcomes). The numerical values represent a measure of the likelihood that a particular outcome will occur. How these numerical values come about has both mathematical and philosophical implications, but once a probability assignment is made (be it by hook or by crook), the rules of the game are strictly mathematical.

Of the many ways in which probabilities can be assigned to outcomes, a particularly important case is the one in which all outcomes have the same probability. When this happens, the critical steps in calculating probabilities revolve around two basic (but not necessarily easy) questions: (1) given a sample space, what is its size? and (2) given an event, what is its size?

In this chapter we have only scratched the surface of what, in reality, is a deep and sophisticated mathematical theory, with important applications in almost every walk of life. When one stops to think how much of life is ruled by fate and chance, the pervasive role of probability is not entirely surprising. The Roman statesman Cicero said, "Probability is the very guide of life." Was he really serious? Probably not, but then again . . . .

KEY CONCEPTS

certain event
counting problem
event
formal probability model
impossible event
independent events
multiplication principle
odds

probability assignment
probability model
probability multiplication
   principle for independent events
random experiment
random variable
sample space

EXERCISES

Walking

1. Consider the random experiment of tossing a coin four times.
   (a) Write out the sample space for this random experiment.
   (b) What is the size of this sample space?
   (c) Write out the event that exactly two of the coin tosses come out heads.
   (d) Assuming that the coin is honest, what is the probability of the event described in (c)?

2. Consider the random experiment of drawing 1 card out of an ordinary deck of 52 cards.
   (a) Describe the sample space for this random experiment.
   (b) What is the size of this sample space?
   (c) Describe the event that the card drawn is a Spade.
   (d) Assuming the deck is honest and the cards are shuffled, what is the probability of the event described in (c)?

3. There are seven players (call them $P_1, P_2, \ldots, P_7$) entered in a tennis tournament. According to one expert, P_1 is twice as likely to win as any of the other players, and $P_2, P_3, \ldots, P_7$ all have an equal chance of winning.
   (a) Write down the sample space and find the probability assignment for the sample space based on this expert's opinion.
   (b) What are the odds that P_1 will win the tournament? How about P_2?

4. There are eight players (call them $P_1, P_2, \ldots, P_8$) entered in a chess tournament. According to an expert, P_1 has a 25% chance of winning the tournament, P_2 has a 15% chance of winning, P_3 has a 5% chance of winning, and all the other players have an equal chance of winning.
   (a) Write down the sample space and find the probability assignment for the sample space based on this expert's opinion.
   (b) What are the odds that P_1 will win the tournament? How about P_2?

5. (a) How many license plates can be made using three letters followed by three digits (0 through 9)?
   (b) How many license plates can be made using three letters followed by three digits if the first digit cannot be a zero?
   (c) How many license plates can be made using three letters followed by three digits if no license plate can have a repeated letter or digit?

6. (a) How many four-letter code words are there? (A code word is any string of letters—it doesn't have to mean anything.)
   (b) How many four-letter code words are there that start with the letter A?
   (c) How many four-letter code words are there that have no repeated letters?

7. A set of reference books consists of eight volumes numbered 1 to 8.
   (a) In how many ways can the eight books be arranged together on a shelf?
   (b) In how many ways can the eight books be arranged on a shelf so that at least one book is out of order?

8. Four men and four women line up at a checkout stand in a grocery store.
   (a) In how many ways can they line up?
   (b) In how many ways can they line up if the first person in line must be a woman?
   (c) In how many ways can they line up if they must alternate woman, man, woman, man, etc.? (A woman is first in line.)

9. A child's spinner has five outcomes: red, blue, yellow, purple, and orange. Experience shows that each primary color (red, blue, or yellow) comes up about 100 times in every 1000 spins and that each nonprimary color has an equal probability of occurring.
   (a) What is the sample space for the random experiment consisting of a single spin?
   (b) Write down the event that the outcome is a primary color.
   (c) Write down the event that the outcome is not a primary color.
   (d) Give a reasonable probability assignment for the sample space given in (a).

10. A child's spinner has five outcomes: 1, 2, 3, 4, and 5.
    (a) What is the sample space for the random experiment consisting of a single spin?
    (b) Write down the event that the outcome of a spin is an odd number.
    (c) Write down the event that the outcome of a spin is an even number.
    (d) If the probability of the event described in (b) is .3, find the probability of the event described in (c).

(e) If the probability of the event described in (b) is .3 and $Pr(1) = Pr(3) = Pr(5)$ and $Pr(2) = Pr(4)$, find the probability assignment for the sample space.

11. Consider the sample space $S = \{A, B, C\}$. Make a list of all the possible events for this sample space. (Remember that an event is any subset of S including $\{\}$ and S itself.)

12. Consider the sample space $S = \{A, B, C, D\}$. Make a list of all the possible events for this sample space. (Remember that an event is any subset of S including $\{\}$ and S itself.)

13. Two friends (Thelma and Mabel) go to the racetrack. There are five horses entered in the first race. Let's call them A, B, C, D, and E. Thelma buys a "to show" ticket (this means that she picks a horse and if her horse comes in either first, second, or third she will win). Mabel buys an exacta ticket (this means that she picks two horses to come in 1-2 and if her two horses come in 1-2 in that order she will win).
    (a) Write down the sample space for Thelma.
    (b) Write down the sample space for Mabel.

14. Two friends (Thelma and Mabel) go to the racetrack. There are eight horses entered in the second race. Let's call them A, B, C, D, E, F, G, and H. Thelma buys a "to place" ticket (this means that she picks a horse and if her horse comes in either first or second she will win). Mabel buys a trifecta ticket (she must pick the first-, second-, and third-place finishers and in exactly that order to win her bet).
    (a) Write down the sample space for Thelma.
    (b) Without writing it all down, determine the size of the sample space for Mabel.

Exercises 15 through 20 concern the random experiment of rolling a pair of dice. The random variable X is the sum of the faces showing. Assume the dice are honest.

15. (a) What is the probability that $X = 10$?
    (b) What are the odds in favor of rolling a 10?
    (c) What are the odds against rolling a 10?
    (d) What is the probability of *not* rolling a 10?

16. (a) What is the probability of rolling snake-eyes (i.e., both dice coming up 1)?
    (b) What are the odds in favor of rolling snake-eyes?
    (c) What is the probability of *not* rolling snake-eyes?
    (d) What are the odds against rolling snake-eyes?

17. (a) Describe the event $X \leq 5$.
    (b) What is the probability of the event described in (a)?
    (c) What are the odds in favor of the event described in (a)?

18. (a) What is the probability that at least one of the two dice comes up 1?
    (b) What are the odds in favor of at least one of the two dice coming up 1?

19. Suppose the pair of dice is rolled twice and that a roll of 7 is considered a success and a roll of anything other than 7 is considered a failure.
    (a) Find the probability that both rolls will result in a success.
    (b) Give a probability assignment for the sample space $S = \{ss, sf, fs, ff\}$.

20. Suppose the pair of dice is rolled twice and that a roll of any number less than 7 is considered a success and a roll of any number larger than or equal to 7 is considered a failure.
  (a) Find the probability that both rolls will result in a failure.
  (b) Give a probability assignment for the sample space $S = \{ss, sf, fs, ff\}$.

21. (a) An honest coin is tossed three times. What is the probability of getting heads at least once?
  (b) An honest coin is tossed ten times. What is the probability of getting heads at least once?

22. A coin is tossed repeatedly until a head appears. Describe a sample space for this experiment.

23. Assuming that boys and girls have an equal chance of being born, what is the probability that in a family of three children,
  (a) all three will be girls
  (b) there will be two girls and one boy?

24. A box contains four times as many red marbles as black marbles. What is the probability that a marble drawn at random from the box is a red marble?

25. A box contains twenty-five red marbles and fifty white marbles. What is the probability that if two marbles are drawn at random from the box, both are red?

Jogging

Exercises 26 through 30 refer to a club that has fifteen members. A delegation of four members must be chosen to represent the club at a convention.

26. How many different four-person delegations are possible?

27. How many different four-person delegations are there that include Mary?

28. How many different four-person delegations are there that do not include Mary?

29. Anna and Bob are a pair, and each has said that they refuse to be on the delegation unless the other one is. How many different four-person delegations are possible if Anna and Bob refuse to be separated?

30. Carol and Luis are not speaking to each other, and each refuses to be in the delegation if the other one is. How many four-person delegations are possible if Carol and Luis will not serve on the same delegation?

Exercises 31 through 35 refer to a standard deck of cards. There are 52 cards in a standard deck, each one having a face value (Ace, 2, 3, 4, 5, 6, 7, 8, 9, 10, Jack, Queen, King), and a suit (Hearts, Diamonds, Spades, Clubs). Hearts and Diamonds are red, Spades and Clubs are black. By a "poker hand" we will mean five cards, randomly drawn from the deck.

31. How many different five-card poker hands are possible?

32. What is the probability of getting all four Aces in a five-card poker hand?

33. What is the probability of getting "four of a kind" in a five-card poker hand?

34. What is the probability of getting all five cards of the same color in a poker hand?

35. What is the probability of getting all five cards of the same suit in a poker hand?

36. Three different colored flags are available. How many different signals can be made on a flagpole, assuming that at least one flag must be used and that the order of the flags on the flagpole makes a difference?

37. A study group of ten students is to be split into two groups of five students each. In how many ways can this be done?

38. Eight points are taken on a circle.
(a) How many chords can be drawn by joining all possible pairs of the points?
(b) How many triangles can be made using these points as vertices?

39. Dolores wants to walk from point A to point B (a total of six blocks) as shown on the street map. Assuming that she always walks toward B, how many different ways can she take this walk?

40. Consider the following game: We roll a pair of honest dice twenty-five times. If we roll "boxcars" (i.e.,) at least once, we will win; otherwise we will lose. What is the probability that we will win?

41. Consider the following game: We roll a pair of honest dice five times. If we roll a total of seven at least once, we will win; otherwise we will lose. What is the probability that we will win?

42. A factory assembles car stereos. From random testing at the factory it is known that on the average one out of every fifty car stereos will be defective (which means that the probability that a car stereo randomly chosen from the assembly line will be defective is .02). After manufacture, car stereos are packaged in boxes of twelve for delivery to the stores.
(a) What is the probability that in a box of twelve there are no defective car stereos? What assumptions are you making?
(b) What is the probability that in a box of twelve there is at most one defective car stereo?

43. In the game of craps, a pair of dice is rolled. A person betting "on the field" is betting that on the next roll the total rolled will be a 2, 3, 4, 9, 10, 11, or 12. What is the probability of winning a field bet?

44. There is a dice game played by some eccentric types called "subtract instead of add." In this game a pair of dice is rolled and bets are made on the difference between the numbers rolled (larger minus smaller). Let the random variable X represent this difference.
(a) What are the possible values of X?
(b) Assuming that both dice are honest, find the probability of the event $X = 1$.
(c) Assuming that both dice are honest, find the probability of the event $X = 0$.

45. A pizza parlor offers six toppings—pepperoni, Canadian bacon, sausage, mushroom, anchovies, and olives—that can be put on their basic cheese pizza. How many different pizzas can be made? (A pizza can have anywhere from no topping to all six toppings.)

46. **(a)** In how many different ways can ten people form a line?
    (b) In how many different ways can ten people hold hands and form a circle? (*Hint:* The answer to [b] is much smaller than the answer to [a]. There are many different ways in which the same circle of ten people can be broken up to form a line. How many?)

47. Two teams (call them X and Y) play in the World Series. The World Series is a best of seven series. This means that the two teams play against each other and the first team to win four games wins the series and the series is over. (Games cannot end in a tie.) We can describe an outcome for the World Series by writing a string of letters that indicate (in order) the winner of each game. For example, the string $XYXXYX$ represents the outcome: X wins game 1, Y wins game 2, X wins game 3, etc.
    (a) Using the notation described above, write the sample space S for the World Series.
    (b) Describe the event "X wins in five games."
    (c) Describe the event "the series lasts seven games."

48. **(a)** Suppose that a coin is tossed eight times. Each time the coin comes up H you win $1.00, each time it comes up T you lose $1.00 (see Example 20). If W is the random variable that describes your total winnings, explain why the possible values of W are $-8, -6, -4, -2, 0, 2, 4, 6, 8$. (Explain why you can't, for example, end up with $W = 3$.)
    (b) Suppose you win $1.50 for every H tossed and you lose $1.75 for every T tossed. What are the possible values of your winnings W?

49. A standard deck of cards is thoroughly shuffled.
    (a) What is the probability that the top card in the deck is an Ace?
    (b) What is the probability that the bottom card in the deck is an Ace?
    (c) What is the probability that the tenth card from the top of the deck is an Ace?

Running

50. How many different "words" (they don't have to mean anything) can be formed using all the letters in
    (a) the word PARSLEY (Note: This one is easy!)
    (b) the word PEPPER. (Note: This one is much harder! *Hint:* Think about the difference between [a] and [b].)

51. In the game of craps, the player's first roll of a pair of dice is very important. If the first roll is 7 or 11, the player wins. If the first roll is 2, 3, or 12, the player loses. If the first roll is any other number (4, 5, 6, 8, 9, 10), this number is called the player's "point." The player then continues to roll until the point reappears, in which case the player wins, or a 7 shows up before the point, in which case the player loses. What is the probability that the player will win? (Assume the dice are honest.)

52. **The birthday problem.** There are thirty people in a room. What is the probability that at least two of these people have the same birthday, that is, have their birthdays on the same day and month?

53. **The Monty Hall problem.** You have just been chosen to be on a game show in which there are three doors, behind one of which is a new sports car. There is nothing behind the other two doors. You know in advance that the game show host will ask you to pick one of the doors, but before the door you picked is opened, the host

will open one of the other two doors with nothing behind it. You will then be given the option of keeping the door you picked or switching to the other closed door. What should you do and why?

54. **The optimal choice problem.** A large box contains 100 tickets. Each ticket has a different number written on it. Other than the fact that the numbers are all different, we know nothing about them. The numbers can be positive or negative, integers or decimals, large or small—anything goes. Of all the tickets in the box there is, of course, one that has the biggest of all the numbers. That's the winning ticket. If we turn that ticket in, we will win a $1000 prize. If we turn any other ticket in, we will get nothing. The ground rules are that we can draw a ticket out of the box, look at it, and if we think it's the winning ticket we can turn it in. If we don't, we get to draw again, but before we do so we must tear the other ticket up—once we pass on a ticket we can't use it again! We can continue drawing tickets this way until we find one we like or run out of tickets.

(a) What is the probability that the first ticket we draw will be the winning ticket?

(b) What's the probability that after we have drawn 50 tickets, the winning ticket will still be in the box?

(c) It is very surprising, but there is a strategy for playing this game that will give us a better than 25% chance of winning. Describe such a strategy.

REFERENCES AND FURTHER READINGS

1. David, F. N., *Games, Gods and Gambling*. Buckinghamshire, England: Charles Griffin & Co., 1962.

2. di Finetti, B., *Theory of Probability*. New York: John Wiley & Sons, Inc., 1970.

3. Freedman, D., R. Pisani, R. Purves, and A. Adhikari, *Statistics* (2nd ed.). New York: W. W. Norton, Inc., 1991, chaps. 13 and 14.

4. Gnedenko, B. V., and A. Y. Khinchin, *An Elementary Introduction to the Theory of Probability*. New York: Dover Publications, Inc., 1962.

5. Keynes, John M., *A Treatise of Probability*. New York: Harper and Row, 1962.

6. Koosis, Donald, *Probability*. New York: John Wiley & Sons, Inc., 1973.

7. Levinson, Horace, *Chance, Luck and Statistics: The Science of Chance*. New York: Dover Publications, Inc., 1963.

8. Levinson, Horace, *The Science of Chance*. New York: Rinehart & Co., 1950.

9. Maistrov, L. E., *Probability Theory: A Historical Perspective*. New York: Academic Press, Inc., 1974.

10. McGervey, John D., *Probabilities in Everyday Life*. New York: Ivy Books, 1986.

11. Trefil, James, "Odds Are Against Your Breaking That Law of Averages," *Smithsonian* (September, 1984), 66–75.

12. Weaver, Warren, *Lady Luck: The Theory of Probability*. New York: Dover Publications, Inc., 1963.

CHAPTER 16

Everything Is Back To Normal (Almost)

Normal Distributions

Variety is the spice of life.
                    ANONYMOUS

Variation is at the heart of statistics. If there were no variation in data, there would be very little for statisticians to do. Fortunately, the common thing for a variable, be it a test score, a grade, a height, etc., is to take many different values throughout the population (that is why it is called a variable!). The pattern of variation of the variable over the entire population is called the variable's **distribution**.

A variable's distribution can follow a very irregular, unpredictable pattern, or it can follow a pattern in the true sense of the word, showing a behavior that is fairly regular and quite predictable. The latter happens surprisingly often, and when it does, we can visualize the distribution as fitting into a certain established "mold." In this case, the bar graphs or histograms that describe the distribution have a predictable *overall shape*—maybe a straight line, maybe a bell shape, maybe something else. A statistician thinks of this nice, regular mold into which the data more or less fit as a mathematical ideal for the way the data are expected to come out. This mathematical ideal is most commonly referred to as a *statistical model* for the data.

Thinking of data in terms of mathematical models may seem to be a form of statistical wishful thinking, but the usefulness of this idea cannot be overstated. The characteristics of a mathematical model are abstract, theoretical facts which can be applied to the specific distribution at hand, where they translate into useful, concrete information about the population.

Of the many different mathematical models for data, by far the most pervasive is the *normal* model. This is a model used to analyze distributions whose bar graphs or histograms follow a "bell-shaped" pattern, and it is the focus of our discussion for this chapter.

APPROXIMATELY NORMAL DISTRIBUTIONS

We start with a pair of examples.

Example 1. We are back to the Stat 101 data set, which we first discussed in Chapter 14 (see Example 1, Chapter 14). Let's look at the distribution of test scores for the exam, but let's throw out the two outliers (Fig. 16-1). (By definition these two scores are exceptional values of the data, and it's OK to take this liberty.)

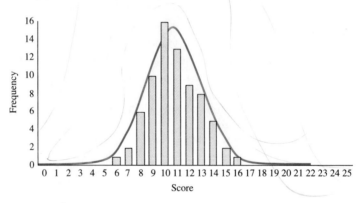

FIGURE 16-1 Distribution of Stat 101 test scores (without outliers).

We can see that the bar graph in Fig. 16-1 has a very distinctive shape—it is that of a bell. The bell-shaped curve superimposed on the bar graph is the mathematical idealization of the distribution of test scores. One shouldn't worry too much about the fact that the fit isn't perfect. As we will see in the next section, having such an idealization still gives us a powerful tool for describing and analyzing the data.

To those readers who are experienced test takers, the fact that the test scores on the Stat 101 midterm follow a bell-shaped pattern is not all that surprising—test scores often do that. (At the same time, we should be wary of generalizations: It is wrong to assume that test scores *always* follow a bell-shaped pattern.)

Example 2. In this example we will consider the distribution of the heights of the top 270 professional basketball players in the United States at the start of the 1993–1994 NBA season. Our data set was obtained as follows: For each of the twenty-seven teams in the National Basketball Association we selected the top

ATLANTIC DIVISION

Team	Top 10 Players' Heights									
Boston	6-0	6-0	6-1	6-5	6-7	6-7	6-9	6-10	6-11	7-0
Miami	6-2	6-5	6-6	6-6	6-8	6-9	6-11	6-11	7-0	7-0
New Jersey	6-1	6-2	6-3	6-4	6-5	6-9	6-10	6-11	7-0	7-2
New York	6-2	6-4	6-5	6-6	6-7	6-7	6-9	6-11	7-0	7-0
Orlando	6-1	6-6	6-6	6-7	6-8	6-8	6-8	6-11	6-11	7-1
Philadelphia	6-2	6-3	6-4	6-4	6-6	6-9	6-9	6-10	6-11	7-5
Washington	5-10	6-3	6-4	6-7	6-7	6-8	6-10	6-10	7-0	7-7

CENTRAL DIVISION

Team	Top 10 Players' Heights									
Atlanta	6-1	6-6	6-7	6-8	6-8	6-11	6-11	7-0	7-0	7-0
Charlotte	5-3	6-0	6-5	6-5	6-5	6-7	6-7	6-7	6-10	6-11
Chicago	6-2	6-2	6-5	6-7	6-8	6-10	6-10	6-11	7-0	7-1
Cleveland	6-0	6-0	6-2	6-4	6-6	6-6	6-10	6-10	6-11	7-0
Detroit	6-1	6-2	6-3	6-4	6-5	6-6	6-6	6-8	6-11	7-0
Indiana	6-1	6-5	6-7	6-7	6-8	6-10	6-11	6-11	7-1	7-4
Milwaukee	6-1	6-4	6-5	6-6	6-8	6-8	6-9	6-11	6-11	6-11

MIDWEST DIVISION

Team	Top 10 Players' Heights									
Dallas	6-3	6-4	6-4	6-4	6-6	6-6	6-8	6-8	6-10	7-0
Denver	6-1	6-4	6-5	6-7	6-7	6-8	6-8	6-9	6-11	7-2
Houston	5-11	6-3	6-3	6-4	6-5	6-9	6-10	6-10	7-0	7-0
Minnesota	6-2	6-3	6-5	6-6	6-8	6-8	6-10	6-11	6-11	7-2
San Antonio	6-3	6-4	6-7	6-7	6-8	6-8	6-9	6-9	6-9	7-1
Utah	6-1	6-3	6-6	6-8	6-9	6-9	6-10	7-2	7-2	7-4

PACIFIC DIVISION

Team	Top 10 Players' Heights									
Golden State	6-0	6-5	6-5	6-5	6-6	6-7	6-9	6-9	6-9	6-10
LA Clippers	6-2	6-3	6-3	6-4	6-6	6-8	6-9	6-9	6-10	7-0
LA Lakers	6-1	6-2	6-4	6-4	6-6	6-9	6-11	7-1	7-1	7-1
Phoenix	6-1	6-1	6-5	6-6	6-6	6-6	6-7	6-9	6-10	7-0
Portland	6-3	6-3	6-7	6-7	6-7	6-8	6-9	6-9	6-10	6-10
Sacramento	5-7	5-11	6-0	6-5	6-7	6-8	6-8	6-8	6-9	7-0
Seattle	5-11	6-4	6-5	6-6	6-7	6-9	6-9	6-10	6-10	7-2

■ Table 16-1   The Heights of the Top 270 National Basketball Association Players at the Start of the 1993–1994 Season Listed by Team (Source: *1994 Complete Handbook of Pro Basketball*, Z. Hollander [ed.].)

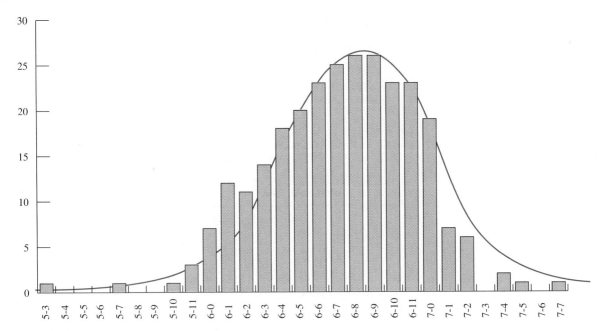

FIGURE 16-2 Heights of 270 players in the National Basketball Association (1993–1994). (Source: *1994 Complete Handbook of Pro Basketball*, Z. Hollander [ed.].)

ten players (the starting five plus the first five substitutes off the bench). Table 16-1 shows the heights (to the nearest inch) of the top ten players on each team.

Figure 16-2 is a bar graph for the 270 heights shown in Table 16-1. Once again there is a rough bell-shaped pattern to the data.[1] A mathematical idealization of the pattern is given by the curve superimposed on the bar graph. Once again the fit isn't perfect, but it is good enough to be useful in practice.

Examples 1 and 2 illustrate two very different situations (different populations and different variables) having one thing in common: Both distributions can best be described as fitting a bell-shaped mold. When this happens, we say that the distribution is **approximately normal**. A **normal distribution** is the idealized mathematical model we use to analyze any distribution that is approximately normal. A normal distribution is described by a bell-shaped curve called a **normal curve**.

[1] Notice, however, the presence of several outliers, to wit: 5ft., 3 in. Tyrone (Muggsy) Bogues of the Charlotte Hornets; 5ft., 7in. Anthony (Spud) Webb of the Sacramento Kings; and, at the other end of the spectrum, 7ft., 7in. Gheorghe Muresan of the Washington Bullets.

The long and the short of it in the NBA: Outliers
"Muggsy" Bogues and Gheorghe Muresan size
each other up. © Time, Inc.)

Because we are using the word "normal" in so many contexts here, let's summarize the three concepts we have just introduced.

■ **Approximately Normal Distribution.** This is a distribution of real-life data with a bar graph or histogram that more or less follows a bell-shaped pattern.

■ **Normal Distribution.** This is a mathematical model, an idealization of how the data would come out in a perfect world.

■ **Normal Curve.** A graphical representation of a normal distribution. It is always a bell-shaped curve.

NORMAL CURVES AND THEIR PROPERTIES

In this section, we will briefly discuss the main mathematical features of normal curves and then see how these facts can be used to obtain important information about any variable with an approximately normal distribution. All the variables that we will discuss from here on will be quantitative variables.

Normal curves (Fig. 16-3) can come in many different packages: Some are tall and skinny, others are short and squat, others are in between. Mathematically speaking, however, there are hardly any differences among normal curves. In fact, whether a normal curve is skinny and tall or short and squat or somewhere in between depends on the way we scale the units on the axes. With the proper choice of scale on the axes, any two normal curves can be made to look like one another.

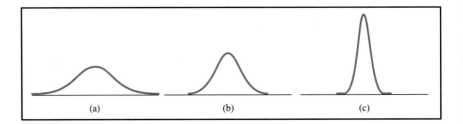

FIGURE 16-3 Three normal curves. (a) Short and squat. (b) In between. (c) Tall and skinny.

What follows is a summary of the most important facts about normal distributions and normal curves.

■ **Symmetry.** Every normal curve is symmetric about a vertical axis (Fig. 16-4). The axis of symmetry splits the bell-shaped region outlined by the curve into two identical halves.

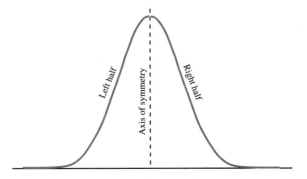

FIGURE 16-4 A normal curve has one vertical reflection symmetry.

■ **Center.** The point where the vertical axis of reflection cuts the horizontal axis is called the **center** of the normal distribution. The data value corresponding to this point is both the *median* and the *average* of the distribution. We will denote the center by the Greek letter μ (mu) (Fig. 16-5). The fact that the median and the average are the same is a consequence of the fact that a normal curve has a vertical reflection symmetry. Any distribution (be it normal or not) that has such a symmetry will have the median equal to the average.

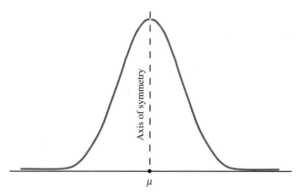

FIGURE 16-5 The median and the average of a normal distribution are equal.

Axis of symmetry

μ

■ **Standard Deviation.** For any normal distribution the standard deviation is the key to measuring spread. We discussed the standard deviation in Chapter 14. Among other things, we saw that calculating the standard deviation for a large distribution of data values can be a great deal of work and that it's something best left to a good calculator or a statistical software package.

The easiest way to describe the standard deviation of a normal distribution is geometrically. Suppose we have a piece of wire we want to bend so that it is shaped exactly like the right half of a normal curve (if we can do the right half, then we can also do the left half since they are symmetric). At the very top we must bend the wire downward (as illustrated in Fig. 16-6[a]) while at the bottom we must bend the wire upward (Fig. 16-6[b]). There is exactly one point P where the curvature of the wire changes from down to up (Fig. 16-6[c]). The point P is called an **inflection point** of the curve. Locating the inflection point P gives us a way to find the standard deviation of the normal distribution—it is the distance from P to the axis of symmetry of the distribution as shown in Fig. 16-7. We will use the Greek letter σ (sigma) to denote the standard deviation.

In general, calculating the standard deviation of a normal distribution requires mathematical tools that go beyond the scope of this book. For the purposes of our discussion we will take it for granted that the standard deviation is known.

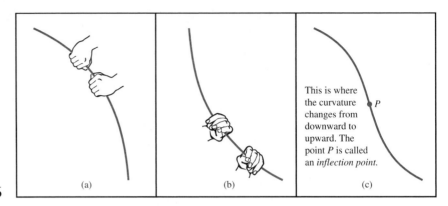

This is where the curvature changes from downward to upward. The point P is called an *inflection point.*

(a)   (b)   (c)

FIGURE 16-6

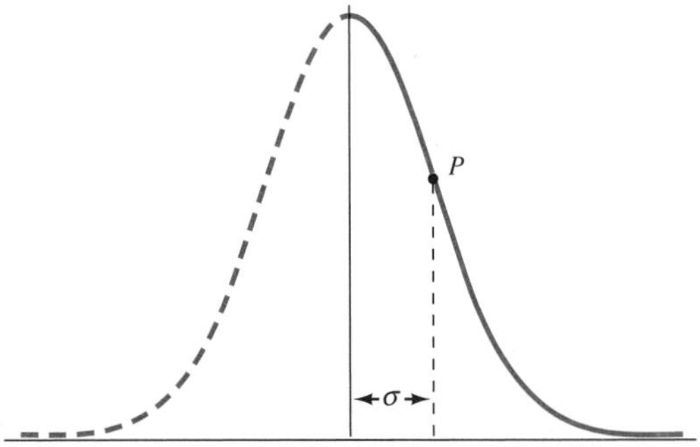

FIGURE 16-7 The standard deviation (σ) is the horizontal distance between the inflection point P and the axis of symmetry of the bell.

■ **Areas under a Normal Curve. The 68-95-99.7 Rule.** For any normal curve the following three facts are always true.

1. The area under the curve and within one standard deviation to the left and right of the center ($\mu \pm \sigma$) equals about 68% of the total area under the curve (Fig. 16-8[a]).

2. The area under the curve and within two standard deviations to the left and right of the center ($\mu \pm 2\sigma$) equals about 95% of the total area under the curve (Fig. 16-8[b]).

3. The area under the curve and within three standard deviations to the left and right of the center ($\mu \pm 3\sigma$) equals about 99.7% of the total area under the curve (Fig. 16-8[c]).

Figure 16-8 illustrates the 68-95-99.7 rule, as well as some of its consequences, in complete detail. The reader is advised to study Fig. 16-8 carefully.

One of the important consequences of the 68-95-99.7 rule is that for all practical purposes almost all the area under the normal curve (99.7%) is concentrated in a section that is plus or minus three standard deviations from the center. Outside that interval there simply isn't that much left of the distribution.

Any normal distribution is completely described by two numbers: the center (μ) and the standard deviation (σ). If someone were to give us two numbers for μ and σ (σ must be a positive number), say $\mu = 16.5$ and $\sigma = 2.3$, we could talk about *the* normal distribution with center 16.5 and standard deviation 2.3, because there is one and only one such beast.

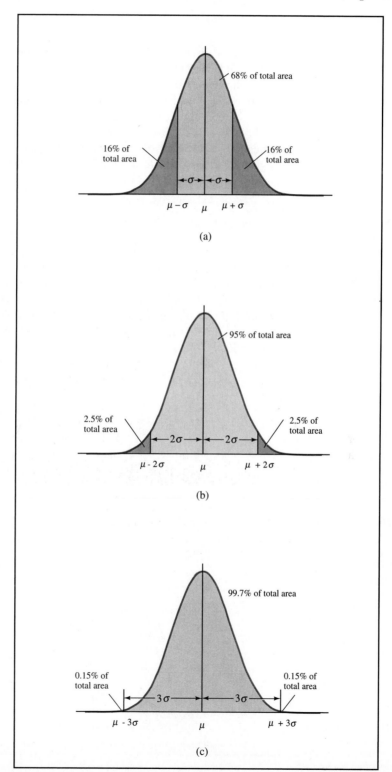

Figure 16-8

AREAS AND PERCENTILES

We will now discuss a critical question: Why should we care about areas under the normal curve? Or, to put it in a different way, what do areas under a normal curve tell us about the associated distribution? Remember that our starting point is some variable which when evaluated over a population produces an approximately normal distribution. The bar graph or histogram for this distribution is roughly bell-shaped, and the normal curve that we associate with it is a mathematical idealization of the bar graph. The curve itself follows (roughly) the heights of the bars, and *the areas of the various segments under the curve correspond to percentages of the population.*

Area under curve = Percentage of population

It all sounds terribly complicated, but it isn't. Let's start with the fact that the total area under a normal curve represents the total population (100%) (Fig. 16-9). Next, let's look at the center μ. We know that the normal curve is symmetric about a vertical axis passing through μ, which tells us that the shaded area in Fig. 16-10 must be one-half the total area under the curve. This tells us that for 50% of the population the variable takes values less than or equal to μ, and for 50% of the population the variable takes values bigger than or equal to μ. This explains why μ is the median of the distribution.

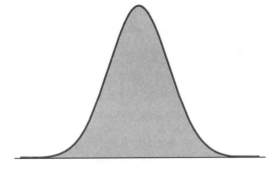

FIGURE 16-9 The entire area under the normal curve represents 100% of the population.

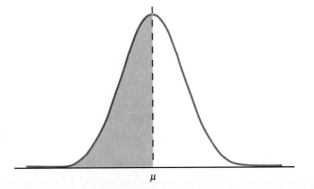

FIGURE 16-10 The vertical axis through μ divides the area under the curve into two pieces of equal area.

How about the quartiles? Following the same line of thinking that we used with the median, we guess that the first quartile (Q_1) must be located at the place where the area to its left and under the curve is 25% of the total area. This is indeed the case, as illustrated in Fig. 16-11(a). Likewise, the third quartile (Q_3) is located so that the area to its left and under the curve is 75% of the total area (Fig. 16-11[b]).

When we know the value of μ and σ, then we can locate the quartiles using the following two rules:

$$Q_3 \approx \mu + (0.675)\sigma$$

$$Q_1 \approx \mu - (0.675)\sigma.$$

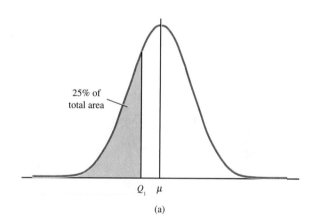

25% of
total area

Q_1 μ

(a)

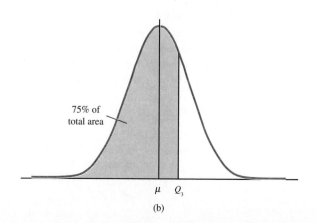

75% of
total area

μ Q_3

(b)

FIGURE 16-11 (a) The area under the curve and to the left of Q_1 is 25% of the total area. (b) The area under the curve and to the left of Q_3 is 75% of the total area.

We won't explain where the magic number 0.675 comes from other than to say that there are tables that give not only the location of the quartiles but also of any other arbitrary **percentile** of the population. (The xth percentile is the value of the variable for which x percent of the population is at or below that value.) For a very simplified version of such a table the reader is referred to Exercises 26 and 27.

We illustrate some of the ideas of this section with a couple of examples.

Example 3. Let's analyze the normal curve with center located at 500 and standard deviation $\sigma = 100$. The median is $\mu = 500$, the first quartile is found by taking $Q_1 \approx 500 - 0.675 \times 100 = 432.5$, and the third quartile is $Q_3 \approx 500 + 0.675 \times 100 = 567.5$. The 68-95-99.7 rule tells us that 99.7% of the area under the curve falls between the values of 200 and 800, that 95% of the area under the curve falls between the values of 300 and 700, and that 68% of the area under the curve falls between the values of 400 and 600 (see Fig. 16-12).

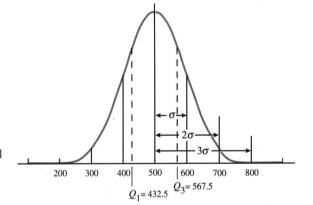

FIGURE 16-12 The normal curve with center $\mu = 500$ and standard deviation $\sigma = 100$.

Example 4. Let's suppose that we are told that the distribution of scores on the mathematics part of the SAT for the entering class at Tasmania State University had an approximately normal distribution with center $\mu = 500$ points and standard deviation $\sigma = 100$ points.[1] We can think of the normal curve shown in Fig. 16-12 as a mathematical model for the distribution of the SAT scores, and we can put all the abstract things we learned in Example 3 to good use (see Fig. 16-13.) Here are some of the things we can say:

---

[1] The reader is warned that these are ficticious (and conveniently chosen) values. In 1993, for example, for the entire national population of college-bound high school seniors, the average SAT mathematics score was 478.

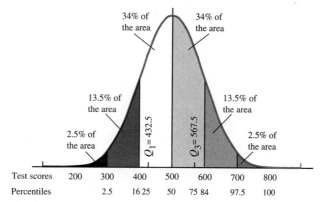

FIGURE 16-13 Scores on the SAT mathematics test and associated percentiles for the entering class at Tasmania State University.

A typical SAT score report. The first row shows SAT-verbal score, the second row shows SAT-mathematics score. The two columns on the right show this student's percentile for two different reference populations: college-bound seniors across the nation as opposed to college-bound seniors from the same state the student is from.

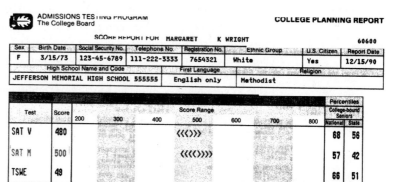

(© 1987 by College Entrance Examination Board and Educational Testing Service)

■ Of all the students taking the SAT-mathematics test, 50% scored 500 points or less and 50% scored 500 points or more.

■ About 25% of all the students scored below $Q_1 = 432.5$ points, and about 25% scored above $Q_3 = 567.5$ points.

■ About 99.7% of all students scored somewhere between 200 and 800 points. Actually, since 200 is the minimum score (they give 200 points just for showing up), and 800 is the maximum score reported, we can go out on a limb and say that 100% of the students scored between 200 and 800 inclusive.

■ About 95% of all students scored between 300 and 700 points. The remaining 5% was split equally between those who scored 300 or below (2.5%) and those who scored 700 or above (2.5%).

■ About 68% of all students scored between 400 and 600 points. By symmetry it follows that the remaining 32% was split equally between those who scored 400 or below (16%) and those who scored 600 or above (16%). ■

Let's put our new formal knowledge to good use by considering the case of three TSU freshpersons (Vanessa, Rudy, and Cleo) who scored 490, 570, and 710 points, respectively, on the mathematics portion of the SAT. In what percentiles of Tasmania State University entering class would these scores roughly place them?

A score of 490 is close to 500, which would place Vanessa a little bit under the 50th percentile—a good guess would be somewhere around the 47th or 48th percentile. Rudy's score of 570 is slightly above the third quartile ($Q_3 = 567.5$). We would guess that Rudy's score places him somewhere around the 75th or 76th percentile. As for Cleo, her score of 710 is above the 97.5th percentile, so we can figure that Cleo placed somewhere around the 98th or 99th percentile. Our guesses are actually very good. The exact percentiles for any score can be calculated by using special statistical tables for normal distributions (or a good piece of statistical software.) It turns out that Vanessa's score of 490 places her in the 46th percentile, Rudy's score of 570 puts him in the 76th percentile, and Cleo's score of 710 puts her in the 98th percentile.

We conclude with a parting comment: When we started Example 4, there was precious little that we had to go on—Just that the scores followed an approximately normal distribution with median 500 and standard deviation 100. From these skimpy facts and the little bit of theory that we learned, we were able to get plenty of mileage. Most remarkably, we were able to do this without looking at the actual data.

NORMAL DISTRIBUTIONS OF RANDOM EVENTS

We are now ready to take up another important aspect of normal distributions— their connection with the behavior of random events. Our starting point is the following important example.

Example 5 (A coin-tossing experiment). We are going to take an honest coin and toss it 100 times. Of these 100 tosses, how many times will the coin come up heads? Fifty? Maybe, maybe not. Forty-five? Why not? Seventy-five? Possible, but it wouldn't be wise to bet on it.

For convenience, we will use X to represent the exact number of heads that come up when we toss the coin 100 times. What does common sense tell us about the possible values of the random variable X? First of all, we cannot predict the exact value of X—in principle it could be anything from $X = 0$ (the 100 tosses come up tails) to $X = 100$ (the 100 tosses come up heads). Of course, some values of X are less likely to occur than others: We suspect that tossing 100 heads is very unlikely; that tossing 55 heads is a lot more likely; and that tossing 50 heads is even more likely.

One way that we can test our intuition about how likely the various values of X are is to repeat the 100 toss experiment many times and check the frequencies of the various outcomes. To do this all one needs is an honest coin, some pencil and paper for record keeping, and lots and lots of time. The South African mathematician John Kerrich was imprisoned by the Germans during World War

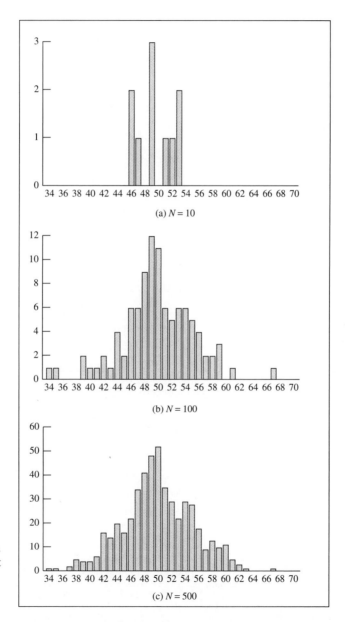

FIGURE 16-14 The 100 toss experiment: Toss an honest coin 100 times and count the number of heads (X). Repeat the experiment N times and chart the frequencies of X. *(Cont'd, on page 534)*

II, and while in prison he carried out such an experiment. He tossed a coin 10,000 times and kept records of the number of heads in each successive set of 100 tosses.

Rather than reproduce Kerrich's data here, we decided to do our own high-tech version of the experiment by letting a computer do the coin tossing as well as the record keeping. (It is very easy to have a computer toss a pretend coin, and the results are just as valid as those we would get tossing a real coin. It is also a lot easier on the thumb!)

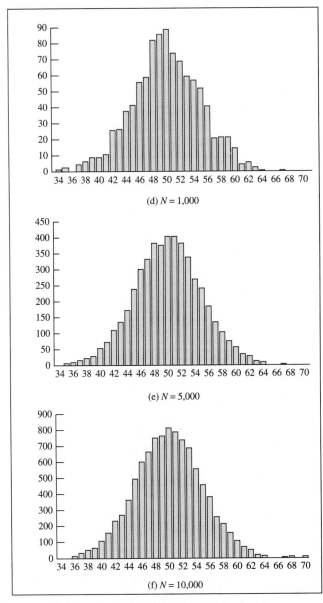

(d) $N = 1,000$

(e) $N = 5,000$

(f) $N = 10,000$

Figure 16-14 *cont.*

Figure 16-14 shows the frequencies for the random variable X when we repeated the 100 toss experiment (a) 10 times, (b) 100 times, (c) 500 times, (d) 1000 times, (e) 5000 times, and (f) 10,000 times.[2]

Figure 16-14 paints a pretty clear picture of what happened. When we repeated the experiment 10 times, we got $X = 46$ twice, $X = 47$ once, $X = 49$ three times, $X = 51$ once, $X = 52$ once, and $X = 53$ twice. A bar graph for these data values is shown in Fig. 16-14(a). This distribution may look a little surprising, but with such a small number of trials just about anything can happen. As we increased the number N of repetitions of the experiment, the picture started to come into sharper focus, and by the time we got to $N = 5000$ the unmistakable shape of a normal distribution showed up loud and clear. When $N = 10,000$ [Fig. 16-14(f)], we got an almost perfect normal distribution!

As the reader can probably guess, what happened in this last example was not an accident. For small values of N things are pretty unpredictable, but as N gets bigger, the normal nature of the distribution is guaranteed to come out. Let's put it this way: Suppose someone else decided to do the same experiment all over, once again tossing an honest coin (be it by hand or by computer) 100 times, counting the value of the random variable X = the number of heads, and then repeating the whole thing N times. For $N = 10$ their results are likely to be quite different from the results we show in Fig. 16-14(a), but as N gets bigger, the results will begin to look more and more alike. By the time they get to $N = 10,000$ their data will be almost identical to the data we show in Fig. 16-14(f). In a sense, we can say that doing these experiments is a waste. For *large values of N we can predict what the distribution of X will be like with an extremely high degree of accuracy without ever tossing a coin*!

The most important conclusion that we can draw from Example 5 is that for sufficiently large values of N, the random variable X has a normal distribution. For a complete grasp of the situation we need two more pieces of information: the values of the center μ and the standard deviation σ of this distribution. Looking at Fig. 16-14(f), we can pretty much see where the center of this normal distribution is—right at 50. The fact that $\mu = 50$ should be no surprise. Since the coin is an honest coin, we know that the likelihood of tossing fewer than 50 heads is the same as that of tossing fewer than 50 tails, which in turn is the same as that of tossing more than 50 heads. Clearly the axis of symmetry of the distribution has to pass through 50.

The value of the standard deviation is less obvious. For now, let's accept the fact that it is $\sigma = 5$. We will explain how we got this value shortly.

Let's summarize what we now know: An honest coin is tossed 100 times. The number of heads in the 100 tosses is a random variable which we call X. If we repeat this experiment a large number of times (N), the random variable X

---

[2] This means $10,000 \times 100$ individual tosses. It would take about 5 minutes (working fast) to toss a real coin 100 times and record the data. If one were to toss coins without a break for 8 hours a day, 7 days a week, it would take about 3 months worth of coin tossing to collect this amount of data by hand.

will have an approximately normal distribution with center $\mu = 50$ and standard deviation $\sigma = 5$, and the larger the value of N, the better this approximation will be.

Collecting the data the hard way. (© Estate of Harold Edgerton, 1965)

The real significance of these facts is that they are valid not just because we went to the trouble (or rather had the computer take the trouble for us) of tossing a coin hundreds of thousands of times. The truth of the matter is that even if we did not toss a coin at all, all of the above would still be true: *For a sufficiently large number of repetitions of the experiment of tossing an honest coin 100 times, the number of heads X is a random variable that has an approximately normal distribution with center $\mu = 50$ and standard deviation $\sigma = 5$.* This is a mathematical fact.

The next step we are going to take is critical. Suppose that we have an honest coin and are going to toss it 100 times. We are going to do this just once—period. Can we make any predictions about the number of heads that are going to come up? The answer is most definitely yes. By taking advantage of the fact that the number of heads that will come up is a random variable with an approximately normal distribution we will be able to make excellent predictions about the outcome of such an experiment. For starters, we can predict that the chances that the number of heads will fall somewhere between 45 and 55 are 68% (it follows from the fact that this is one standard deviation to the left and right of the center), the chances that the number of heads will fall somewhere between 40 and 60 are 95%, and the chances that the number of heads will fall somwhere between 35 and 65 are a whopping 99.7%. All we are using here is the knowledge that we are dealing with an approximately normal distribution with center $\mu = 50$ and standard deviation $\sigma = 5$, together with some of the things that we learned about normal distributions (such as the 68-95-99.7 rule).

THE HONEST AND DISHONEST COIN PRINCIPLES

In Example 5 everything was based on tossing an honest coin 100 times. What if we were to toss a coin 500 times? Or 1000 times? Or n times? Not surprisingly, everything we discovered in Example 5 would still be true, except that the values of μ and σ would be different. Specifically, we can make the following general statement for which we have coined the name the "honest coin principle."

THE HONEST COIN PRINCIPLE

An experiment consists of tossing an honest coin n times and counting the number of heads (X). If we repeat this experiment a large number of times (N), then

1. The random variable X has an approximately normal distribution, and the larger N is, the closer the distribution is to a normal distribution.

2. The center of the distribution is $\mu = n/2$.

3. The standard deviation of the distribution is $\sigma = \sqrt{n}/2$.

(This is why for $n = 100$ we found that the center is $\mu = \frac{100}{2} = 50$ and the standard deviation is $\sigma = \sqrt{100}/2 = \frac{10}{2} = 5$.)

Example 6. An honest coin is going to be tossed 256 times. Before this is done there is an opportunity to make some bets. Let's say that we can make a bet (with even odds) that if the number of heads falls somewhere between 120 and 136 we will win, otherwise we will lose. Should we make such a bet?

By the honest coin principle we know that the number of heads in 256 tosses of an honest coin is a random variable having a distribution that is approximately normal with center $\mu = \frac{256}{2} = 128$ and standard deviation $\sigma = \sqrt{256}/2 = \frac{16}{2} = 8$. The values 120 to 136 are exactly one standard deviation to the right and left of 128, and by the 68-95-99.7 rule there is a 68% chance that the number of heads will fall somewhere between 120 and 136. We should indeed make this bet!

By a similar calculation we can argue that there is a 95% chance that the number of heads will fall somewhere between 112 and 144, and the chance that the number of heads will fall somewhere between 104 and 152 is a whopping 99.7%.

Our next step is to see if we can discover a **dishonest coin principle**. (We can't always count on an even break!) Would the ideas that we have seen so far apply if we have a dishonest coin? The answer is yes. In fact, the nature of the coin only plays a role in calculating the center and the standard deviation.

THE DISHONEST COIN PRINCIPLE

An experiment consists of tossing a dishonest coin n times and counting the number of heads (X). The coin has a probability of landing heads equal to p (and therefore the probability of landing tails is $1 - p$). If we repeat this experiment a large number of times (N), then

1. The random variable X has an approximately normal distribution, and the larger N is, the closer the distribution is to a normal distribution.

2. The center of the distribution is $\mu = n \cdot p$.

3. The standard deviation of the distribution is $\sigma = \sqrt{n \cdot p \cdot (1 - p)}$.

Example 7. A coin is rigged so that it comes up heads only 20% of the time (i.e., $p = .20$). The coin is tossed 100 times ($n = 100$).

According to the dishonest coin principle, the distribution of the number of heads is approximately normal with center $\mu = 100 \times .20 = 20$ and standard deviation $\sigma = \sqrt{100 \times .20 \times .80} = 4$.

Note that in this case heads and tails are no longer symmetric, but the dishonest coin principle will work just as well for tails as it does for heads. The distribution for the number of tails is approximately normal with center $\mu = 100 \times .80 = 80$ and standard deviation $\sigma = \sqrt{100 \times .80 \times .20} = 4$.

Based on these facts we can now make the following assertions:

(a) There is a 68% chance that the number of heads will fall somewhere between 16 and 24.

(b) There is a 95% chance that the number of heads will fall somewhere between 12 and 28.

(c) The number of heads is almost guaranteed (a 99.7% chance) to fall somewhere between 8 and 32.

The dishonest coin principle can be applied to any coin, even one that is fair ($p = \frac{1}{2}$). In the case $p = \frac{1}{2}$ the honest and dishonest coin principles say the same thing (Exercise 25).

The dishonest coin principle is a down-to-earth version of one of the most important facts in all of statistics, known by the somewhat intimidating name of the *central limit theorem*. We will now briefly illustrate why the importance of the dishonest coin principle goes beyond the tossing of coins.

SAMPLING AND THE DISHONEST COIN PRINCIPLE

Example 8. A large container (sometimes called an urn) is filled with 100,000 beads, 20,000 of which are red and 80,000 of which are white. The beads in the urn are thoroughly mixed. We will now draw a sample from this urn as follows: (1) We stick our hand in the urn and randomly pick a bead, (2) we check and record the color of the bead, (3) we put the bead back in the urn. We repeat this process $n = 100$ times. Let the random variable X represent the number of red beads counted in the 100 draws. What can we say about X?

A moment's reflection will show that statistically this example is identical to Example 7—each red bead drawn (probability .20) can be identified with the coin coming up heads (also probability .20). It follows that the dishonest coin principle can be applied to make the following assertions.

(a) There is a 68% chance that the number of red beads drawn will fall somewhere between 16 and 24.

(b) There is a 95% chance that the number of red beads drawn will fall somewhere between 12 and 28.

(c) The number of red beads drawn is almost guaranteed (a 99.7% chance) to fall somewhere between 8 and 32.

Probably the most important point of all this is that each of the above facts can be rephrased in terms of sample errors, a concept we first discussed in Chapter 13. Suppose that the purpose of drawing the 100 beads from the urn is to use the number of red beads in the sample to estimate the percentage of red beads in the entire population. Let's say, for the sake of argument, that we draw 24 red beads in the sample of 100. If this statistic (24%) were used to estimate the percent of red beads in the urn, then the sample error would be 4% (the estimate is 24% and the exact value of the parameter is 20%). By the same token, if we had drawn 16 red beads, the sample error would be −4%. Since the standard deviation of $\sigma = 4$ beads (which we computed in Example 7) is exactly 4% of the sample size ($n = 100$) we can rephrase assertions a through c in terms of sample errors as follows:

(a) When estimating the proportion of red beads in the urn using a sample of 100 beads, the chance that the sample error will fall somewhere between −4% and 4% is 68%.

(b) When estimating the proportion of red beads in the urn using a sample of 100 beads, the chance that the sample error will fall somewhere between −8% and 8% is 95%.

(c) We can guarantee with great confidence (99.7% certainty) that, when estimating the proportion of red beads in the urn using a sample of 100 beads, the sample error will fall somewhere between −12% and 12%. ▬

Example 9. Suppose we have the same urn as in Example 8, but this time we are going to draw a sample of size $n = 1600$. Before we even count the number of red beads in the sample, let's see how much mileage we can get out of the dishonest coin principle. Theoretically, we are doing the same thing as tossing a dishonest coin (with $p = .2$) a total of 1600 times. The standard deviation here is $\sqrt{1600 \times .2 \times .8} = 16$, which is exactly 1% of the sample. This means that when we estimate the proportion of red beads in the urn using this sample, we can once again use the dishonest coin principle to make certain qualified assurances about the size of the sample error, namely,

(a) We can say with some confidence (68%) that the sample error will fall somewhere between -1% and 1%.

(b) We can say with a lot of confidence (95%) that the sample error will fall somewhere between -2% and 2%.

(c) We can say with a tremendous amount of confidence (99.7%) that the sample error will fall somewhere between -3% and 3%.

The ideas discussed in this section are fundamental to an understanding of how to estimate the reliability of any sample. We end this section, however, with a disclaimer: When using a random sample to estimate a population proportion out there in the real world, the standard deviation σ given by the dishonest coin principle cannot be calculated directly (the way we did in Examples 8 and 9) because we don't know p. If we did, we wouldn't have to use a sample! It is possible, however, to get a very good estimate of σ and to use it to make the same types of qualified assurances we gave in the last set of statements a through c. (See Exercise 30.)

CONCLUSION

The wheels of science do not always spin forward. In previous chapters we learned about the wonderful things that we can do with data. In a sense, this chapter was about all the wonderful things that we can do without data. The process of drawing conclusions with limited information is an essential aspect of statistics. The way the process works is by looking at a mathematical idealization of the data. In many real-life situations one can safely predict ahead of time what the overall pattern of the data will be like—this overall pattern is what we call a *mathematical model* for the data. A good understanding of the mathematical properties of a model is an invaluable tool for analyzing the real data.

Perhaps the most commonly used mathematical model of real-life data is the normal distribution. Contrary to what many people think, the normal distribution is not guaranteed to always be the right model for looking at real-life data. It is nevertheless true that when we look at almost any variable that measures a characteristic of a population (height, IQ, blood pressure, etc.), it is often the case that its distribution is approximately normal. When this happens, we can use the mathematical theory of normal distributions to gain a better understanding of the data.

A second context in which normal distributions came up in this chapter was in connection with random experiments. Many types of random experiments exhibit exactly the same long-run behavior as that shown by the repeated tossing of a coin. In these cases the *dishonest coin principle* (which includes the *honest coin principle* as a special case) tells us that this long-run behavior can be described by a normal distribution. The importance of the dishonest coin principle in real-life applications of statistics cannot be overstated. It is, among other things, the mathematical principle that allows us to use random sampling to draw reliable conclusions about a population, and, in a more sobering vein, it is also the principle that guarantees that in the long run gamblers will always lose and casinos will always win.

KEY CONCEPTS

68-95-99.7 rule
approximately normal distribution
center
dishonest coin principle
distribution
honest coin principle

inflection point
normal curve
normal distribution
percentile
standard deviation

EXERCISES

Walking

Exercises 1 through 4 refer to the following: 250 students take a college entrance exam. The scores on the exam have an approximately normal distribution with center $\mu = 52$ points and the standard deviation $\sigma = 11$ points.

1. (a) Estimate the average score on the exam.
   (b) Estimate what percent of the students scored 52 points or more.
   (c) Estimate what percent of the students scored between 41 and 63 points.
   (d) Estimate what percent of the students scored 63 points or more.

2. (a) Estimate how many students scored between 30 and 74 points.
   (b) Estimate how many students scored 74 points or more.
   (c) Estimate how many students scored 85 points or more.

3. (a) Estimate the first-quartile score for this exam.
   (b) Estimate the third-quartile score for this exam.
   (c) Estimate the interquartile range for this exam.

4. For each of the following scores, estimate in what percentile of the students taking the exam the score would place you.
   (a) 51
   (b) 64

(c) 60

(d) 85.

Exercises 5 through 8 refer to the following: As part of a research project, the blood pressures of 2000 patients in a hospital are recorded. The systolic blood pressures (given in millimeters) have an approximately normal distribution with center $\mu = 125$ and $\sigma = 13$.

5. (a) Estimate the number of patients whose blood pressure was between 99 and 151 millimeters.

   (b) Estimate the number of patients whose blood pressure was 99 millimeters or less.

6. (a) Estimate the third quartile (Q_3) for the distribution of blood pressures.

   (b) Estimate the interquartile range for the distribution of blood pressures.

7. For each of the following blood pressures, estimate the percentile of the patient population to which they correspond.

   (a) 100 millimeters

   (b) 112 millimeters

   (c) 115 millimeters

   (d) 138 millimeters

   (e) 164 millimeters.

8. (a) Estimate the value of the lowest (Min) and the highest (Max) blood pressures. (Assume there were no outliers and use the 68-95-99.7 rule.)

   (b) Assuming there were no outliers, give an estimate of the five-number summary (Min, Q_1, μ, Q_3, Max) for the distribution of blood pressures.

Exercises 9 through 12 refer to the following: Packaged foods sold at supermarkets are not always the weight indicated on the package. Variability always crops up in the manufacturing and packaging process. Suppose that the exact weight of a "12-ounce" bag of potato chips is a random variable that has an approximately normal distribution with center $\mu = 12$ ounces and standard deviation $\sigma = 0.5$ ounces.

9. If a "12-ounce" bag of potato chips is chosen at random, what are the chances that

   (a) it weighs somewhere between 11 and 13 ounces

   (b) it weighs somewhere between 12 and 13 ounces

   (c) it weighs more than 11 ounces?

10. If a "12-ounce" bag of potato chips is chosen at random, what are the chances that

   (a) it weighs somewhere between 11.5 and 12.5 ounces

   (b) it weighs somewhere between 12 and 12.5 ounces

   (c) it weighs more than 12.5 ounces?

11. Suppose that 500 bags of potato chips are chosen at random. Estimate the number of bags with weight

   (a) 11 ounces or less

   (b) 11.5 ounces or less

   (c) 12 ounces or less

   (d 12.5 ounces or less

   (e) 13 ounces or less

   (f) 13.5 ounces or less.

12. Suppose that 1500 bags of potato chips are chosen at random. Estimate the number of bags of potato chips with weight
   (a) between 11 and 11.5 ounces
   (b) between 11.5 and 12 ounces
   (c) between 12 and 12.5 ounces
   (d) between 12.5 and 13 ounces
   (e) between 13 and 13.5 ounces.

13. Find μ and σ for the normal curve shown in the figure. (P and P' are the inflection points of the normal curve—see Fig. 16-7.)

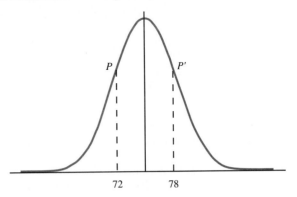

14. Find μ and σ for the normal curve shown in the figure.

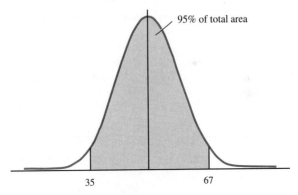

15. Find μ and σ for the normal curve shown in the figure.

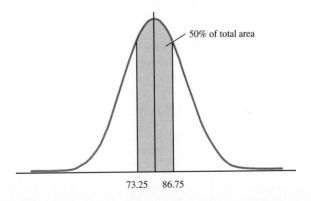

16. Find μ and σ for the normal curve shown in the figure.

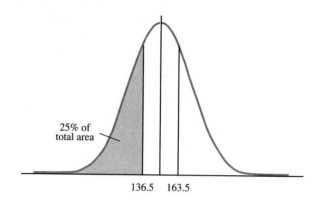

Exercises 17 through 20 refer to the following: *The distribution of weights for children of a given age and sex is approximately normal. This fact allows a doctor or nurse to find from a child's weight the weight percentile of the population (all children of the same age and sex) to which the child belongs. Typically, this is done using special charts provided to the doctor or nurse, but these percentiles can also be computed using facts about approximately normal distributions such as the ones we learned in this chapter. (The figures in these examples are approximate values taken from tables produced by the National Center for Health Statistics, U.S. Department of Health and Human Services.)*

17. The distribution of weights for six-month-old baby boys is approximately normal with center $\mu = 17.25$ pounds and standard deviation $\sigma = 2$ pounds.
   (a) Suppose that a six-month-old boy weighs 15.25 pounds. Approximately what weight percentile is he in?
   (b) Suppose that a six-month-old boy weighs 21.25 pounds. Approximately what weight percentile is he in?
   (c) Suppose that a six-month-old boy is in the 75th percentile in weight. Estimate his weight.

18. The distribution of weights for twelve-month-old baby girls is approximately normal with center $\mu = 21$ pounds and standard deviation $\sigma = 2.2$ pounds.
   (a) Suppose that a twelve-month-old girl weighs 16.6 pounds. Approximately what weight percentile is she in?
   (b) Suppose that a twelve-month-old girl weighs 18.8 pounds. Approximately what weight percentile is she in?
   (c) Suppose that a twelve-month-old girl is in the 75th percentile in weight. Estimate her weight.

19. The distribution of weights for one-month-old baby girls is approximately normal with center $\mu = 8.75$ pounds and standard deviation $\sigma = 1.1$ pounds.
   (a) Suppose that a one-month-old girl weighs 11 pounds. Approximately what weight percentile is she in?
   (b) Suppose that a one-month-old girl weighs 12 pounds. Approximately what weight percentile is she in?
   (c) Suppose that a one-month-old girl is in the 25th percentile in weight. Estimate her weight.

20. The distribution of weights for twelve-month-old baby boys is approximately normal with center $\mu = 22.5$ pounds and standard deviation $\sigma = 2.2$ pounds.
   (a) Suppose that a twelve-month-old boy weighs 24 pounds. Approximately what weight percentile is he in?
   (b) Suppose that a twelve-month-old boy weighs 21 pounds. Approximately what weight percentile is he in?
   (c) Suppose that a twelve-month-old boy is in the 84th percentile in weight. Estimate his weight.

Jogging

21. Over the years, the eggs produced by the Fibonacci Egg Company have been known to have weights with an approximately normal distribution whose center is $\mu = 1.5$ ounces and standard deviation is $\sigma = 0.35$ ounces. Eggs are classified by weight as follows:

■ Extra large: more than 2.2 ounces

■ Large: between 1.5 and 2.2 ounces

■ Discount (sold to wholesalers): less than 1.5 ounces.

Out of 5000 eggs, approximately how many would you expect would fall in each of the three categories?

22. An honest coin is tossed $n = 3600$ times. Let the random variable Y denote the number of tails tossed.
   (a) Find the center μ and the standard deviation σ for the distribution of the random variable Y.
   (b) What are the chances that the value of Y will fall somewhere between 1770 and 1830?
   (c) What are the chances that the value of Y will fall somewhere between 1800 and 1830?
   (d) What are the chances that the value of Y will fall somewhere between 1830 and 1860?

23. An honest die is rolled. If the roll comes out even (2, 4, or 6), you will win \$1; if the roll comes out odd (1, 3, or 5), you will lose \$1. Suppose that in one evening you play this game $n = 2500$ times in a row.
   (a) What is the probability that by the end of the evening you will not have lost any money?
   (b) What is the probability that the number of even rolls will fall between 1250 and 1300?
   (c) What is the probability that you will win \$100 or more?
   (d) What is the probability that you will win exactly \$101?

24. A dishonest coin with probability of heads $p = .4$ is tossed $n = 600$ times. Let the random variable X represent the number of times the coin comes up heads.
   (a) Find the center and standard deviation for the distribution of X.
   (b) Find the first and third quartiles for the distribution of X.
   (c) Suppose that you could choose between one of the following bets (\$1 to win \$1):
      (i) The number of heads will fall somewhere between 230 and 250.
      (ii) The number of heads will be less than or equal to 230 or more than or equal to 250.
   Which of the two bets would you choose? Explain your answer.

25. Explain why when the dishonest coin principle is applied with an honest coin, we get the honest coin principle.

Exercises 26 and 27 refer to the following table. It is a simplified version of a more elaborate statistical table that gives, for every value of x, the approximate location of the xth percentile for a normal distribution with center μ and standard deviation σ.

Percentile	Approximate Location	Percentile	Approximate Location
99th	$\mu + 2.33\sigma$	1st	$\mu - 2.33\sigma$
95th	$\mu + 1.65\sigma$	5th	$\mu - 1.65\sigma$
90th	$\mu + 1.28\sigma$	10th	$\mu - 1.28\sigma$
80th	$\mu + 0.84\sigma$	20th	$\mu - 0.84\sigma$
75th	$\mu + 0.675\sigma$	25th	$\mu - 0.675\sigma$
70th	$\mu + 0.52\sigma$	30th	$\mu - 0.52\sigma$
60th	$\mu + 0.25\sigma$	40th	$\mu - 0.25\sigma$
50th	μ		

26. The distribution of weights for six-month-old baby boys is approximately normal with center $\mu = 17.25$ pounds and standard deviation $\sigma = 2$ pounds.
   (a) Suppose that a six-month-old baby boy weighs in the 95th percentile of his age group. Find his weight in pounds approximated to two decimal places.
   (b) Suppose that a six-month-old boy weighs 17.75 pounds. Determine in what percentile of the weight distribution for his age group this child is in.
   (c) Suppose that a six-month-old baby boy weighs 15.2 pounds. Estimate in what percentile of the weight distribution for his age group this child is in.

27. Five-thousand students took a college entrance exam. The scores on the exam have an approximately normal distribution with center $\mu = 55$ points and standard deviation $\sigma = 12$ points.
   (a) Suppose a student's score places her in the 60th percentile. Find her score on the exam.
   (b) Suppose that a student scored 45 points on the exam. Estimate the percentile in which this score places him.
   (c) Suppose that a student scored 83 points on the exam. Estimate the percentile in which this score places her.
   (d) Approximately how many students scored 83 points or more?
   (e) Approximately how many students scored 52 points or less?

Running

28. On an American roulette wheel, there are 18 red numbers and 18 black numbers, plus 2 green numbers (0 and a 00). Thus, the probability of a red number coming up on a spin of the wheel is $p = \frac{18}{38} \approx .47$. Suppose that we go on a binge and we bet $1 on red 10,000 times in a row. (A $1 bet wins $1 if red comes up, otherwise we lose the $1.)

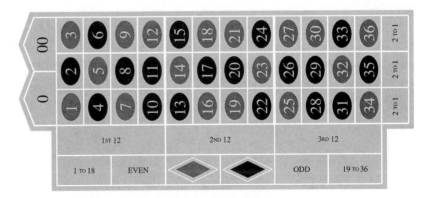

(a) Let Y represent the number of times we lose (i.e., red does not come up). Use the dishonest coin principle to describe the distribution of the random variable Y.

(b) Approximately what are the chances that we will lose 5300 times or more?

(c) Approximately what are the chances that we will lose somewhere between 5150 and 5450 times?

(d) Explain why the chances that we will break even or win in this situation are essentially zero.

29. An urn contains 10,000 beads, 20% of which are red and the rest white. Suppose that we draw a sample of size $n = 400$ (drawing a bead and each time replacing it in the urn before drawing again).

(a) If Y is the number of white beads in the sample, find the center and standard deviation for the distribution of Y.

(b) What are the chances that the number of white beads in the sample will fall somewhere between 304 and 332?

(c) When estimating the proportion of white beads in the urn using a sample of 400 beads, what are the chances that the sample error will fall somewhere between -4% and 4%?

30. An urn contains a large number of beads which are either red or white. Suppose that we draw a sample of 1200 beads, of which 300 (25%) are red. We want to use the statistic $25\% = .25$ as an estimate of the number of red beads in the urn. To estimate the sample error we can use the dishonest coin principle and use $\hat{p} = .25$ as an approximation for the exact value of p.

Using this value of $\hat{p}$ and the dishonest coin principle, explain why we can say with a lot of confidence (95%) that our estimate of 25% is within $\pm 2.5\%$ of the exact value of p.

REFERENCES AND FURTHER READINGS

1. Berry, D. A., and B. W. Lindgren, *Statistics: Theory and Methods.* Pacific Grove, CA: Brooks/Cole Publishing Co., 1990, chap. 8.

2. Converse, P. E., and M. W. Traugott, "Assessing the Accuracy of Polls and Surveys," *Science*, 234 (1986), 1094–1098.

3. Freedman, D., R. Pisani, R. Purves, and A. Adhikari, *Statistics* (2nd ed.). New York: W. W. Norton, Inc., 1991, chaps. 16 and 18.

4. Groeneveld, Richard A., *Introductory Statistical Methods.* Boston: PWS-Kent Publishing Co., 1988, chap 5.

5. Larsen, J., and D. F. Stroup, *Statistics in the Real World.* New York: Macmillan Publishing Co., Inc., 1976.

6. Mosteller, F., W. Kruskal, et al., *Statistics by Example: Detecting Patterns.* Reading, MA: Addison-Wesley Publishing Co., Inc., 1973.

7. Sincich, Terry, *Statistics by Example.* San Francisco, CA: Dellen Publishing Co., 1990, chaps. 6 and 7.

8. Tanner, Martin, *Investigations for a Course in Statistics.* New York: Macmillan Publishing Co., Inc., 1990.

Answers to Selected Problems

CHAPTER 1

Walking

1. (a)

Number of voters	6	2	1	3
1st choice	A	C	B	C
2nd choice	B	D	D	B
3rd choice	C	B	C	D
4th choice	D	A	A	A

(b) no **(c)** The Atrium (A) **(d)** Blair's Kitchen (B)

3. (a) 51 **(b)** A

5. B

7. C

9. (a)

Number of voters	8	5	3	2	4
1st choice	A	C	C	C	B
2nd choice	B	E	B	B	E
3rd choice	C	A	E	A	A
4th choice	E	B	A	E	C

(b) A and B tie

11. Winner: A. Second place: C and D (tied). Fourth place: B. Last place: E.

13. Winner: C. Second place: A and B (tied). Fourth place: D. Last place: E.

15. H

17. Winner: R. Second place: H. Third place: C. Last place: O and S (tied).

19. Winner: H. Second place: R. Third place: C. Fourth place: S. Last place: O.

21. (a) B **(b)** No. No candidate has a majority of the first-place votes.

23. Winner: D. Second place: B. Third place: C. Fourth place: E. Last place: A.

25. (a) B **(b)** No. No candidate has a majority of the first-place votes.

27. Winner: B. Second place: A. Third place: C. Fourth place: D. Last place: E.

29. (a) D **(b)** No. No candidate has a majority of the first-place votes.

31. Winner: B. Second place: C. Third place: D. Last place: A.

33. Winner: A. Second place: B. Third place: C. Fourth place: D. Last place: E.

35. Winner: A and B (tied). Third place: D. Fourth place: E. Last place: C.

37. Winner: B. Second place: A. Third place: C. Fourth place: D. Last place: E.

39. Winner: C. Second place: B. Third place: A. Fourth place: D. Last place: E.

Jogging

41. (a) 1225 **(b)** 2450

43. (a) Since there are only two columns in the preference schedule, one of the columns must represent the votes of 11 or more voters and so the first choice in that column is the first choice of more than half of the voters.

(b) The argument given in (a) can be repeated provided there are an odd number of voters since then one of the two columns must represent the votes of more than half of the voters (there cannot be a tie).

45. Suppose that the results of the election under the Borda count method (first place: 5 points, second place: 4 points, third place: 3 points, fourth place: 2 points, and fifth place: 1 point) were

Candidate	A	B	C	D	E
Points	a	b	c	d	e

Under the revised scheme (first place: 4 points, second place: 3 points, third place: 2 points, fourth place, 1 point, and fifth place: 0 points), A loses a point for every voter, so does B, etc. It follows that the result of the election under the revised scheme must be

Candidate	A	B	C	D	E
Points	$a - 21$	$b - 21$	$c - 21$	$d - 21$	$e - 21$

Since these numbers have the same values relative to each other as the original numbers, the outcome of the election is still the same.

47. If there is a candidate that is the first choice of a majority of the voters then using the plurality-with-elimination method, that candidate will be declared the winner of the election in the first round.

49. If there is a candidate that is the first choice of a majority of the voters then that candidate will win every one-to-one comparison with any other candidate and so will be a Condorcet candidate. If a voting method violates the majority criterion, then there is an election for which a candidate is the first choice of a majority of the voters (and hence a Condorcet candidate), and yet is not the winner of the election. Consequently, the voting method also violates the Condorcet criterion.

51. (a) $1523 + 1494 + 1447 = 62 \cdot (25 + 24 + 23)$

(b)

	2nd-place votes	3rd-place votes
Florida State	25	1
Notre Dame	18	19
Nebraska	19	42

CHAPTER 2

Walking

1. (a) 3   **(b)** 10   **(c)** 10   **(d)** $\{P_1, P_2\}, \{P_1, P_3\}, \{P_1, P_2, P_3\}$
   (e) P_1 only   **(f)** $P_1 : \frac{3}{5}; P_2 : \frac{1}{5}, P_3 : \frac{1}{5}.$

3. (a) $P_1 : \frac{3}{5}; P_2 : \frac{1}{5}; P_3 : \frac{1}{5}$ **(b)** $P_1 : \frac{1}{2}; P_2 : \frac{1}{2}; P_3 : 0.$

5. $P_1 : \frac{1}{3}; P_2 : \frac{1}{3}; P_3 : \frac{1}{3}; P_4 : 0; P_5 : 0.$

7. (a) $\langle P_1, P_2, P_3 \rangle, \langle P_1, P_3, P_2 \rangle, \langle P_2, P_1, P_3 \rangle, \langle P_2, P_3, P_1 \rangle, \langle P_3, P_1, P_2 \rangle, \langle P_3, P_2, P_1 \rangle.$
   (b) $\langle P_1, \underline{P_2}, P_3 \rangle, \langle P_1, \underline{P_3}, P_2 \rangle, \langle P_2, \underline{P_1}, P_3 \rangle, \langle P_2, P_3, \underline{P_1} \rangle, \langle P_3, \underline{P_1}, P_2 \rangle, \langle P_3, P_2, \underline{P_1} \rangle.$
   (c) $P_1 : \frac{2}{3}; P_2 : \frac{1}{6}; P_3 : \frac{1}{6}$

9. (a) $P_1 : \frac{2}{3}; P_2 : \frac{1}{6}; P_3 : \frac{1}{6}$ **(b)** $P_1 : \frac{1}{2}; P_2 : \frac{1}{2}; P_3 : 0.$

11. (a) $P_1 : \frac{1}{2}; P_2 : \frac{3}{10}; P_3 : \frac{1}{10}; P_4 : \frac{1}{10}.$ **(b)** $P_1 : \frac{7}{12}; P_2 : \frac{3}{12}; P_3 : \frac{1}{12}; P_4 : \frac{1}{12}.$

13. (a) P_1 and P_2 have veto power; P_3 is a dummy.
   (b) P_1 and P_2 have veto power; P_3 is a dummy.
   (c) There is no dictator, no one has veto power, and no one is a dummy.

15. (a) P_1 and P_2 have veto power; P_5 is a dummy.
   (b) P_1 is a dictator; P_2, P_3, P_4 are dummies.
   (c) All four players have veto power.
   (d) P_1 and P_2 have veto power; P_3 and P_4 are dummies.

17. (a) 16   **(b)** 31   **(c)** 31   **(d)** 120

19. (a) $P_1 : 1; P_2 : 0; P_3 : 0.$ **(b)** $P_1 : \frac{2}{3}; P_2 : \frac{1}{6}; P_3 : \frac{1}{6}$ **(c)** $P_1 : \frac{1}{2}; P_2 : \frac{1}{2}; P_3 : 0.$
   (d) $P_1 : \frac{1}{2}; P_2 : \frac{1}{2}; P_3 : 0.$ **(e)** $P_1 : \frac{1}{3}; P_2 : \frac{1}{3}; P_3 : \frac{1}{3}$

21. (a) 40,320   **(b)** 479,001,600   **(c)** 6,227,020,800
   (d) 39,916,800   **(e)** $13! = 6,227,020,800$

23. $A : \frac{1}{3}; B : \frac{1}{3}; C : \frac{1}{3}; D : 0.$

25. $A : \frac{7}{17}; B : \frac{7}{17}; C : \frac{1}{17}; D : \frac{1}{17}; E : \frac{1}{17}.$

Jogging

27. $P_1 : \frac{15}{52}; \quad P_2 : \frac{13}{52}; P_3 : \frac{11}{52}; P_4 : \frac{9}{52}; P_5 : \frac{3}{52}; P_6 : \frac{1}{52}.$

29. (a) 720
   (b) The player must be the last (sixth) player in the sequential coalition.
   (c) 120   **(d)** $\frac{120}{720} = \frac{1}{6}$
   (e) $\frac{1}{6}$ (Each player is the last player in 120 of the 720 sequential coalitions.)
   (f) If the quota equals the sum of all the weights ($q = w_1 + w_2 + \cdots + w_N$) then the only way a player can be pivotal is for the player to be the last player in the sequential coalition. Since every player will be the last player in the same number of sequential coalitions, all players must have the same Shapley-Shubik power index. It follows that each of the N players has Shapley-Shubik power index of $\frac{1}{N}$.

31. (a) [7: 6,3,2,1,1]. Shapley-Shubik power distribution:
   $P_1 : \frac{3}{5}; P_2 : \frac{1}{10}; P_3 : \frac{1}{10}; P_4 : \frac{1}{10}; P_5 : \frac{1}{10}.$
   (b) [9: 6,3,1,1,1]. Shapley-Shubik power distribution:
   $P_1 : \frac{11}{20}; P_2 : \frac{6}{20}; P_3 : \frac{1}{20}; P_4 : \frac{1}{20}; P_5 : \frac{1}{20}.$

    (c) [10: 6,3,2,1,1,1]. Shapley-Shubik power distribution:
$$P_1 : \tfrac{1}{2};\ P_2 : \tfrac{1}{4};\ P_3 : \tfrac{1}{12};\ P_4 : \tfrac{1}{12};\ P_5 : \tfrac{1}{12}.$$

    (d) [13: 6,3,2,1,1,1]. Shapley-Shubik power distribution:
$$P_1 : \tfrac{1}{5};\ P_2 : \tfrac{1}{5};\ P_3 : \tfrac{1}{5};\ P_4 : \tfrac{1}{5};\ P_5 : \tfrac{1}{5}.$$

33. (a) Both have Banzhaf power distribution $P_1 : \tfrac{2}{5};\ P_2 : \tfrac{1}{5};\ P_3 : \tfrac{1}{5};\ P_4 : \tfrac{1}{5}.$

    (b) In the weighted voting system $[q : w_1, w_2, ..., w_N]$, P_k is critical in a coalition means that the sum of the weights of all the players in the coalition (including P_k) is at least q but the sum of the weights of all the players in the coalition except P_k is less than q. Consequently, if the weights of all the players are multiplied by $c > 0$ ($c \le 0$ would make no sense), then the sum of the weights of all the players in the coalition (including P_k) is at least cq but the sum of the weights of all the players in the coalition except P_k is less than cq. Therefore P_k is critical in the same coalition in the weighted voting system $[cq : cw_1, cw_2, ..., cw_N]$. Since the critical players are the same in both weighted voting systems, the Banzhaf power distributions will be the same.

35. (a) If a player X has Banzhaf power index 0 then X is not critical in any coalition and so the addition or deletion of X to or from any coalition will never change the coalition from losing to winning or winning to losing. It follows that X can never be pivotal in any sequential coalition and so X must have Shapley-Shubik power index 0.

    (b) If a player X has Shapley-Shubik power index 0 then X is not pivotal in any sequential coalition and so X can never be added to a losing coalition and turn it into a winning coalition. It follows that X can never be critical in any coalition and so X has Banzhaf power index 0.

37. You should buy your vote from P_1. The following table explains why.

Buying a vote from	Resulting weighted voting system	Resulting Banzhaf power distribution	Your power
P_1	[6: 3,2,2,2,2]	$P_1 : \tfrac{1}{5};\ P_2 : \tfrac{1}{5};\ P_3 : \tfrac{1}{5};\ P_4 : \tfrac{1}{5};\ P_5 : \tfrac{1}{5}$	$\tfrac{1}{5}$
P_2	[6: 4,1,2,2,2]	$P_1 : \tfrac{1}{2};\ P_2 : 0;\ P_3 : \tfrac{1}{6};\ P_4 : \tfrac{1}{6};\ P_5 : \tfrac{1}{6}$	$\tfrac{1}{6}$
P_3	[6: 4,2,1,2,2]	$P_1 : \tfrac{1}{2};\ P_2 : \tfrac{1}{6};\ P_3 : 0;\ P_4 : \tfrac{1}{6};\ P_5 : \tfrac{1}{6}$	$\tfrac{1}{6}$
P_4	[6: 4,2,2,1,2]	$P_1 : \tfrac{1}{2};\ P_2 : \tfrac{1}{6};\ P_3 : \tfrac{1}{6};\ P_4 : 0;\ P_5 : \tfrac{1}{6}$	$\tfrac{1}{6}$

39. (a) You should buy your vote from P_2. The following table explains why.

Buying a vote from	Resulting weighted voting system	Resulting Banzhaf power distribution	Your power
P_1	[18: 9,8,6,4,3]	$P_1 : \tfrac{4}{13};\ P_2 : \tfrac{3}{13};\ P_3 : \tfrac{3}{13};\ P_4 : \tfrac{2}{13};\ P_5 : \tfrac{1}{13}$	$\tfrac{1}{13}$
P_2	[18: 10,7,6,4,3]	$P_1 : \tfrac{9}{25};\ P_2 : \tfrac{1}{5};\ P_3 : \tfrac{1}{5};\ P_4 : \tfrac{3}{25};\ P_5 : \tfrac{3}{25}$	$\tfrac{3}{25}$
P_3	[18: 10,8,5,4,3]	$P_1 : \tfrac{5}{12};\ P_2 : \tfrac{1}{4};\ P_3 : \tfrac{1}{6};\ P_4 : \tfrac{1}{12};\ P_5 : \tfrac{1}{12}$	$\tfrac{1}{12}$
P_4	[18: 10,8,6,3,3]	$P_1 : \tfrac{5}{12};\ P_2 : \tfrac{1}{4};\ P_3 : \tfrac{1}{6};\ P_4 : \tfrac{1}{12};\ P_5 : \tfrac{1}{12}$	$\tfrac{1}{12}$

(b) You should buy from P_2 since this gives you the largest power index $\left(\frac{2}{13}\right)$.

(c) Buying a single vote from P_2 raises your power from $\frac{1}{25} = 4\%$ to $\frac{3}{25} = 12\%$. Buying a second vote from P_2 raises your power to $\frac{2}{13} \approx 15.4\%$. The increase in power is less with the second vote, but if you value power over money, it might still be worth it to you to buy that second vote.

CHAPTER 3

Walking

1. (a) $9.00 **(b)** $3.00 **(c)** $3.00

3. (a) $9.00 **(b)** $6.00 **(c)** $3.00 **(d)**

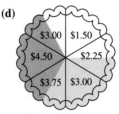

5. (a) only (iii) **(b)** (i): **I**; (ii): **I**; (iii): **I**; (iv) either

7. (a) There are three possible answers as shown in the following table:

Chooser 1	Chooser 2	Divider
s_2	s_1	s_3
s_3	s_1	s_2
s_3	s_2	s_1

(b) There are two possible answers as shown in the following table:

Chooser 1	Chooser 2	Divider
s_1	s_3	s_2
s_2	s_3	s_1

(c) The only possible fair division is

Chooser 1	Chooser 2	Divider
s_1	s_3	s_2

(d) The divider can pick between s_1 and s_3—let's say the divider picks s_1. Then s_2 and s_3 can be combined again into a cake that may then be divided between Chooser 1 and Chooser 2 using the divider-chooser method.

9. (a) One possible fair division of the cake is

Chooser 1	Chooser 2	Chooser 3	Divider
s_2	s_4	s_1	s_3

(b) Another fair division of the cake is

Chooser 1	Chooser 2	Chooser 3	Divider
s_4	s_1	s_2	s_3

(c) Since none of the choosers have chosen s_3, s_3 can only be given to the divider.

11. (a) A fair division of the cake is

Chooser 1	Chooser 2	Chooser 3	Chooser 4	Divider
s_3	s_4	s_2	s_5	s_1

(b) Another fair division of the cake is

Chooser 1	Chooser 2	Chooser 3	Chooser 4	Divider
s_4	s_3	s_2	s_5	s_1

(c) Since none of the choosers chose s_1, s_1 can only be given to the divider.

13. (a) A fair division of the cake is

Chooser 1	Chooser 2	Chooser 3	Chooser 4	Chooser 5	Divider
s_2	s_5	s_3	s_1	s_6	s_4

(b) Chooser 5 must get s_6 which forces chooser 4 to get s_1. This leaves only s_5 for chooser 2 which in turn leaves only s_3 for chooser 3. Consequently only s_2 is left for chooser 1 and the divider gets the leftover (s_4).

15. (a) Chooser 1: $\{s_3, s_4\}$; chooser 2: $\{s_1, s_3, s_4\}$; chooser 3: $\{s_3\}$.

(b) A fair division of the cake is

Chooser 1	Chooser 2	Chooser 3	Divider
s_4	s_1	s_3	s_2

17. The value of the parts of the cake (as a percentage of the total) as seen by each person is shown in the figure.

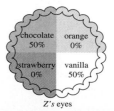

(a) One possible fair division is

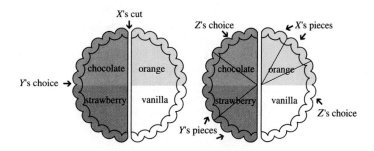

In this division the value of X's final share in X's own eyes is $\frac{2}{3} \cdot 50\% = 33\frac{1}{3}\%$, the value of Y's final share in Y's own eyes is $\frac{2}{3} \cdot 100\% = 66\frac{2}{3}\%$, and the value of Z's final share in Z's own eyes is $\frac{2}{3} \cdot 50\% + 50\% = 83\frac{1}{3}\%$.

(b) One possible fair division is

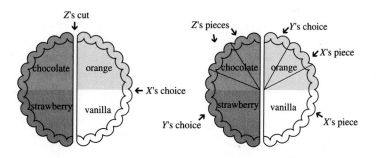

In this division the value of X's final share in X's own eyes is $\frac{2}{3} \cdot 50\% = 33\frac{1}{3}\%$, the value of Y's final share in Y's own eyes is $\frac{1}{3} \cdot 50\% + 50\% + 0\% = 66\frac{2}{3}\%$, and the value of Z's final share in Z's own eyes is $\frac{2}{3} \cdot 50\% = 33\frac{1}{3}\%$.

(c) One possible fair division is

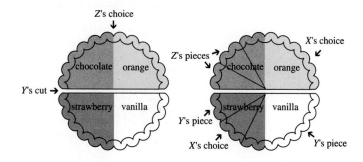

In this division the value of X's final share in X's own eyes is $\frac{1}{3} \cdot 50\% + 50\% = 66\frac{2}{3}\%$, the value of Y's final share in Y's own eyes is $\frac{2}{3} \cdot 50\% = 33\frac{1}{3}\%$, and the value of Z's final share in Z's own eyes is $\frac{2}{3} \cdot 50\% = 33\frac{1}{3}\%$.

19. **(a)** P_{10} **(b)** P_1 **(c)** P_{15} **(d)** P_3 **(e)** P_1 **(f)** P_{15}
  (g) P_1 **(h)** 14

21. *A* gets the desk and the tapestry and must pay \$295.55; *B* gets the dresser and receives \$111.11; *C* gets the vanity and receives \$184.44.

23. **(a)** Jane gets the partnership and pays \$170,000.
  (b) Bob gets \$90,000 and Ann gets \$80,000.

25. *A* ends up with all four items and must pay \$173,777.78; *B* ends up with \$89,888.89; *C* ends up with \$83,888.89.

27. *A* ends up with item 4 and pays \$226; *B* ends up with \$609; *C* ends up with items 1 and 3 and pays \$264: *D* ends up with item 5 and \$120; *E* ends up with items 2 and 6 and pays \$239.

29. **(a)** P_1 gets items 10, 11, 12, 13; P_2 gets items 1, 2, 3; P_3 gets items 5, 6, 7.
  (b) Items 4, 8, and 9 are left over.

31. **(a)** P_1 gets items 1, 2; P_2 gets items 10, 11, 12; P_3 gets items 4, 5, 6, 7.
  (b) Items 3, 8, and 9 are left over.

33. **(a)** P_1 gets items 19, 20; P_2 gets items 15, 16, 17; P_3 gets items 1, 2, 3; P_4 gets items 11, 12, 13; P_5 gets items 5, 6, 7, 8.
  (b) Items 4, 9, 10, 14, and 18 are left over

35. **(a)** P_1 gets items 4, 5; P_2 gets item 10; P_3 gets item 15; P_4 gets items 1, 2.
  (b) Items 3, 6, 7, 8, 9, 11, 12, 13, and 14 are left over.

Jogging

37. **(a)** The total area is 30,000 m^2 and the area of *C* is only 8000 m^2. Since P_2 and P_3 value the land uniformly, each thinks that a fair share must have an area of at least 10,000 m^2.
  (b) Since there are 22,000 m^2 left, any cut that divides the remaining property in parts of 11,000 m^2 will work. For example

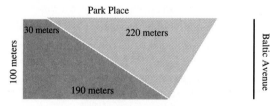

  (c) The cut parallel to Baltic Avenue which divides the parcel in half is

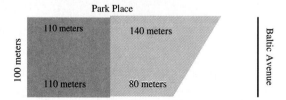

39. **Move 1 (Bidding).** Each player makes a sealed bid giving his or her honest assessment of the dollar value of each of the items in the estate. **Move 2 (Original Allocation).** Each item goes to the highest bidder for that item. (In case of ties, a predetermined tie-breaking procedure should be invoked.) Each player's fair share is calculated by multiplying the total of that player's bids by the percentage that player is entitled to. (Multiply the total of P_1's bids by $r_1/100$, etc.) Each player puts in or takes out from a common pot (the estate) the difference between his or her fair share and the total value of the items allocated to that player. **Move 3 (Dividing the surplus).** After the original allocations are completed there may be a surplus of cash in the estate. This surplus is divided among the players according to the percentage each is entitled to. (P_1 gets $r_1/100$ of the surplus, etc.)

41. (a)

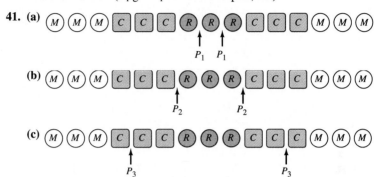

(d) P_1 gets 1 Reese's Piece; P_2 gets 3 caramels and 3 mints; P_3 gets 1 caramel and 3 mints.

(e) 2 caramels and 2 Reese's Pieces.

43. Think of this problem as an ordinary fair division problem with ten players where seven of the players (P_1, P_2, P_3. P_4, P_5, P_6, P_7) are clones of A and the three other players (P_8, P_9, P_{10}) are clones of B. Using the lone divider method, A will receive seven pieces each of which is worth (to A) at least one-tenth of the cake and B will get three pieces each of which is worth (to B) at least one-tenth of the cake.

45. (a) No. The chooser is not assured a fair share in her own value system.

(b) The divider. (The divider has the advantage that he can assure himself a fair share of the cake if he divides it appropriately. The chooser is essentially at the mercy of the laws of chance.)

CHAPTER 4

Walking

1. (a) 50,000   **(b)** A: 26.6; B: 53.4; C: 14.2; D: 65.8
(c) A: 27; B: 53; C: 14; D: 66

3. (a) A: 26.233; B:52.663; C: 14.004; D:64.892
(b) A: 27; B: 53; C: 15; D: 65

5. (a) The standard divisor (50,000) works.   **(b)** A: 27; B:53; C:14; D:66

7. A: 12; B: 39; C: 20; D: 31; E: 34; F: 14

9. A: 13; B: 38; C: 20; D: 31; E: 33; F: 15

11. A: 82; B: 66; C: 48; D: 29

13. A: 82; B: 66; C: 48; D: 29

15. (a) 124   **(b)** 200,000
(c) A: 5,052,000; B: 3,664,000; C: 516,000; D: 7,432,000; E: 8,136,000

17. *A*: 25; *B*: 18; *C:* 2; *D*: 38; *E*: 41

19. *A*: 251 *B*: 19; *C*: 3; *D*: 37; *E*: 40

21. Agriculture: 82; Business: 44; Education: 96; Humanities: 114; Science: 164

23. Agriculture: 82; Business: 44; Education: 96; Humanities: 114; Science: 164

25. (a) Bob: 8; Peter: 3; Ron: 0.   **(b)** Bob: 8; Peter: 2; Ron 1.
   (c) Yes. For studying an extra 2 minutes (an increase of 3.70%) Ron gets a piece of candy while Peter who studies an extra 12 minutes (an increase of 4.94%) has to give up a piece. This is an example of the population paradox.

Jogging

27. Answers will vary. One such example is: Apportion 10 seats among the four states *A*, *B*, *C*, and *D* with populations given in the following table

State	*A*	*B*	*C*	*D*
Population (in millions)	2.24	2.71	2.13	2.92

29. (a) The standard quotas add up to 100. Rounding these standard quotas in the usual way gives *A*: 11; *B*: 24; *C*: 8; *D*: 36; and *E*: 20. These integers add up to 99. Consequently, we must choose a divisor that is *smaller* than the standard divisor so that we can obtain modified quotas that are slightly larger.

   (b) The standard quotas add up to 100. Rounding these standard quotas in the usual way gives *A*: 12; *B*: 25; *C*: 8; *D*: 36; and *E*: 20. These integers add up to 101. Consequently, we must choose a divisor that is *bigger* than the standard divisor so that we can obtain modified quotas that are slightly smaller.

   (c) If the standard quotas rounded in the conventional way (to the nearest integer) add up to *M* then the standard divisor works as an appropriate divisor for Webster's method.

31. (a) *A*: 5; *B*: 10; *C*: 15; *D*: 21.

   (b) For $D = 100$ the modified quotas are *A*: 5, *B*: 10, *C*: 15, *D*: 20. For $D < 100$, each of the modified quotas above will increase and so rounding upward will give at least *A*: 6, *B*: 11, *C*: 16, *D*: 21 or a total of at least 54. For $D > 100$, each of the modified quotas above will decrease and so rounding upward will give at most *A*: 5, *B*: 10, *C*: 15, *D*: 20 or a total of at most 50.

   (c) From (b) we see that there is no divisor such that after rounding the modified quotas upward, the total is 51.

33. (a)

CN	DE	GA	KY	MD	MA	NH	NJ	NY	NC	PA	RI	SC	VT	VA
7	2	2	2	8	14	4	5	10	10	13	2	6	2	18

   (b)

CN	DE	GA	KY	MD	MA	NH	NJ	NY	NC	PA	RI	SC	VT	VA
7	1	2	2	8	14	4	5	10	10	13	2	6	2	19

   (c) Virginia; Delaware

35. (a) In Jefferson's method the modified quotas are larger than the standard quotas and so rounding downward will give each state at least the integer part of the standard quota for that state.

(b) In Adams' method the modified quotas are smaller than the standard quota and so rounding upward will give each state at most one more than the integer part of the standard quota for that state.

(c) If there are only two states, an upper quota violation for one state results in a lower quota violation for the other state (and vice versa). Since neither Jefferson's nor Adams' method can have both upper and lower violations of the quota rule, neither can violate the quota rule when there are only two states.

37. (a) Take for example $q_1 = 3.9$ and $q_2 = 10.1$ ($M = 14$). Under both Hamilton's method and Lowndes' method, A gets 4 seats and B gets 10 seats.

(b) Take for example $q_1 = 3.4$ and $q_2 = 10.6$ ($M = 14$). Under Hamilton's method, A gets 3 seats and B gets 11 seats. Under Lowndes' method, A gets 4 seats and B gets 10 seats.

(c) If $f_1 > f_2$, then under Hamilton's method the surplus seat goes to A. Under Lowndes' method, the surplus seat would go to B if

$$\frac{f_2}{q_2 - f_2} > \frac{f_2}{q_1 - f_1}.$$

CHAPTER 5

Walking

1. (a) Vertices: A, B, C, D; Edges: AB, AC, AD, BD.
(b) Vertices: A, B, C; Edges: none.
(c) Vertices: V, W, X, Y, Z; Edges: $XX, XY, XZ, XV, XW, WY, YZ$.

3. (a)

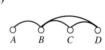

(b)

  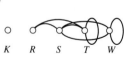

5. (a) $\deg(A) = 3$; $\deg(B) = 2$; $\deg(C) = 1$; $\deg(D) = 2$.
(b) $\deg(A) = 0$; $\deg(B) = 0$; $\deg(C) = 0$.
(c) $\deg(X) = 6$; $\deg(Y) = 3$; $\deg(Z) = 2$; $\deg(V) = 1$; $\deg(W) = 2$.

7. (a) Both graphs have four vertices $A, B, C,$ and D and (the same) edges AB, AC, AD, BD.

(b)

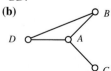

9. (a) 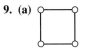 **(b)**

11. (a)   **(b)** 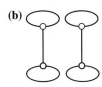 **(c)**

(d)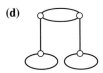

13. (a) *C, B, A, H, F* **(b)** *C, B, D, A, H, F*
  (c) 4 (*C, B, A; C, D, A; C, B, D, A; C, D, B, A*)
  (d) 3 (*H, F; H, G, F; H, G, G, F*)
  (e) 12 (Any one of the paths in [c] followed by *AH*, followed by any one of the paths in [d].)

15. (a) *D, C, B, A, D*
  (b) 6 (*D, C, B, D; D, B, C, D; D, A, B, D; D, B, A, D; D, C, B, A, D; D, A, B, C, D*)
  (c) *HA* and *FE*

17. (a) None of them. **(b)**

19. (a) Has an Euler circuit since all vertices have even degree.
  (b) Has no Euler circuit, but has an Euler path since there are exactly two vertices of odd degree.
  (c) Has neither an Euler circuit nor an Euler path since there are four vertices of odd degree.

21. (a) Has an Euler circuit since all vertices have even degree.
  (b) Has noEuler circuit, but has an Euler path since there are exactly two vertices of odd degree.
  (c) Has no Euler circuit, but has an Euler path since there are exactly two vertices of odd degree.

23. (a) Can't be done since there are more than two vertices of odd degree.
  (b) 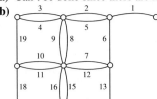 **(c)**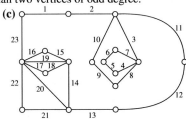

25. (a)

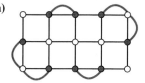

(b)

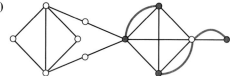

27.

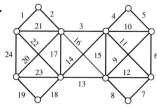

29. *A, B, C, D, E, F, G, A, C, E, G, B, D, F, A, D, G, C, F, B, E, A*

31.

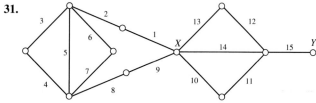

Jogging

33. (a)

Graph	Number of edges	Sum of the degrees of all vertices
Exercise 1(a)	4	$3 + 2 + 1 + 2 = 8$
Exercise 1(b)	0	0
Exercise 1(c)	7	$6 + 2 + 3 + 1 + 2 = 14$
Exercise 2(a)	4	$3 + 2 + 1 + 2 = 8$
Exercise 2(b)	10	$4 + 4 + 4 + 4 + 4 = 20$
Exercise 2(c)	7	$1 + 3 + 2 + 1 + 5 + 2 = 14$

(b) Each edge contributes 2 (1 at each end of the edge) to the sum of the degrees of all the vertices.

(c) If there were an odd number of vertices of odd degree, the sum of the degrees of all the vertices would be odd.

35. (a)

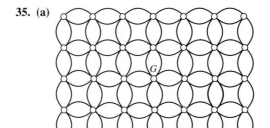

(b)

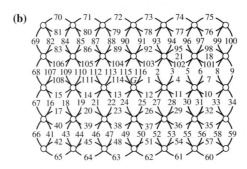

37. (a)

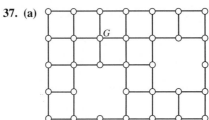

(b)

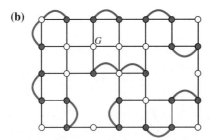

(c)

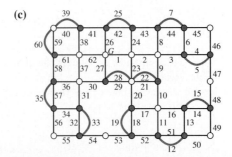

39.

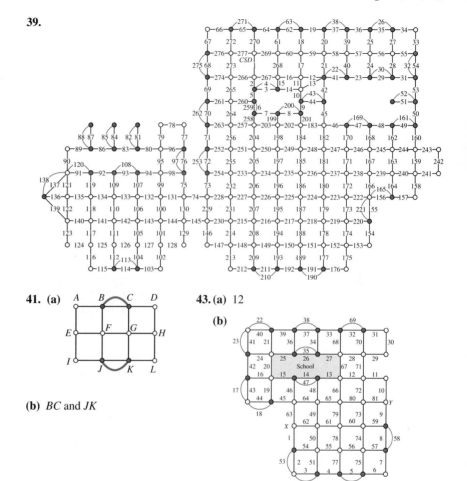

41. (a)

```
A   B   C   D
O---O===O---O
|   |   |   |
E---O---O---H
|   F   G   |
O---O---O---O
I   |   |   |
O---O===O---O
    J   K   L
```

(b) *BC* and *JK*

43. (a) 12

(b)

45. (a) The office complex can be represented by a graph (where each vertex represents a location and each edge a door).

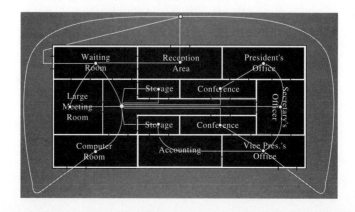

Since there are vertices of odd degree (for example the secretary's office has degree 3), there is no Euler circuit.

(b) Since there are exactly 2 vertices of odd degree (the secretary's office has degree 3 and the hall has degree 9), there is an Euler path starting at either the secretary's office or the hall and ending at the other.

(c) If the door from the secretary's office to the hall is removed (i.e., the edge between the secretary's office and the hall is removed), then every vertex will have even degree and so there will be an Euler circuit. Consequently, it would be possible to start at any location, walk through every door exactly once and end up at the starting location.

CHAPTER 6

Walking

1. *A, D, B, E, C, F, G, A* and *A, G, B, D, C, E, F, A*

3. *A, B, C, D, E, F, G, A* | *A, G, F, E, D, C, B, A*
A, B, E, D, C, F, G, A | *A, G, F, C, D, E, B, A*
A, F, C, D, E, B, G, A | *A, G, B, E, D, C, F, A*
A, F, E, D, C, B, G, A | *A, G, B, C, D, E, F, A*

5. **(a)** *A, B, C, D, E, F, A* | *A, F, E, D, C, B, A*
A, B, E, D, C, F, A | *A, F, C, D, E, B, A*

(b) *D, E, F, A, B, C, D* | *D, C, B, A, F, E, D*
D, C, F, A, B, E, D | *D, E, B, A, F, C, D*

7. **(a)** 6   **(b)** 4   **(c)** *A, B, C, D, E, A* (weight 32)
(d) *A, D, B, C, E, A* (weight 27)

9. **(a)** 11   **(b)** *A, B, C, F, E, D, A* (weight 37)
(c) *A, D, F, E, B, C, A* (weight 41)

11. **(a)** $6! = 5! \times 6 = 120 \times 6 = 720$ **(b)** $9! = \dfrac{10!}{10} = 362{,}880$
(c) $9! = 362{,}880$

13. **(a)** *A, C, B, D, A* (weight 62)   **(b)** *A, D, C, B, A* (weight 80)
(c) *A, B, D, C, A* (weight 74)
(d) The solution obtained in (b) is 29.0% longer than the optimal solution found in (a)—relative error = 0.290. The solution obtained in (c) is 19.4% longer than the optimal solution—relative error = 0.194.

15. **(a)** *A, B, D, C, A* (15 minutes)   **(b)** *A, D, D, C, A* (12.5 minutes)
(c) *A, D, C, B, A* (11.5 minutes)

17. *B, C, E, A, D, B*

19. **(a)** *A, E, B, C, D, A*
(b) No. The circuit in (a) uses up the 5 cheapest edges in the graph.

21. **(a)** *A, E, C, D, B, A* (weight 91)   **(b)** Yes; *A, C, B, D, E, A* (weight 56)

23. *C, D, E, A, B, C* (weight 9.8)

25. *A, B, F, C, D, E, A*

27. *A, B, E, D, C, F, A*

29. (a)

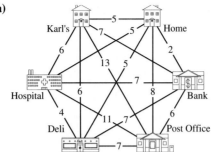

(b) Home, Bank, Post Office, Deli, Hospital, Karl's, Home. The total length of the trip is 30 miles.

Jogging

31.

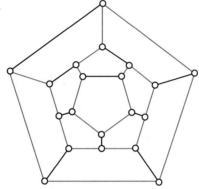

33. Any circuit passing through vertex B must contain edge AB. Any circuit passing through vertex D must contain edge AD. Any circuit passing through vertex G must contain edge AG. Consequently, any circuit passing through vertices B, D, and G must contain at least three edges meeting at A and hence would pass through vertex A more than once.

35. $A, B, C, D, J, I, F, G, E, H$

37. (a) $2^3 = 8 > 6 = 3!$ **(b)** $2^4 = 16 < 24 = 4!$

(c) $N!$ is bigger.

$2^5 = 2 \cdot 2^4 < 2 \cdot 4! < 5 \cdot 4! = 5!$

$2^6 = 2 \cdot 2^5 < 2 \cdot 5! < 6 \cdot 5! = 6!$

$2^7 = 2 \cdot 2^6 < 2 \cdot 6! < 7 \cdot 6! = 7!$

$\vdots$

$2^{k+1} = 2 \cdot 2^k < 2 \cdot k! < (k+1) \cdot k! = (k+1)!$

$\vdots$

In other words, as k increases by 1, 2^k increases by a factor of 2, but $k!$ increases by a factor of $(k+1)$.

39. Dallas, Houston, Memphis, Louisville, Columbus, Chicago, Kansas City, Denver, Atlanta, Buffalo, Boston, Dallas.

CHAPTER 7

Walking

1. (a) tree **(b)** not a tree (has a circuit, is not connected)
(c) not a tree (has a circuit) **(d)** tree

3.

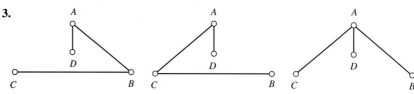

5. (a) **(b)**

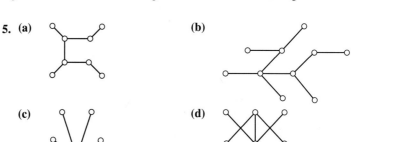

(c) **(d)**

7.

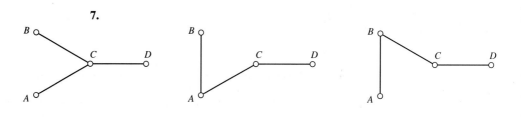

9. (a) 3   **(b)** 1

11.

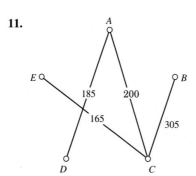

13.

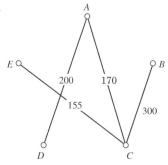

15.

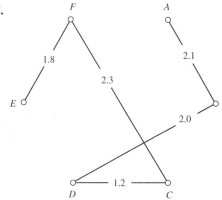

17. **(a)** *CE + ED + EB* is larger since *CD + DB* is the shortest network connecting the points *C, D,* and *B.*

(b) *CD + DB* is the shortest network connecting the points *C, D,* and *B* since angle *CDB* is 120° and so the shortest network is the same as the minimum spanning tree.

(c) *CE + EB* is the shortest network connecting the points *C, E,* and *B* since angle *CEB* is more than 120° and so the shortest network is the same as the minimum spanning tree.

19. 88.2

21. 50.6

23. Call angle *SAB = x.* Then angle *SBA* = 60° − *x*; angle *SBC* = 60° − (60° − *x*) = *x* and angle *SCB* = 60° − *x.* Also, *AB = BC.* It follows that triangles *ASB* and *BSC* are congruent. An identical argument shows that triangles *BSC* and *CSA* are congruent.

25. **(a)** From Exercise 24, it follows that triangle *ABJ* is a 30°-60°-90° triangle with hypotenuse *AB* = 500. It follows that *BJ* = 250 and *JA* = 250 $\sqrt{3} \approx$ 433.0.

(b) Triangle *BSJ* is also a 30°-60°-90° triangle with longer leg *BJ* = 250. It follows that the shorter leg *SJ* = 250$\sqrt{3}$ and the hypotenuse *BS* = 500/$\sqrt{3}$. Since *BS* = *SA*, we have *SA* = 500/$\sqrt{3}$ = 500 $\sqrt{3}$/3 ≈ 288.7.

Jogging

27. 230 miles

29. Each step of Kruskal's algorithm produces exactly one edge in a spanning tree of G. Since G has N vertices, the number of edges in a spanning tree of G must be $N - 1$.

31. **(a)** 27   **(b)** 7   **(c)** 56

33. **(a)** No. A tree with four vertices must have three edges and hence the sum of the degrees of all the vertices is 6.
     (b) 6   **(c)** 8
     (d) $2N - 2$ (This follows from the fact that a tree has $N = 1$ edges and that in any graph the sum of the degrees of all the vertices is twice the number of edges [Exercise 33, Chapter 5].)

35. **(a)** 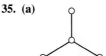   **(b)** 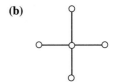   **(c)**

37. The length of the network is $4x$ (see figure).

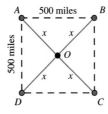

Since the diagonals of a square are perpendicular, we have $x^2 + x^2 = 500^2$ which gives $2x^2 = 500^2$ or $x = \dfrac{500}{\sqrt{2}} = \dfrac{500\sqrt{2}}{2} = 250\sqrt{2}$. Thus $4x = 1000\sqrt{2} \approx 1414$.

39. **(a)** The length of the network is $4x + (300 - x) = 3x + 300$, where $200^2 + \left(\dfrac{x}{2}\right)^2 = x^2$. (See figure.)

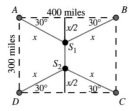

Solving the equation $x = \dfrac{400\sqrt{3}}{3}$, and so the length of the network is $400\sqrt{3} + 300 \approx 993$.

(b) The length of the network is $4x + (400 - x) = 3x + 400$, where $150^2 + \left(\dfrac{x}{2}\right)^2 = x^2$. (See figure.)

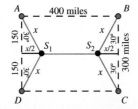

Solving the equation gives $x = \dfrac{300\sqrt{3}}{3}$, and so the length of the network is $300\sqrt{3} + 400 \approx 919.6$.

CHAPTER 8

Walking

1. (a)

Vertex	Degree	Indegree	Outdegree	Vertex is incident to	Vertex is incident from
A	3	2	1	C	B, D
B	2	0	2	A, D	—
C	1	1	0	—	A
D	2	2	1	A	B

(b)

Vertex	Degree	Indegree	Outdegree	Vertex is incident to	Vertex is incident from
A	3	2	1	C	B, D
B	2	0	2	A, D	—
C	4	1	3	A, D, E	A
D	3	3	0	—	B, C, E
E	2	1	1	D	C

(c)

Vertex	Degree	Indegree	Outdegree	Vertex is incident to	Vertex is incident from
A	1	1	0	—	B
B	3	2	1	A	E
C	2	1	1	F	E
D	1	1	0	—	E
E	5	0	5	B, C, D, F	—
F	2	2	0	—	C, E

3. (a)   **(b)**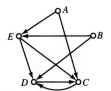

5.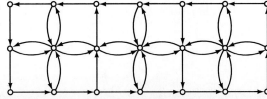

7. (a) **(b)**

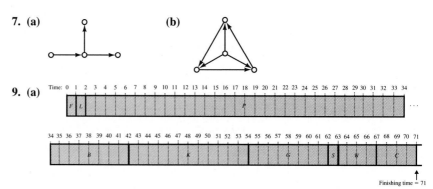

9. (a)

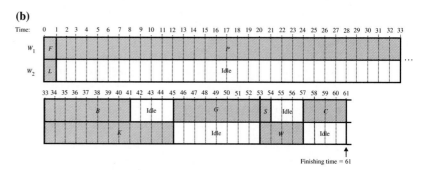

(b)

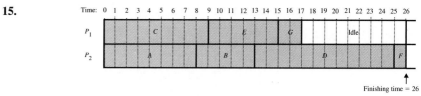

11. According to the precedence relations, G cannot be started until K is completed.

13. According to the precedence relations, G cannot be started until both K and B are completed.

15.

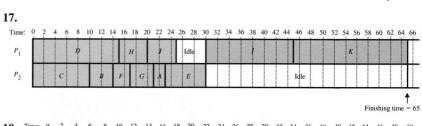

17.

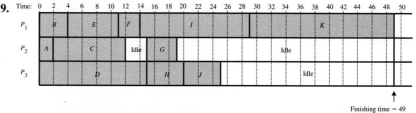

19.

21. Since there is a total of 25 hours of work to be done by 2 processors, the work cannot be completed in less than 12.5 hours. But the times for all the jobs are whole numbers, so the work cannot be completed in less than 13 hours.

23.

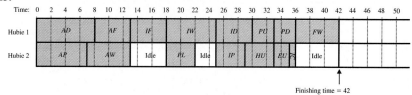

Finishing time = 42

25.

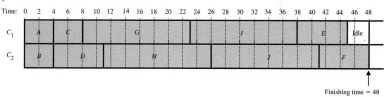

27. (a)

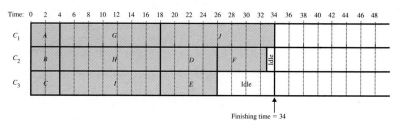

Finishing time = 48

(b)

Finishing time = 34

(c)

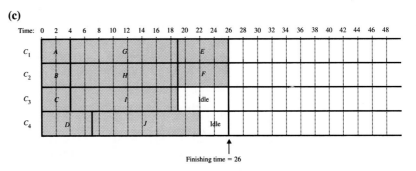

Finishing time = 26

29. (a)

Finishing time = 66

(b)

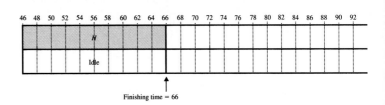

Finishing time = 42

31. (a)

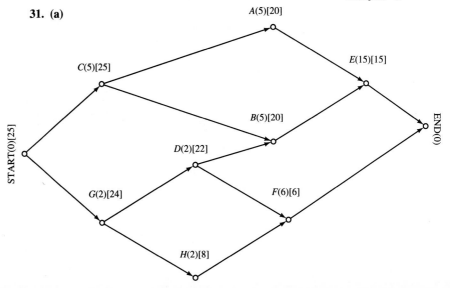

(b)

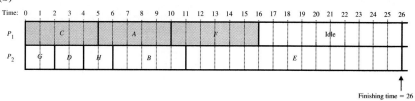

Finishing time = 26

Jogging

33. Each arc of the graph contributes 1 to the sum of the indegrees and 1 to the sum of the outdegrees.

35.

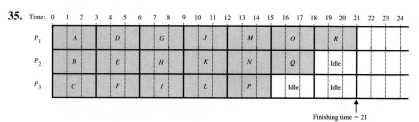

Finishing time = 21

37. (a)

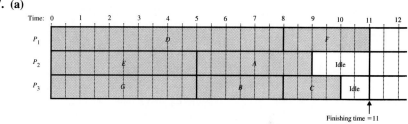

Finishing time = 11

(b)

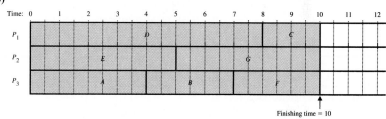

Finishing time = 10

39. (a)

Time: 0   5   10   15   20   25   30   35   40   45   50   55   60   65

P_1	A	F	G	H				I	
P_2		D		B	Idle	
P_3		E		C	Idle	

Finishing time = 65

(b)

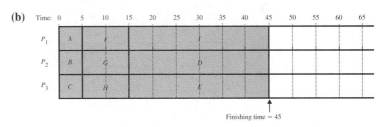

Finishing time = 45

41. (a)

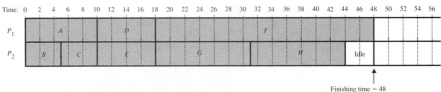

Finishing time = 48

(b)

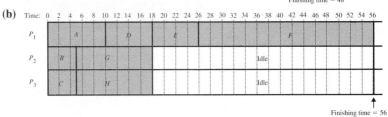

Finishing time = 56

(c) The critical path for this project has length 48 and so the job cannot be completed any sooner. Consequently, the critical path algorithm with 2 processors produced an optimal schedule. With 3 processors the critical path algorithm produced a longer schedule. This paradoxical situation is a consequence of the requirement of the critical path algorithm that a processor cannot remain idle if there is an available task. Tasks D and E need to be completed early so that the long task F can be started, but processor 2 and processor 3 were forced to start other jobs and so were not available to start D or E when those jobs were available.

CHAPTER 9

Walking

1. $F_{15} = 610$, $F_{16} = 987$, $F_{17} = 1597$, $F_{18} = 2584$
3. (a) $F_{38} = 39088169$ (b) $F_{35} = 9227465$
5. $\dfrac{F_{14}}{F_{13}} = 1.6180258$, $\dfrac{F_{16}}{F_{15}} = 1.6180328$, $\dfrac{F_{18}}{F_{17}} = 1.6180338$. These numbers are increasing and getting closer to the golden ratio Φ.
7. $N = 12$ and $M = 12$. ($F_{12} = 12^2$)
9. (a) $47 = 13 + 34$ (b) $48 = 1 + 13 + 34$
   (c) $207 = 8 + 55 + 144$ (d) $210 = 3 + 8 + 55 + 144$

11. (a) $(F_1 + F_2 + F_3 + F_4) + 1 = (1 + 1 + 2 + 3) + 1 = 8 = F_6 = F_{4+2}$
   (b) $(F_1 + F_2 + F_3 + F_4 + F_5) + 1 = (1 + 1 + 2 + 3 + 5) + 1 = 13 = F_7 = F_{5+2}$
   (c) $(F_1 + F_2 + F_3 + \cdots + F_{10}) + 1 = (1 + 1 + 2 + 3 + 5 + 8 + 13 + 21 + 34 +$
   $55\,) + 1 = 144 = F_{12} = F_{10+2}$
   (d) $(F_1 + F_2 + F_3 + \cdots + F_{11}) + 1 = (1 + 1 + 2 + 3 + 5 + 8 + 13 + 21 + 34 +$
   $55 + 89) + 1 = 233 = F_{13} = F_{11+2}$

13. (a) $2F_{N+2} - F_{N+3} = F_N$
   (b) $2F_{1+2} - F_{1+3} = 2F_3 - F_4 = 2 \cdot 2 - 3 = 1 = F_1$
   (c) $2F_{4+2} - F_{4+3} = 2F_6 - F_7 = 2 \cdot 8 - 13 = 3 = F_4$
   (d) $2F_{8+2} - F_{8+3} = 2F_{10} - F_{11} = 2 \cdot 55 - 89 = 21 = F_8$

15. (a) 122.9919 (rounded to four decimal places).
   (b) 0.0081 (rounded to four decimal places).
   (c) 55.0000 (rounded to four decimal places).

17. $x = 1 + \sqrt{2} \approx 2.414, x = 1 - \sqrt{2} \approx -0.414$

19. $x = \dfrac{8}{3} \approx 2.66667$

21. $x = \dfrac{1 + \sqrt{5}}{2} = \Phi$

23 $x = \dfrac{-1 + \sqrt{17}}{8} \approx 0.39039$

25. 20 by 30
27. $c = 25 \;[3:9 = 9:(c + 3)]$
29. $x = 4$
31. $x = 12, y = 10 \quad [3:4:5 = 9:x:(5 + y)]$

Jogging

33. $x = 6, y = 12, z = 10$
35. $x = 3, y = 5$

37. If $\Phi^N = a\Phi + b$ then $\Phi^{N+1} = (a\Phi + b)\Phi = a\Phi^2 + b\Phi = a(\Phi + 1) + b\Phi = (a + b)\Phi + a$. (Remember $\Phi^2 = \Phi + 1$.)

39. We must have $\dfrac{b + y}{b} = \dfrac{h + x}{h}$ or equivalently $1 + \dfrac{y}{b} = 1 + \dfrac{x}{h}$. This gives $\dfrac{y}{b} = \dfrac{x}{h}$, or, equivalently, $\dfrac{y}{x} = \dfrac{b}{h}$

41. (a) 55.
   (b) Any discrepancy is due to round-off error in the calculator.
43. $A_N = 5F_N$
45. From elementary geometry angle $AEF \cong$ angle DBA. Consequently triangle AEF is similar to triangle DBA (since they are both right triangles and so have all their corresponding angles congruent). So $AF: FE = DA : AB$ which shows rectangle $ADEF$ is similar to rectangle $ABCD$.

CHAPTER 10

Walking

1. (a) $P_4 = 26$ **(b)** $P_6 = 68$ **(c)** No. The sum of two even numbers is even.

3. (a) $P_N = P_{N-1} + 5; P_1 = 3$ **(b)** $P_N = 3 + 5(N - 1) = 5N - 2$
    (c) $P_{300} = 1498$

5. (a) $P_2 = 205, P_3 = 330, P_4 = 455$ **(b)** $P_{100} = 12,455$
    (c) $P_N = 80 + 125(N - 1) = 125N - 45$

7. (a) $d = 3$ **(b)** $P_{51} = 158$ **(c)** $P_N = 8 + 3(N - 1) = 3N + 5$

9. 24, 950

11. 16,050

13. (a) 3,519,500   **(b)** 3,482,550

15. (a) 213   **(b)** $137 + 2N$ **(c)** $7124   **(d)** $2652

17. (a) $40.50   **(b)** 40.5%   **(c)** 40.5%

19. 39.15%

21. $4587.64

23. (a) $9083.48   **(b)** 12.6825%

25. The Great Bulldog Bank: 6%; The First Northern Bank: $\approx$ 5.9%; The Bank of Wonderland: $\approx$ 5.65%

27. $\approx$ $1133.56

29. (a) $6209.21   **(b)** $6102.71   **(c)** $6077.89

31. (a) $P_2 = 13.75$ **(b)** $P_{10} = 11 \times 1.25^9 \approx 81.956$ **(c)** $P_N = 11 \times 1.25^{N-1}$

33. (a) $3(2^{99})$ **(b)** $3(2^{N-1})$ **(c)** $3(2^{100} - 1)$ **(d)** $3(2^{49})(2^{51} - 1)$

35. (a) $p_2 = 0.357$ **(b)** $p_3 = 0.6427428$ **(c)** $p_5 \approx 0.64278397$

37. $p_2 = 0.3825$, $p_3 = 0.70858125$, $p_4 = 0.619481586$, $p_5 = 0.707172452$, $p_6 = 0.621238725$, $p_7 = 0.705903515$, $p_8 = 0.622811228$, $p_9 = 0.704752207$, $p_{10} = 0.624229601$ (Answers rounded to 9 decimal places after each transition.)

Jogging

39. 100%

41. $10,737,418.23

43. $\approx$ 14,619 snails

45. $\approx$ $105,006

47. 6425

CHAPTER 11

Walking

1.

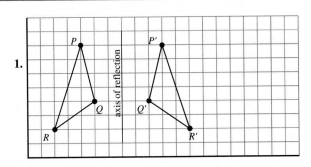

3. (a) 140°   **(b)** 279°

5.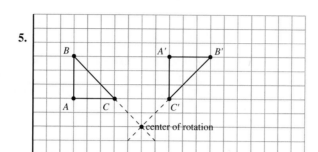

7.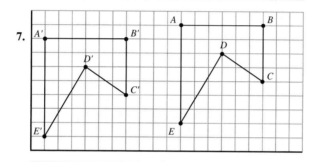

9.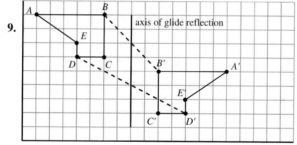

11. (a) b A T **(b)** M I T **(c)** b A d **(d)** b o Y
13. (a) vertical reflection **(b)** horizontal reflection
  (c) horizontal reflection, vertical reflection, reflection about an axis in the northwest direction, reflection about an axis in the northeast direction, rotation of 90° (180°, 270°, etc.)
  (d) 180° rotation **(e)** identity only
15. (a) C **(b)** V **(c)** I **(d)** S **(e)** J
17. (a) **(b)**

19. (a) horizontal reflection, vertical reflection, 180° rotation
  (b) horizontal reflection, vertical reflection, reflections about the diagonals of the center white square, rotation of 90° (180°, 270°, etc.)
  (c) reflection about any axis that passes through the tips of opposite petals, rotation about the center of 120° (240°, etc)

21. (a) translation, vertical reflection   **(b)** translation, horizontal reflection
     (c) translation, 180° rotation   **(d)** translation
23. (a) translation   **(b)** translation, vertical reflection
     (c) translation, horizontal reflection   **(d)** translation, 180° rotation
25. Since every rigid motion is equivalent to either a reflection, rotation, translation, or glide reflection, and a rotation has only one fixed point and translations and glide reflections have no fixed points, the specified rigid motion must be equivalent to a reflection.

Jogging

27. (a)

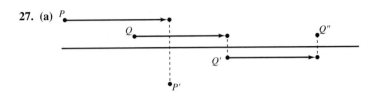

(b) The result of applying the same glide reflection twice is equivalent to a translation in the direction of the glide and twice the amount of the original glide. (See figure in [a].)

29. (a)

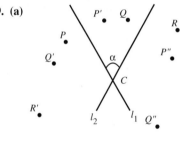

(b) The result of applying reflection 1 followed by reflection 2 is a clockwise rotation with center C and angle of rotation $\gamma + \gamma + \beta + \beta = 2(\gamma + \beta) = 2\alpha$. (See figure)

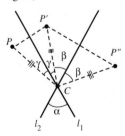

(c) The result of applying reflection 2 followed by reflection 1 is a counterclockwise rotation with center C and angle 2α.
31. (a) There are five different reflection symetries—the axes are the perpendicular bisectors of the five sides.
     (b) There are six different reflection symmetries. Three of the axes are the perpendicular bisectors of the sides and the other three axes pass through opposite vertices of the hexagon.

(c) There are N different reflection symmetries. If N is odd, the axes are the perpendicular bisectors of the sides. If N is even, half of the axes pass through opposite vertices of the regular polygon and the other half are the perpendicular bisectors of the sides of the regular polygon.

33. **(a)** translation, 180° rotation   **(b)** translation, 180° rotation
     (c) translation, vertical reflection, 180° rotation, glide reflection

35. **(a)** Rotations and translations are proper rigid motions, and hence preserve clockwise-counterclockwise orientations. The given motion is an improper rigid motion (it reverses the clockwise-counterclockwise orientation).

   (b) If the rigid motion was a reflection, then PP', RR', and QQ' would all be perpendicular to the axis of reflection and hence would all be parallel.

   (c) It must be a glide reflection (the only rigid motion left).

37. a rotation

39. a glide reflection

CHAPTER 12

Walking

1. **(a)**, **(b)**, **(c)** A line segment joining the midpoints of two sides of a triangle is parallel to the third side and has length one-half the third side. This implies that all the triangles are congruent and therefore the area of each one is $\frac{1}{4}X$.

   (d) $\frac{3}{4}X$ (See [a], [b], [c] above).

3. **(a)** $\frac{9}{16}X$ **(b)** $\frac{27}{64}X$ **(c)** $\left(\frac{3}{4}\right)^N X$

   (d) $\left(\frac{3}{4}\right)^N$ gets closer and closer to 0 as N gets bigger and bigger.

5. $6^2 = 36, 6^9, 6^{N-1}$

7. **(a)** $\frac{1}{3}P$ **(b)** $2P$ **(c)** $4P$

9.

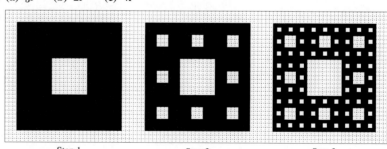

Step 1   Step 2   Step 3

11. **(a)** 512   **(b)** $\left(\frac{8}{9}\right)^3 X = \frac{512}{729}X$ **(c)** 8^N **(d)** $\left(\frac{8}{9}\right)^N X$

   (e) $\left(\frac{8}{9}\right)^N$ gets closer and closer to 0 as N gets bigger and bigger.

13. Start with a ■. Wherever you see a _____ , replace it with a _■_.

15. **(a)** $\frac{20}{3}, \frac{100}{9}$ **(b)** $4\left(\frac{5}{3}\right)^N$

17.

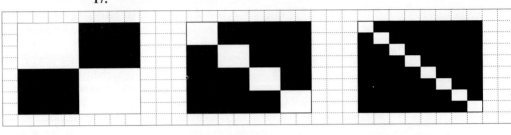

Step 1   Step 2   Step 3

19. $\dfrac{2^N - 1}{2^N} X$

21.

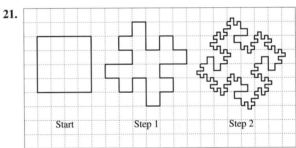

Start   Step 1   Step 2

23. (a) Start: Perimeter $= P$
Step 1: Perimeter $= 2P$
Step 2: Perimeter $= 4P$
(*Note:* This can best be seen by observing that ⌐⌐⌐ is exactly twice as long
as ——.)
(b) Step N: Perimeter $= 2^N P$

25.

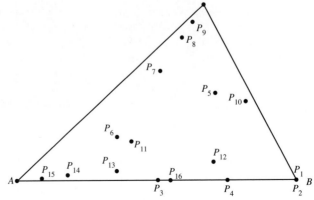

27.

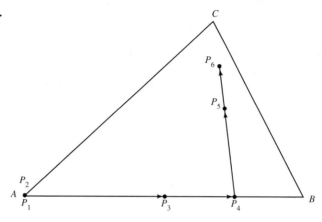

29. (a) 2, 2, 2, 2, 2   **(b)** stable at 2

Jogging

31. At each step, one new white triangle is introduced for each black triangle and the number of black triangles is tripled.

	White Δ's	Black Δ's
Step 1	1	3
Step 2	$1 + 3$	3^2
Step 3	$1 + 3 + 3^2$	3^3
Step 4	$1 + 3 + 3^2 + 3^3$	3^4

We see that at the Nth step there will be

$$1 + 3 + 3^2 + 3^3 + \cdots + 3^{N-1} = \frac{1 - 3^N}{1 - 3} = \frac{3^N - 1}{2}$$

white triangles.

33. There will be infinitely many points left. For example, the 3 vertices of the original triangle will be left as well as the vertices of every black triangle that occurs at each step of the construction.

35. Step 99: -0.4299; step 100: -0.5652. Long-term behavior, oscillates above and below -0.5 and gets closer and closer to -0.5.

37. Step 99: 0.4906; step 100: 0.4907. Long-term behavior, gets closer and closer to 0.5.

39. Step 1: $2 + \sqrt{2}$. Long-term behavior, grows without bound.

CHAPTER 13

Walking

1. (a) Answer 1: All married people. Answer 2: All married people who read Dear Abby's column. Note: While both answers are acceptable, it is clear that Dear Abby was trying to draw conclusions about married people at large so answer 1 is better. Also note that from the wording of her conclusion ("...there are far more faithfully wed couples than I surmised") one can assume that she was primarily interested in currently married people as opposed to divorcees, widows, and widowers.

    (b) 210,336   **(c)** self selection

    (d) 85% is a statistic, since it is based on data taken from a sample.

3. (a) 74.0%   **(b)** 81.8%

    (c) Not very accurate. The sample was far from being representatives of the entire population.

5. (a) The citizens of Cleansburg.   **(b)** 475

7. (a) The choice of street corner could make a great deal of difference in the responses collected.

    (b) *D*. (We are making the assumption that people who live or work downtown are much more likely to answer yes than people in other parts of town.)

    (c) Yes, for two main reasons: (i) people out on the street between 4:00 P.M. and 6:00 P.M. are not representative of the population at large. For example, office and white-collar workers are much more likely to be in the sample than homemakers and school teachers. (ii) The five street corners were chosen by the interviewers and the passersby are unlikely to represent a cross section of the city.

    (d) No. No attempt was made to use quotas to get a representative cross section of the population.

9. (a) All undergraduates at Tasmania State University.   **(b)** $N = 15,000$

11. (a) No. The sample was chosen by a random method.

    (b) In simple random sampling, any two members of the population have as much chance of both being in the sample as any other two. But in this sample, two people with the same last name—say Len Euler and Linda Euler—have no chance of both being in the sample. (By the way, the sampling method described in this exercise is frequently used and goes by the technical name of **systematic sampling**.)

13. (a) Anyone who could have a cold and would consider buying vitamin *X* (i.e., pretty much all adults.)

    (b) Presumably they volunteered. (We cound infer this from the fact that they are being paid.)

    (c) $n = 500$ **(d)** No. There was no control group.

15. ■ Using college students. (College students are not a representative cross section of the population in terms of age and therefore in terms of how they would respond to the treatment.)

    ■ Using subjects only from the San Diego area.

    ■ Offering money as an incentive to participate. (Bad idea!)

    ■ Allowing self-reporting (the subjects themselves determine when their colds are over) is a very unreliable way to collect data and is especially bad when the subjects are paid volunteers.

17. Anyone who could potentially suffer from arteriosclerosis.

19. (a) There was a treatment group (the ones getting the beta carotene pill) and there was a control group. The control group received a placebo pill. These two elements make it a controlled placebo experiment.

    (b) The group that received the beta carotene pills.

    (c) Both the treatment and control groups were chosen by random selection.

21. The professor was conducting a clinical study because he was, after all, trying to establish the connection between a cause (10 milligrams of caffeine a day) and an effect (improved performance in college courses). Other than that, the experiment had little going for it: it was not controlled (no control group) and not randomized (the subjects were chosen because of their poor grades); no placebo was used and consequently the study was not double-blinded.

23. **(i)** A regular visit to the professor's office could in itself be a boost to a student's self-confidence and help improve his or her grades.

    (ii) The "individualized tutoring" that took place during office meetings could also be the reason for improved performance.

    (iii) The students selected for the study all got F's on their first midterm, making them likely candidates to show some improvement.

Jogging

25. **(a)** (i) the entire sky; (ii) all the coffee in the cup; (iii) all the blood in Carla's body.

    (b) In none of the three examples is the sample random.

    (c) (i) In some situations one can have a good idea as to whether it will rain or not by seeing only a small section of the sky, but in many other situations rain clouds can be patchy and one might draw the wrong conclusions by just peeking out the window. (ii) If the coffee is burning hot on top, it is likely to be pretty hot throughout, so Betty's conclusion is likely to be valid. (iii) Because of the constant circulation of blood in our system, the 5 ml. of blood drawn out of Carla's right arm is representative of all the blood in her body, so the lab's results are very likely to be valid.

27. **(a)** The question was worded in a way that made it almost impossible to answer yes.

    (b) "Will you support some form of tax increase if it can be proven that such a tax increase is justified?" is better, but still not neutral. "Do you support or oppose some form of tax increase?" is bland but probably as neutral as one can get.

29. **(a)** Under method 1, people whose phone numbers are unlisted are automatically ruled out from the sample. At the same time, method 1 is cheaper and easier to implement than method 2.

    (b) For this particular situation, method 2 is likely to produce much more reliable data than method 1. The two main reasons are: (i) People with unlisted phone numbers are very likely to be the same kind of people that would seriously consider buying a burglar alarm, and (ii) the listing bias is more likely to be significant in a place like New York City. (People with unlisted phone numbers make up a much higher percentage of the population in a large city such as New York than in a small town or rural area. Interestingly enough, the largest percentage of unlisted phone numbers for any American city is in Las Vegas, Nevada.)

31. **(a)** 2000   **(b)** $N = \dfrac{n_2}{k} \cdot n_1$

    (c) Both samples should be a representative cross section of the same population. In particular, it is essential that the first sample, after being released, be allowed to disperse evenly throughout the population, and that the population should not change between the time of the capture and the time of the recapture.

    (d) It is possible (especially when dealing with elusive types of animals) that the very fact that the animals in the first sample allowed themselves to be captured makes such a sample biased (they could represent a slower, less cunning group). This type of bias is compounded with the animals that get captured the second time around. A second problem is the effect that the first capture can have on the captured animals. Sometimes the animal may be hurt (physically or emotionally) making it more (or less) likely to be captured the second time around. A third source of bias is the possibility that some of the tags will come off.

    (e) 20,500

CHAPTER 14

Walking

1. (a)

Score	10	50	60	70	80	100
Frequency	1	3	7	6	5	2

(b)   **(c)**

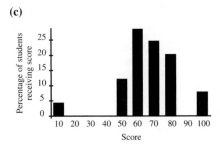

3. (a) Min $= 10$, $Q_1 = 60$, $M = 70$, $Q_3 = 80$, Max $= 100$

(b)

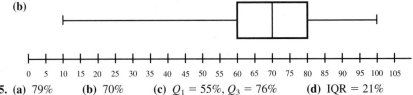

5. (a) 79%   **(b)** 70%   **(c)** $Q_1 = 55\%$, $Q_3 = 76\%$ **(d)** IQR $= 21\%$

7. (a) $\approx 65.32\%$ **(b)** $\approx 15.76\%$

9. (a)

Distance to School (miles)	0.0	0.5	1.0	1.5	2.0	2.5	3.0	3.5	8.5
Frequency	5	3	4	6	3	2	1	1	1

(b)

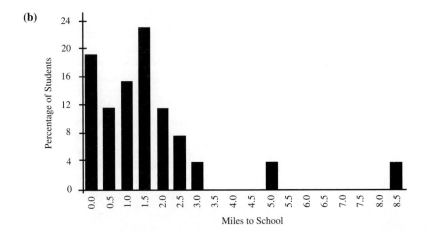

11. (a) 30

(b)

Score	3	4	5	6	7	8	9	10
Frequency	2	5	6	4	4	5	3	1

    (c) 6.17   **(d)** 6

13. Asian: 39.6°; Hispanic: 68.4°; African-American: 86.4°; Caucasian: 140.4°; Other: 25.2°.

15. (a) 132   **(b)** 9

    (c)

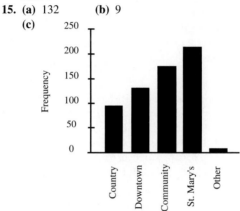

17. (a) 12 ounces

    (b) The third class interval: "more than 72 ounces and less than or equal to 84 ounces." Values that fall exactly on the boundary between two class intervals belong to the class interval to the left.

19.

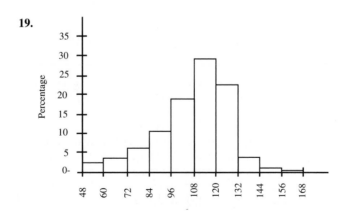

21. Average = 5, SD = 0; Average = 5, SD ≈ 3.54; Average = 5, SD ≈ 11.73. The averages are all the same but the standard deviations get larger as the numbers are more spread out.

23. (a) 4.5   **(b)** 4.5   **(c)** ≈ 2.87

25. (a) 2   **(b)** 7   **(c)** 5

27. (a) 50.5   **(b)** 50.5

29. (a) Min = 9, P_1 = 13, M = 16, P_3 = 18, Max = 25. Average ≈ 15.49 standard deviation ≈ 3.11.

(b)

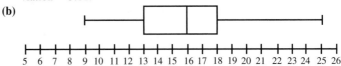

```
5  6  7  8  9 10 11 12 13 14 15 16 17 18 19 20 21 22 23 24 25 26
```

31. (a) $35,000 is the first quartile of engineering salaries, so 459 engineering majors made $35,000 or more.

(b) $25,000 is the first quartile of agriculture salaries, so 240 agriculture majors made $25,000 or less. Note that the exact number that made strictly less than $25,000 cannot be determined from the given information.

33. 58 points

Jogging

35. An example is Ramon gets 85 out of 100 on each of the first four exams and 60 out of 100 on the fifth exam while Josh gets 80 out of 100 on all five of the exams.

37. (a) {1, 1, 1, 1, 6, 6, 6, 6, 6, 6} Average = 4; Median = 6.

(b) {1, 1, 1, 1, 1, 1, 6, 6, 6, 6} Average = 3; Median = 1.

(c) {1, 1, 6, 6, 6, 6, 6, 6, 6, 6} Average = 5; Q_1 = 6.

(d) {1, 1, 1, 1, 1, 1, 1, 1, 6, 6} Average = 2; Q_3 = 1.

39. (a) The five-number summary for the original scores was Min = 1, Q_1 = 9; M = 11, Q_3 = 12, and Max = 24. When 2 points are added to each test score, the five-number summary will also have 2 points added to each of its numbers (i.e., Min = 3, Q_1 = 11, M = 13, Q_3 = 14, and Max = 26).

(b) When 10% is added to each score (i.e., each score is multiplied by 1.1) then each number in the five-number summary will also be multiplied by 1.1 (i.e., Min = 1.1, Q_1 = 9.9, M = 12.1, Q_3 = 13.2, and Max = 26.4).

41. (a) To represent 50% of the population, the column over the interval 30–35 should be 4 units high. (Column area/10 = 50%/25%, so the column area is 20 and the column height is 4.)

(b) To represent 10% of the population, the column over the interval 35–40 should be 0.4 units high. (Column area/10 = 10%/25%, so the column area is 4 and the column height is 0.4.)

(c) To represent 15% of the population, the column over the interval 45–60 should be 0.4 units high. (Column area/10 = 15%/25%, so the column area is 6 and the column height is 0.4.)

43. (a) 10%, 20%   **(b)** 80%, 90%

(c) The figures for both schools were combined. A total of 830 males were admitted out of a total of 1200 that applied—giving approximately 68.3% of the total number of males that applied were admitted. Similarly, a total of 460 females were admitted out of a total of 900 that applied—giving approximately 51.1% of the total number of females that applied were admitted.

(d) In this example, females have a higher percentage $\left(\frac{100}{500} = 20\%\right)$ than males $\left(\frac{20}{200} = 10\%\right)$ for admissions to the School of Architecture and also a higher percentage $\left(\frac{360}{400} = 90\%\right)$ than males $\left(\frac{800}{1000} = 80\%\right)$ for admissions to the School of Engineering. When the numbers are combined, however, females have a lower percentage $\left(\frac{100+360}{500+400} \approx 51.1\%\right)$ than males $\left(\frac{20+800}{200+1000} \approx 68.3\%\right)$ in total admissions. The reason that this apparent paradox can occur is purely a matter of arithmetic: Just because $\frac{a_1}{a_2} > \frac{b_1}{b_2}$ and $\frac{c_1}{c_2} > \frac{d_1}{d_2}$ it does not necessarily follow that

$$\frac{a_1 + c_1}{a_2 + c_2} > \frac{b_1 + d_1}{b_2 + d_2}$$

CHAPTER 15

Walking

1. (a) {*HHHH, HHHT, HHTH, HHTT, HTHH, HTHT, HTTH, HTTT, THHH, THHT, THTH, THTT, TTHH, TTHT, TTTH, TTTT*}
   (b) 16   (c) {*HHTT, HTHT, HTTH, THHT, THTH, TTHH*}   (d) $\frac{6}{16} = \frac{3}{8}$
3. (a) $\{P_1, P_2, P_3, P_4, P_5, P_6, P_7\}$; $\Pr(P_1) = .25$; $\Pr(P_2) = \Pr(P_3) = \Pr(P_4) = \Pr(P_5) = \Pr(P_6) = \Pr(P_7) = .125$
   (b) The odds for P_1 winning the tournament are 1 to 3.
   The odds for P_2 winning the tournament are 1 to 7.
5. (a) 17,576,000   (b) 15,818,400   (c) 11,232,000
7. (a) 40,320   (b) 40,319
9. (a) {red, blue, yellow, purple, orange}   (b) {red, blue, yellow}
   (c) {purple, orange}
   (d) $\Pr(\text{red}) = \Pr(\text{blue}) = \Pr(\text{yellow}) = 0.1$; $\Pr(\text{purple}) = \Pr(\text{orange}) = .35$
11. { }, {*A*}, {*B*}, {*C*}, {*A, B*}, {*A, C*}, {*B,C*}, {*A, B, C*}
13. (a) {*A, B, C, D, E*}
    (b) {*AB, AC, AD, AE, BA, BC, BD, BE, CA, CB, CD, CE, DA, DB, DC, DE, EA, EB, EC, ED*}
15. (a) $\frac{1}{12}$ (b) 1 to 11   (c) 11 to 1   (d) $\frac{11}{12}$
17. (a) { ⊡⊡, ⊡⊡, ⊡⊡, ⊡⊡, ⊡⊡, ⊡⊡, ⊡⊡, ⊡⊡, ⊡⊡ }
    (b) $\frac{5}{18}$ (c) 5 to 13
19. (a) $\frac{1}{36}$ (b) $\Pr(ss) = \frac{1}{36}$; $\Pr(sf) = \frac{5}{36}$; $\Pr(fs) = \frac{5}{36}$; $\Pr(ff) = \frac{25}{36}$
21. (a) $\frac{7}{8}$ (b) $\frac{1023}{1024}$
23. (a) $\frac{1}{8}$ (b) $\frac{3}{8}$
25. $\frac{4}{37}$

Jogging

27. 364
29. 793
31. 2,598,960
33. $\frac{1}{4.165}$
35. $\frac{33}{16,660}$
37. 252
39. 20
41. $\Pr(\text{not rolling a total of 7 in one roll}) = \frac{5}{6}$
    $\Pr(\text{lose}) = \Pr(\text{not rolling a 7 in five rolls}) = \left(\frac{5}{6}\right)^5$
    $\Pr(\text{win}) = 1 - \left(\frac{5}{6}\right)^5 \approx 0.6$
43. $\Pr(\text{rolling a } 2, 3, 4, 9, 10, \text{ or } 12) = \frac{1}{36} + \frac{2}{36} + \frac{3}{36} + \frac{4}{36} + \frac{3}{36} + \frac{2}{36} + \frac{1}{36} = \frac{16}{36} = \frac{4}{9}$

45. $2^6 = 64$

47. **(a)** **(b)** **(c)** There are 70 different outcomes in S as shown in the table below.

	X wins	Y wins
4-game series	*XXXX*	*YYYY*
5-game series	*YXXXX, XYXXX, XXYXX, XXXYX*	*XYYYY, YXYYY, YYXYY, YYYXY*
6-game series	*YYXXXX, YXYXXX, YXXYXX, YXXXYX, XYYXXX, XYXYXX, XYXXYX, XXYYXX, XXYXYX, XXXYYX*	*XXYYYY, XYXYYY, XYYXYY, XYYYXY, YXXYYY, YXYXYY, YXYYXY, YYXXYY, YYXYXY, YYYXXY*
7-game series	*YYYXXXX, YYXYXXX, YYXXYXX, YYXXXYX, YXYYXXX, YXYXYXX, YXYXXYX, YXXYYXX, YXXYXYX, YXXXYYX, XYYYXXX, XYYXYXX, XYYXXYX, XYXYYXX, XYXYXYX, XYXXYYX, XXYYYXX, XXYYXYX, XXYXYYX, XXXYYYX*	*XXXYYYY, XXYXYYY, XXYYXYY, XXYYYXY, XYXXYYY, XYXYXYY, XYXYYXY, XYYXXYY, XYYXYXY, XYYYXXY, YXXXYYY, YXXYXYY, YXXYYXY, YXYXXYY, YXYXYXY, YXYYXXY, YYXXXYY, YYXXYXY, YYXYXXY, YYYXXXY*

49. **(a)** $\frac{1}{13}$ **(b)** $\frac{1}{13}$ **(c)** $\frac{1}{13}$

CHAPTER 16

Walking

1. **(a)** 52   **(b)** 50%   **(c)** 68%   **(d)** 16%

3. **(a)** 44.6   **(b)** 59.4   **(c)** 14.8

5. **(a)** 1900   **(b)** 50

7. **(a)** approximately the 3rd percentile   **(b)** the 16th percentile
(c) around the 22nd or the 23rd percentile   **(d)** the 84th percentile
(e) the 99.85th percentile

9. **(a)** 95%   **(b)** 47.5%   **(c)** 97.5%

11. **(a)** 13   **(b)** 80   **(c)** 250   **(d)** 420   **(e)** 488   **(f)** 499

13. $\mu = 75, \sigma = 3$

15. $\mu = 80, \sigma = 10$

17. **(a)** 16th percentile   **(b)** 97.5th percentile   **(c)** 18.6 lbs

19. **(a)** 97.5th percentile   **(b)** 99.85th percentile   **(c)** 8 lbs

21. Extra large: 125; large: 2375; discount: 2500.

23. **(a)** .5   **(b)** .475   **(c)** .025   **(d)** 0

25. For an honest coin $p = 1/2$. Using the dishonest coin principle with $p = 1/2$, one gets
$\mu = n \cdot (1/2) = n/2$ and $\sigma = \sqrt{n \cdot (1/2) \cdot (1/2)} = \sqrt{n/4} = \sqrt{n}/2$.

27. **(a)** $\mu + 0.25\sigma = 55 + 0.25 \times 12 = 58$ points
(b) 45 points corresponds to the 20th percentile. ($55 - 0.84 \times 12 \approx 45$)
(c) 83 points corresponds to the 99th percentile. ($55 + 2.33 \times 12 \approx 83$)
(d) Approximately 50 students (1%)
(e) Approximately 2000 students. (52 points corresponds to the 40th percentile.)

Index